Hydraulic Systems Volume 7

Modeling and Simulation for Application Engineers

Dr. Medhat Kamel Bahr Khalil, Ph.D, CFPHS, CFPAI.
Director of Professional Education and Research Development,
Applied Technology Center, Milwaukee School of Engineering,
Milwaukee, WI, USA.

CompuDraulic LLC
www.CompuDraulic.com

CompuDraulic LLC

Hydraulic Systems Volume 7

Modeling and Simulation for Application Engineers

ISBN: 978-0-9977634-3-0

Printed in the United States of America
First Published by June 2020
Revised by September 2023

3850 Scenic Way, Franksville, WI, 53126 USA.
www.compudraulic.com

Disclaimer

It is always advisable to review the relevant standards and the recommendations from the system manufacturer. However, the content of this book provides guidelines based on the author's experience.

Any portion of information presented in this book might not be suitable for some applications due to various reasons. Since errors can occur in circuits, tables, and text, the author/publisher assumes no liability for the safe and/or satisfactory operation of any system designed based on the information in this book.

The author/publisher does not endorse or recommend any brand name product by including such brand name products in this book. Conversely the author/publisher does not disapprove any brand name product not included in this book. The publisher obtained data from catalogs, literatures, and material from hydraulic components and systems manufacturers based on their permissions. The author/publisher welcomes additional data from other sources for future editions. This disclaimer is applicable for the workbook (if available) for this textbook.

Hydraulic Systems Volume 7
Modeling and Simulations for Application Engineers

Chapter 9: Modeling of Hydraulic Cylinders, 276

Chapter 10: Modeling of Hydraulic Valves, 293

PREFACE

Modeling and simulation techniques are essential tools for dynamic systems design and production. This book introduces an overview of common mathematical modeling techniques in t-domain and s-domain, various types of physical systems, and challenges of modeling them.

This book is targeting industry professionals who oversee modeling machine at large rather than modeling a single component. This book is also a great resource for mechanical engineering graduate students for their research work.

The book adopted lumped modeling technique, using Matlab-Simulink, to model discrete hydraulic components that can be recharacterized and used repeatedly in system models. The book isn't intended to present a model for every hydraulic component, it rather applies the lumped modeling concept on hydraulic fluids, transmission lines, pumps, motors, cylinders, pressure relief valves, flow control valves, proportional valves, and servo valves. This book uses the component lumped models to assemble an electrohydraulic cylinder position control system and an electrohydraulic motor speed control as case studies.

More than 60 models are presented. This book provides a comprehensive explanation on how these models are structured, validated, and used for analyzing system performance. These models are available to download when you purchase the book.

The author is working hard to finish his goal of supporting fluid power professional education by developing the following series of volumes and relevant software:

- Hydraulic Systems Volume 1: Introduction to Hydraulics for Industry Professionals.
- Hydraulic Systems Volume 2: Electro-Hydraulic Components and Systems.
- Hydraulic Systems Volume 3: Hydraulic Fluids and Contamination Control.
- Hydraulic Systems Volume 4: Hydraulic Fluids Conditioning. Under Development.
- Hydraulic Systems Volume 5: Safety and Maintenance.
- Hydraulic Systems Volume 6: Troubleshooting and Failure Analysis.
- Hydraulic Systems Volume 7: Modeling and Simulation for Application Engineers.

Dr. Medhat Kamel Bahr Khalil

ACKNOWLEDGEMENT

This book was written during the hardship of Covid-19 Virus.

All praise is to Allah who granted me the knowledge, resources and health to finish this work

To the soul of my parents who taught me the values of ISLAM

To my wife who offered me all the best she can to make this work complete

To my family: wife, sons, daughters in law, and grandson "Adam"

To my best teachers and supervisors

The author also thanks the following gentlemen for their effective support in developing this book:

- Kamara Sheku, Dean of Applied Researches at Milwaukee School of Engineering.
- Tom Wanke, CFPE, Director of Fluid Power Industrial Consortium and Industry Relations at Milwaukee School of Engineering.

ABOUT THE BOOK

Book Description:

This book is targeting system design engineers who oversee hydraulic control system design whether for industrial or mobile applications. This book introduces conceptual methodology to build lumped models for hydraulic components and assemble them to form a system. This book is also a great resource for mechanical engineering graduate students for their research work. The book presents models for hydraulic fluids, transmission lines, pumps, motors, cylinders, pressure relief valves, flow control valves, proportional valves, and servo valves.

This book is colored and has the size of standard A4. The book is associated with a separate-colored workbook. The workbook contains printed power point slides, chapter reviews and assignments. This book is the seventh in a series that the author plans to publish to offer complete and comprehensive teaching references for the fluid power industry. The book contains a total of eleven chapters distributed over 320 pages with very demonstrative figures and tables. The contents of the book are brand non-biased and intends to introduce the latest technologies related to the subject of the book.

Book Objectives:

Chapter 1: Introduction to Physical Systems Modeling and Simulation

Modeling and simulation are essential tools in today's system design process. This chapter introduces the subject matter overviewing, the importance, historic background, and the challenges in physical systems modeling and simulation. The chapter also provides a brief overview of the common techniques used for mathematical modeling of physical systems in t-domain and s-domain. The chapter also presents the typical forcing functions used to simulate physical systems performance analysis under various load conditions or commands.

Chapter 2: Modeling and Simulation of First-Order Dynamic Systems

In this chapter, methods and theories presented in Chapter 1 are applied to First-Order dynamic systems. The chapter presents mathematical modeling for first-order systems in t-domain and s-domain. This chapter also presents the response of first-order systems to the typical forcing functions including, step, ramp, and harmonic inputs. The chapter discusses the measured characteristics of first-order step response and how to develop the transfer function of the system based on existing dynamic characteristics.

Chapter 3: Modeling and Simulation of Second-Order Dynamic Systems

In this chapter, methods and theories presented in Chapter 1 are applied for Second-Order dynamic system. The chapter presents mathematical modeling for second-order systems in t-domain and s-domain. This chapter also presents the response of second-order systems to the typical forcing functions including, step, ramp, and harmonic inputs. The chapter discusses measured characteristics of second-order step response and how to develop the transfer function of the system based on existing dynamic characteristics.

Chapter 4: Modeling Approaches for Hydraulic Components and Systems

This chapter explores the different approaches when modeling a hydraulic component versus modeling a hydraulic system at large. The chapter presents the basic idea and the structure of lumped modeling, an adopted modeling approach for application engineers.

Chapter 5: Modeling of Fluid Properties

This chapter presents different techniques to model hydraulic fluid properties based on available information. Properties considered in this chapter are bulk modulus, density, specific gravity and viscosity. In modeling such properties, effects of working temperature and pressure are considered. Case studies are presented, and Matlab-Simulink models were built and validated based on given information.

Chapter 6: Modeling of Hydraulic Transmission Lines

This chapter presents modeling transmission lines, fittings and orifices. Model for a transmission line considers compressible fluid so that effect of line capacitance can be investigated. Developed models were validated based on other software.

Chapter 7: Modeling of Hydraulic Pumps

This chapter presents the lumped modeling concept as applied for fixed and variable displacement pumps. This chapter considers situations where a pump works under a constant or variable pressure and driving speed. Models for pressure-compensated, displacement-controlled, and torque-limited pumps are developed.

Chapter 8: Modeling of Hydraulic Motors

This chapter presents the lumped modeling concept as applied for fixed and variable displacement motors. This chapter considers situations where a motor works under a constant or variable torque and inlet flow. Models for two-position control, proportional control, and torque-limited motors are developed.

Chapter 9: Modeling of Hydraulic Cylinders

This chapter presents the lumped modeling concept as applied for double-acting hydraulic cylinders. This chapter considers situations where a cylinder works under a constant or variable external load and inlet flow. Calculations for cylinder slowing due to leakage, cylinder drift due to oil compressibility, and pressure increase due to thermal expansion are presented.

Chapter 10: Modeling of Hydraulic Valves

This chapter presents the lumped modeling concept as applied for hydraulic valves. This chapter considers modeling at least one pressure control valve, one flow control valve, and one directional control valves. Models for electro-hydraulic proportional and servo valves are also developed.

Chapter 11: Modeling of Hydraulic Control Systems

This chapter utilizes the lumped component models previously built to build system models. In this chapter, there is no additional math models to be developed. This chapter assembles system models from component models. After validating system models, they can be used as reference models for purposes of system design or investigating effects of operating conditions on system performance. This chapter presents models for electrohydraulic cylinder position control, electrohydraulic motor speed control, and hydraulic loading system.

Book Statistics:

The table shown below contains interesting statistical data about the textbook:

Chapter #	Pages	Figures	Models	Equations	Tables	Lines	Words	Characters
Chapter 1	38	19	0	12	5	390	8218	46847
Chapter2	29	27	9	16	0	192	4044	23057
Chapter 3	45	36	13	27	0	245	5172	29484
Chapter 4	13	7	0	2	0	131	2757	15720
Chapter 5	23	19	5	20	3	189	3990	22749
Chapter 6	20	20	4	15	0	119	2520	14369
Chapter 7	57	68	12	7	0	344	7244	41292
Chapter 8	36	41	7	6	0	234	4926	28080
Chapter 9	17	18	1	12	1	130	2752	15691
Chapter 10	24	28	6	7	0	157	3323	18946
Chapter 11	16	21	5	0	0	98	2074	11826
Other	39	0	0	0	0	0	0	0
Total	357	304	62	124	9	2229	47020	268061

ABOUT THE AUTHOR

Medhat Khalil, Ph.D. is Director of Professional Education & Research Development at the Applied Technology Center, Milwaukee School of Engineering, Milwaukee, WI, USA. Medhat has consistently been working on his academic development through the years, starting from bachelor's and master's Degrees in Mechanical Engineering in Cairo Egypt and proceeding with his Ph.D. in Mechanical Engineering and Post-Doctoral Industrial Research Fellowship at Concordia University in Montreal, Quebec, Canada. He has been certified and is a member of many institutions such as: Certified Fluid Power Hydraulic Specialist (CFPHS) by the International Fluid Power Society (IFPS); Certified Fluid Power Accredited Instructor (CFPAI) by the International Fluid Power Society (IFPS); Member of Center for Compact and Efficient Fluid Power Engineering Research Center (CCEFP); Listed Fluid Power Consultant by the National Fluid Power Association (NFPA); and Listed Professional Instructor by the American Society of Mechanical Engineers (ASME). Medhat has balanced academic and industrial experience. Medhat has vast working experience in Fluid Power teaching courses for industry professionals. Being quite aware of the technological developments in the field of fluid power,

Medhat had worked for several world-wide recognized industrial organizations such as Rexroth in Egypt and CAE in Canada. Medhat had designed several hydraulic systems and developed several analytical and educational software. Medhat also has considerable experience in modeling and simulation of dynamic systems using Matlab-Simulink. Medhat has been selected among the inductees for Pioneers in fluid Power by NFPA (2012) and Hall of Fame in fluid Power by IFPS (2021).

Chapter 1
Introduction to Physical Systems Modeling and Simulation

Objectives

Modeling and simulation are essential tools in today's system design process. This chapter introduces the subject matter overviewing, the importance, historic background, and the challenges in physical systems modeling and simulation. The chapter also provides a brief overview of the common techniques used for mathematical modeling of physical systems in t-domain and s-domain. The chapter also presents the typical forcing functions used to simulate physical systems performance analysis under various load conditions or commands.

Brief Contents

Chapter 1: Introduction to Physical Systems Modeling and Simulation

1.1- Importance of Physical Systems Modeling and Simulation

As shown in Fig. 1.1, various industry sectors consider modeling and simulation in system design strategy for the following advantages:

- Optimize the design of a physical system to make it more competitive such as making it more cost-effective, more efficient, last longer, etc.
- Investigate the static and dynamic performance of a physical system under various operating conditions.
- Reduced time & cost for product prototyping.
- Design and prototyping better controllers using hardware-in-the-loop techniques.

Fig. 1.1 – Various Industry Sectors

1.2- History of Physical Systems Modeling and Simulation

Five to eight decades ago, system designers relied mostly on mathematicians to find mathematical solutions to size and predict the static and dynamic characteristics of physical dynamic systems. During that time, system designers used to consider too many assumptions for the missing parameters. Due to the lack of computational tools, they also used to follow simplified calculation techniques such as system linearization. As a result, system designers used to spend much time, money, and production resources in a physical trial and error design strategy before the system achieved acceptable performance.

Today, microprocessors can perform any computational requirements for system design in a very fast way, even in *Real-Time*. In parallel with the existence of computers, the term "*Modeling and Simulation*" was created. As shown in Fig. 1.2, the term "Modeling" means using the system physics to develop the set of mathematical equations that represent the static and dynamic characteristics of the physical system. The term "Simulation" is to upload these equations into a software platform that can solve them simultaneously to simulate the system performance. So, the trial-and-error design strategy are now applied on a virtual system.

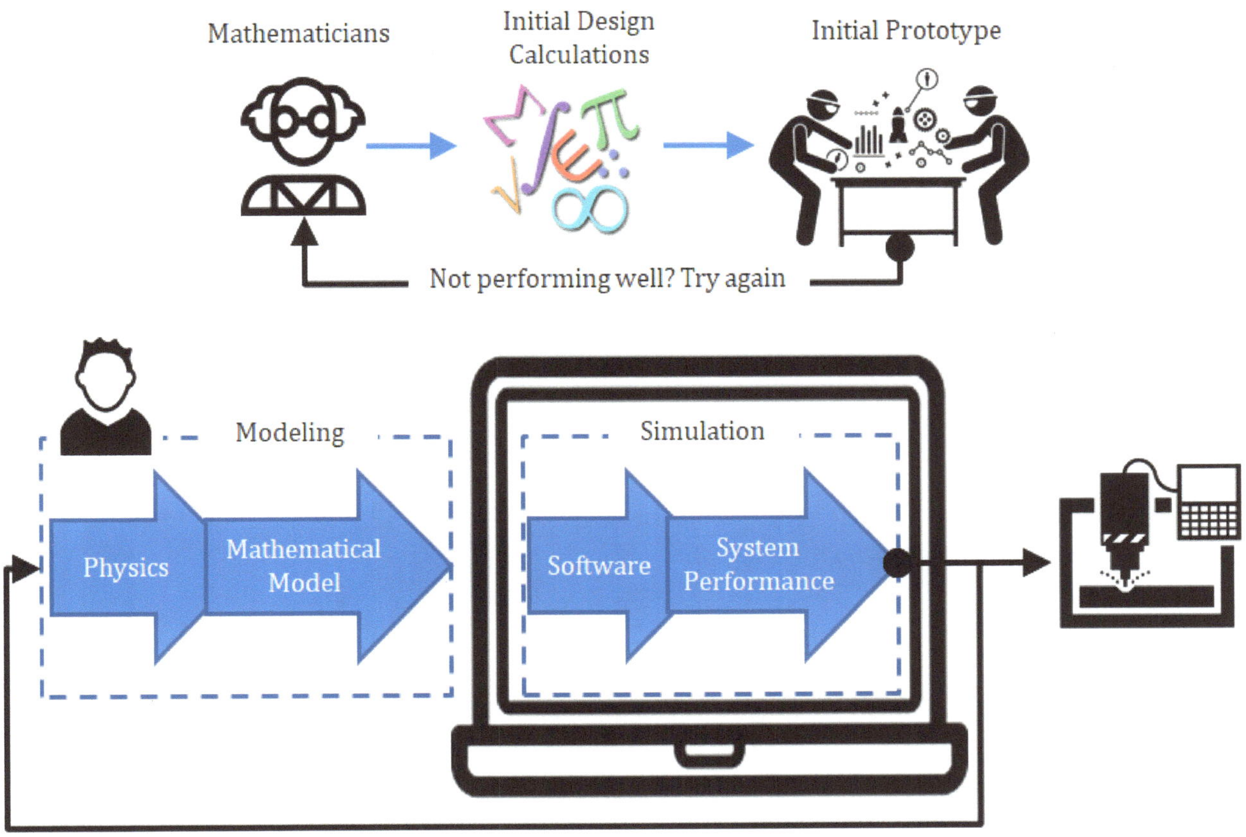

Fig. 1.2 – Physical Systems Design Methodology

1.3- Product Development Cycle

As a result, as shown in Fig. 1.3, system modeling and simulation became a crucial step in system design and product development cycle was changed to include the following steps:

1- **Physical Systems Identification:** Thoroughly understand the physical systems including the way it works, design parameter, kinematics, dynamics, etc.
2- **Mathematical Modeling:** Develop a solvable mathematical model that represents the system static and dynamic characteristics.
3- **Simulation Modeling:** Upload the mathematical model into a proper software platform that can solve the mathematical model that meets the simulation requirements.
4- **Performance Simulation**: Simulate and capture the system response in time-domain as it is exposed to one of the standard time-variant forcing functions; and in frequency domain as it is exposed to periodic load.
5- **Performance Analysis:** Use the simulation results to analyze and investigate the system performance based on the standard measures of acceptable response and stability.
6- **Model Validation:** If the model simulation shows acceptable performance, it must be validated before relying on it as a reference model for further design process. Model *validation* increases the level of fidelity in the model. Model is validated based on a second source of information, usually experimental measurements.
7- **Prototyping and Field Tests:** Build a system prototype and perform field testing.

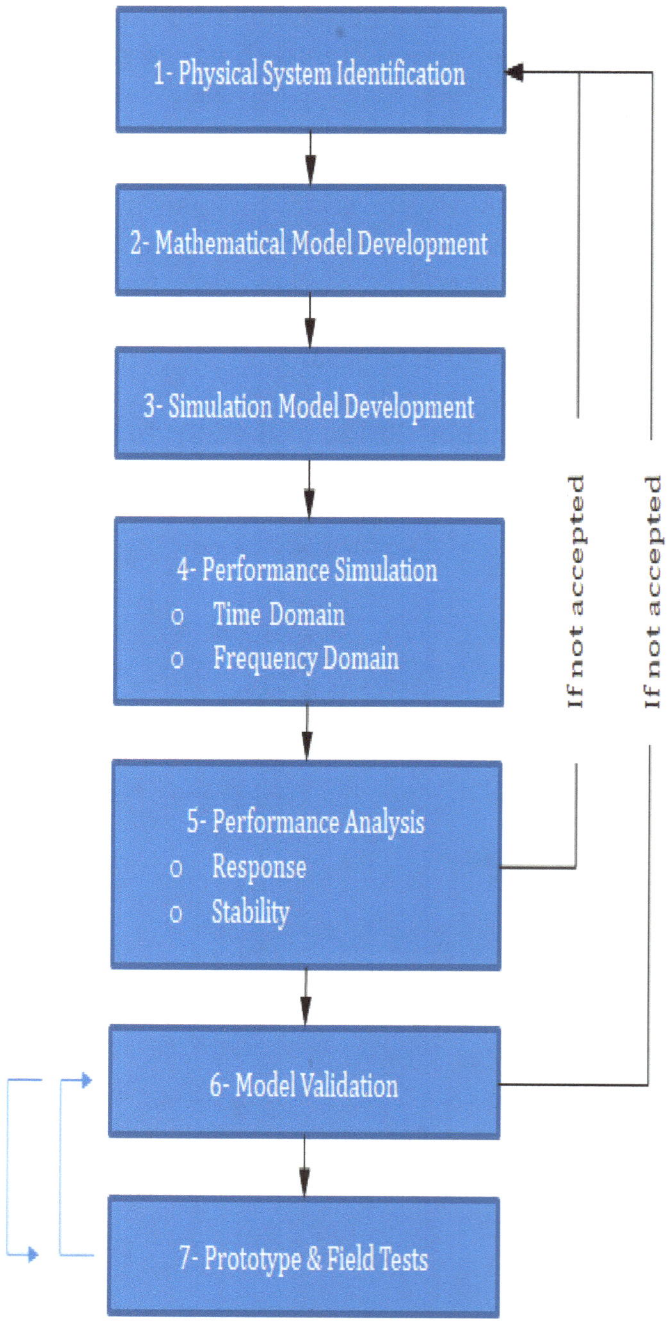

Fig. 1.3 – Product Development Cycle

1.4- Physical Systems Identification

As shown in Fig. 1.4, performance of a physical system can vary with the space, time, and input. Modeling a physical system should start by identifying the features of the system variations.

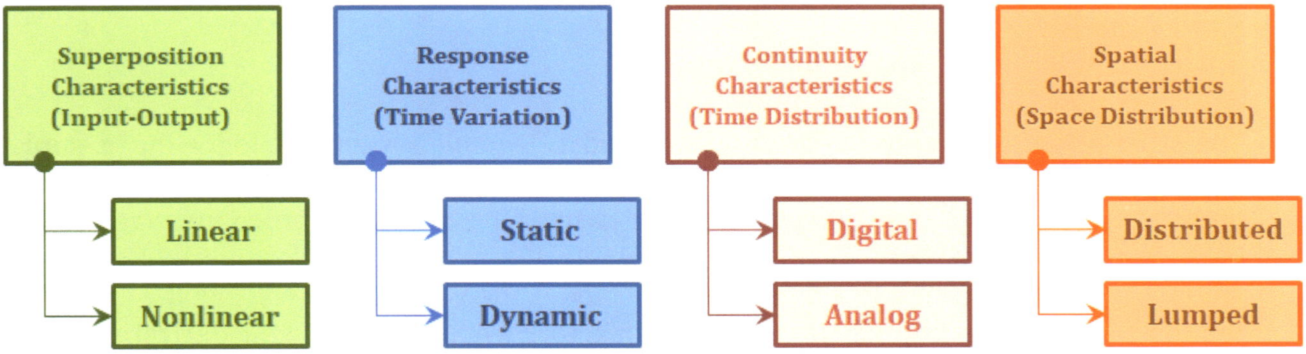

Fig. 1.4 – Features of Physical Systems Performance

1.4.1- Linear versus Nonlinear Systems

A system response to an input, in the steady state conditions, is referred to as the *Static Characteristics* of the system. Based on the nature of the physical system, its static characteristics can be identified as *Linear* or *Nonlinear*. As shown in Fig. 1.5, differential pressure **ΔP** across a hydraulic transmission line is linearly proportional to the input flow **Q** and varies nonlinearly across an orifice.

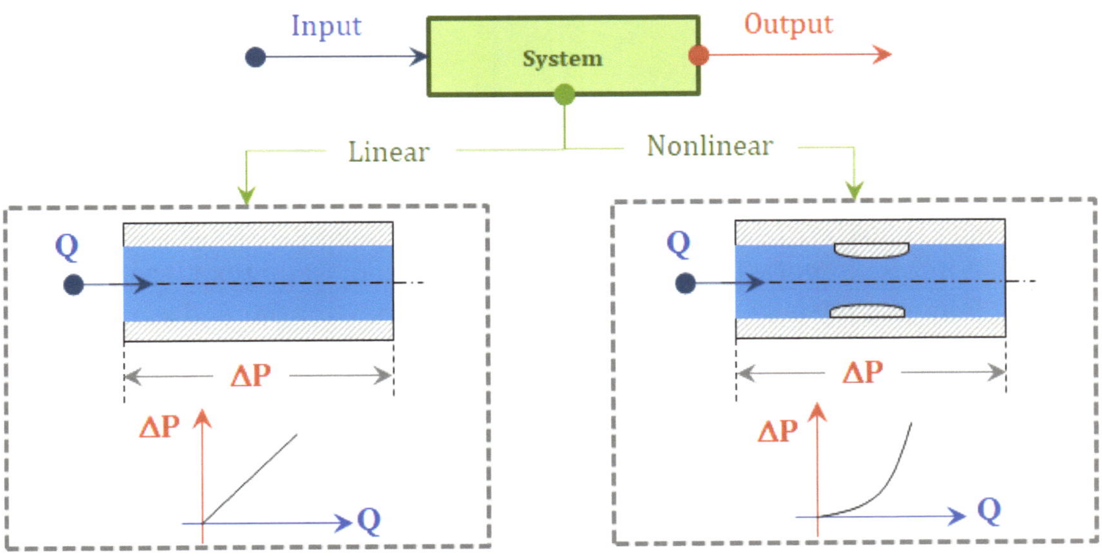

Fig. 1.5 – Linear Versus Nonlinear Systems

In math and physics, the term "linear" means simple because it complies with superposition characteristics and mathematical solutions for linear equations are well developed. The following equations show what is meant by superposition characteristics:

$$\mathbf{f(x1+x2) = f(x1) + f(x2)\ \&\ f(kx) = kf(x)}$$

1.4.2- Static versus Dynamic Systems

To Elaborate the difference between a *Static* and a *Dynamic* System, the following example of a proportional electro-hydraulic valve shown in Table 1.1 is considered. However, when modeling a system at large, a system modeler must be careful when connecting models for static and dynamic systems together. For example, if a pump model considered the pump as a static system is connected to a dynamic valve model, the valve model will see the change in the pump flow as a surge flow which may result in errors when calculating the derivatives.

Static	Dynamic
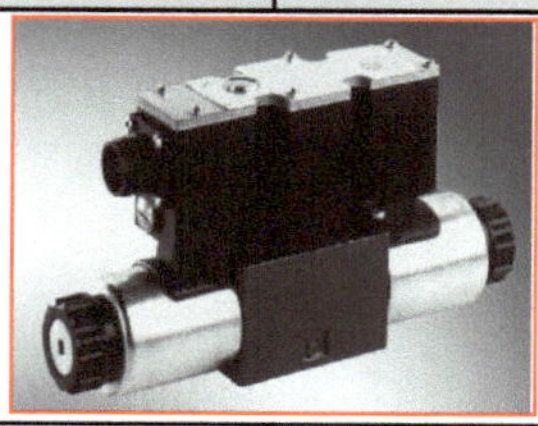	
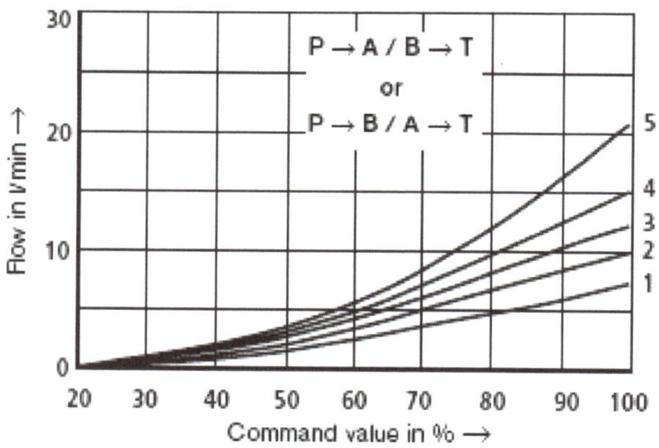	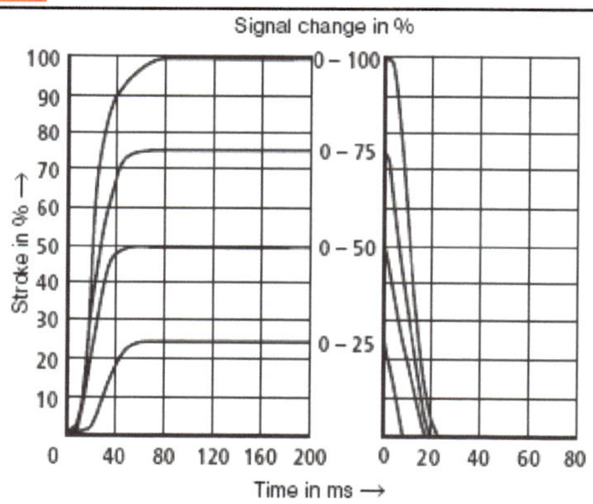
Valve responds at no time to a command value just based on its static characteristics.	Valve dynamic response contains transient and steady state response.
Valve stops immediately once the input stops.	Valve may continue to move after the input command stops.
Valve response can be mathematically modeled using algebraic equations or lookup tables.	Valve response can be mathematically modeled using differential equations or transfer functions.
Static system is considered inherently stable.	Stability of dynamic systems must be investigated.

Table 1.1 – Static versus Dynamic Systems

1.4.3- Digital versus Analog Systems

Based on the distribution of a dynamic system response within time domain, as shown in Table 1.2, physical systems are identified as follows:

Digital (Binary) System: Time/Value = Continuous/Binary:
Today the term "*Digital*" is conventionally used to indicate microprocessor-controlled systems. However, it has been used here to indicate that the system response has *Binary* value like True or False, Zero or One, On or Off, etc. An example of that is the spool position of a solenoid-operated switching directional control hydraulic valve. In such valves, the spool jumps to the alternative position without being able to set in intermediate position. The valve is modeled based on the need to know the spool position at any moment of the valve operation.

Analog (Continuous) System: Time/Value = Continuous/Continuous:
Today nowadays the term "analog" is conventionally used to indicate a control system that is built based on physical electrical parts. However, it has been used here to indicate that the system response has *continuous value.* An example of that is the spool position of a proportional directional control hydraulic valve. In such valves, the spool travels smoothly from one position to other and can take infinite number of intermediate positions. The valve is modeled based on the need to know the spool position at any moment of the valve operation.

Discrete System: Time/Value = Discrete/Continuous:
Systems in between digital and analog are called *Discrete* systems. In such systems, the response varies in analog fashion. However, because the system is slow, there is no sense to monitor the system variation on every single moment. It may be enough to take sample readings every minute, hour, etc. depends on what is needed. An example of this is the working temperature of a hydraulic system.

Hybrid System:
When a machine at large is modeled, it is considered a Hybrid system because it contains a mix of the above-mentioned systems.

Digital (Binary)	Analog (Continuous)	Discreet
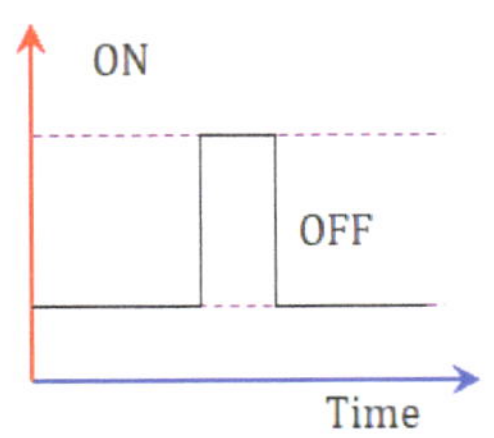	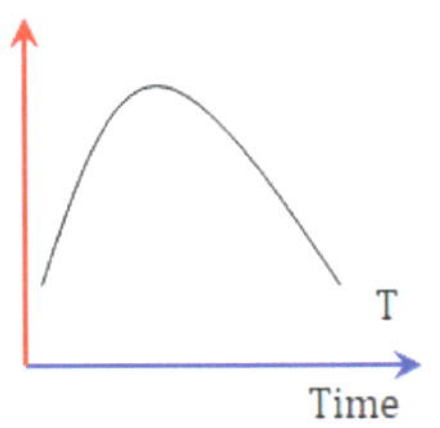	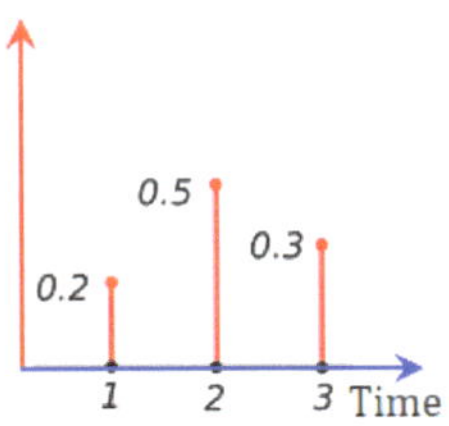
e.g. spool position of switching directional valve	e.g. spool position of Proportional directional valve	e.g. working temperature of a hydraulic system

Table 1.2 – Digital versus Analog Systems

1.4.4- Distributed versus Lumped Systems

Space-wise, systems may be modeled to know the distribution of the system response over the system space. Such systems are referred to as *Distributed* systems. An example of this, as shown in Fig. 1.6 (left side), a system may be modeled to simulate the pressure distribution within the pipeline to investigate water hammer or pressure spikes. It could be also useful for stress or thermal analysis.

In other systems, it may be enough to know what is going on at the boundaries of the system rather than knowing the details within the space of the system. Such systems are referred to *Lumped* systems based on which the technique of *Finite Elements* has been developed. An example of this, as shown in Fig. 1.6 (right side), a system may be modeled to simulate the differential pressure across a transmission line no matter what the pressure distribution within the line is.

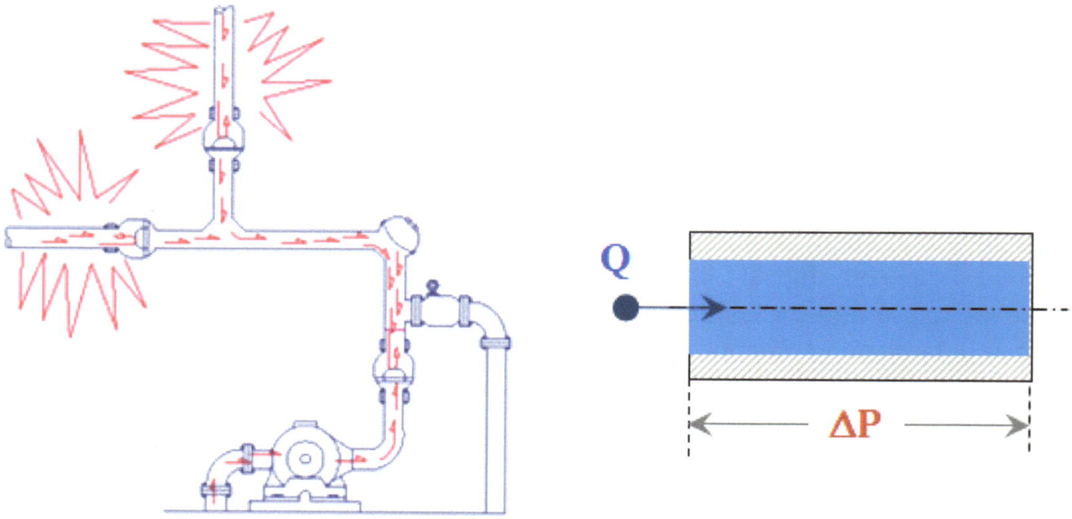

Fig. 1.6 – Distributed Versus Lumped Systems

1.4.5- Design Parameters and Assumptions

Identify the system design parameters such as moving mass, spring stiffness, dimensional parameters, etc.. The unidentifiable or unmeasurable design parameters must be reasonably, and feasibly assumed. A system modeler won't be able to make such assumptions unless he is experience enough to thoroughly understand the physical system. In order to develop a mathematical model, kinematics and dynamics of the system must also be clearly understood.

1.5- Physical Systems Mathematical Modeling

Physical systems can be mathematically modeled using various techniques. However, historically the following couple of methods are the most common.

Physical Systems Mathematical Modeling in t-Domain: Physical systems can be mathematically modeled in *t-Domain* using *Differential Equations* (DE). Modeling a physical system in t-Domain means that the system performance is presented versus time. Differential Equations vary depending on the physical nature of the system.

Physical Systems Mathematical Modeling in s-Domain: Physical systems can also be mathematically modeled in *s-Domain* using *Laplace Transform* mathematical technique. Modeling a physical system in s-Domain means that the system is represented by a *Transfer Function* (TF) and the system performance is presented on a *Complex Plane* that contains a *Real* and *Imaginary* axis.

Other Methods: Physical systems can be mathematically modeled using different techniques such as:

- **State-Space Representation:** in the method of *State-Space* representation, the model is built in form of a *Matrix* and therefore, matrix rules are applied.

- **Bond-Graph Modeling:** Along with an ever-increasing model complexity, a so-called *object-oriented* approach to physical systems modeling has become more and more popular throughout the last few years. The *Bond-Graph* method is a graphical approach to modeling in which component energy ports are connected by bonds that specify the transfer of energy between system components.

However, this textbook considers only the first two common mathematical modeling methods.

1.6- Physical Systems Modeling in t-Domain using Differential Equations

Physical systems can be modeled in *Time-Domain* using *Differential Equations* (DE). Modeling a physical system in T-Domain means that the system performance is presented versus time. Differential Equations vary depending on the physical nature of the system.

1.6.1- Classifications of Differential Equations

1.6.1.1- Ordinary versus Partial Differential Equations

Ordinary DE: An *Ordinary Differential Equation* represents a system in which a dependent variable **y** is function of only one independent variable. In dynamic systems, the independent variable is time **t**.

Example of an Ordinary DE:

$$\frac{dy}{dt} = 3 \sin(t)$$

A practical example from hydraulic systems could be the temperature at the outlet of a hydraulic pump is function of time.

Partial DE: A *Partial Differential Equation* represents a system in which a dependent variable is function of time and other independent variables.

Example of a Partial DE:

$$\frac{\partial y}{\partial t} + \frac{\partial y}{\partial x} = at + bx$$

Figure 1.7 shows a practical example from hydraulic systems. The figure shows the pressure distribution over a length of a transmission line. Pressure also varies with the time as the flow passing through the line is time-variant.

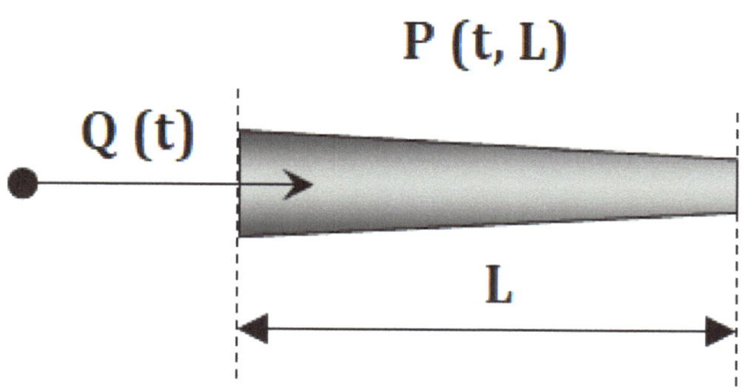

Fig. 1.7 – Potential Energy for Hydraulic Cylinders

1.6.1.2- Linear versus Nonlinear Differential Equations

Linear DE: In a *Linear Differential Equation,* all variables and their derivatives appear with first power. In math and physics, the term "linear" means simple because mathematical solutions for linear equations are well developed.

Example of an Ordinary Linear DE:

$$\mathbf{ax'' + bx' + cx = F(t)}$$

Nonlinear DE: In a *Nonlinear Differential Equation,* at least one of the variables or its derivative is raised to power more than one OR contains trigonometric function.

Examples of Ordinary Nonlinear DE:

$$\mathbf{y' + \frac{1}{y} = 0\ ,\ y' + y^2 = 0,\ y'' + \sin(y) = 0, \text{and } yy' = 1}$$

In math and physics, the term "nonlinear" means complex because mathematical solutions for nonlinear equations are difficult. That is why, in the past before developing the current simulation software, system designers relied on *System Linearization* technique.

For example, if the original system is mathematically modeled as follows:

$$\mathbf{y'' + 2y' + e^y = 5t}$$

The operator $\mathbf{e^y = 1 + y + \frac{y^2}{2} + \frac{y^3}{6} + etc.}$

Then, by removing the nonlinear parts $\rightarrow$ $\mathbf{e^y = 1 + y}$

Then the linearized system is mathematically modeled as follows:

$$\mathbf{y'' + 2y' + 1 + y = 5t}$$

The fact is, by using the linearized model, some of the system minor details will not be captured. This means that the linearized model will be less accurate in representing the real physical system.

The question now is "do system designers still need to develop linearized models anymore?". The answer is "not necessarily" because there are software tools that can solve the system model as is, no matter what the linearity state of the model is. Linearized techniques may still be used to apply simple design methods and predict an approximate system response.

1.6.1.3- Homogeneous versus Non-Homogeneous Differential Equations

Non-Homogeneous DE: The shown below ordinary linear DE is referred to as *Non-Homogeneous* DE because it contains the *Forcing Function* **F(t).**

$$\mathbf{aX'' + bX' + cX = F(t)}$$

Homogeneous DE: By equating the Left-Hand-side (LHS) of the non-homogeneous DE to zero, it is referred to as *Homogeneous* DE or the *Characteristic Equation*.

$$\mathbf{aX'' + bX' + cX = 0}$$

1.6.1.4- First Order versus Second Order Differential Equations

The order of a differential equation means the number of the derivative of the dependent variable of a differential equation.

First Order DE: Since Eq. 1.1 contains only the first derivative, it is considered a *First Order* DE.

$$\mathbf{ax' + bx = F(t)} \qquad \mathbf{1.1}$$

Second Order DE: Since Eq. 1.2 contains the second derivative, it is considered a *Second Order* DE. Such a system contains the three basic physical elements (inductive, capacitive, and resistive).

$$\mathbf{ax'' + bx' + cx = F(t)} \qquad \mathbf{1.2}$$

Higher Order DE: Since the Eq. 1.3 contains more than the second derivative, it is considered a *Higher Order* DE. An example is a third-order system in which a mass is accelerated by variable acceleration acting jerky.

$$\mathbf{ax''' + bx'' + cx' + dx = F(t)} \qquad \mathbf{1.3}$$

1.6.2- Mathematical Solution in t-Domain for a Physical System Response

Figure 1.8 shows the traditional methodology for designing a physical system that is based on developing a mathematical solution for the DEs that represent the system. As shown in the figure, mathematical solution method contains the following parts:

Transient (Free) Response: *Transient (Free) Response* $\mathbf{X_H(t)}$ is developed by finding the *Roots* of the characteristic equation. Value of the roots indicates the system response and stability.

Steady State (Forced) Response: Second step is to find the *Steady Sate (Forced) Response* $\mathbf{X_{SS}}$ of the system out of the *Non-Homogeneous DE.*

Complete Response: Third step is to find the complete solution by applying the *Initial Conditions* of the system.

As it has been discussed previously, since the response of a physical system nowadays can be simulated by uploading the mathematical model to a simulation software to solve for the system response, the method of developing mathematical solution won't be discussed in this textbook.

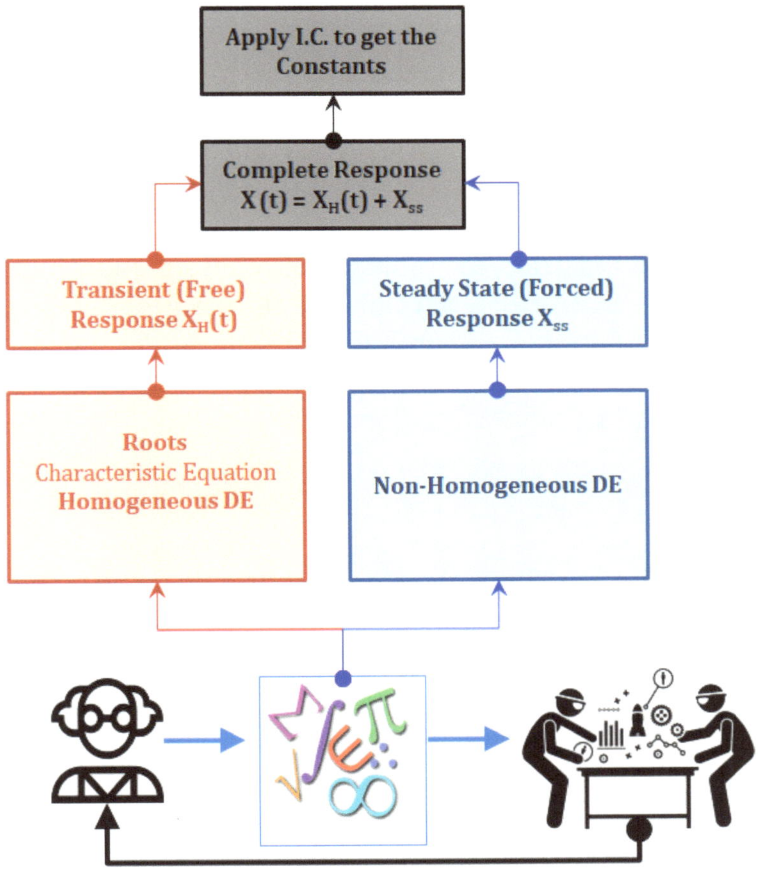

Fig. 1.8 – Mathematical Solutions for Physical Systems

1.7- Physical Systems Modeling in s-Domain using Laplace Transform

Physical systems can also be modeled in *s-Domain* using *Laplace Transform* mathematical technique. Modeling a physical system in s-Domain means that the system is represented by a *Transfer Function* (TF) and the system performance is presented on a *Complex Plane* that contains a *Real* and *Imaginary* axis.

The background of this mathematical modeling dates to when there were no computers to help simulate physical systems and perform the required computational efforts for that. So, mathematicians were looking for a method to easily solve the differential equations and predict system performance. The conceptual idea of Laplace Transform is to convert a differential equation into an algebraic equation that can be solved easily, then the solution can be reconverted using *Inverse Laplace Transform*.

The question now is, in view of having computers and simulation software that can solve differential equations, do we still need to model systems using LT in form of TF?

The answer is it has been left to the system modeler based on his preference. In large systems, probably this technique is helpful because subsystems can be capsulated in just one TF. Fortunately, simulation software can accept either or combination of t-domain and s-Domain models on the same layer of the system model.

1.7.1- Laplace Transform

Converting a differential equation in t-domain into an equation in s-Domain is conceptually based on replacing a derivative of a variable by an operator "**s**". Since the integration is the inverse of the differentiation, then integral of a variable is replaced by the reciprocal of s "$\mathbf{1/s}$". In that sense, as shown in Eq. 1.4, applying the Laplace Transform operator on a t-domain function **f(t)** converts it to s-Domain function **F(s).**

$$\mathcal{L}\{\mathbf{F(t)}\} = \mathbf{F(s)} \qquad \mathbf{1.4}$$

From there, lots of tables have been developed to list the conversion of the typical t-domain functions to equivalent s-domain functions. This textbook is not intended to teach this method. However, such conversion tables can easily be found in lots of mathematics textbooks.

1.7.2- Block Diagram Algebra

Along with the use of the Laplace Transform method, a set of rules has been developed using graphical block diagram to help in developing the mathematical model. The following set of examples gives a brief overview of the block diagram algebra used with Laplace Transforms.

Example 1 (multiplying by a constant):

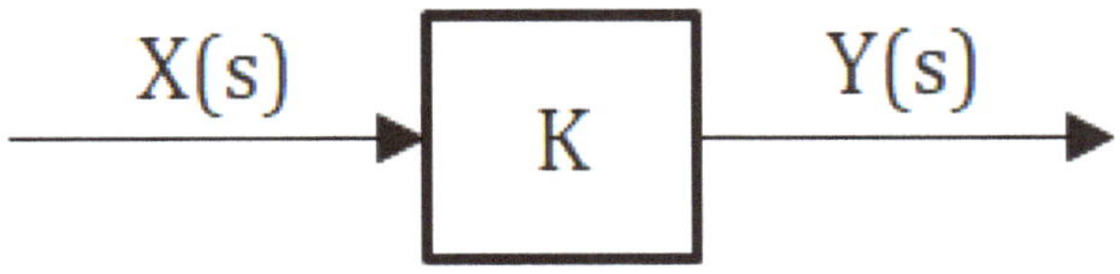

$$Y(s) = K\,X(s)$$

Example 2 (Derivative):

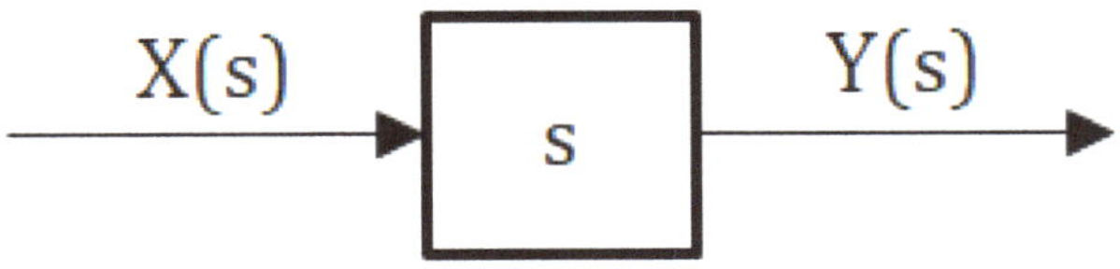

$$Y(s) = s\,X(s)$$

Example 3 (Integral):

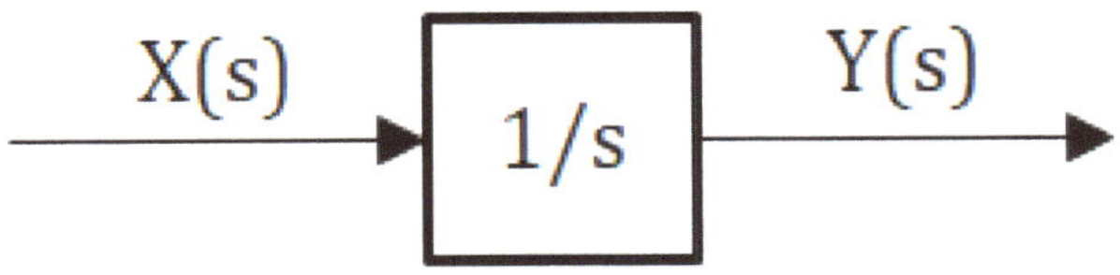

$$Y(s) = 1/s\,X(s)$$

Example 4 (Addition and Subtraction):

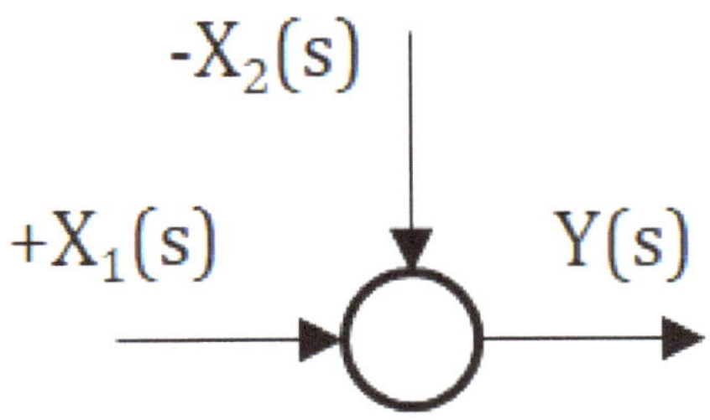

$$Y(s) = X_1(s) - X_2(s)$$

Example 5 (Associative and Commutative Properties):

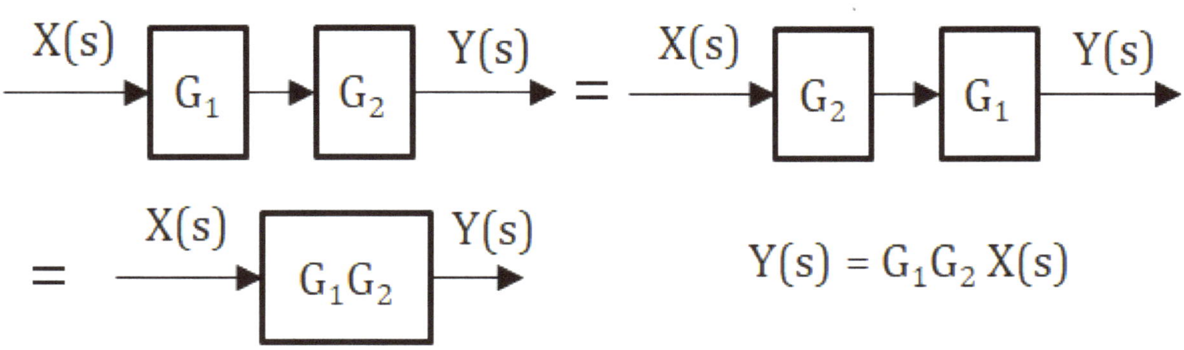

$$Y(s) = G_1 G_2\, X(s)$$

Example 6 (Distributive Property):

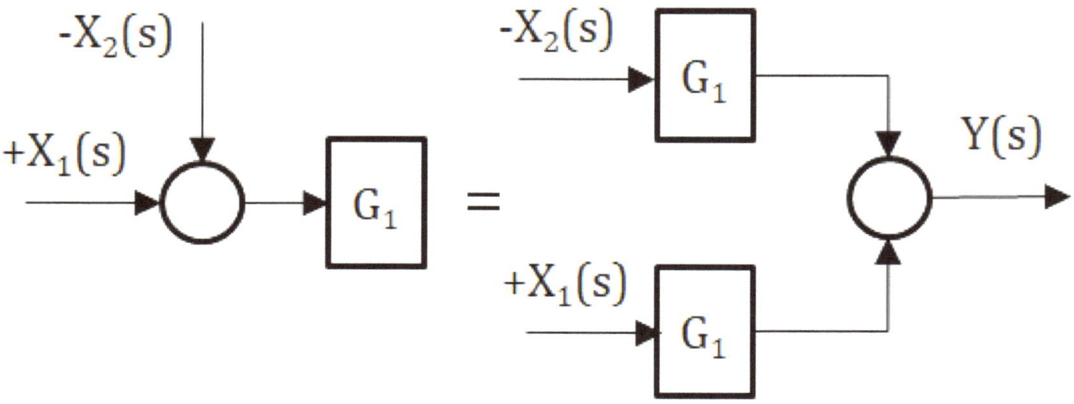

$$Y(s) = G_1 X_1(s) - G_1 X_2(s)$$

Example 7 (Blocks in Parallel):

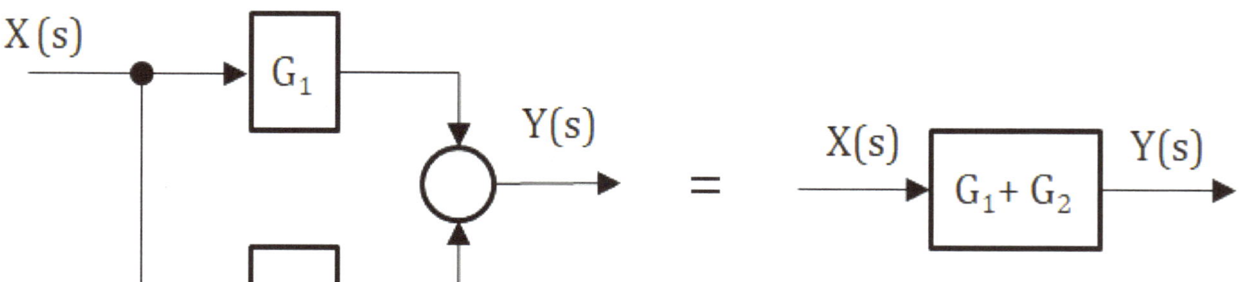

$$Y(s) = (G_1 + G_2) X(s) = G_1 X(s) + G_2 X(s)$$

1.7.3- Transfer Function

Along with block diagram algebra, the concept of *Transfer Function* is developed. A transfer function is the ratio (output/input) of a block diagram when the whole diagram is reduced to just one block. The following two examples show what is meant by transfer function.

Example 1: Positive Feedback Loop:

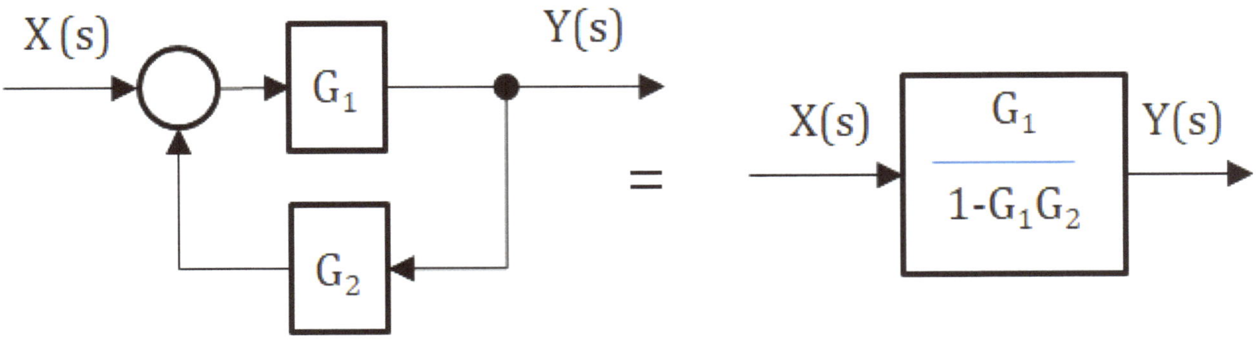

$$[\mathbf{X(s)} + \mathbf{G_2 Y(s)}]\mathbf{G_1} = \mathbf{Y(s)} \rightarrow \mathbf{TF} = \frac{\mathbf{Y(s)}}{\mathbf{X(s)}} = \frac{\mathbf{G_1}}{\mathbf{1 - G_1 G_{21}}}$$

Example 2: Negative Feedback Loop:

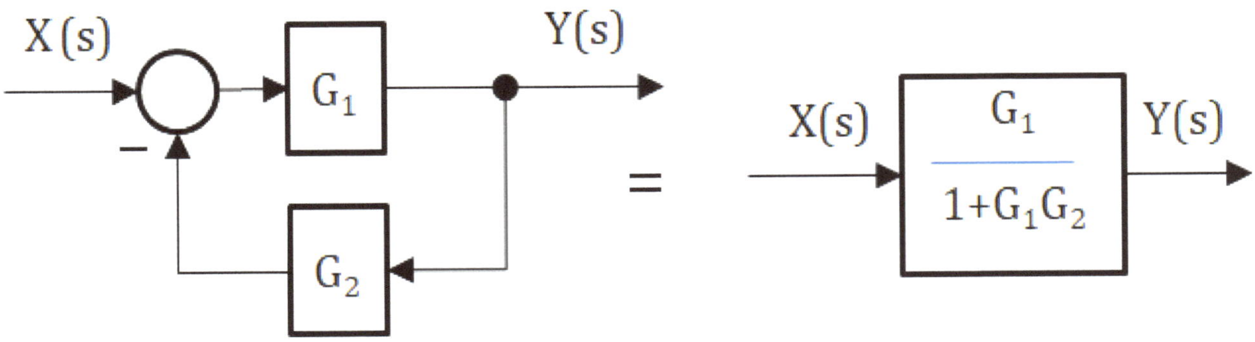

$$[\mathbf{X(s)} - \mathbf{G_2 Y(s)}]\mathbf{G_1} = \mathbf{Y(s)} \rightarrow \mathbf{TF} = \frac{\mathbf{Y(s)}}{\mathbf{X(s)}} = \frac{\mathbf{G_1}}{\mathbf{1 + G_1 G_{21}}}$$

1.8- t-Domain versus s-Domain Physical Systems Modeling

The main question here is "does the domain in which a physical system is modeled affect the system response"? The answer is "NO" as long as the same design parameters, assumptions, and system features are considered.

As it has been stated previously, fortunately, Matlab-Simulink accepts t-domain and s-domain models on the same layer. So, it is totally up to the system modeler to use his/her preferred modeling method. However, the following set of bullets conclude the t-domain versus s-domain modeling approaches:

- **Use for Complex Systems:**
  - **t-Domain Model:** Is not recommended because the model contains a large set of differential equations. Any change in the layout of the components that form the system requires a change in sequence of solving the equations. For example, if a hydraulic valve is placed downstream rather than upstream of a hydraulic motor, then the relevant DE must be placed in the same sequence.
  - **s-Domain Model:** Is recommended because models for components are built as transfer functions that can easily connect or change their layout.

- **Optimize Design Parameters:**
  - **t-Domain Model:** Recommended because each design parameter is individually represented in the DE. For example, the spool mass and spring stiffness of a directional valve is presented in its differential equation.
  - **s-Domain Model:** Is not recommended because models for components are built as transfer functions where the design parameters are hidden inside other parameters. For example, the spring stiffness is hidden inside the time constant of a first order system and inside the natural frequency and damping ratio of a second order system. Therefore, every time the spring stiffness is changed, then the relevant parameters of the transfer function must be recalculated.

- **System Performance Prediction:**
  - **t-Domain Model:** Is not recommended because system performance can't be predicted without simulating the system performance.
  - **s-Domain Model:** Is recommended because design parameter of the transfer functions indicates the system response. For example, small time constant indicates a first order system with faster response.

1.9- Development of Simulation Models for Physical Systems

After a mathematical model is developed for a physical system, it must be converted into a simulation model to be solved by a software. As shown in Fig. 1.9, based on the time-base and the contents of the software and hardware in a simulation model, the following simulation approaches are used

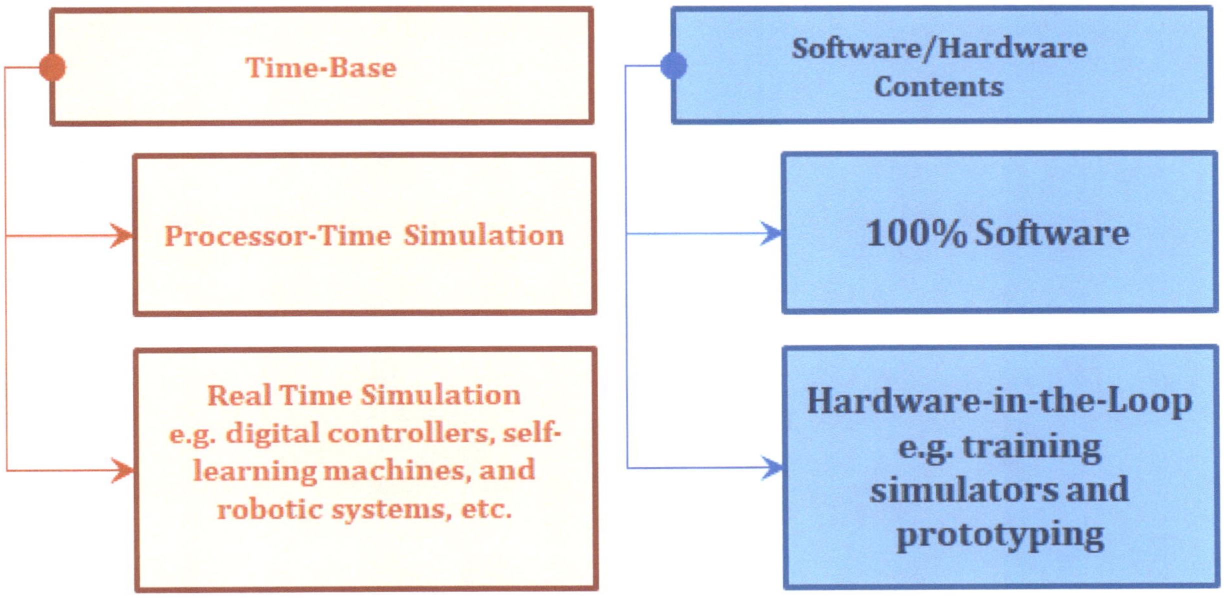

Fig. 1.9 – Configurations of Simulation Models

1.9.1- Processor-Time Simulation

In this simulation model, the processor takes whatever time needed to finish the computational requirements to simulate the performance of the physical system. So, simulation time basically depends on the microprocessors speed and most likely it takes much more time than the real time of the physical system response. In the past, when the processors were relatively slow, simulating just a minute of a physical system response may take an hour.

1.9.2- Real-Time Simulation

Today, the job of system designers is much easier because microprocessors became significantly fast. So, performance of a physical system can be simulated in Real-Time or even faster. Such capability helped in developing digital controllers, self-learning machines, and robotic systems, etc.

1.9.3- Hardware-in-the-Loop Simulation

Simulation models can be developed as 100% software or they can be configured as interfaces between software and hardware. Such a mix is referred to as *Hardware-in-the-Loop*. As shown in Fig. 1.10, this configuration was needed when building training simulators such as Flight Simulators. It is also helpful in prototyping a system. For example, as shown in the figure, a physical controller is interfaced with a virtual plant to test the overall closed loop control system performance. Or a physical system is interfaced with a virtual controller for prototyping the controller itself.

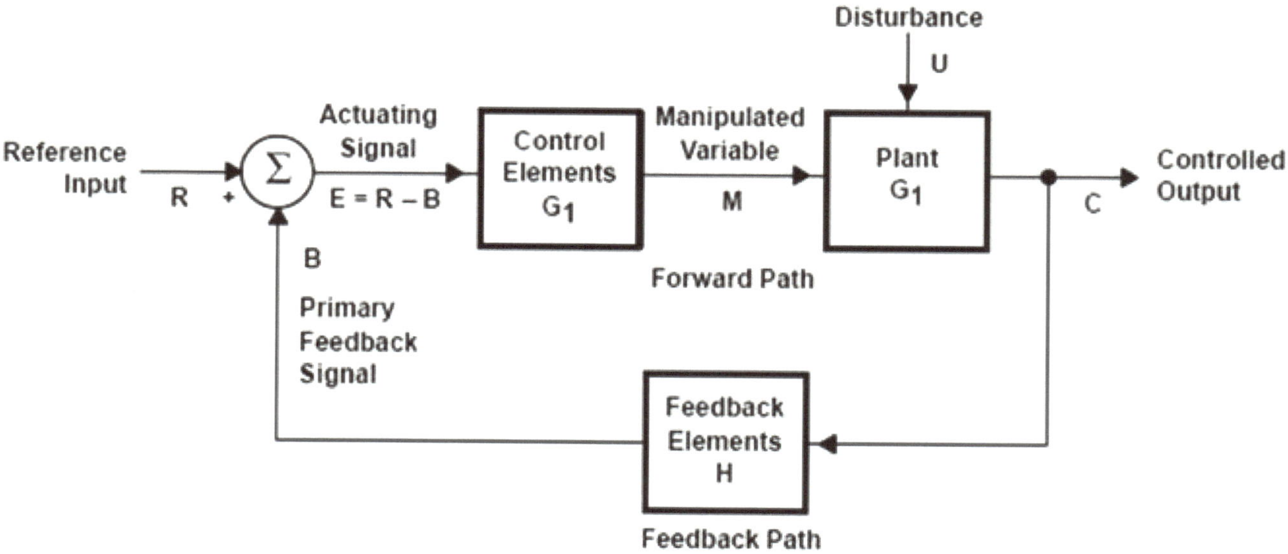

Fig. 1.10 – Hardware-in-the-Loop Simulation

1.10- Physical Systems Performance Simulation

After a simulation model is developed for a physical system, in order to investigate the system static and dynamic characteristics, it must be exposed to input forcing functions. As shown in Table 1.3, such input forcing functions represents typical and realistic loading or command scenarios.

Name	Physical Interpretations	F(t)	F(s)	Graphical Representation
Impulse	Represents shock loads or commands that occur in a very short time, such as pressure spikes or opening and closing an on/off valve.	**δ(t)**	**k**	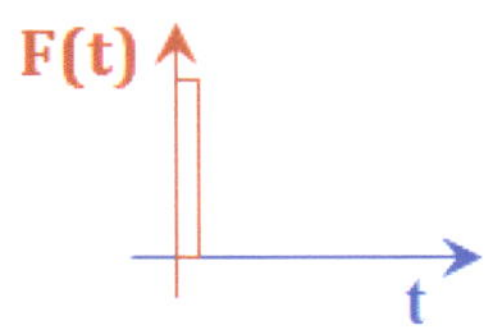
Step	Represents sudden loading or commands to a constant value, such as requesting a variable pump to go from zero to stay at 75% displacement in a very short time.	**k**	**k/s**	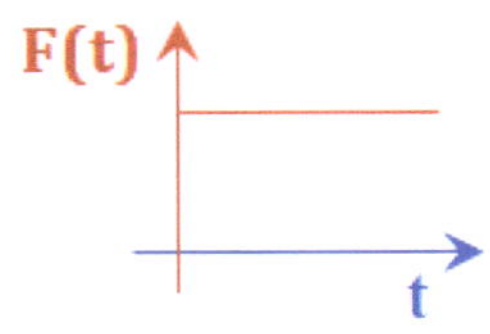
Ramp	Represents gradual loading or commands with constant rate, such as opening of a proportional DCV over a relatively long time e.g. 3 seconds. Usually used to develop static characteristic of a system.	**kt**	**k/s^2**	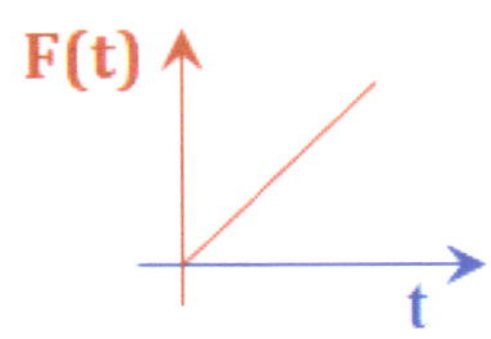
Harmonic	Physical system must be tested for periodic loads. Pure harmonic signal is a result of Fourier Analysis of periodic load. If a system passes the test under harmonic signal, it should pass the test under periodic load.	**Sin(ωt)** & **Cos(ωt)**	**$\omega/(s^2+\omega^2)$** & **$s/(s^2+\omega^2)$**	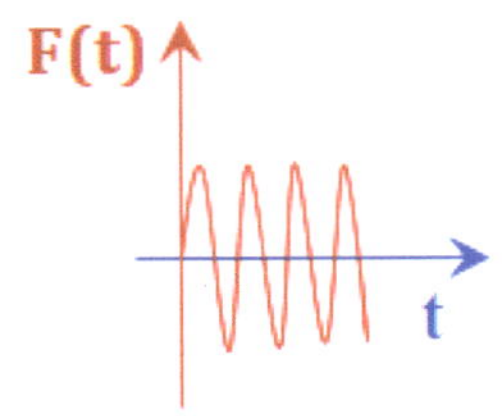

Table 1.3 – Typical Forcing Functions

1.11- Physical Systems Performance Analysis

After performance of a physical system is simulated, system performance should be investigated to see if it meets the design requirements. There are standard measures for evaluating the dynamic performance of physical systems. These measures are based on the *Transient Response* to input signal and the *System Stability*. the speed of response and the degree for system stability depends on the order of the system (such as 1st or 2nd order) and will be discussed in the relevant chapter.

1.11.1- Impulse Response

Figure 1.11 shows a typical *Impulse Response* of a physical system in t-domain. When a physical system is exposed to shock load., if the system is stable, it will return to its initial state.

1.11.2- Step Response

Figure 1.12 shows a typical *Step Response* of a physical system in t-domain. When a physical system is exposed to sudden loading or command to a constant value, if the system is stable, it will stabilize around a steady state value with a *Steady-State Error* (SSE).

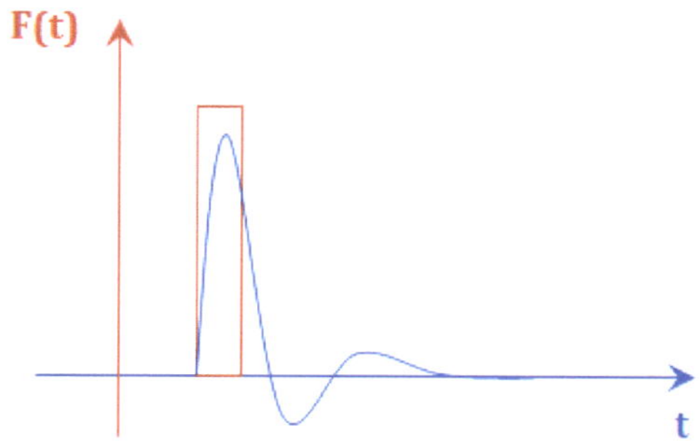

Fig. 1.11 – Impulse Response of Physical Systems

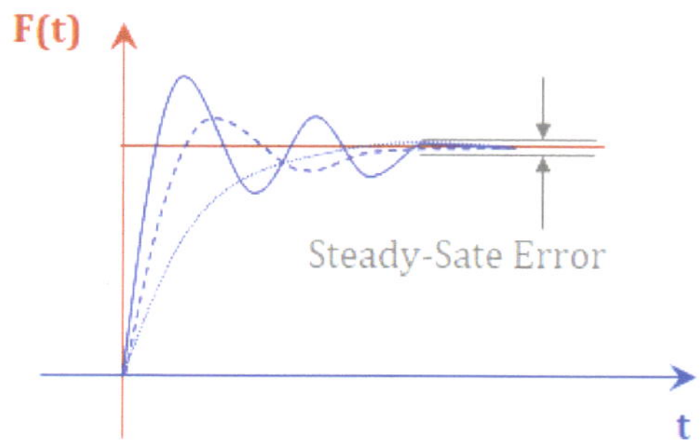

Fig. 1.12 – Step Response of Physical Systems

1.11.3- Ramp Response

Figure 1.13 shows a typical *Ramp Response* of a physical system in t-domain. When a physical system is exposed to gradual loading or command with constant rate, if the system is stable, it will stabilize around a steady state value with certain *Following Error*.

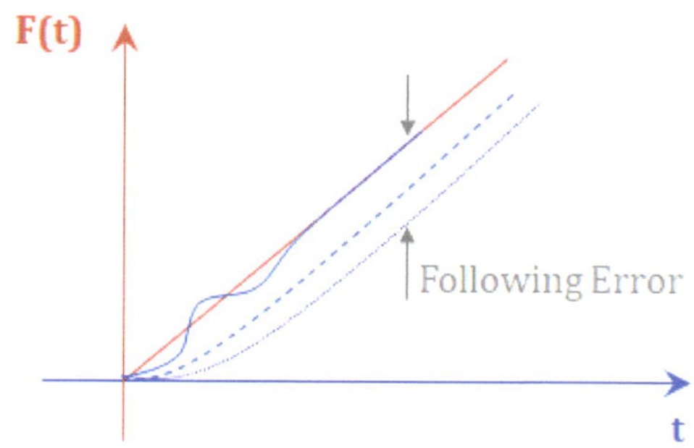

Fig. 1.13 – Ramp Response of Physical Systems

1.11.4- Frequency Response

Figure 1.14 shows a typical response of a physical system, in t-domain, to a harmonic input with certain amplitude and frequency. The figure shows that the amplitude of the system response is attenuated and is lagging as compared to the exciting signal. Amplitude attenuation is quantified by the *Amplitude Ratio* (AR = A/B).

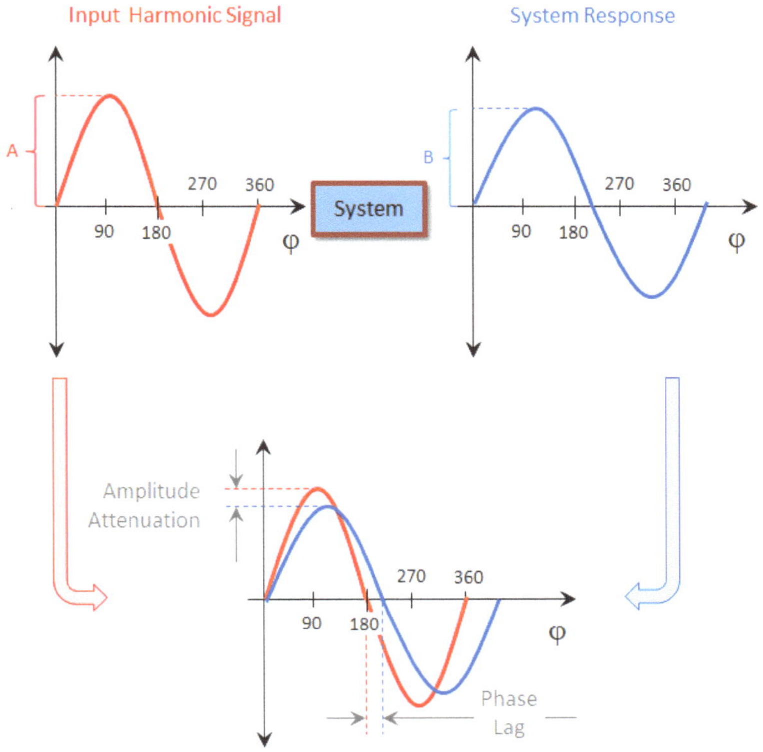

Fig. 1.14 – Response of a Physical System to a Harmonic Input Signal

In order to develop the *Frequency Response* of a physical system, both the *Amplitude Ratio* and the *Phase Lag* are presented against frequency in what is called a *Bode Plot* or Bode Diagram. Figure 1.15 shows a generic Bode diagram that contains the following parts.

Frequency: Because the frequency range is large, a logarithmic scale is used. Frequency can be presented in units of *Hertz* (Hz) or *Radians* per seconds (rad/s)

Gain Curve: *Gain Curve* shows the change of AR with the increase of the exciting frequency. As shown in the figure, at low frequency, AR = 100% on the normalized scale that means no amplitude attenuation. If the gain curve goes above 100%, this means overshooting of the system response. With increasing exciting frequency, AR decreases. At high exciting frequency, the physical system is not even able to see the input signal. For example, a human ear can't hear ultrasonic sounds and a human eye can see lights only within certain band. As shown in Eq. 1.5, AR is also presented in units of *Decibel* [dB].

$$\mathbf{AR[dB] = 20\log\left[\frac{B}{A}\right]} \quad \mathbf{1.5}$$

Because log 1 = 0, then at AR = 100%, AR = 0 [dB]. When the system overshoots, AR will have positive [dB] units because log any value >1 is positive. When the system amplitude starts to decay, AR will have negative [dB] because log any value <1 is negative. The frequency at which AR = 70% = -3dB is defined as the *Bandwidth* of the system.

Phase Curve: *Phase Curve* shows the change of *Phase Angle* with the increase of the exciting frequency. As shown in the figure, at low frequencies, the system cope with the exciting frequency with no lage. The system response lags more with the increase of the input frequency. So, this phase curve is referred to as "*Phase Angle*". Alternatively, phase curve can be plotted showing descending (negative) phase angle. In this case, the curve is titled "*Phase Lag*".

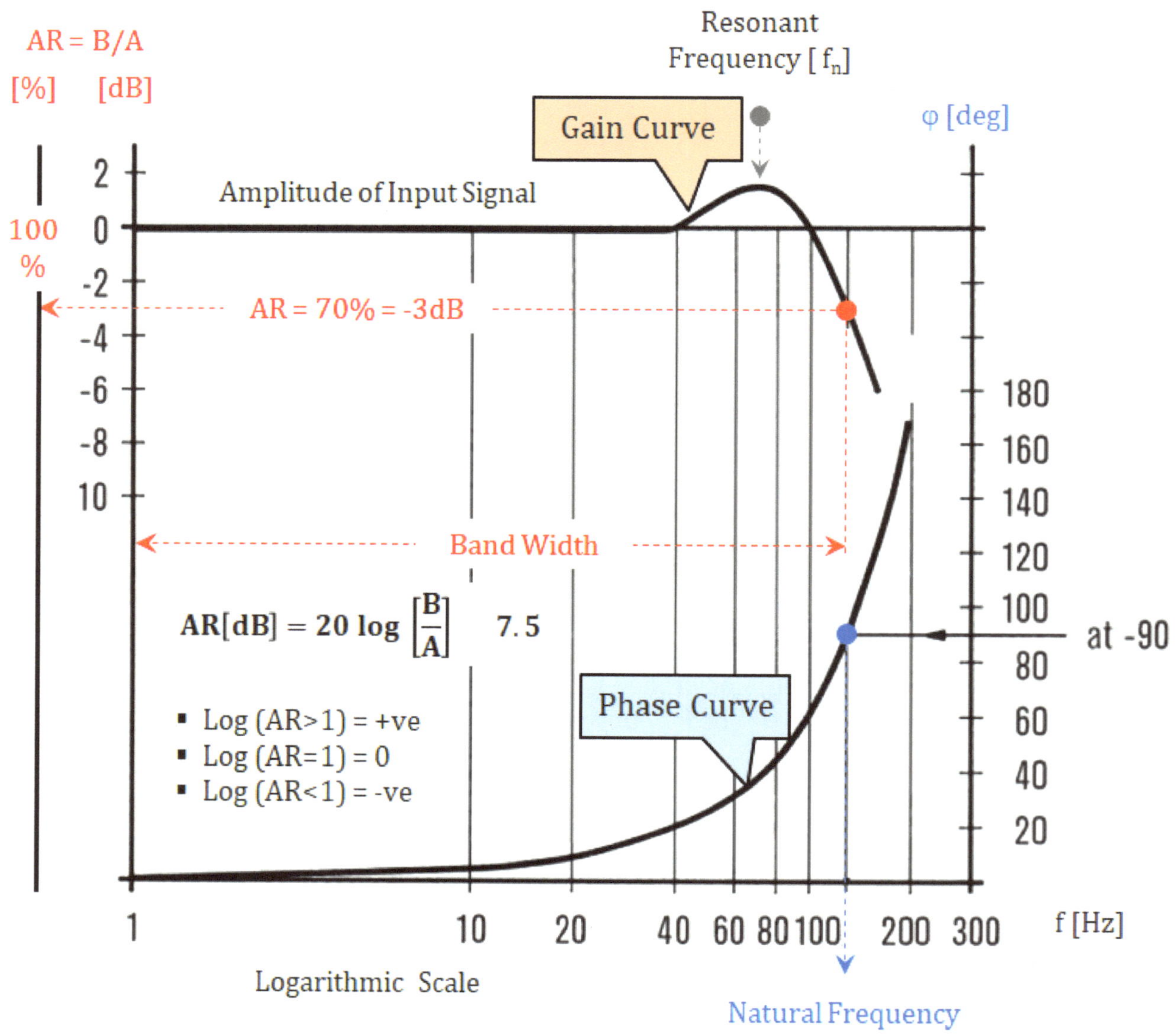

Fig. 1.15 – Generic Bode Plot

1.12- Challenges of Physical Systems Modeling and Simulation

Based on what has been previously discussed, physical systems modeling, and simulation has challenges that can be described as follows:

1.12.1- Challenges in Physical System Identification

- **Features of the Physical System:** A system modeler must clearly understand the features of the system, linear or nonlinear, static or dynamic, digital or analog, and distributed or lumped.

- **Component Design Parameters:** Component design parameters such as dimensional parameters, clearances, masses, friction coefficients, spring constants, etc. are required for the modeling process. Getting access to the specifications of a physical component may be a challenge because these specifications are likely considered classified design parameters for the component manufacturer.

- **Assumptions:** It is almost impossible to know in advance all the design parameters about the modeled physical system. Therefore, feasible and reasonable assumptions must be made to compensate for the lack of information or to define unmeasurable parameters. These assumptions must be made based on experienced understanding of the physical system.

1.12.2- Challenges in Mathematical Model Development

- **Level of Model Complexity:** The main challenge of modeling a physical system is that there is no one method that is applicable for all cases. It is rather a case-by-case, or in other words, a very tailoring process. A mathematical model shouldn't be tight (simplified) to the level of not being able to capture necessary details of system performance, and at the same time, it shouldn't be large (exaugurated) to the level of giving unnecessary details wasting valuable resources of time and cost.

- **Mathematical Modeling Technique:** Deciding to use *Differential Equations* (DE) or *Transfer Functions* (TF) or any other technique is another challenge because they are all available and it depends on the know-how of the system modeler.

- **Solvable Model:** Develop a solvable mathematical model "set of equations" that represent the system characteristics is another challenge. That is because basically the number of equations must be equal to the number of unknowns. Also, *Algebraic Loops* and *Overflow* cases must be avoided. Algebraic loop is when the calculations fall in a circle without coming to an end. Overflow is when dividing by zero.

1.12.3- Challenges in Simulation Model Development

- **Input/output Parameters:** from the very beginning, a system modeler must understand what the available input information to the model is and what the required output information from the model is.

- **Models for Static vs. Dynamics:** A system modeler should be careful when combining models for static components with models for dynamic components. An example is modeling a pump as just a constant source of flow dumped into a model for cylinder dynamics. At the first iteration of the model, the cylinder will surge flow from the pump as input flow at no time! That may result in hypothetical very high pressure or the model will result in calculation error.

- **Programming Language:** Since there is a wide variety of software platforms available in the market, selecting adequate software to solve the model is a challenge.

- **Artwork:** In addition to the engineering basics, physical system modeling requires some sort of artwork and creativity to build flexible and visually enhanced models.

1.12.4- Challenges in Performance Simulation

- **Sampling Rate:** Modeling software solves the mathematical model using iterative techniques based on a *Sampling Rate.* Sampling rate is the incremental time change for each iteration. Considering a large sampling rate will result in missing details of system dynamics in the transient periods. Considering a small sampling rate will increase actual time of simulation and the required computational efforts. Therefore, sampling rate must be compromised. Matlab-Simulink offers selecting fixed or variable sampling rate.

- **Input Signals:** Deciding what type of input signals should be used to simulate the system performance requires clear understanding of the system operation, i.e. will this system likely expose to steady, shock, or periodic loads.

- **Model Debugging:** If a system simulation generated did give unrealistic results, a system modeler must be able to debug the model to know the source of error. A modeler should be aware by the debugging tools that are offered by the software.

1.12.5- Challenges in Performance Analysis

- **Model Results:** A system modeler should have a good level of understanding of the physical system to assess how close the simulation results in representing the actual system characteristics. A system modeler should not surrender as a slave to the math! He/She should continue to work as a master and do not believe in the model unless the it is validated and generates results that make sense.

1.13- Basic Elements of Physical Systems

A hydraulic system that drives an external load is composed of one or a combination of the basic physical systems:

- Mechanical Translational System.
- Mechanical Rotational System.
- Fluidic System.
- Electrical System.
- Thermal System.

As shown in Table 1.4, each one of the above mentioned basic physical systems can be broken down to the following basic element:

- An *Inductive* Element: This is the inertial element (not applicable for thermal system).
- A *Resistive* Element: This element dissipates the input power in form of heat.
- A *Capacitive* Element: This is the element in which energy can be stored.

The contents of the table are as follows:

❖ Mechanical Translational System:
- Inductive Element is a *Translational Mass* **m**.
- Resistive Element is a *Liner Damper* with a *Damping Coefficient* $\mathbf{k_f}$.
- Capacitive Element is a *Linear Spring* with a *Capacitance* **C** = 1/ *Spring Stiffness*($\mathbf{k_s}$).

❖ Mechanical Rotational System:
- Inductive Element is a *Rotational Mass* with a *Mass Moment of Inertia* **J**.
- Resistive Element is a *Rotational Damper* with a *Damping Coefficient* $\mathbf{k_f}$.
- Capacitive Element is a *Torsion Spring* with a *Capacitance* **C** = 1/*Spring Stiffness*($\mathbf{k_s}$).

❖ Hydraulic System:
- Inductive Element is the inertial effect **I** of the moving fluid mass.
- Resistive Elements is the element that resists fluid flow (such as transmission lines and valves) with *Fluid Flow Resistivity* **R**.
- Capacitive Element *Capacitance* **C** = Initial Volume ($\mathbf{V_o}$)/*Fluid Bulk Modulus* (**β**).

❖ Electrical System:
- Inductive Element is the *Electrical Inductor* with *Electrical Inductance* **L**. It is much like the inertial effect of the moving electrons that resists the change in the flow of electrons.
- Resistive Element is an element that resists flow of electrons (such as Electrical Resistors) with *Electric Resistivity* **R**.
- Capacitive Element is the *Electrical Capacitor* with *Electrical Capacitance* **C**.

❖ Thermal System:
- Inductive Element is not applicable because there is no moving mass in thermal systems.
- Resistive Element is *Thermal Resistor* (such as *Heat Shield*) with *Thermal Resistivity* **R**.
- Capacitive Element is a *Heat Sink* with *Thermal Capacitance* **C**.

Elements→ System↓	**Inductive "Inertial"**	**Resistive "Energy Dissipative"**	**Capacitive "Energy Storage"**
Mechanical Translational 	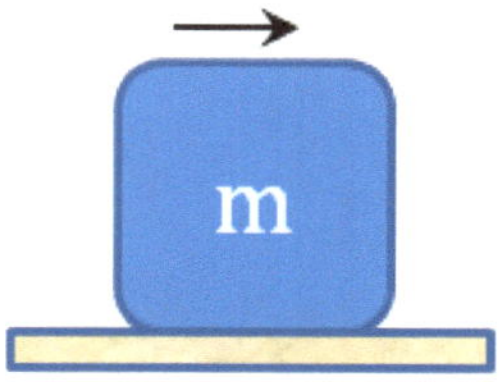 Translational Mass (m)	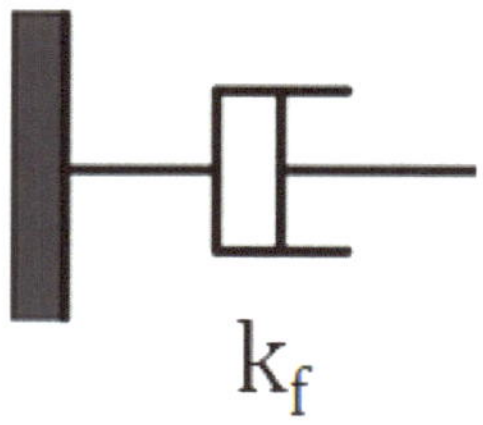 Linear Damper Damping Coefficient (k_f)	 Linear Spring Capacitance (C) = 1/Spring Stiffness (k_s)
Mechanical Rotational 	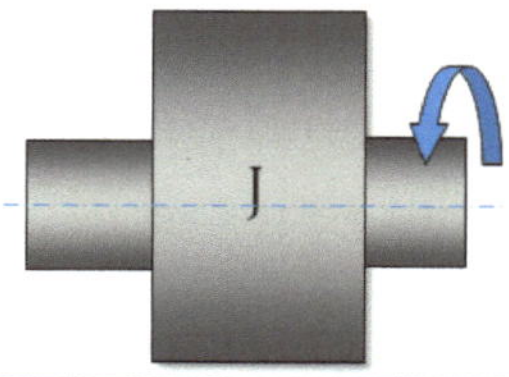 Rotational Mass Mass Moment of Inertia (J)	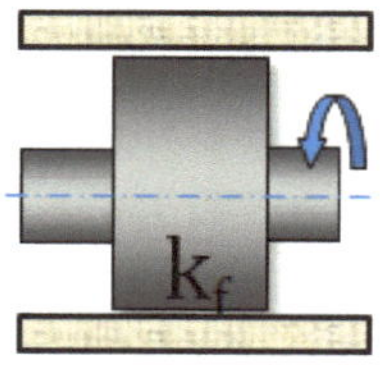 Torsional Damper Damping Coefficient (k_f)	 Torsion Spring Capacitance (C) = 1/Spring Stiffness (k_s)
Hydraulic 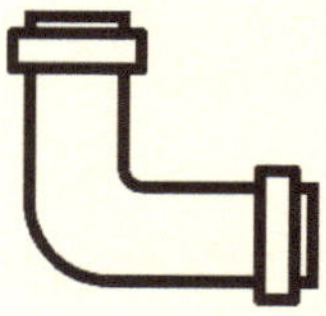	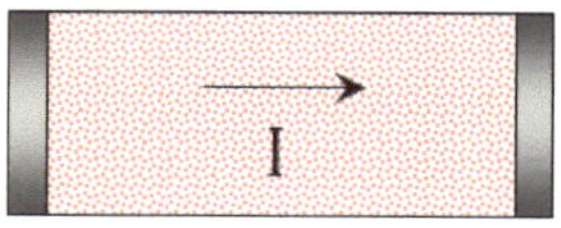 Fluid Line Inductance (I)	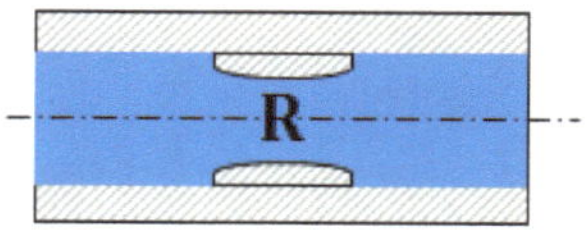 Fluid Line Resistance (R)	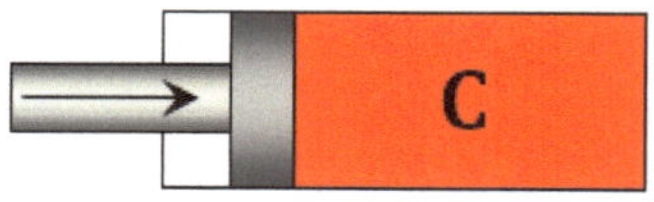 Fluid Line Capacitance Capacitance (C) = V_0 /Bulk Modulus (β)
Electrical 	 Electric Inductor Inductance (L)	 Electric Resistor Resistance (R)	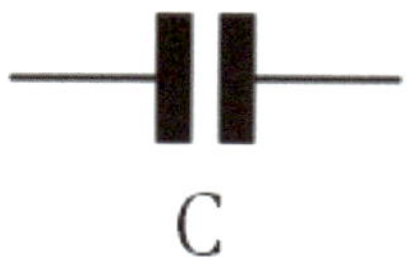 Electrical Capacitor Capacitance (C)
Thermal 	Not Applicable	 Heat Resistor Resistance (R)	 Heat Sink Capacitance (C)

Table 1.4 – Basic Elements of Physical Systems

1.14- Effort and Flow Variables of Physical Systems

As shown in Table 1.5, each of the previously mentioned basic physical systems has an *Effort Variable* and a *Flow Variable*. Physical systems behave similarly so that the following one statement represents the flow-effort variables relationship.

If |Effort Variable| > 0 → |Flow Variable| ≠ Constant
If |Effort Variable| = 0 → |Flow Variable|= 0 OR = Constant
Charge Variable = ∫(Flow Variable)

For a better understanding of the relationship between the effort and flow variables:

- Mechanical Translational System:
  - Effort Variable is the *Resultant Force* **F** and Flow Variable is *Linear Speed* **v**.
  - If $|\Sigma \mathbf{F}| > 0 \rightarrow |\mathbf{v}| \neq$ Constant
  - It means that the Mass **m** is either accelerating or decelerating depends on the sign of **ΣF**.
  - If $|\Sigma \mathbf{F}| = 0 \rightarrow \mathbf{v} = 0$ OR $|\mathbf{v}| =$ Constant
  - It means that the Mass **m** is either standstill or translating with constant **v**.
  - Charge Variable is *Linear Displacement* $\mathbf{x} = \int \mathbf{v}dt$.

- Mechanical Rotational System:
  - Effort Variable is the *Resultant Torque* **T** and Flow Variable is *Angular Speed* **ω**.
  - If $|\Sigma \mathbf{T}| > 0 \rightarrow |\boldsymbol{\omega}| \neq$ Constant
  - It means that the Moment of Inertia **J** is either accelerating or decelerating depends on the sign of **ΣT**.
  - If $|\Sigma \mathbf{T}| = 0 \rightarrow \boldsymbol{\omega} = 0$ OR $|\boldsymbol{\omega}| =$ Constant
  - It means that the moment of inertia **J** is either standstill or rotating with constant **ω**.
  - Charge Variable is *Angular Displacement* $\boldsymbol{\Theta} = \int \boldsymbol{\omega}dt$.

- Hydraulic System:
  - Effort Variable is the *Differential Pressure* **ΔP** and Flow Variable is *Fluid Flow* **Q**.
  - If $|\Delta \mathbf{P}| > 0 \rightarrow$ Fluid Flow **Q** moves in a direction depends on the sign of **ΔP**.
  - Charge Variable is *Fluid Volume* $\Delta \mathbf{V} = \int \mathbf{Q}dt$.

- Electrical System:
  - Effort Variable is the *Voltage Difference* **Δv** and Flow Variable is *Electrical Current* **i**.
  - If $|\Delta \mathbf{v}| > 0 \rightarrow$ Electrical Current **i** moves in a direction depends on the sign of **Δv**.
  - Charge Variable is *Electrical Charge* $\mathbf{Q} = \int \mathbf{i}dt$.

- Thermal System:
  - Effort Variable is the *Temperature Difference* **ΔT** and Flow Variable is *Heat Flow* **q**.
  - If $|\Delta \mathbf{T}| > 0 \rightarrow$ Heat Flow **q** moves in a direction depends on the sign of **ΔT**.
  - Charge Variable is *Heat* $\mathbf{Q} = \int \mathbf{q}dt$.

Variables→ Systems↓	Effort Variable	Flow Variable	Charge Variable	
Mechanical Translational 	Resultant Force **F**	Linear Velocity **v**	Linear Displace-ment **x**	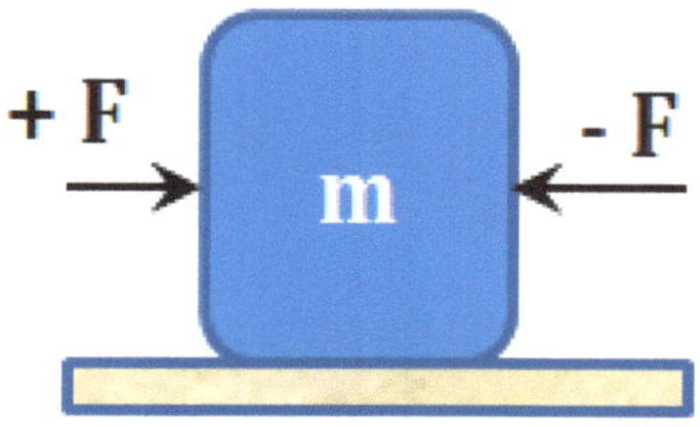
Mechanical Rotational 	Resultant Torque **T**	Angular Velocity ω	Angular Displace-ment θ	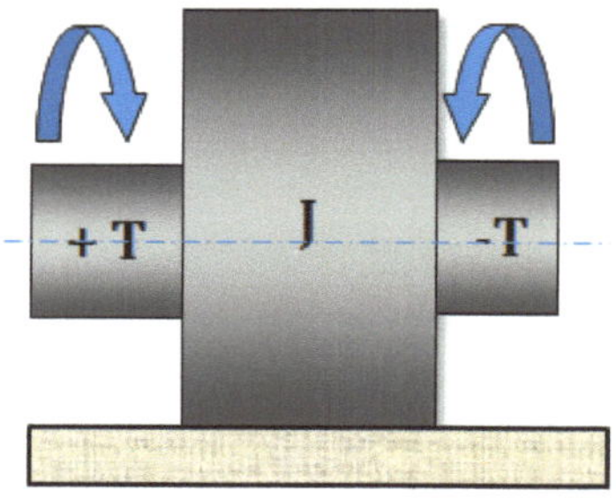
Hydraulic 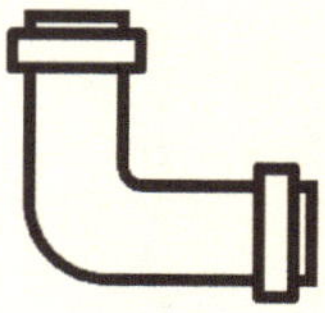	Differential Pressure **ΔP**	Fluid Flow **Q**	Fluid Volume **ΔV**	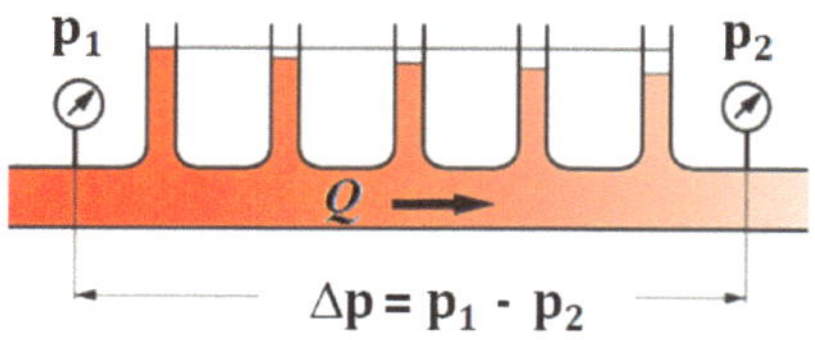
Electrical 	Voltage Difference **Δv**	Electrical Current **i**	Electrical Charge **Q**	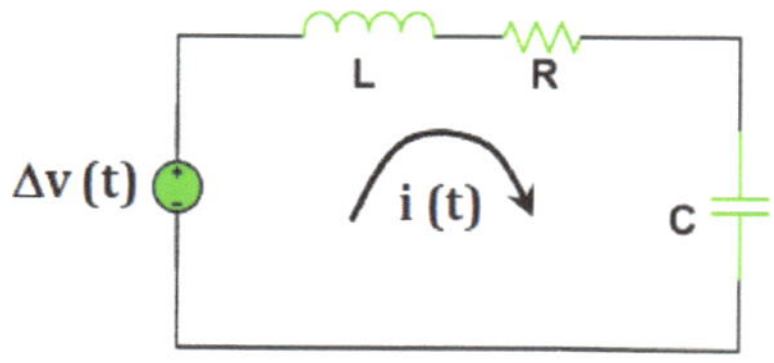
Thermal 	Temperature Difference **ΔT**	Heat Flow **q**	Heat **Q**	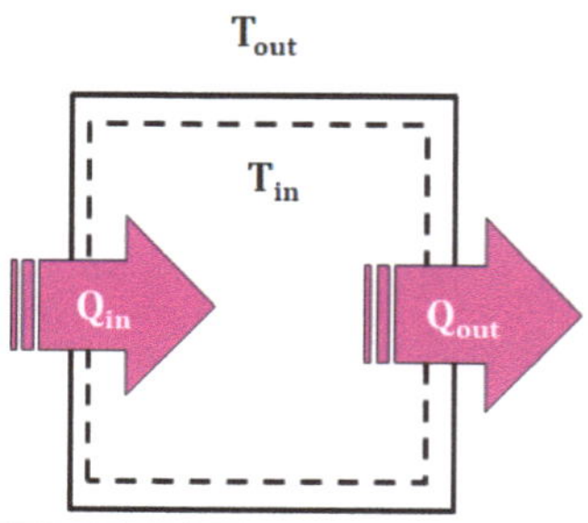

Table 1.5 - Effort and Flow Variables for Physical Systems

1.15- Power Calculation for Physical Systems

In order to quantify the power requirements for physical systems, the following set of equations shows one common equation is used:

$$\textbf{Power} = \textbf{Effort Variable} \times \textbf{Flow Variable} \quad ----- \quad \textbf{1.5}$$

- For a Mechanical Translational System:

$$\textbf{Power, H} = \textbf{F} \times \textbf{v} \qquad \textbf{1.5A}$$

- For a Mechanical Rotational System:

$$\textbf{Power, H} = \textbf{T} \times \boldsymbol{\omega} \qquad \textbf{1.5B}$$

- For a Hydraulic System:

$$\textbf{Power, H} = \boldsymbol{\Delta}\textbf{P} \times \textbf{Q} \qquad \textbf{1.5C}$$

- For an Electrical System:

$$\textbf{Power, H} = \boldsymbol{\Delta}\textbf{v} \times \textbf{i} \qquad \textbf{1.5D}$$

Note: As shown in Fig. 1.16, same equation (Eq. 1.5C) is used to quantify the hydraulic power whether the power is gained by a pump, consumed to do useful work by a hydraulic actuator, or wasted as heat in an element.

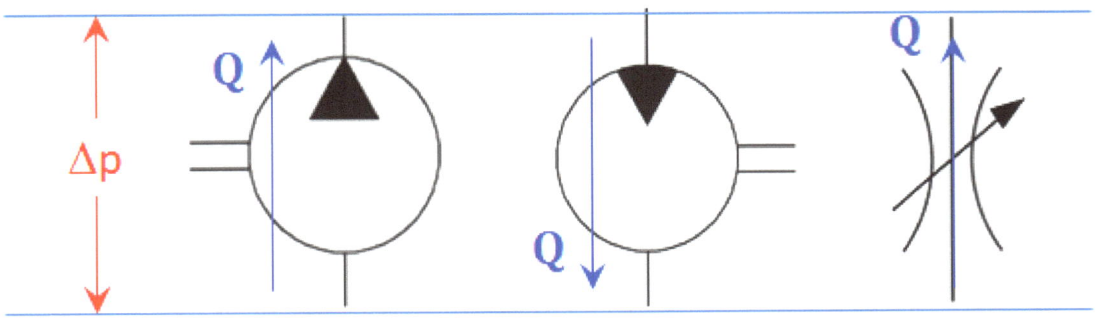

Fig. 1.16 – Power Calculation for Hydraulic Components

1.16- Mathematical Representation of Inductive Elements

In order to quantify the effort required to accelerate an inductive element, the following set of equations shows one common equation is used:

$$\textbf{Effort Variable} = \textbf{Inductance} \times \textbf{Rate of change of Flow Variable} \quad ---- \; 1.6$$

❖ For a Mechanical Translational System:

$$F = m \times \frac{dv}{dt} \qquad 1.6A$$

❖ For a Mechanical Rotational System:

$$T = J \times \frac{d\omega}{dt} = J \times \alpha \qquad 1.6B$$

❖ For an Electrical System:

$$\Delta v = L \times \frac{di}{dt} \qquad 1.6C$$

Inductance **L** of an electrical inductor is a given value depending on the design of the inductor.

❖ For a Hydraulic System:

$$\Delta P = I \times \frac{dQ}{dt} \qquad 1.6D$$

Inductance **I** of a hydraulic transmission line can be derived as follows. As shown in Fig. 1.17, differential pressure that is required to accelerate certain mass of fluid is

$$\Delta P = \frac{\text{Fluid Inertia}}{\text{Line Area (A)}} = \frac{\text{Fluid Mass (m)}}{\text{Line Area (A)}} \times \frac{d\,[\text{Fluid Velocity (v)}]}{dt}$$

Since Fluid Flow Q = v x A, and Fluid Mass = Fluid Density ρ x Fluid Volume V, then

$$\Delta P = \frac{m}{A} \times \frac{1}{A} \times \frac{dQ}{dt} = \frac{\rho \times V}{A} \times \frac{1}{A} \times \frac{dQ}{dt} = \frac{\rho \times A \times L}{A} \times \frac{1}{A} \times \frac{dQ}{dt} = \frac{\rho L}{A}\frac{dQ}{dt} = I\frac{dQ}{dt}$$

$$\text{Then } I = \frac{\rho L}{A} \qquad 1.7$$

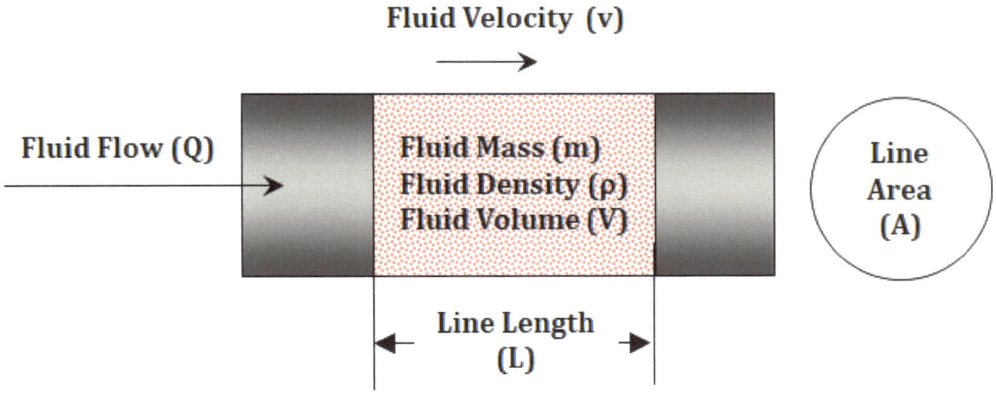

Fig. 1.17 – Inductance of a Hydraulic Transmission Line

Kinetic Energy (associated with inductive elements): In physics, *Kinetic Energy* of an object is the energy that it possesses due to its motion. It is defined as the work needed to accelerate a body of a given mass from one velocity to other velocity. In order to quantify the kinetic energy of an inductive element, the following set of equations shows one common equation is used:

$$\textbf{Kinetic Energy} = \tfrac{1}{2}\,\textbf{Inductance} \times (\textbf{Flow Variable})^2 \quad ----- \quad 1.8$$

- ❖ For a Mechanical Translational System:

$$E_K = \frac{1}{2}mv^2 \qquad 1.8A$$

- ❖ For a Mechanical Rotational System:

$$E_K = \frac{1}{2}J\omega^2 \qquad 1.8B$$

- ❖ For an Electrical System:

$$E_K = \frac{1}{2}Li^2 \qquad 1.8C$$

- ❖ For a Hydraulic System:

$$E_K = \frac{1}{2}mv^2 = \frac{1}{2}(\rho V)\left(\frac{Q}{A}\right)^2 = \frac{1}{2}(\rho AL)\left(\frac{Q}{A}\right)^2 = \frac{1}{2}\left(\frac{\rho L}{A}\right)Q^2 = \frac{1}{2}IQ^2 \qquad 1.8D$$

1.17- Mathematical Representation of Resistive Elements

In order to quantify the effort required to dissipate energy in a resistive element for physical systems, the following set of equations shows one common equation is used:

$$\textbf{Effort Variable} = \textbf{Resistivity Factor} \times \textbf{Flow Variable} \quad ---- \; 1.9$$

- For a Mechanical Translational System (Hook's Law):

$$\mathbf{F = k_f v} \qquad \mathbf{1.9A}$$

- For a Mechanical Rotational System:

$$\mathbf{T = k_f \omega} \qquad \mathbf{1.9B}$$

- For an Electrical System:

$$\mathbf{\Delta v = Ri} \qquad \mathbf{1.9C}$$

- For a Hydraulic Transmission Lines:

$$\mathbf{\Delta P = RQ} \qquad \mathbf{1.9D}$$

- For a Hydraulic Valves (Nonlinear):

$$\mathbf{\Delta P = RQ^2} \qquad \mathbf{1.9E}$$

Where:

k_f = the linear/angular *Friction Coefficient.*

R = the electric/hydraulic *Resistivity Factor.*

Hydraulic resistivity for transmission lines and valves will be discussed in Chapter 6.

Dissipated (Wasted) Energy (associated with resistive elements): Energy losses are relative because energy can't vanish. In physics, *Dissipated Energy* means energy that is converted into a non-useful form. Energy losses in hydraulic components, such as energy wasted in transmission lines and valves, are converted into heat. In an electrical heater, any energy that does not converted into heat is considered wasted energy!

However, in order to quantify the potential energy of a capacitive element, the following set of equations shows one common equation is used:

$$\textbf{Wasted Energy} = \textbf{Wasted Power} \times \textbf{Time}$$

Equation 1.5 →

$$\textbf{Wasted Energy} = \textbf{Effort Variable} \times \textbf{Flow Variable} \times \textbf{Time}$$

Then

$$\textbf{Wasted Energy} = \textbf{Effort Variable} \times \textbf{Charge Variable} ---- \quad \textbf{1.10}$$

- For a Mechanical Translational System (Hook's Law):

$$\mathbf{E_W = Fx = k_f v x} \qquad \textbf{1.10A}$$

- For a Mechanical Rotational System:

$$\mathbf{E_W = T\theta = k_f \omega \theta} \qquad \textbf{1.10B}$$

- For an Electrical System:

$$\mathbf{E_W = \Delta v Q = R i Q} \qquad \textbf{1.10C}$$

- For a Hydraulic Transmission Lines:

$$\mathbf{E_W = \Delta P \Delta V = R Q \Delta V} \qquad \textbf{1.10D}$$

- For a Hydraulic Valves (Nonlinear):

$$\mathbf{E_W = \Delta P \Delta V = R Q^2 \Delta V} \qquad \textbf{1.10E}$$

1.18- Mathematical Representation of Capacitive Elements

In order to quantify the effort required to store energy in a capacitive element for physical systems, the following set of equations shows one common equation is used:

$$\textbf{Effort Variable} = \frac{\textbf{1}}{\textbf{Capacitnace}} \int (\textbf{Flow Variable})\ \textbf{dt} = \frac{\textbf{Charge Variable}}{\textbf{Capacitnace}} \quad ----\ \textbf{1.11}$$

❖ For a Mechanical Translational System (Hook's Law):

$$\mathbf{F} = \frac{\mathbf{1}}{\mathbf{C}} \int \mathbf{v\ dt} = \mathbf{k_s x} \qquad \mathbf{1.11A}$$

Where $\mathbf{k_s}$ is the spring stiffness (also referred as *Spring Constant*) and **x** is the linear deformation (compression or tensile length) of a linear spring.

❖ For a Mechanical Rotational System:

$$\mathbf{T} = \frac{\mathbf{1}}{\mathbf{C}} \int \boldsymbol{\omega}\ \mathbf{dt} = \mathbf{k_s} \boldsymbol{\theta} \qquad \mathbf{1.11B}$$

Where $\mathbf{k_s}$ is the spring stiffness and **θ** is the angular deformation of a torsion spring.

❖ For an Electrical System:

$$\Delta \mathbf{v} = \frac{\mathbf{1}}{\mathbf{C}} \int \mathbf{i\ dt} = \frac{\textbf{Erlecrtic Charge (Q)}}{\mathbf{C}} \qquad \mathbf{1.11C}$$

❖ For a Hydraulic System:

$$\Delta \mathbf{P} = \frac{\mathbf{1}}{\mathbf{C}} \int \mathbf{Q\ dt} = \frac{\textbf{Fluid Bulck Modulus } (\boldsymbol{\beta})}{\textbf{Container Volume } (\mathbf{V_O})} \int \mathbf{Q\ dt} = \frac{\boldsymbol{\beta} \Delta \mathbf{V}}{\mathbf{V_O}} \qquad \mathbf{1.11D}$$

The interpretation of this equation is, for an initial container of volume $\mathbf{V_O}$, if the sum of flow-in is more than the sum of flow-out, pressure increases, and vice versa. Therefore, this equation can be used to investigate generating pressure spikes and cavitaion developing in hydraulic components. More examples will be discussed in Chapter 6.

Potential (stored) Energy (associated with capacitive elements): In physics, *Potential Energy* of an object is the energy that s stored in the object because of its altitude relative to other objects, elastic deformation, fluid compressibility, or electric charge. It is defined as the work needed to make such a change in the object.

In order to quantify the potential energy of a capacitive element, the following set of equations shows one common equation is used:

$$\textbf{Potential Energy} = \frac{1}{2}\textbf{Effort Variable} \times \textbf{Charge Variable} = \frac{(\textbf{Charge Variable})^2}{2 \times \textbf{Capacitnace}} \qquad \textbf{1.12}$$

❖ For a Mechanical Translational System:

Elastic potential energy is potential energy stored as a result of deformation of an elastic object, such as the stretching or compressing a linear spring. It is equal to the work done to deform the spring, which depends upon the spring constant $\mathbf{k_s}$ as well as the deformation distance.

Equation 1.5A shows that the force **F** to deform a linear spring varies linearly to the deformation length **x**. Then, as shown in Fig. 1.18, the work done equals area under the curve.

$$\mathbf{E_P} = \frac{1}{2}\mathbf{Fx} = \frac{\mathbf{x}}{2}\left[\frac{1}{C}\int \mathbf{v\,dt}\right] = \frac{\mathbf{x^2}}{\mathbf{2C}} = \frac{1}{2}\mathbf{k_s x^2} \qquad \mathbf{1.12A}$$

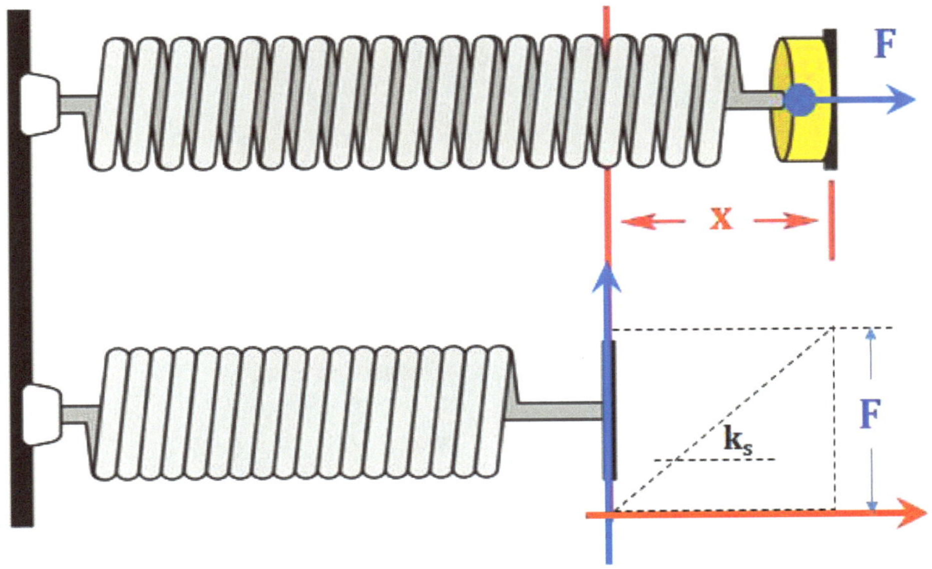

Fig. 1.18 – Potential Energy for Linear Spring

❖ For a Mechanical Rotational System:
Similar to linear springs, torsion springs can store energy as follow:

$$\mathbf{E_P = \frac{1}{2}T\theta = \frac{\theta}{2}\left[\frac{1}{C}\int \omega\, dt\right] = \frac{\theta^2}{2C} = \frac{1}{2}k_s\theta^2} \qquad \mathbf{1.12B}$$

❖ For an Electrical System:

$$\mathbf{E_P = \frac{1}{2}\Delta vQ = \frac{Q}{2}\left[\frac{1}{C}\int i\, dt\right] = \frac{1}{2}\frac{1}{C}\, Q^2} \qquad \mathbf{1.12C}$$

Where **Q** is the electrical charge and **C** is the capacitance of the capacitor.

❖ For a Hydraulic System:

$$\mathbf{E_P = \frac{1}{2}\Delta p\Delta V = \frac{\Delta V}{2}\left[\frac{1}{C}\int Q\, dt\ \right] = \frac{1}{2}\frac{1}{C}\,(\Delta V)^2 = \frac{1}{2}\frac{\beta}{V_O}(\Delta V)^2} \qquad \mathbf{1.12D.1}$$

$$\mathbf{E_P = \frac{1}{2}\Delta p\Delta V = \frac{1}{2}\Delta pAx = \frac{1}{2}Fx} \qquad \mathbf{1.12D.2}$$

As shown in Fig. 1.19, **ΔV** is the change in the oil volume inside a container of an initial volume **Vo**. By definition, **β = Δp / (ΔV/ Vo)**. Then substituting the value β in Eq. 1.6D proves the calculation method of the hydraulic stored energy.

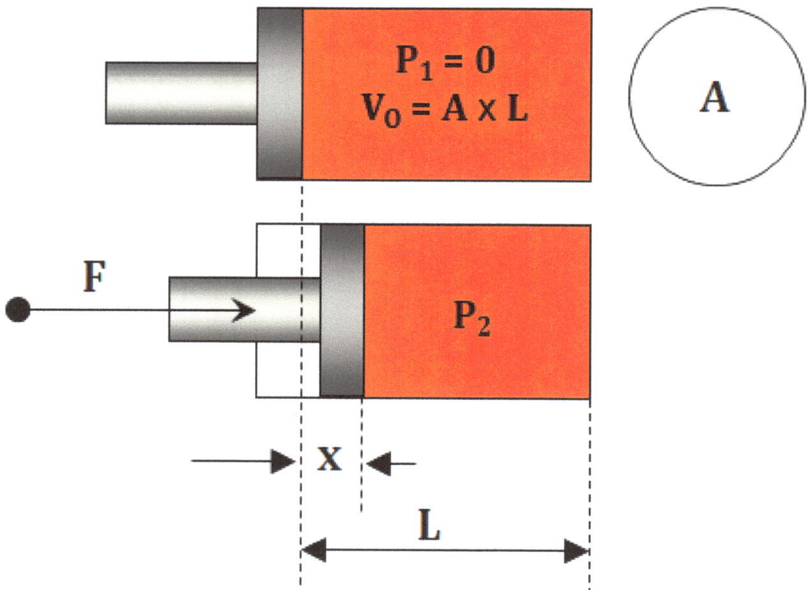

Fig. 1.19 – Potential Energy of Hydraulic Fluids

Chapter 2
Modeling and Simulation of First-Order Dynamic Systems

Objectives

In this chapter, methods and theories presented in Chapter 1 are applied to First-Order dynamic systems. The chapter presents mathematical modeling for first-order systems in t-domain and s-domain. This chapter also presents the response of first-order systems to the typical forcing functions including, step, ramp, and harmonic inputs. The chapter discusses the measured characteristics of first-order step response and how to develop the transfer function of the system based on existing dynamic characteristics.

Brief Contents

Chapter 2: Modeling and Simulation of First-Order Dynamic Systems

A *First-Order* dynamic system is the simplest among the dynamic systems because it contains only first derivative of the investigated system variables and its mathematical solutions are well-known.

2.1- First-Order Physical System Identification

As it has been discussed in chapter 1, first step in a product cycle development is to identify the physical system. As shown in Fig. 2.1, to simplify the discussion, two examples for mechanical translational first-order systems are considered. It is to be noted that a first-order system contains only two elements (inductive & resistive) OR (capacitive and resistive). Resistive element must be there for a system to be stable.

Example 1 (proportional solenoid drives a negligible spool mass against a spring and friction): In the first example (left), the system consists of a spring and a damper that are connected to a moving point with negligible mass. In this example, a model is developed to solve for the displacement **x** of the connecting point. Therefore, the spring stiffness $\mathbf{k_s}$ and the friction coefficient $\mathbf{k_f}$ must be known. The spring stiffness $\mathbf{k_s}$ is a measurable design parameter.

Example 2 (Pressure force is applied in a hydraulic cylinder to drive an equivalent moving mass against friction): In the second example (right), the system consists of a considerable mass connected to a damper. In this example, a model is developed to solve for the velocity **v** of the moving mass. Therefore, the mass **m**, and the friction coefficient $\mathbf{k_f}$ must be known. The mass **m** is a measurable design parameter.

In both examples, since $\mathbf{k_f}$ is not an easy to measure design parameter, it can be reasonably assumed and optimized during system design.

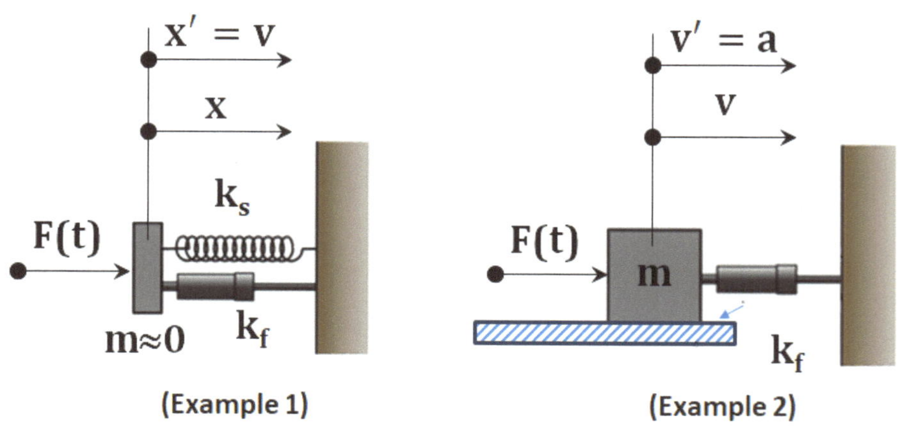

Fig. 2.1 – Examples of Mechanical Translational First-Order Systems

2.2- First-Order System Mathematical Modeling

As it has been discussed in chapter 1, after identifying the system, a mathematical model can be developed.

2.2.1- First-Order System Mathematical Model in t-domain

In Chapter 1, Eq. 1.1 shows the differential equation that represents a generic first-order system.

Example 1: The system contains a resistive and a capacitive element. So, referring to Eq.1.1, Eq.1.9A and 1.11A, the system in example 1 can be represented as follows.

$$\mathbf{k_f v + k_s x = F(t)} \qquad \mathbf{2.1}$$

This system solves for the displacement **x**. The steady state value of **x** is when the rate of change of **x** with the time, i.e. the velocity **v**, equals zero. Then $\mathbf{x_{ss} = F(t)/k_s}$.

If the input forcing function is a step input **F**, then $\mathbf{x_{ss} = F/k_s}$. is the displacement at which the driving force is balanced with the spring force and system stops moving (Hooke's law).

Example 2: The system contains an inductive and a resistive element. So, referring to Eq.1.1, Eq.1.6A and 1.9A, the system in example 2 can be represented as follows.

$$\mathbf{ma + k_f v = F(t)} \qquad \mathbf{2.2}$$

This system solves for the speed **v**. The steady state value of **v** is when the rate of change of **v** with the time, i.e. the acceleration **a**, equals zero. Then $\mathbf{v_{ss} = F(t)/k_f}$.

If the input forcing function is a step input **F**, $\mathbf{v_{ss} = F/k_f}$ is the velocity at which the driving force is balanced with the viscous friction force and the system moves at constant velocity.

Where:

m = Equivalent Moving Mass.
$\mathbf{k_f}$ = Viscous Friction Coefficient.
$\mathbf{k_s}$ = Spring Constant.

2.2.2- First-Order System Mathematical Model in s-domain

By applying the Laplace Transform to the generic D.E. 1.1, a first-order system is represented as follows:

$$\mathbf{(as + b)\,x(s) = F(s)} \qquad \mathbf{2.3}$$

By dividing the whole equation by **b**, and considering the definition of time constant, Eq. 2.3 takes the following form:

$$\mathbf{(\tau s + 1)\,x(s) = \frac{F(s)}{b}} \qquad \mathbf{2.4}$$

Where the ratio **a/b** is called the *Time Constant* (**τ**) that is used to identify the system response as it will be explained later.

Then the Transfer Function that represents a first-order system is as follows:

$$\mathbf{TF = \frac{Output}{Input} = \frac{x(s)}{F(s)} = \frac{1/b}{\tau s + 1}} \qquad \mathbf{2.5}$$

So, this TF can be uploaded to simulation software and it supposed to result in the same system response as if the differential equation is used. Then the system modeler is free to choose how to mathematically represent a first-order system.

However, referring to Fig. 2.1 TF for example #1 is as follows:

$$\mathbf{TF = \frac{Output}{Input} = \frac{x(s)}{F(s)} = \frac{1/k_s}{\tau s + 1}} \qquad \mathbf{2.6}$$

Where the time constant for this system is $\boldsymbol{\tau = k_f/k_s}$

TF for example #2 is as follows:

$$\mathbf{TF = \frac{Output}{Input} = \frac{v(s)}{F(s)} = \frac{1/k_f}{\tau s + 1}} \qquad \mathbf{2.7}$$

Where the time constant for this system is $\boldsymbol{\tau = m/k_f}$

To calculate the steady state value in s-domain, the *Final Value Theorem* applies as follows:

$$\mathbf{x_{ss} = lim_{s\to 0}\; s.x(s) = lim_{s\to 0} \frac{s.F(s)/b}{\tau s + 1}} \qquad \mathbf{2.8}$$

2.2.3- First-Order System Normalized Transfer Function

It is to be reminded here that, if the magnitude of the design parameter **b** in a first order system equals unity, then the numerator of the NTF = 1 and the magnitude of the steady state variable **x** equals the magnitude of the forcing function **F**. The form of a first-order transfer function that has **b = 1** is named in this textbook the *Normalized Transfer Function* (NTF) as follows:

$$\mathbf{NTF} = = \frac{\mathbf{1}}{\boldsymbol{\tau}\mathbf{s} + \mathbf{1}} \qquad \mathbf{2.9}$$

Figure 2.2 shows a simulation model (HSV7CH02Model01) that has been constructed to represent the spring-damper system shown previously as example 1. Assuming $\mathbf{k_f = 2}$, $\mathbf{k_s = 1}$, and a step input **F(t) = 100**, the figure shows clearly that the system is stabilized at a steady state value equal to 100.

This NTF is equivalent to a numerical value of 1 but the output caries the dynamic pattern of the system.

The NTF concept will be used to develop mathematical models out of existing measured dynamics as it will be explained later.

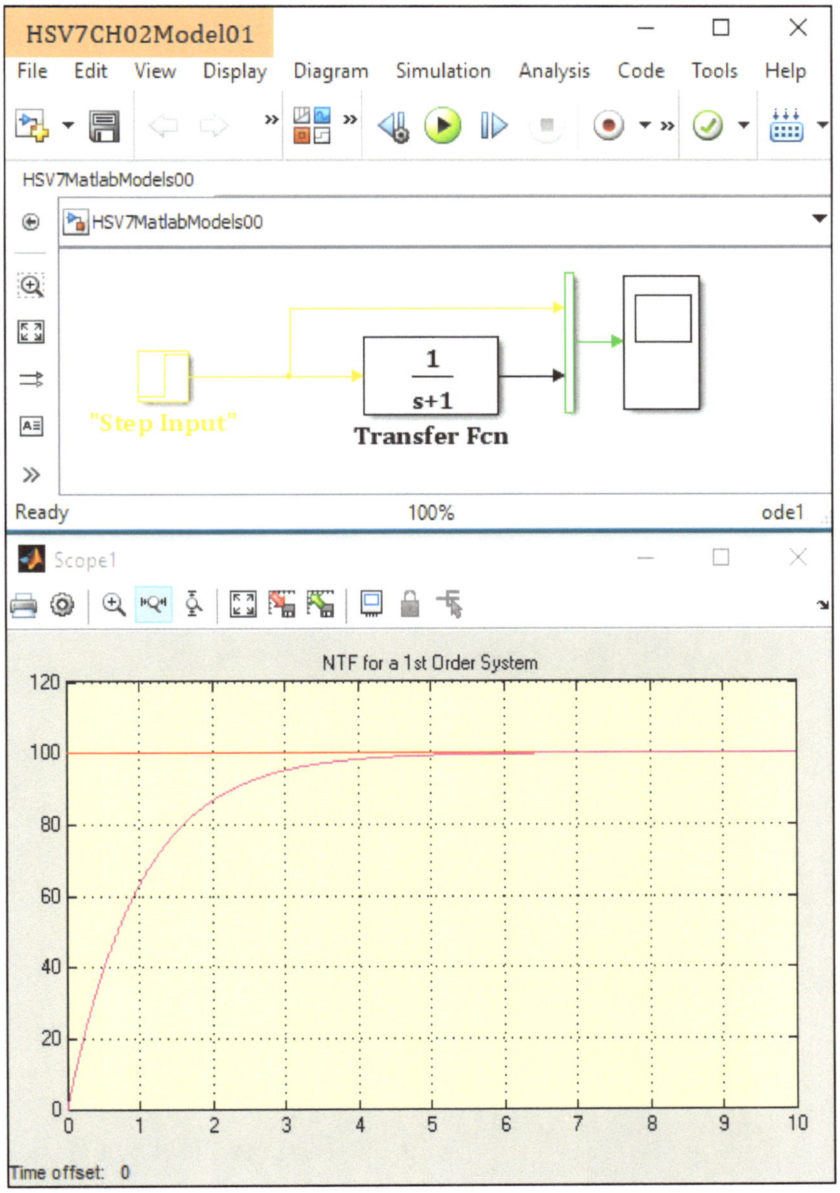

Fig. 2.2 – Response of a First-Order System Represented by NTF

2.3- Simulation Model Development of First-Order Systems

As it has been discussed in chapter 1, after a mathematical model is developed for a physical system, it must be converted into a simulation model to be solved by software. Matlab-Simulink will be considered as the software platform.

As it has been discussed in the previous section, a first order system has been mathematically presented either in t-domain using DE or in s-domain using TF.

Now, the objective is to upload these two mathematical models into a software package to solve for the system response to various exciting forcing functions. This book isn't intended to teach how to use Matlab-Simulink. However, some brief explanation will be provided as needed.

The first challenge is structuring a Simulink Model, is how to upload the system's design parameters to the model. Figure 2.3 shows multiple methods to do that.

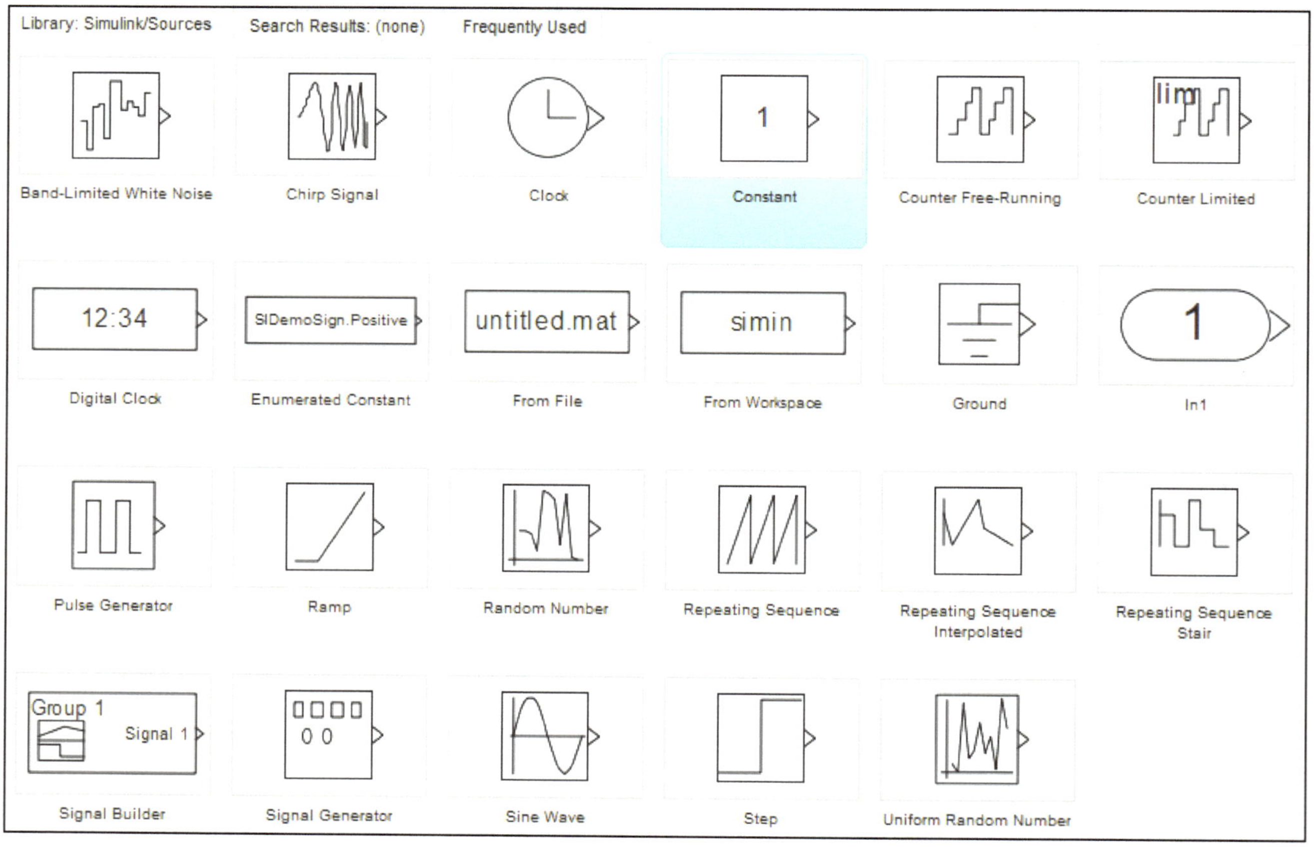

Fig. 2.3 – Source Blocks in Matlab-Simulaink

Figure 2.4 shows a common method to reduce linking lines between blocks while feeding in the system's design parameters. This method uses a combination of a *Constant Block* as a source block and a *GOTO Block* as a sink block. variables can be assigned as Local (for the current layer of the model) or Global to be seen by all model layers. The figure shows that a value (32) is assigned for a global variable name (Viscosity). Value of this variable name is seen by any part of the model and will be updated interactively if it is changed.

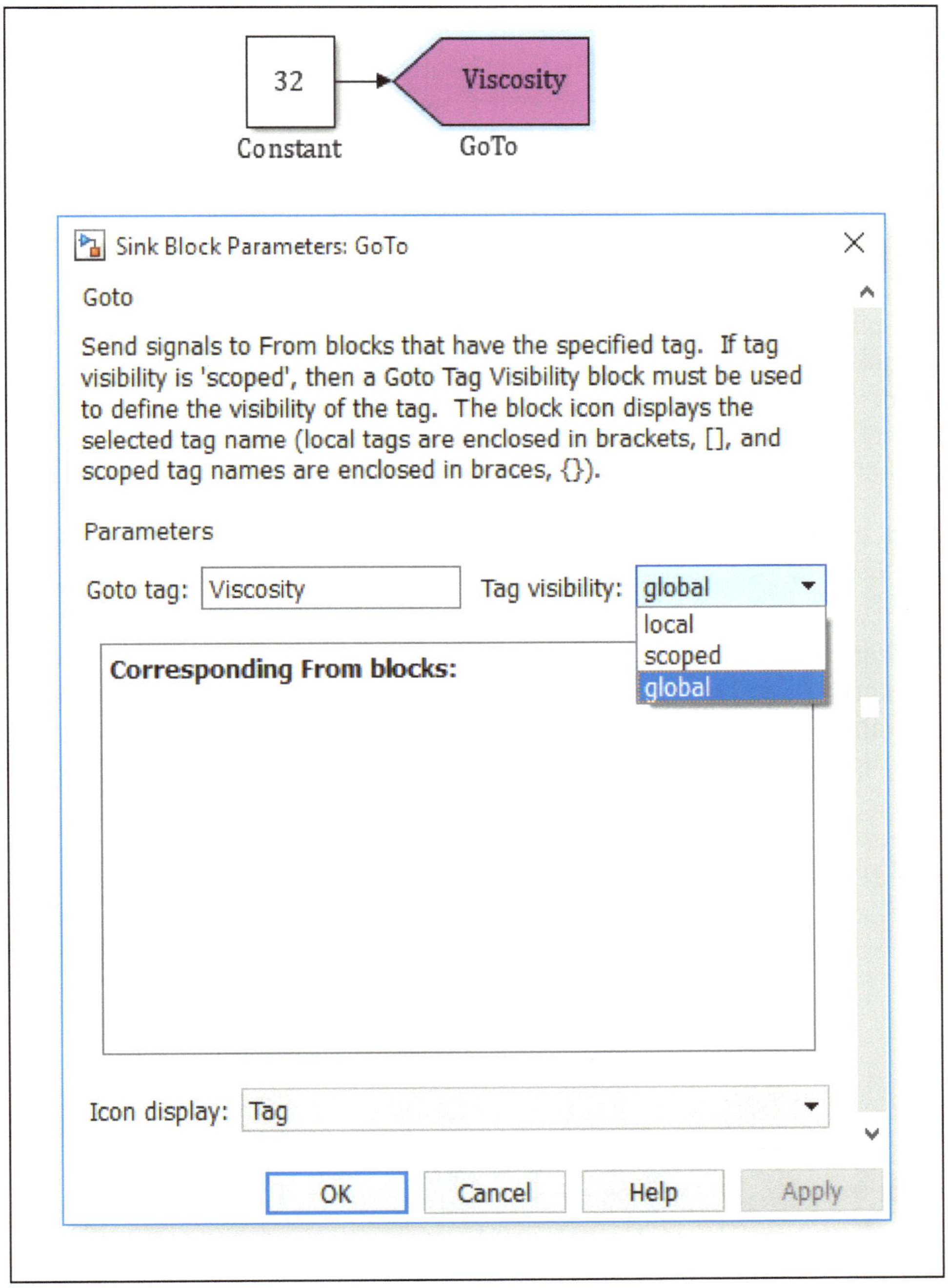

Fig. 2.4 – Sink Blocks in Matlab-Simulaink

One other easy way to upload the design parameters to the model is to type it in the form of named variables in the model or subsystems properties/Callbacks/InitFcn.

Now, the model itself is to be structured. Referring to Example 1 in Fig. 2.1, a simulation model has been developed for the system. Fig. 2.5 shows that simulation model (HSV7CH02Model02) is structured based on the DE shown in Eq. 2.1.

The upper part of the Fig. 2.5 shows that the model compares the response of the system to a Step input F(t). Usually Simulink models are composed of multiple blocks. Some of these blocks perform iterative calculations based on *Sampling Time*. Sampling time must be unified for all blocks in the same model, otherwise an error is generated. Therefore, as shown in Fig. 2.6, sampling time and the value of the forcing function F(t) are assigned in the model properties so that they can be read by the model before any calculations begin.

The lower part of Fig. 2.5 shows that Simulink blocks are arranged to solve for the displacement **x(t).** The figure shows also that saturation limits for both the speed and displacement can be added. These saturation blocks are to limit the calculation based on the physical maximum speed and maximum displacement. All these blocks are capsulated in one *Subsystem* that is shown in the upper part with the relevant masking image.

As shown in Fig. 2.7, design parameters for this system are assigned in the subsystem block properties. It is to be noted that the time constant and the steady state value of the displacement are calculated in the block properties based on the given design parameters. The spring constant considered in this example is an impractical. The reason in using such impractical spring constant (1N/m) is to simplify the calculations and convey the idea. So, this spring constant isn't really representing a linear spring, it represents an elastic cable that the mass is moving against.

For the same system represented by Example 1, the second model (HSV7CH02Model03) is structured based on the TF shown in Eq. 2.6. Figure 2.8 shows just one transfer function block that represents the system. That is why this method is recommended for large systems. The figure shows also the design parameters assigned in the TF block properties.

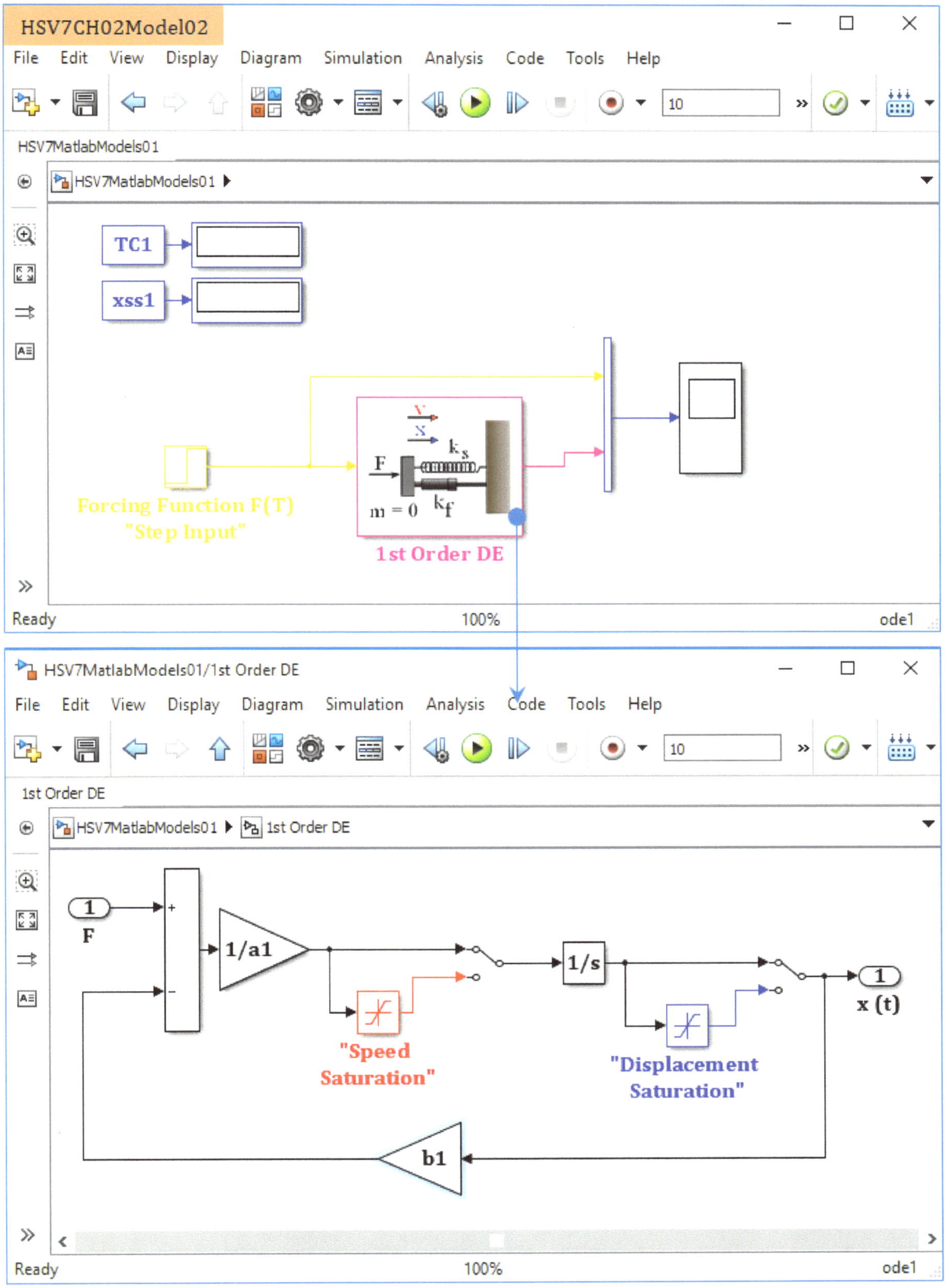

Fig. 2.5 – Simulation Model for First-Order System Based on Differential Equation

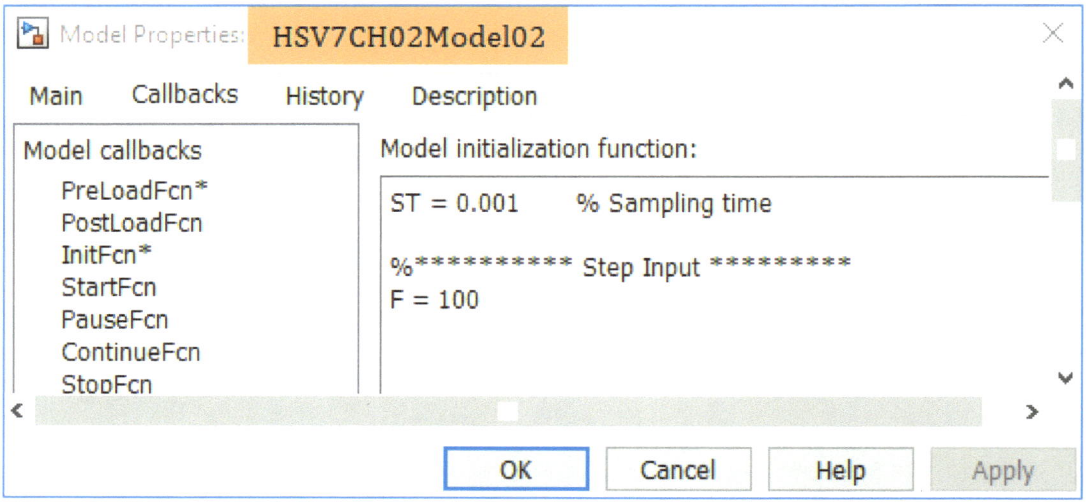

Fig. 2.6 –Properties for the Previous Model

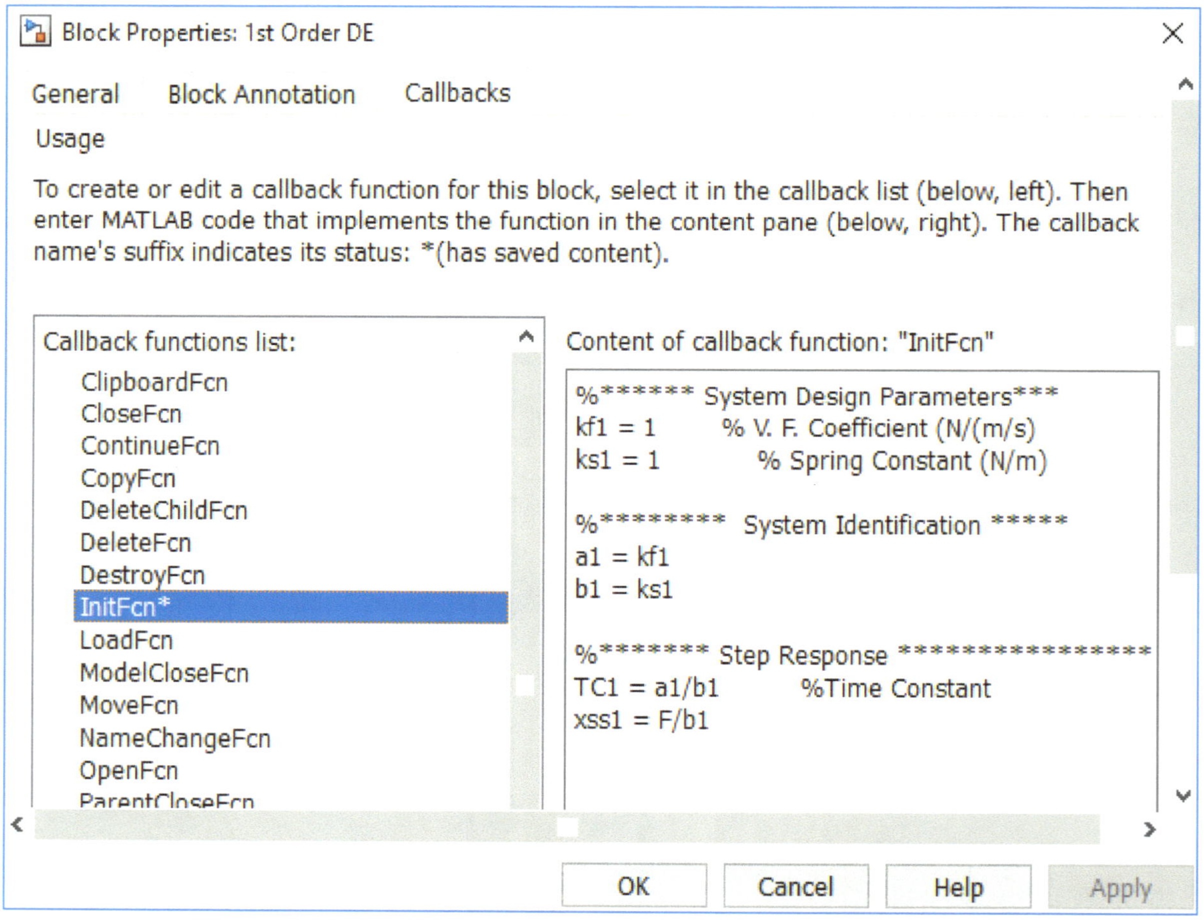

Fig. 2.7 – Design Parameters for the Previous Model

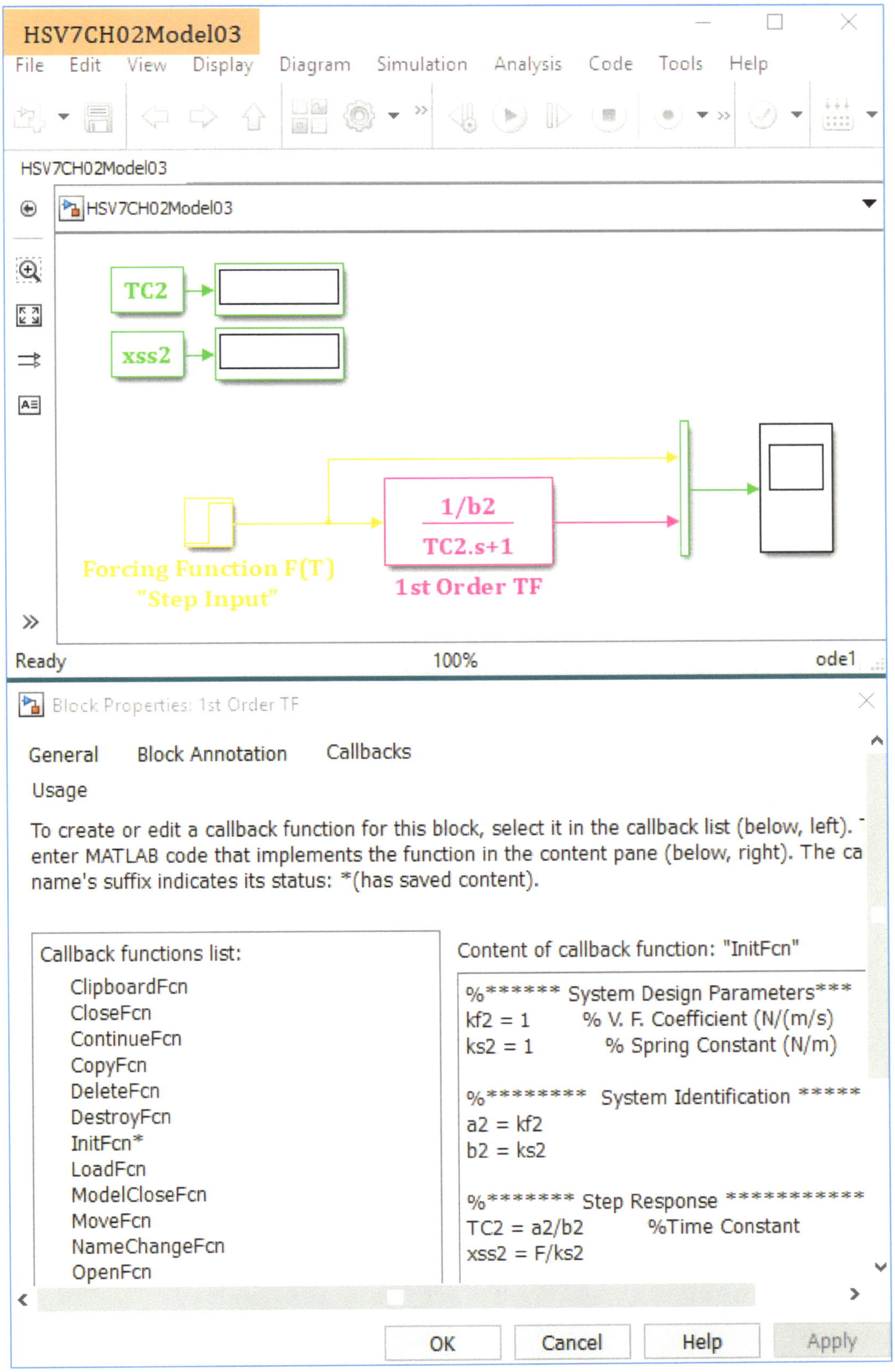

Fig. 2.8 – Simulation Model for First-Order System Based on Transfer Function

2.4- Performance Simulation of First-Order Systems

As it has been previously stated, no matter how a dynamic system mathematically is modeled and no matter what the input signal is, the simulation results should be the same as long as the same design parameters are used. Figures, 2.9 shows a new simulation model (HSV7CH02Model04) that contains the model in t-domain and the model in s-domain. The figure shows that a *Sampling Time* of 1 ms. has been assigned as a global variable in the model properties. A step input **F = 100** is also assigned in the model properties.

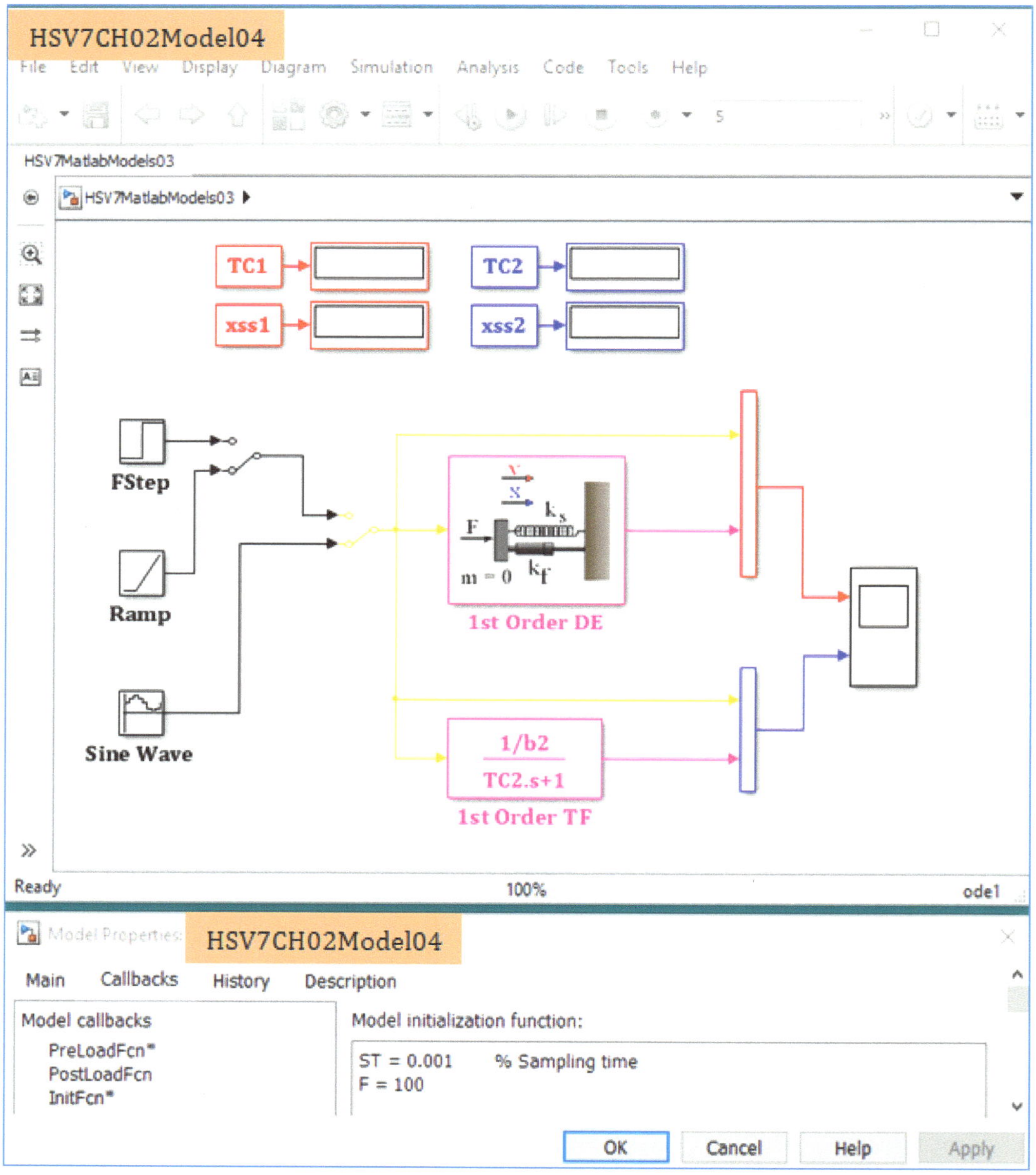

Fig. 2.9 – Comparison of Simulation Models for First-Order System

Figure 2.10 shows the response of a first-order system to a *Step Input*. The following observations can be seen out of the simulation results:

Note 1: Dynamic responses of two models (including the dynamic pattern, Time Constant, and Steady State Value) are identical since same design parameters are assigned in the two models.

Note 2: Since the design parameter **b = 1**, the steady state value in the step response [x (m)] equals the same magnitude of the input step function [F(N)].

Note 3: Steady state value of the system response is when the external force is balanced with the spring force. This is seen whether the system is represented by a DE or a TF. To calculate the SS value when a TF is subjected to a step input, considering b = 1 and F(s) = 100/s, final value theorem shown in Eq. 2.8 is applied as follows.

$$x_{ss} = \lim_{s \to 0} s.x(s)$$

$$= \lim_{s \to 0} \frac{s.F(s)/b}{\tau s + 1}$$

$$= \lim_{s \to 0} \frac{s\,\frac{100}{s}/b}{\tau s + 1} = 100$$

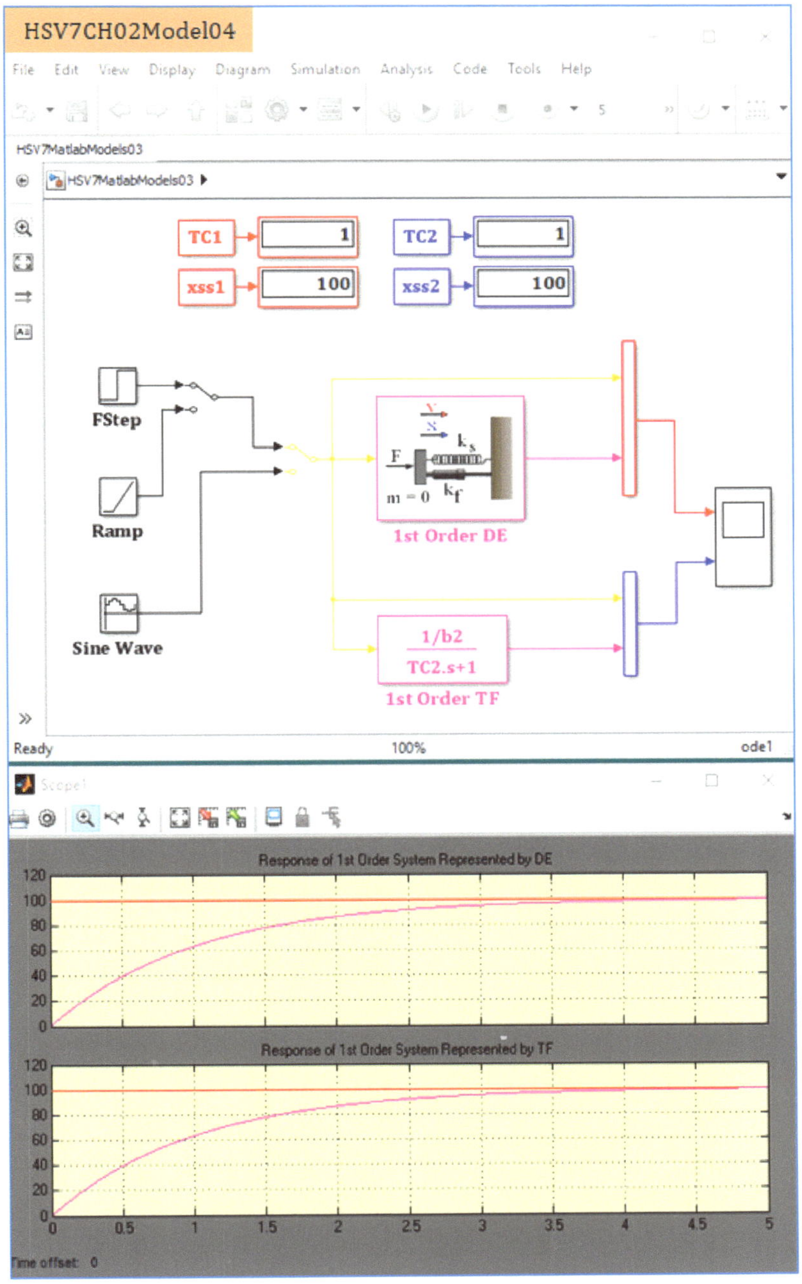

Fig. 2.10 – Step Response of a First-Order System Based on Two Models

Figures, 2.11 shows the response of a first-order system to a *Ramp Input*, as it is subjected to gradual loading or commanding with constant rate. Again, simulation results show identical response despite using different model structures.

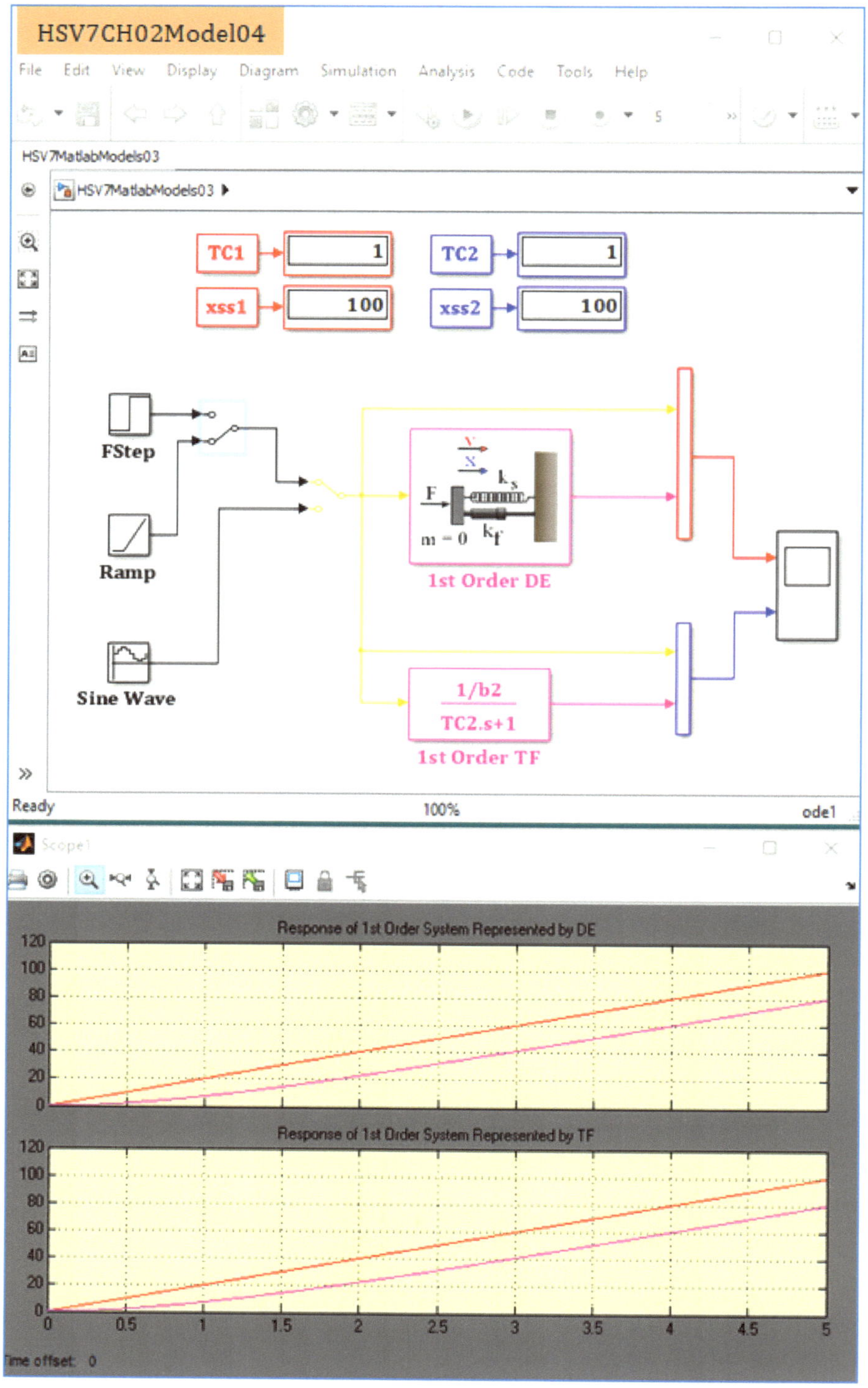

Fig. 2.11 – Ramp Response of a First-Order System Based on Two Models

Figures, 2.12 shows the response of a first-order system to a *Sinusoidal Input*, as it is subjected to periodic loading or commanding. Results show same response no matter the structure of the model.

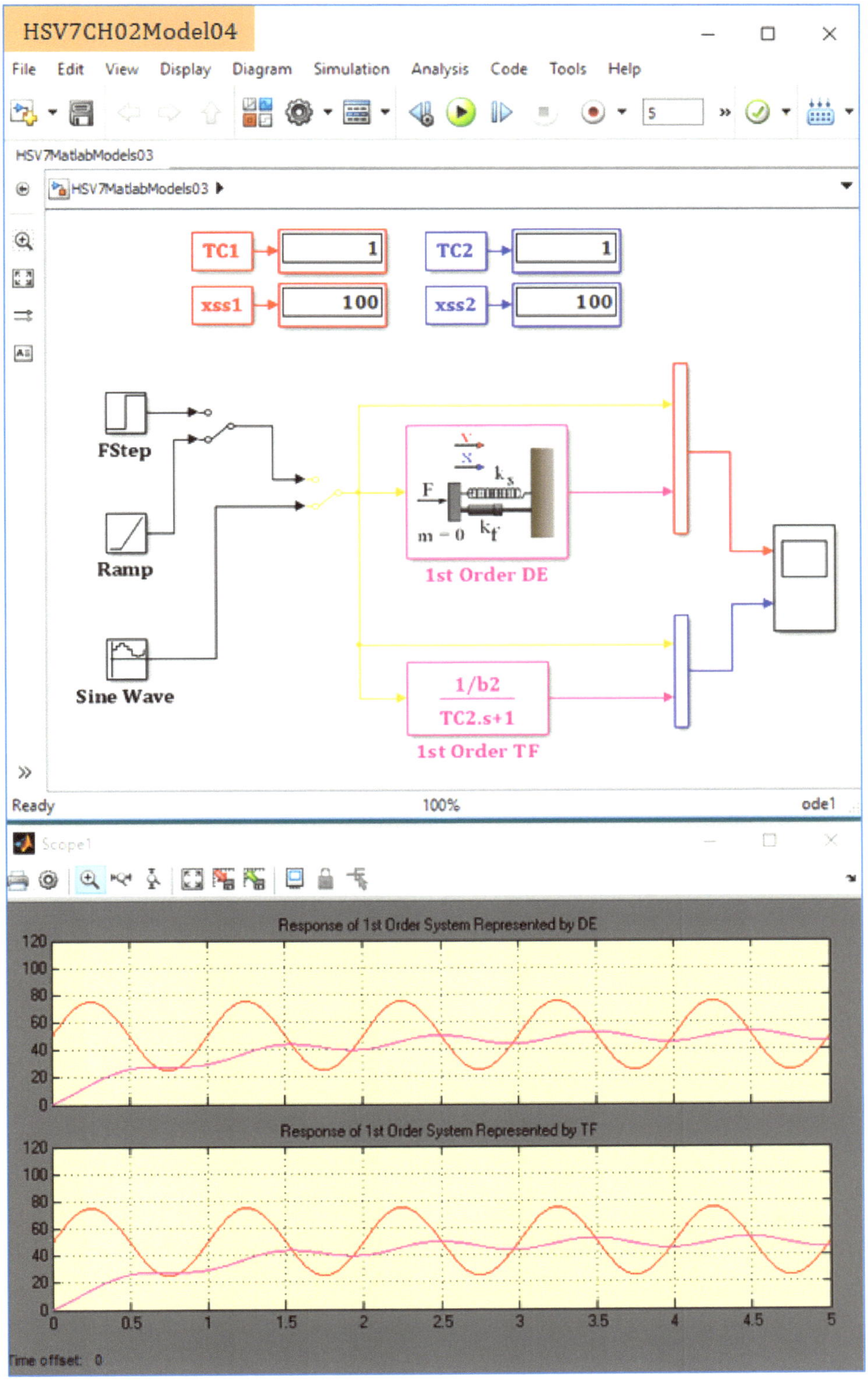

Fig. 2.12 – Harmonic Response of a First-Order System Based on Two Models

2.5- Step Response Analysis of First-Order Systems

2.5.1- Identification of First-Order Systems Based on Step Response

Simulation model (HSV7CH02Model03) has been retrofitted to generate the model (HSV7CH02Model05) shown in Fig. 2.13. The new model has a time constant = 1 sec. and a steady state value = 1. By analyzing the response, the following observations have been found applicable to all first-order systems.

Settling Time:
Settling Time is where the system has approximately settled around a steady state value.

Time Constant τ = 1/5 of the Settling Time
Response of a 1st – Order system is identified by its *Time Constant*. The figure shows that the system settles approximately at a time equal to five times the time constant regardless its value. *Settling Time* is the time the system takes to reach the steady state value. The figure shows also, at time = **τ**, the system reaches 63.2% of its SS value.

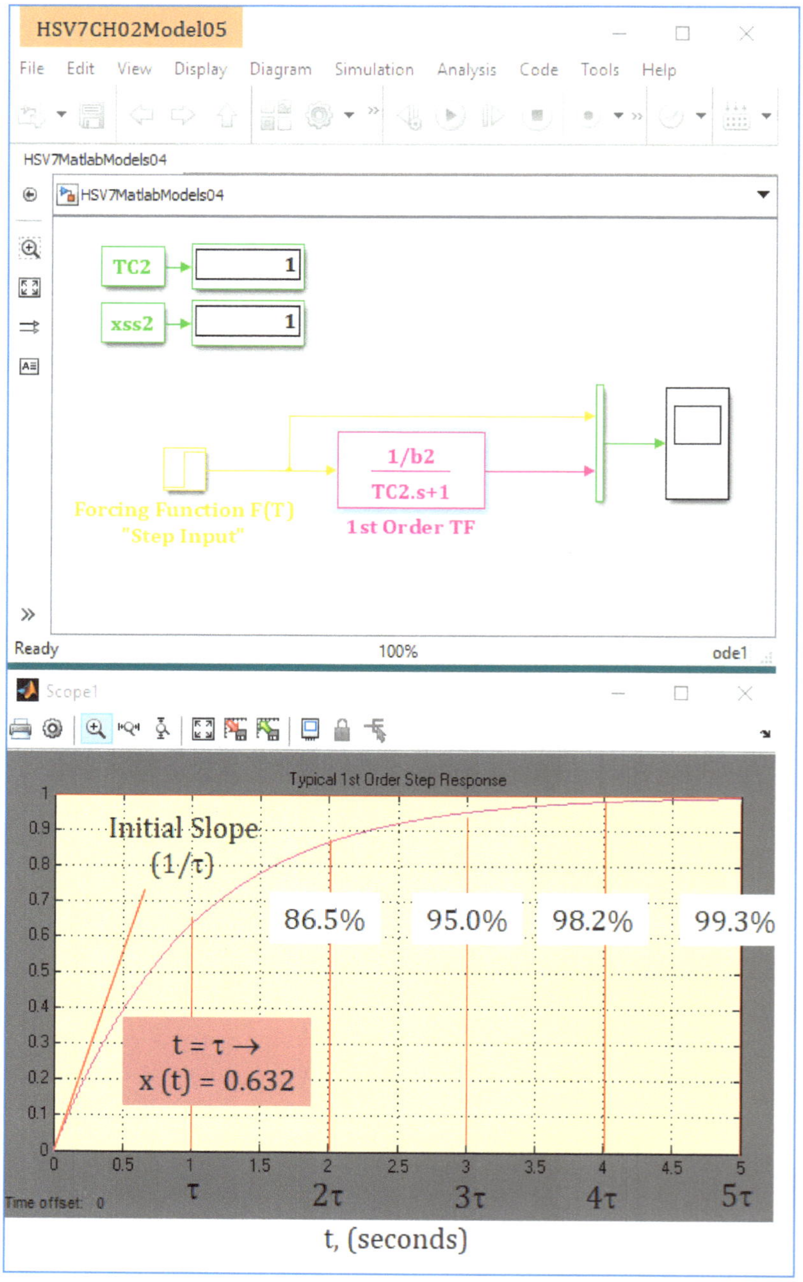

Fig. 2.13 – Step Response Analysis for First-Order Systems

2.5.2- Effect of Design Parameters on First-Order System Step Response

2.5.2.1-Effect of Time Constant

Effect of Time Constant (τ↑ → Settling Time↑ i.e. System Speed of Response↓):
Figure 2.14A shows a simulation model (HSV7CH02Model06) that contains three TFs. In Fig. 2.14B, simulation results show evidently that when the time constant increases, the system's speed of response is reduced. This means that the time constant is inversely proportional to the system's speed of response. So, systems with very small time constant are "nervous" and systems with large time constant are "sluggish".

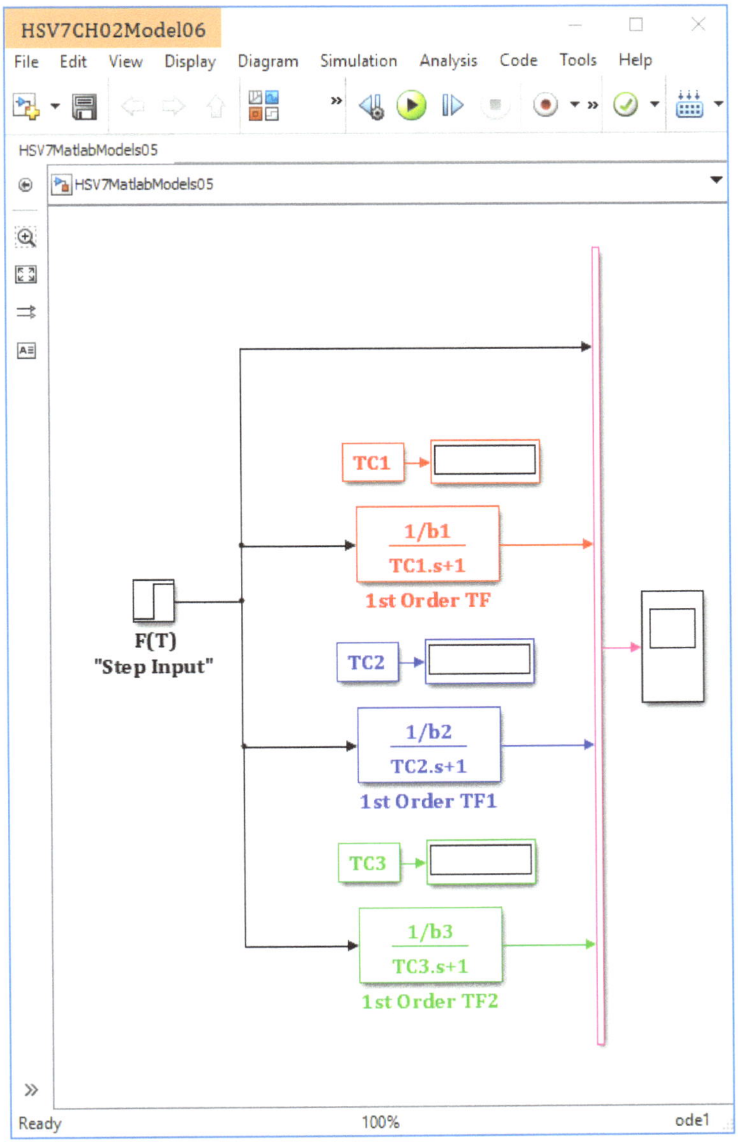

Fig. 2.14A – Simulation Model to Investigate Effect of Time Constant on Step Response of First Order Systems

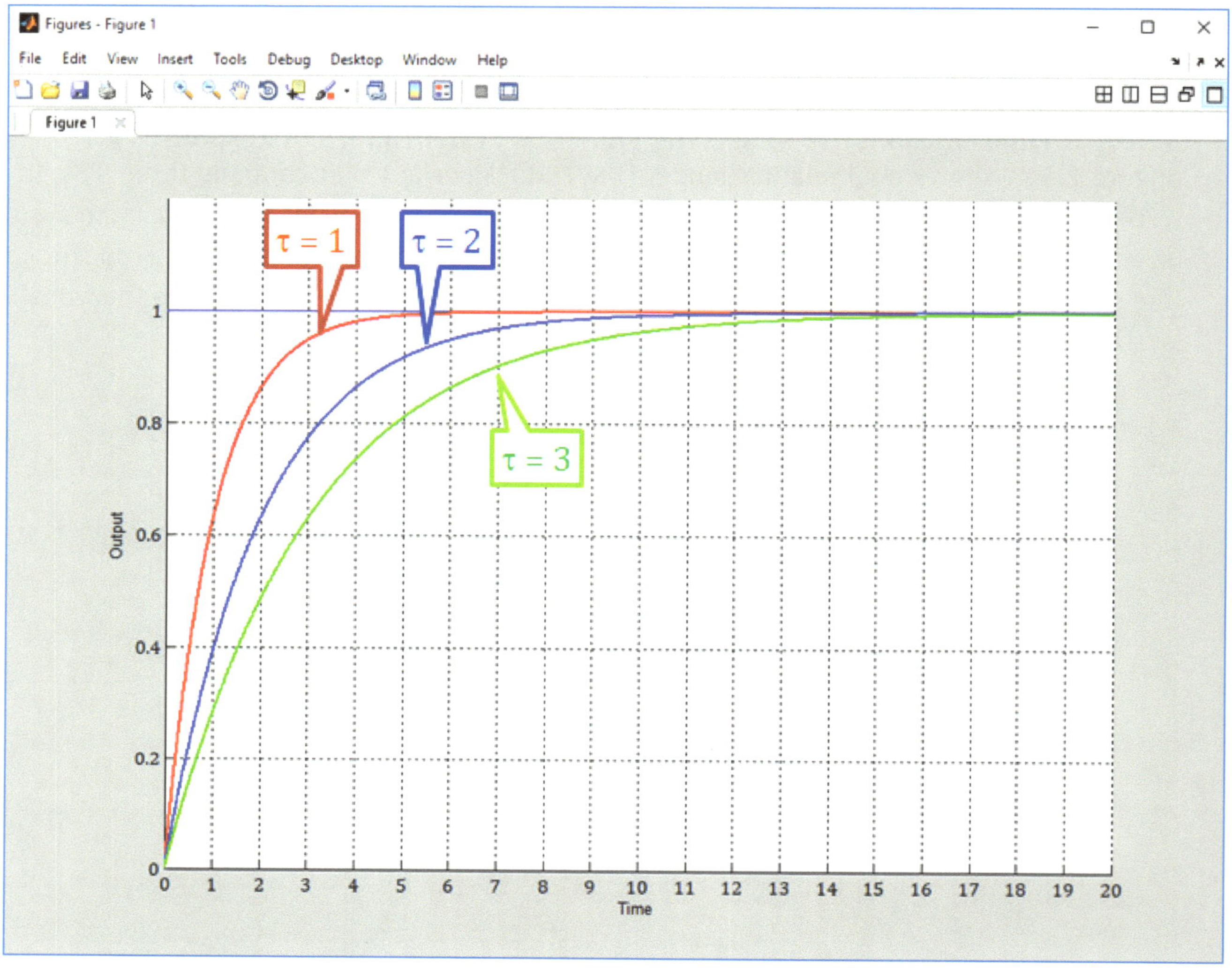

Fig. 2.14B- Effect of Time Constant on 1st Order System Step Response

2.5.2.2-Effect of Friction Coefficient

Effect of Friction Coefficient k_f (is not the same on all systems).

In order to study the effect of damping coefficient on different systems, response of two different systems must be compared while varying the damping coefficient. Referring to Example 2 in Fig. 2.1, a simulation model is structured based on the DE shown in Eq. 2.1. Figure 2.15 shows simulation model (HSV7CH02Model07) that combines example 1 and 2 in the same model. Damping coefficient $\mathbf{k_f}$ has been doubled for both models. It has been found that changing $\mathbf{k_f}$ may have different effects depending on the system. Figure 2.15 shows the interpretation of that as follows:

- **Referring to Example 1:** the system's time constant $\tau = k_f/k_s$ and $x_{ss} = F/k_s$.
- This system solves for the displacement.
- x_{ss} = (Step input forcing function / Spring Stiffness).
- x_{ss} is the displacement at which the driving force is balanced with the spring force and the system stops moving (Hooke's law).
- $k_f \uparrow \rightarrow \tau \uparrow \rightarrow$ system speed of response $\downarrow$, i.e. slower response.
- $k_s \uparrow \rightarrow \tau \downarrow \rightarrow$ system speed of response $\uparrow$, i.e. system reaches x_{ss} faster.

- **Referring to Example 2:** the system's time constant $\tau = m/k_f$ and $v_{ss} = F/k_f$.
- This system solves for the mass velocity.
- v_{ss} = (Step input forcing function / Viscous Friction Coefficient)
- v_{ss} is the velocity at which the driving force is balanced with the viscous friction force and system moves at constant velocity.
- $m \uparrow \rightarrow \tau \uparrow \rightarrow$ system speed of response $\downarrow$, i.e. slower response.
- $k_f \uparrow \rightarrow \tau \downarrow \rightarrow$ system speed of response $\uparrow$, i.e. system reaches to v_{ss} faster.

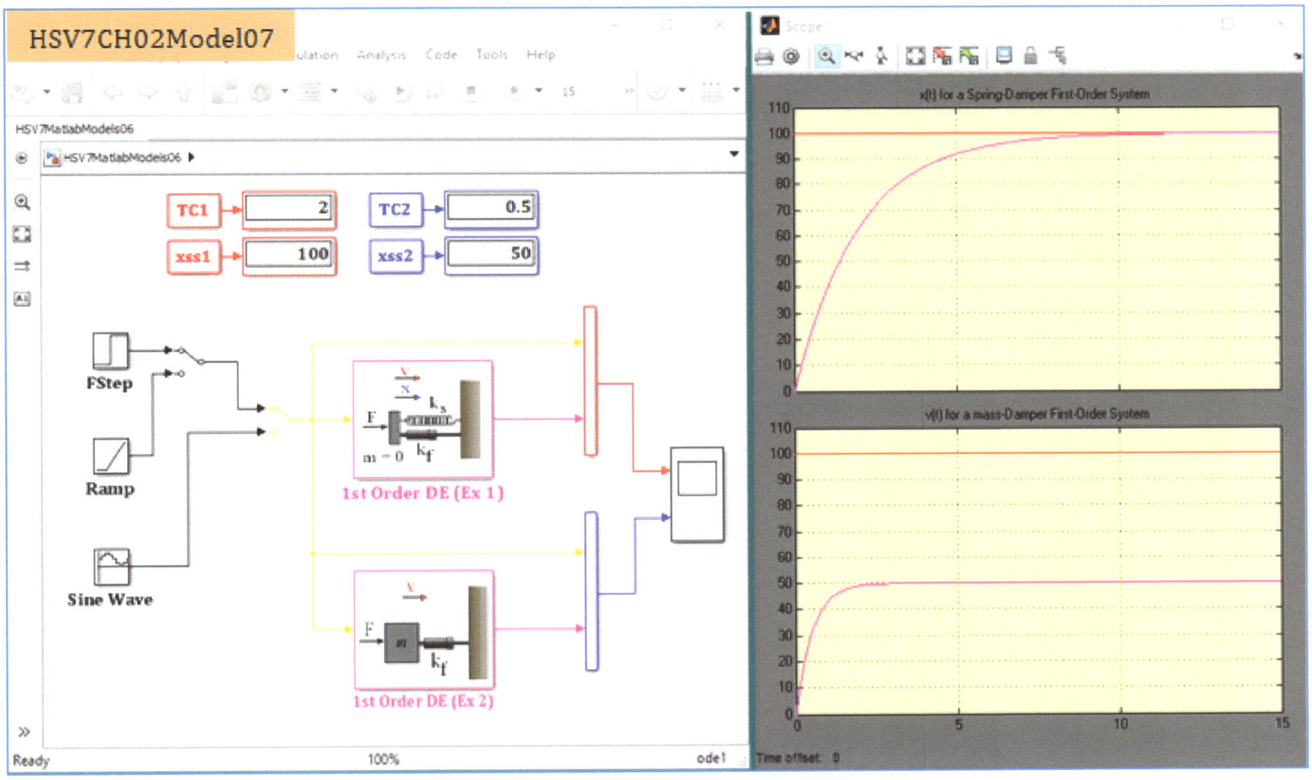

Fig. 2.15 – Effect of Damping Coefficient on 1st Order System Step Response

2.6- First-Order System Modeling Based on Step Response

As shown in Fig. 2.16, if a step response of a first order system exists, a model to represent the dynamics of the system can be developed as follows.

Step 1: Define the Settling Time:
Settling Time is reported graphically or numerically by a system manufacturer. It can also be found experimentally by exposing the system to a step input.

Step 2: Calculate the Time Constant:
Time constant τ equals one fifth of the settling time.

Step 3A: Develop the NTF (in cases where physical design parameters are not known):
Due to lack of knowledge of the design parameters, at least the pattern of the system dynamics can be represented by a *Normalized Transfer Function* (Fig. 2.17). It will be discussed later how to use the NTF with the static characteristics of the system to build a complete system model.

Step 3B (Option 1): Develop the TF (in cases where physical design parameters are known): Figure 2.18 shows the system dynamics are represented by a conventional TF. To do so, as shown in Fig. 2.19, design parameter **b** must be found with the units being balanced.

Step 3C (Option 2): Developing DE (in cases where physical design parameters are known): An alternative method to the previous one, since parameter **b** is known, is to develop the differential equation that represents the system dynamics. Units must be balanced.

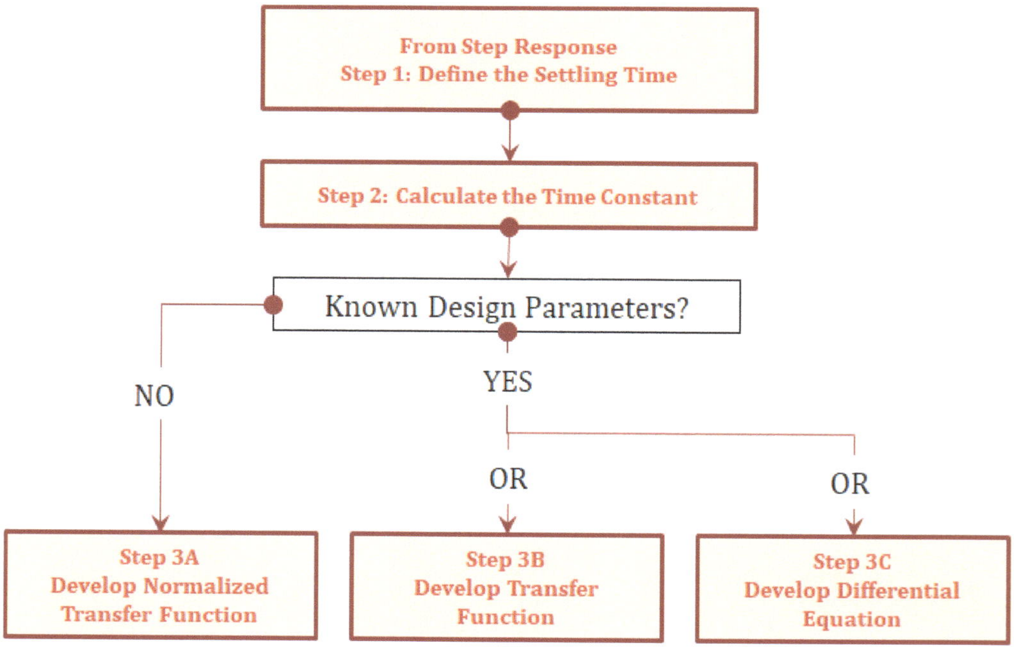

Fig. 2.16 – Road Map to Develop a First-Order Model based on Step Response

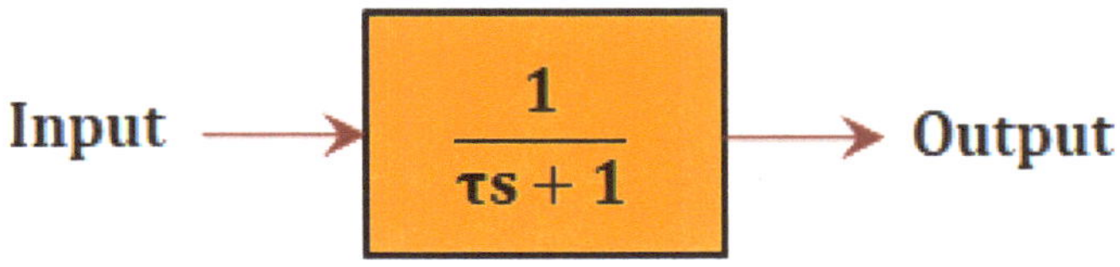

Fig. 2.17 – First-Order Model based on Normalized Transfer Function

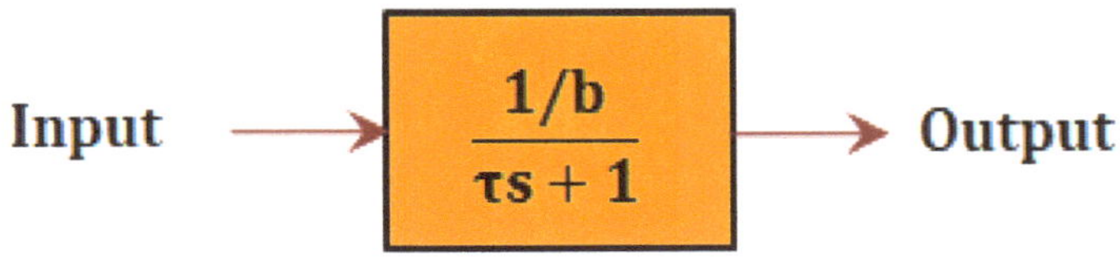

Fig. 2.18 – First-Order Model based on Transfer Function

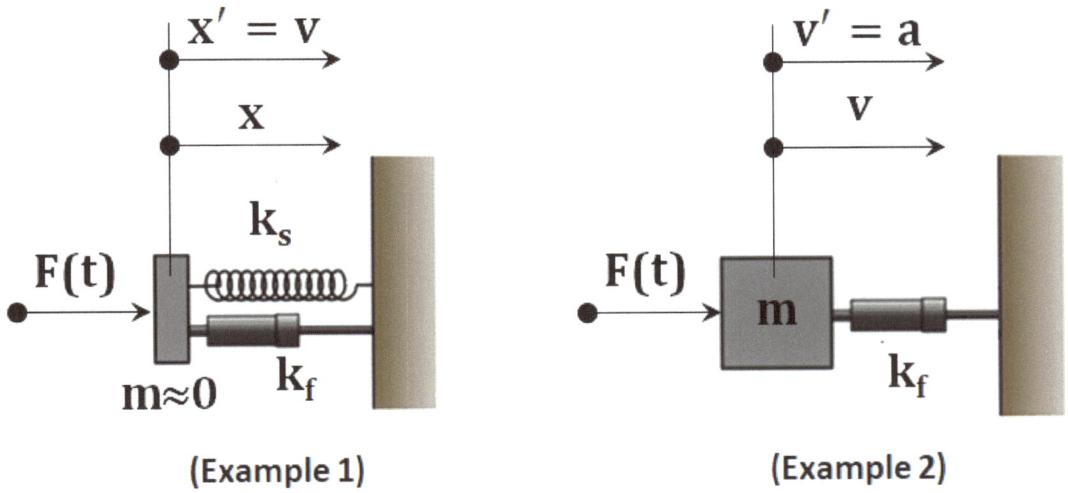

Example 1	Example 2
$k_f v + k_s x = F(t)$	$ma + k_f v = F(t)$
$a = k_f$ "not measurable"	$a = m$ "measurable"
$b = k_s$ "measurable"	$b = k_f$ "not measurable"
$\tau = a/b = k_f / k_s$	$\tau = a/b = m / k_f$
Then $k_f = k_s \times \tau$	Then $k_f = m/\tau$

Fig. 2.19 – First-Order Model based on Differential Equation

Case Study:

Step 1: Define the Settling Time: The hydraulic cylinder, shown below in Fig. 2.20, has an equivalent moving mass of 2000 kg. The cylinder has been extended suddenly by a 50 N force. Dynamics of the cylinder has been captured. The cylinder was found performing like a first-order system and reached a constant velocity in 10 seconds.

Step 2: Calculate the Time Constant: Time constant τ equals one fifth of the settling. Time Since settling time = 10 sec, then time constant **τ = 10/5 = 2 sec.**

Step 3B: Construct the Model based on TF + Measurable Design Parameter: Figure 2.20 shows that the cylinder is equivalent to the mass-damper system. Then Since $\tau = m/k_f$, then $k_f = m/\tau = 2000/2 = 1000$ **(kg/s).** Damping coefficient can also be calculated if the force and the constant velocity are measured. However, with the knowledge of the damping coefficient and Eq. 2.7, system model is constructed as follows:

$$\mathbf{TF} = \frac{\mathbf{Output}}{\mathbf{Input}} = \frac{\mathbf{v(s)}}{\mathbf{F(s)}} = \frac{\mathbf{1/k_f}}{\boldsymbol{\tau}\mathbf{s+1}} = \frac{\mathbf{1/1000}}{\mathbf{2s+1}} \qquad \mathbf{2.10}$$

OR Step 3C: Construct the Model based on Developing DE: Referring to Eq. 2.2, System can also be represented by DE with the units being balanced, as follows:

$$\mathbf{m[kg]} \times \mathbf{a}[\tfrac{\mathbf{m}}{\mathbf{s^2}}] + \mathbf{k_f}\left[\frac{\mathbf{kg}}{\mathbf{s}}\right] \times \mathbf{v}\left[\frac{\mathbf{m}}{\mathbf{s}}\right] = \mathbf{F(t)\ [N]} \qquad \mathbf{2.11}$$

Then, with the knowledge of the damping coefficient, Eq. 2.11 can be rewritten as follow:

$$\mathbf{2000\ a}[\tfrac{\mathbf{m}}{\mathbf{s^2}}] + \mathbf{1000\ v}\left[\frac{\mathbf{m}}{\mathbf{s}}\right] = \mathbf{F(t)\ [N]} \qquad \mathbf{2.12}$$

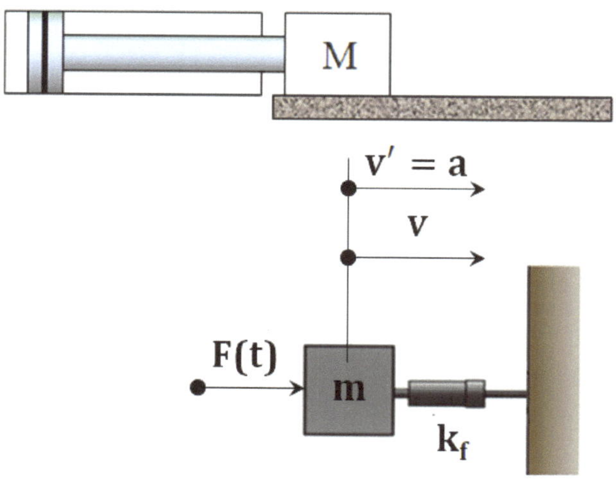

Fig. 2.20 – First-Order Model Development based on Existing Dynamics

As shown in Fig. 2.21, model (HSV7MatlabModel07) runs for simulating the case study assuming a sudden pressure force of 50 N acts on the piston side to extend the cylinder. Results were found identical from both models. Simulation results shows that the cylinder settles at a constant velocity of 0.05 m/s (5 cm/s) after 10 seconds when it is subjected 50 N sudden force. That confirms the calculation of time constant = 2 sec. and a damping coefficient = 1000 kg/s.

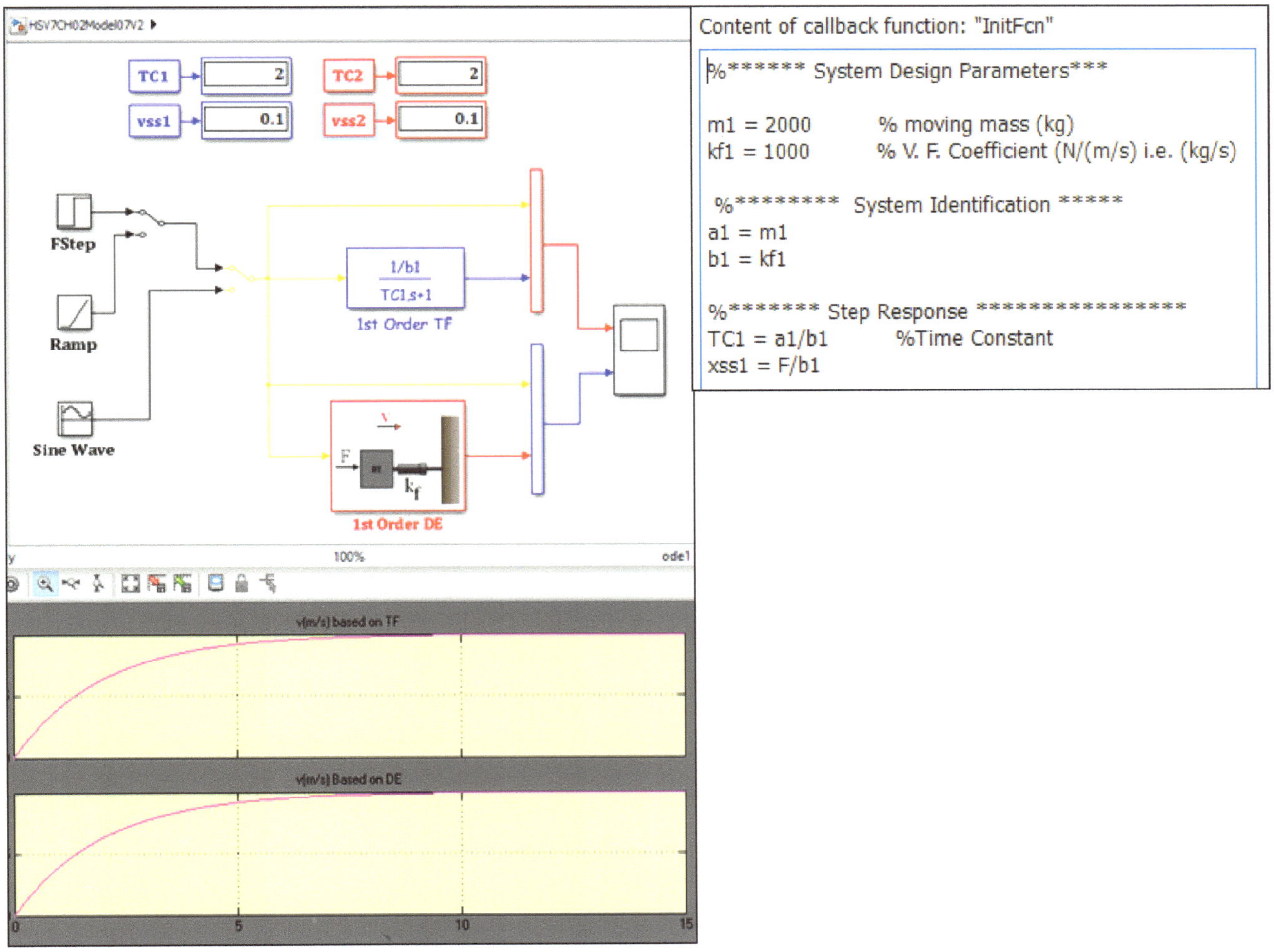

Fig. 2.21 – Simulation Results for the Case Study

2.7- Frequency Response Analysis of First-Order Systems

2.7.1- Identification of First-Order Systems Based on Frequency Response

Mathematical solutions for first-order frequency response are as follows:

$$\mathbf{x(t) = B\sin(\omega t + \varphi)} \tag{2.13}$$

$$\mathbf{B = \frac{A}{\sqrt{1+\omega^2\tau^2}}} \tag{2.14}$$

$$\mathbf{\varphi = \tan^{-1}(-\omega\tau)} \tag{2.15}$$

$$\mathbf{AR\,(dB) = 20\log\frac{B}{A} = 20\log\frac{1}{\sqrt{1+\omega^2\tau^2}}} \tag{2.16}$$

Where:
A = Amplitude of the harmonic input (exciting) signal.
B = Amplitude of the system response.
AR = Amplitude Ratio of the system response.
ω (rad/s) = Frequency of the harmonic input (exciting) signal.
τ = Time constant of the system response.
φ (deg) = Phase lag of the system response.

As shown in Fig. 2.22, *Bode Plot* for a first-order system can be drawn as follows.

- ω < 0.1/τ → AR ≈ 0 dB and φ = 0 deg.
- ω = 1/τ = Critical Frequency → AR = -3 dB (called Bandwidth) and φ = -45 deg.
- ω >= Critical Frequency → AR decays 20 dB/decade of exciting frequency.
- ω = 10/τ → AR = -20 dB.
- ω => 10/τ → φ = -90 deg.

Straight lines show the approximate frequency response. Asymptotic line represents the accurate frequency response.

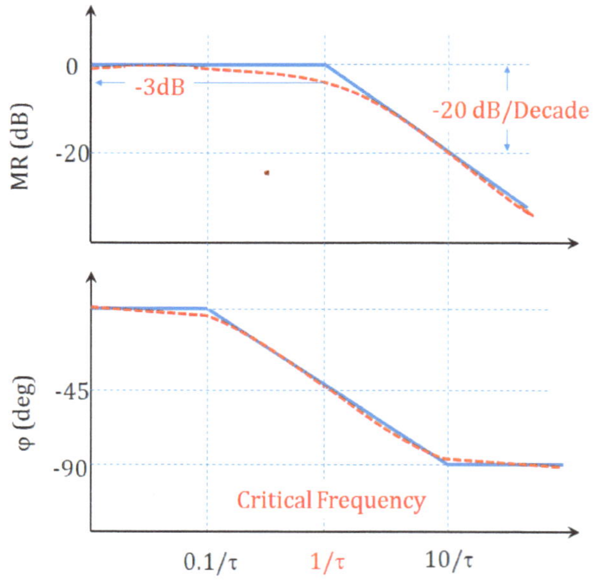

Fig. 2.22 – Bode Plot of a 1st Order System

2.7.2- Effect of Design Parameters on First-Order System Frequency Response

2.7.2.1- Effect of Time Constant

As shown in Fig. 2.23, to confirm the previous analysis, a model (HSV7CH02Model08) was constructed that contains three transfer functions of first order systems. The three systems have been excited by the same sinusoidal function that has exciting frequency equal 1 rad/s.

First System (Red): τ = 0.1 and ω = 1 → ω = 0.1/τ. Therefore, the system appears very responsive and copes with the exciting frequency very well.

Second System (Blue): τ = 1 and ω = 1 → ω = 1/τ = Critical Frequency (1/τ). Therefore, the system shows **AR = 70%** with phase lag **φ = -45 deg.**

Third System (Green): τ = 10 and ω = 1 → ω = 10/τ. Therefore, the system appears very sluggish and not able to respond to the input function.

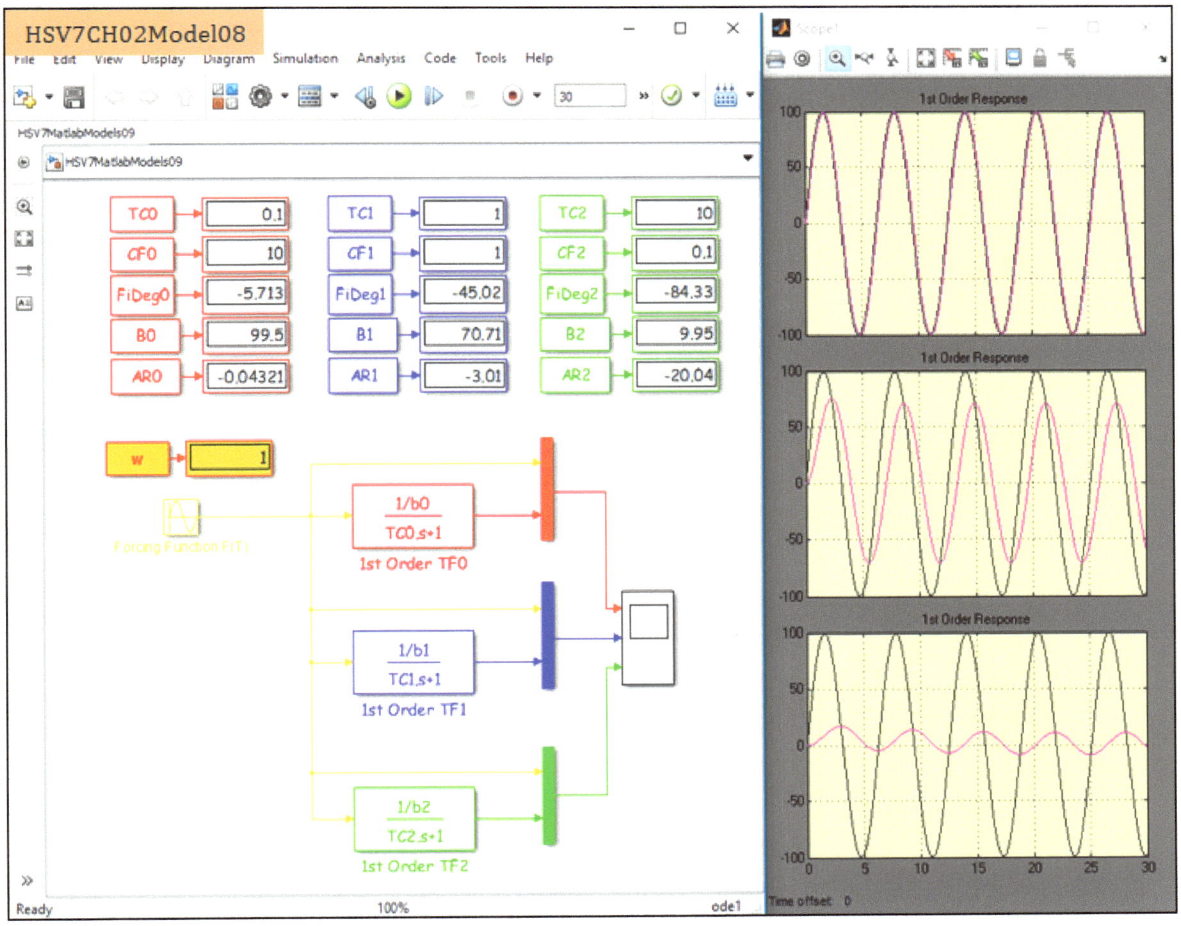

Fig. 2.23 – Effect of Time Constant on 1st Order System Harmonic System

2.7.2.2- Effect of Exciting Frequency

As shown in Fig. 2.24, another model (HSV7CH02Model09) was constructed to contain three transfer functions of same first order system. Design parameters have been selected so that the three systems have a time constant equal 1 and a critical frequency = 1. Each transfer function has been excited by harmonic input with different frequency.

The first system (Red) subjected to low input frequency, therefore **AR** and **φ** are very small. By increasing the frequency to the critical frequency, the second system (blue), shows **AR = 70%** with phase lag **φ = -45 deg.** By increasing the exciting frequency, the system may not able to see the exciting signal.

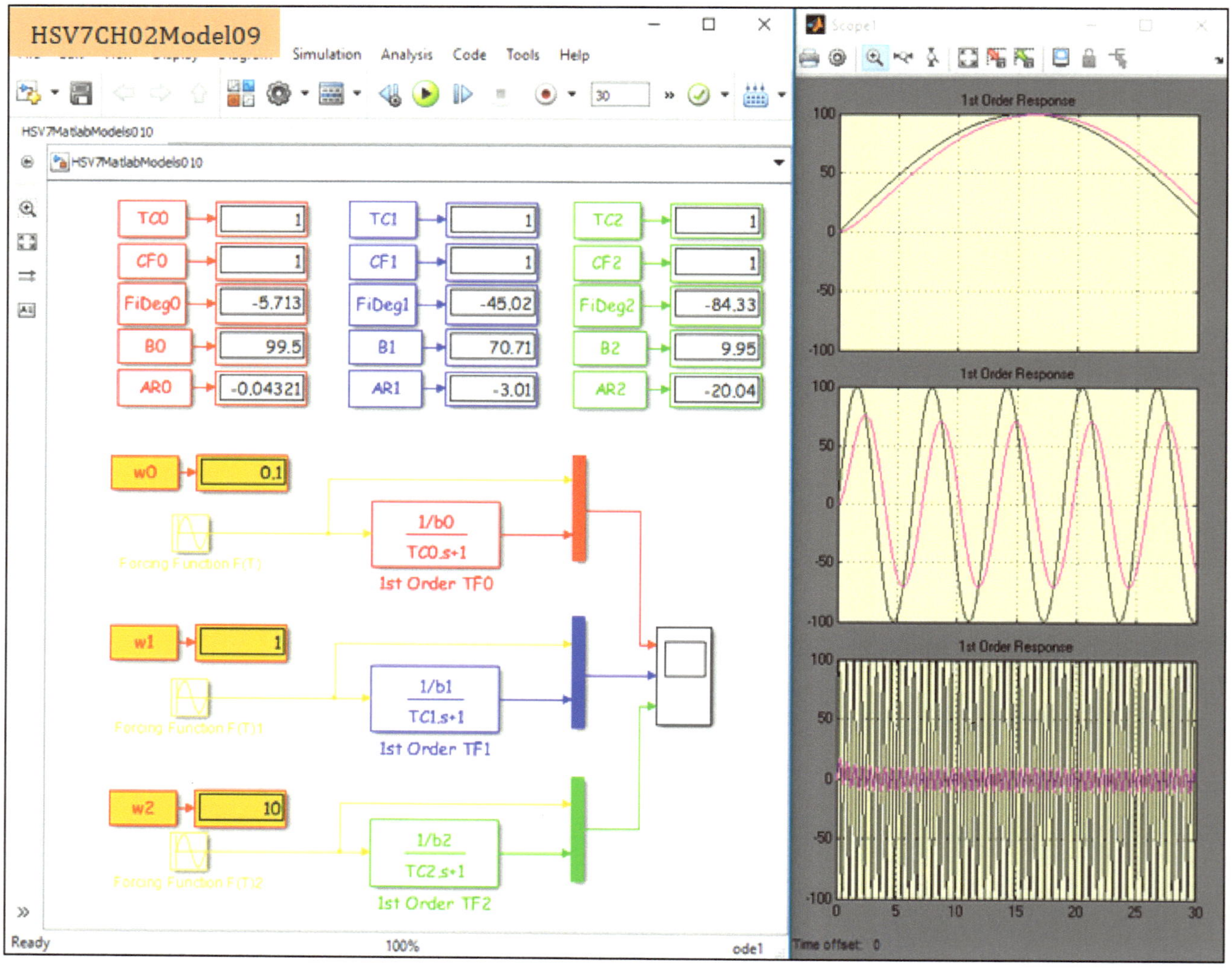

Fig. 2.24 – Effect of Exciting Frequency on 1st Order System Harmonic System

2.8- First-Order System Modeling Based on Frequency Response

As shown in Fig. 2.25, if a frequency or harmonic response of a first order exists, the transfer function of the system can be derived as follows.

Step 1A: Find the Critical Frequency from a Bode Plot: At **AR = - 3 dB** or at **φ = -45 Deg.,** pick the value of the critical frequency.

Step 1B: Find the Phase Lag φ from Harmonic Response: Measure φ as % of the complete cycle time.

Step 2: Calculate the Time Constant:

- Having the critical frequency (from Step 1A), calculate **τ = 1/critical frequency**.
- OR Having the phase lag (from Step 1B), using Eq. 2.15, calculate **τ = - [tan (φ)]/ω**.

Step 3 Construct the Model: Since the time constant has been calculated, follow any of the previously mentioned steps (3A, 3B, or 3C) to construct the model.

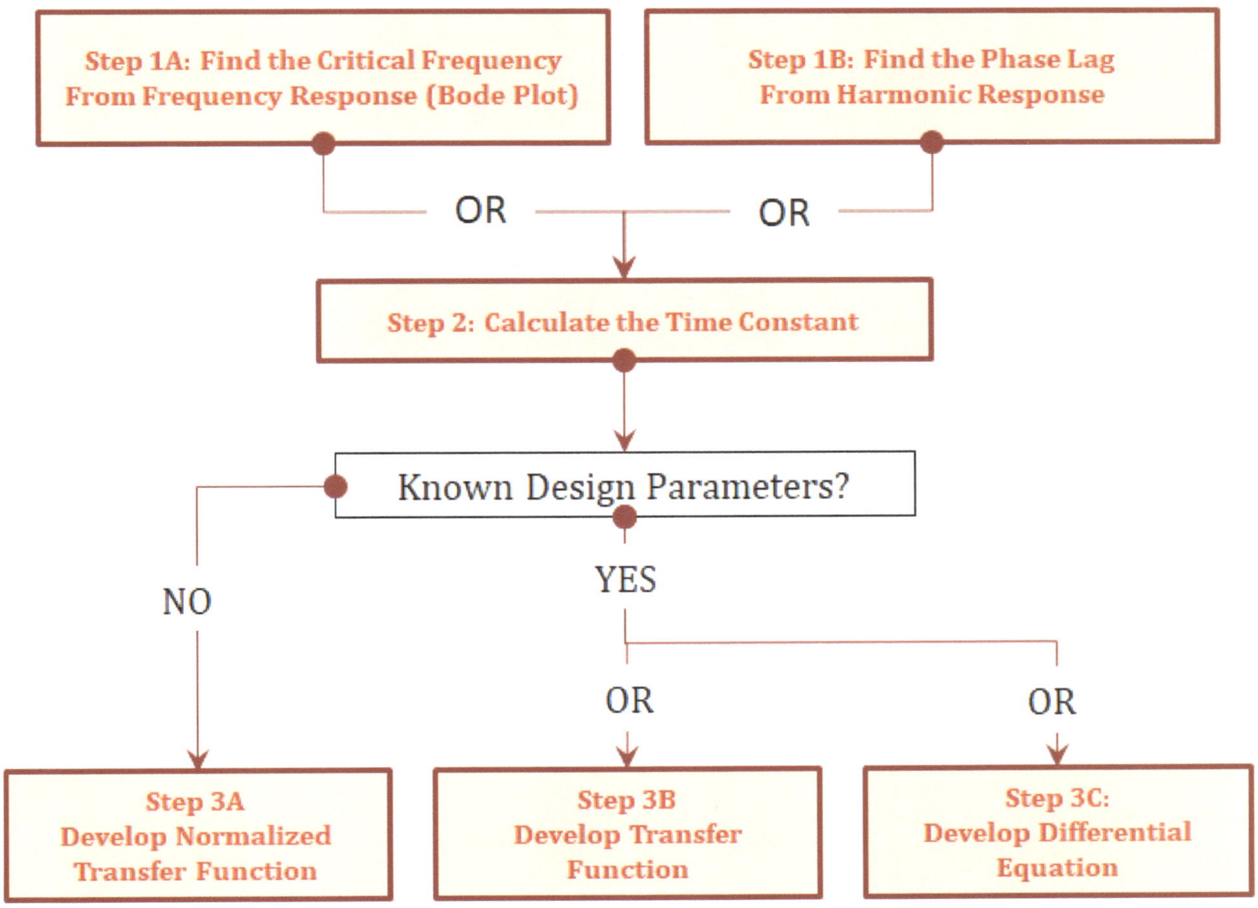

Fig. 2.25 – Road Map to Develop a First-Order Model based on Frequency Response

Case Study 1:

Step 1A: Find the Critical Frequency from a Bode Plot:
Assuming a system has the Bode Plot shown in Fig. 2.26, then the critical frequency = 2 rad/s. If a frequency response does not exist, then excite the system with increasing frequency until the amplitude of the system response becomes 70% of the input amplitude. Then this will be the critical frequency of the system

Step 2: Calculate the Time Constant:
Having the critical frequency (from Step 1A), calculate **τ = 1/critical frequency**. Then the time constant **τ = ½ sec.**

Step 3 Construct the Model:
Since the time constant has been calculated, follow any of the previously mentioned steps (3A, 3B, or 3C) to construct the model.

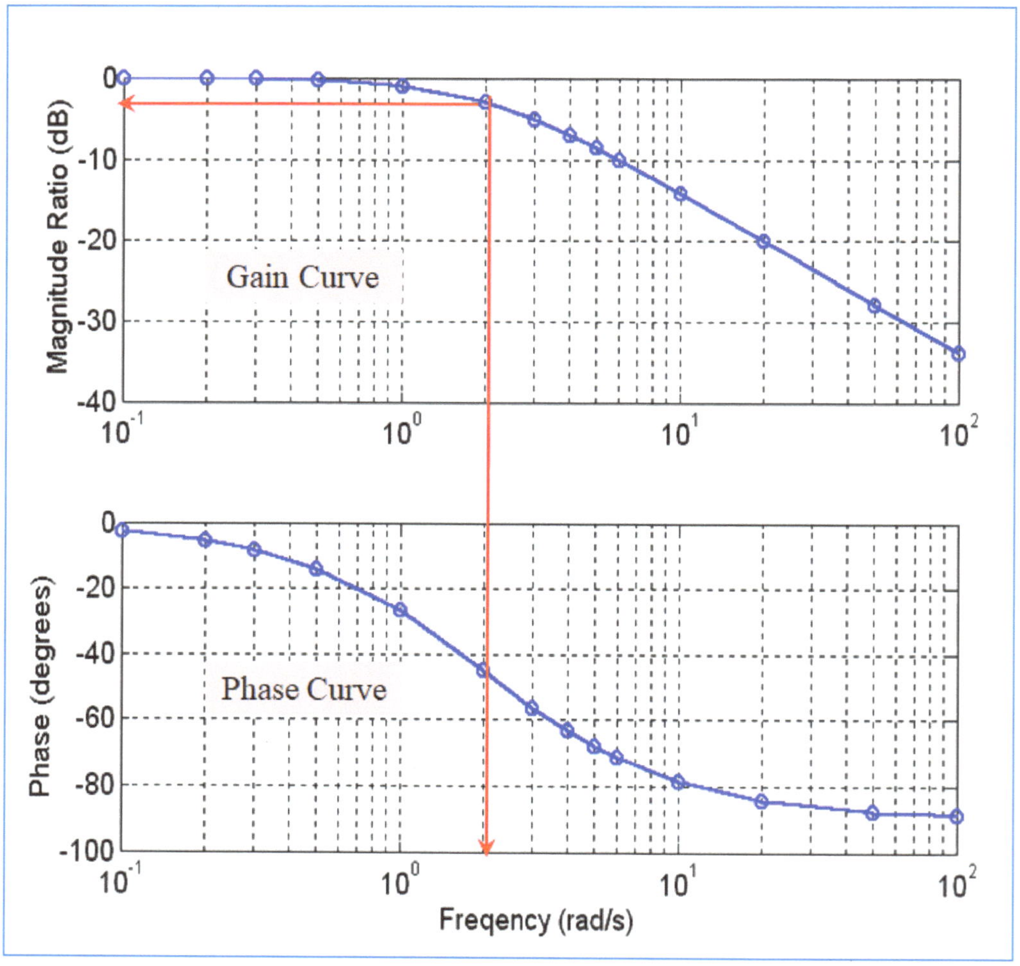

Fig. 2.26 – Frequency Response (Bode Plot) of a First-Order System

Case Study 2:

Step 1B: Find the Phase Lag φ from Harmonic Response:

- Assuming a system has the harmonic response shown in Fig. 2.27, complete cycle time = 2 sec. This means the exciting frequency ω= ½ Hz. = 3.14 rad/s.

- Phase lag is equivalent to -0.25 sec out of a complete cycle time of 2 sec. Since the complete cycle is performed in 360 degrees, the phase lag φ = (-0.25/2) 360 = -45 deg.

Step 2: Calculate the Time Constant:
Having the phase lag (from step 1B), Eq. 2.15 →

$$\tau = -[\tan(\varphi)]/\omega = -[\tan(-45)]/3.14 = 0.516.$$

Step 3 Construct the Model:
Since the time constant has been calculated, follow any of the previously mentioned steps (3A, 3B, or 3C) to construct the model.

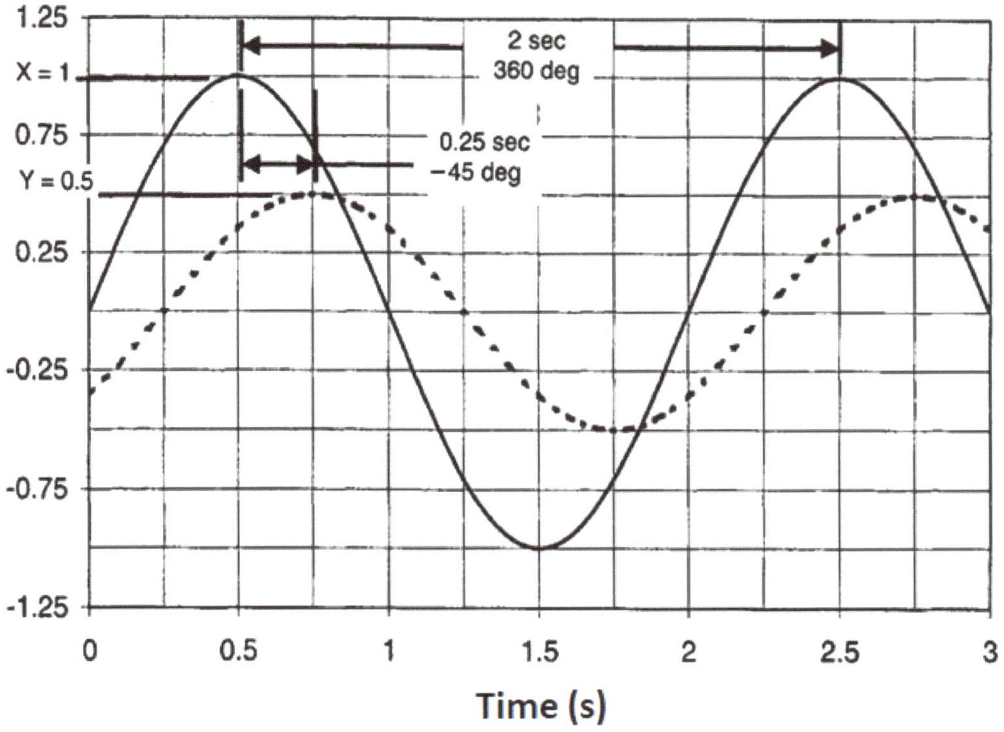

Fig. 2.27 – Harmonic Response of a First-Order System

Chapter 3
Modeling and Simulation of Second-Order Dynamic Systems

Objectives

In this chapter, methods and theories presented in Chapter 1 are applied for Second-Order dynamic system. The chapter presents mathematical modeling for second-order systems in t-domain and s-domain. This chapter also presents the response of second-order systems to the typical forcing functions including, step, ramp, and harmonic inputs. The chapter discusses measured characteristics of second-order step response and how to develop the transfer function of the system based on existing dynamic characteristics.

Brief Contents

Chapter 3: Modeling and Simulation of Second-Order Dynamic Systems

A *Second-Order* dynamic system is a fair approximation for many dynamic systems that contain the three basic physical elements; inductive (inertial), capacitive (energy storage), and resistive (energy dissipative). Mathematical solutions for second-order systems are also well-developed.

3.1- Second-Order Physical System Identification

As it has been discussed in chapter 1, first step in a product cycle development is to identify the physical system. As shown in Fig. 3.1, to simplify the discussion, mechanical translational second-order system i considered.

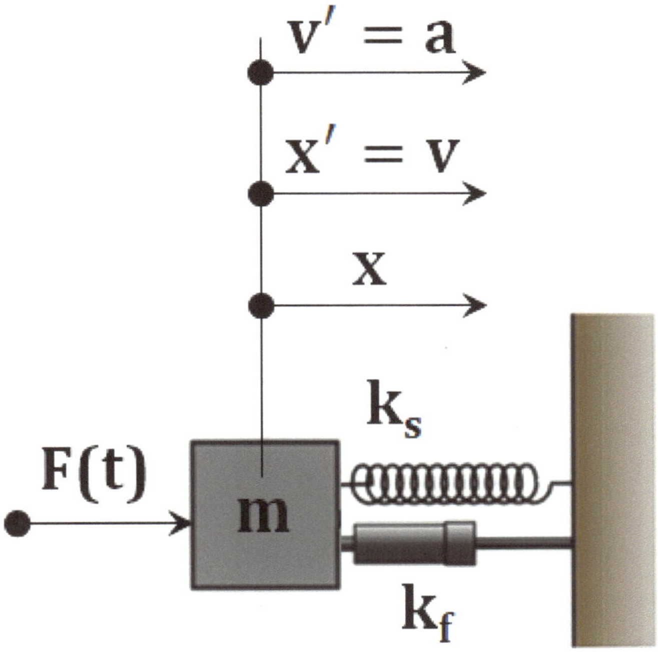

Fig. 3.1 – Mechanical Translational Second-Order Systems

3.2- Second-Order System Mathematical Modeling

As it has been discussed in chapter 1, after identifying the system, a mathematical model is to be developed.

3.2.1- Second-Order System Mathematical Model in t-domain

In Chapter 1, Eq. 1.2 shows the differential equation that represents a generic second-order system. Since a second order system contains three physical elements (**inductive**, **resistive**, **and** capacitive), then referring to Eq.1.6, Eq.1.9, and 1.11, the following four equations represent the basic second-order physical systems.

❖ For a Mechanical Translational System:

$$\mathbf{F(t)} = \mathbf{m}\frac{\mathbf{dv}}{\mathbf{dt}} + \mathbf{k_f\,v} + \mathbf{k_s}\int \mathbf{v\,dt} \qquad \mathbf{3.1}$$

❖ For a Mechanical Rotational System:

$$\mathbf{T(t)} = \mathbf{j}\frac{\mathbf{d\omega}}{\mathbf{dt}} + \mathbf{k_f\,v} + \mathbf{k_s}\int \boldsymbol{\omega}\,\mathbf{dt} \qquad \mathbf{3.2}$$

❖ For an Electrical System:

$$\mathbf{e(t)} = \mathbf{L}\frac{\mathbf{di}}{\mathbf{dt}} + \mathbf{R\,i} + \frac{\mathbf{1}}{\mathbf{C}}\int \mathbf{i\,dt} \qquad \mathbf{3.3}$$

❖ For a Hydraulic System:

$$\mathbf{\Delta P(t)} = \mathbf{I}\frac{\mathbf{dQ}}{\mathbf{dt}} + \mathbf{R\,v} + \frac{\mathbf{1}}{\mathbf{C}}\int \mathbf{Q\,dt} \qquad \mathbf{3.4}$$

Where

- **m** = Equivalent moving mass.
- **v** = Velocity of translational mass.
- **ω** = Angular Speed of a rotational mass.
- $\mathbf{k_f}$ = Viscous frictional coefficient for translational/rotational system.
- $\mathbf{k_s}$ = Linear/rotational spring constant.
- **L** = Electrical inductance.
- **I** = Fluid line inductance.
- **R** = Electrical/hydraulic line resistance.
- **C** = Electrical/hydraulic capacitance.

Mechanical translational system will be considered for the rest of the book. Equation 3.1 can be rewritten in a more familiar form as follows.

$$\mathbf{ma + k_f v + k_s x = F(t)} \qquad 3.5$$

This system solves for the displacement **x**. The steady state value of **x** is when the rate of change of **x** with the time, i.e. the velocity **v = 0 and a = 0**. Then $\mathbf{x_{ss}}$ = $\mathbf{F(t)/k_s}$. If the input forcing function is a step input **F**, then $\mathbf{x_{ss}}$ = $\mathbf{F/k_s}$. is the displacement at which the driving force is balanced with the spring force and system stops moving (Hooke's law).

3.2.2- Second-Order System Mathematical Model in s-domain

By applying the Laplace Transform to the D.E. 3.1, a second-order system is represented as follows:

$$\mathbf{(ms^2 + k_f s + k_s)\, x(s) = F(s)} \qquad 3.6$$

By dividing the whole equation by **m**, Eq. 3.6 will take the following form:

$$\mathbf{(s^2 + \frac{k_f}{m} s + \frac{k_s}{m})\, x(s) = \frac{F(s)}{m}} \qquad 3.7$$

Considering the definition of *Damping Ratio* and *Undamped Natural Frequency* as the frequency at which the system oscillates at no damping **(ξ = 0).**

$$\mathbf{Natural\ Frequency = \omega_n[rad/s] = \sqrt{\frac{k_s}{m}}} \qquad 3.8$$

$$\mathbf{Damping\ Ratio = \xi = \frac{k_f}{2\sqrt{k_s m}}} \qquad 3.9$$

Then the Transfer Function that represents a second-order system is as follows:

$$\mathbf{TF = \frac{Output}{Input} = \frac{x(s)}{F(s)} = \frac{1/m}{s^2 + 2\xi\omega_n s + \omega_n^2}} \qquad 3.10$$

So, this TF can be uploaded to simulation software and will result in the same system response as if the differential equation is used. Then the system modeler is free to choose how to mathematically represent the second-order system.

To calculate the steady state value in s-domain, *Final Value Theorem* applies as follows:

$$\mathbf{x_{ss} = lim_{s\to 0}\, s.x(s) = lim_{s\to 0} \frac{sF(s)/m}{s^2 + 2\xi\omega_n s + \omega_n^2}} \qquad 3.11$$

3.2.3- Second-Order System Normalized Transfer Function

If the magnitude of the design parameter **c** in the generic equation 1.2 (that is the spring constant $\mathbf{k_s}$ in a translational system represented by Eq. 3.5 and Eq. 3.6) equal unity, then the magnitude of the steady state variable **x** equals the magnitude of the forcing function **F**. The form of a second-order transfer function that has **c = 1** is named in this textbook the *Normalized Transfer Function* (NTF).

When $\mathbf{k_s} = 1$, Eq. 3.8 → $1/m = \omega^2$ → then Eq. 3.10 can be rewritten in the following NTF form where the numerator of the NTF = $\boldsymbol{\omega^2}$.

$$\mathbf{NTF} = \frac{\mathbf{Output}}{\mathbf{Input}} = \frac{\mathbf{x(s)}}{\mathbf{F(s)}} = \frac{\boldsymbol{\omega_n^2}}{\mathbf{s^2 + 2\xi\omega_n s + \omega_n^2}} \qquad \mathbf{3.12}$$

As shown in Fig. 3.2, model (HSV7CH03Model01) contains NTF for 1st- Order and 2nd Order systems.

In both systems it was assumed that $\mathbf{k_f = 1}$, $\mathbf{k_s = 1}$, and a step input **F(t) = 100**. In the 2nd-Order system, the mass was assumed **m = 2**.

As shown in the figure, the two systems stabilized at a steady state value equal 100.

Obviously the 2nd Order system has different dynamic response than the 1st Order system and will be discussed later in this chapter.

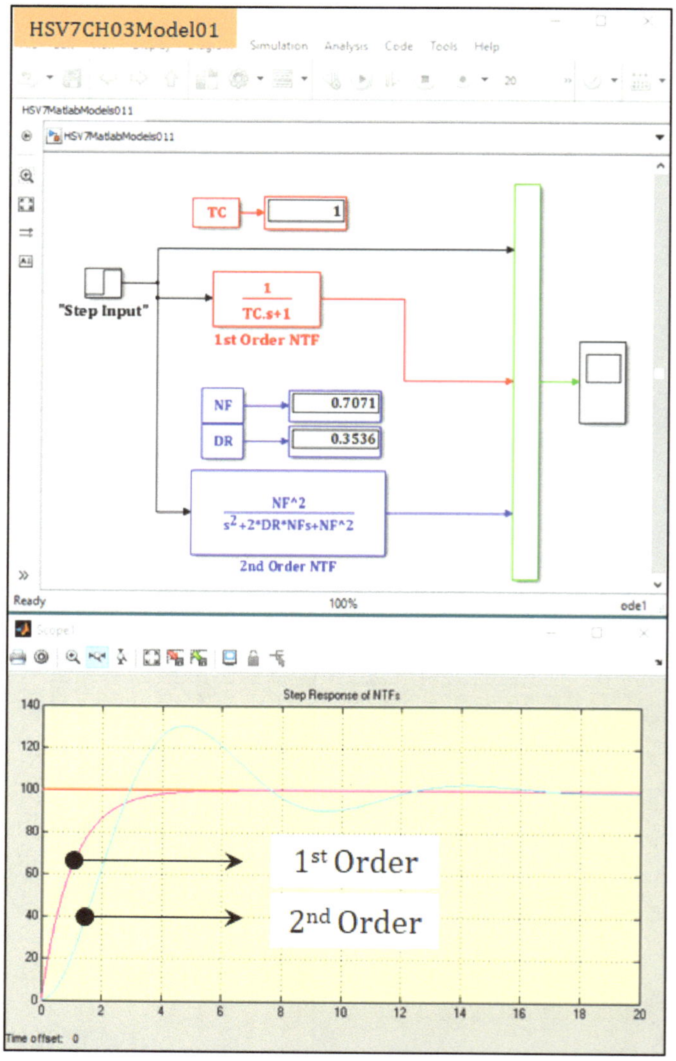

Fig. 3.2 – Response of Normalized Transfer Functions

3.3- Simulation Model Development of Second-Order Systems

As it has been discussed in chapter 1, after a mathematical model is developed for a physical system, it must be converted into a simulation model to be solved by software. As it has been discussed in the previous section, a second order system has been mathematically presented either in t-domain using DE or in s-domain using TF. Now, the objective is to upload these two mathematical models into software to solve for the system response to various exciting forcing functions.

As shown in Fig. 3.3, referring to the DE Eq. 3.6, a simulation model (HSV7CH03Model02) has been developed. The upper part of the figure shows that the model compares the response of the system to a Step input F(t). The lower part of the figure shows that Simulink blocks are arranged to solve for the displacement **x(t).**

The figure shows also that saturation limits for both the speed and displacement can be added. These saturation blocks are to limit the calculation based on the physical maximum speed and maximum displacement. All these blocks are capsulated in one *Subsystem* that is shown in the upper part with the relevant masking image.

As shown in Fig. 3.4, design parameters for this system are assigned in the subsystem block properties. It is to be noted that the *Natural Frequency* and the *Damping Ratio* of the displacement are calculated in the block properties based on the given design parameters.

Sampling Time and magnitude of the Forcing Function F(t) are assigned in the model properties.

As shown in Fig. 3.5, for the same second-order system, another model (HSV7CH03Model03) is structured based on the TF. The figure shows just one transfer function block that represents the system. That is why this method is recommended for large systems. The figure also shows the design parameters assigned in the TF block properties.

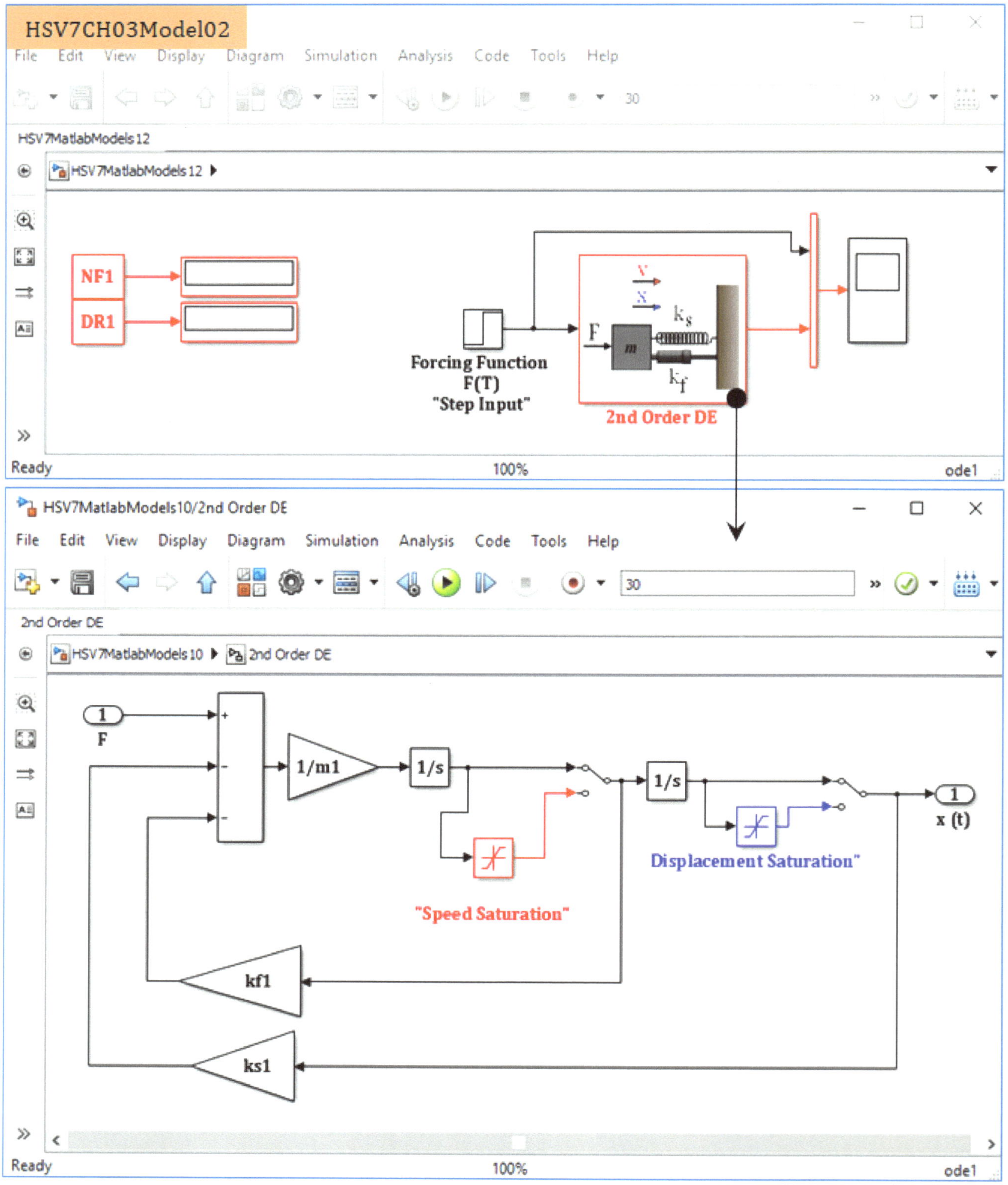

Fig. 3.3 – Simulation Model for Second-Order System Based on Differential Equation

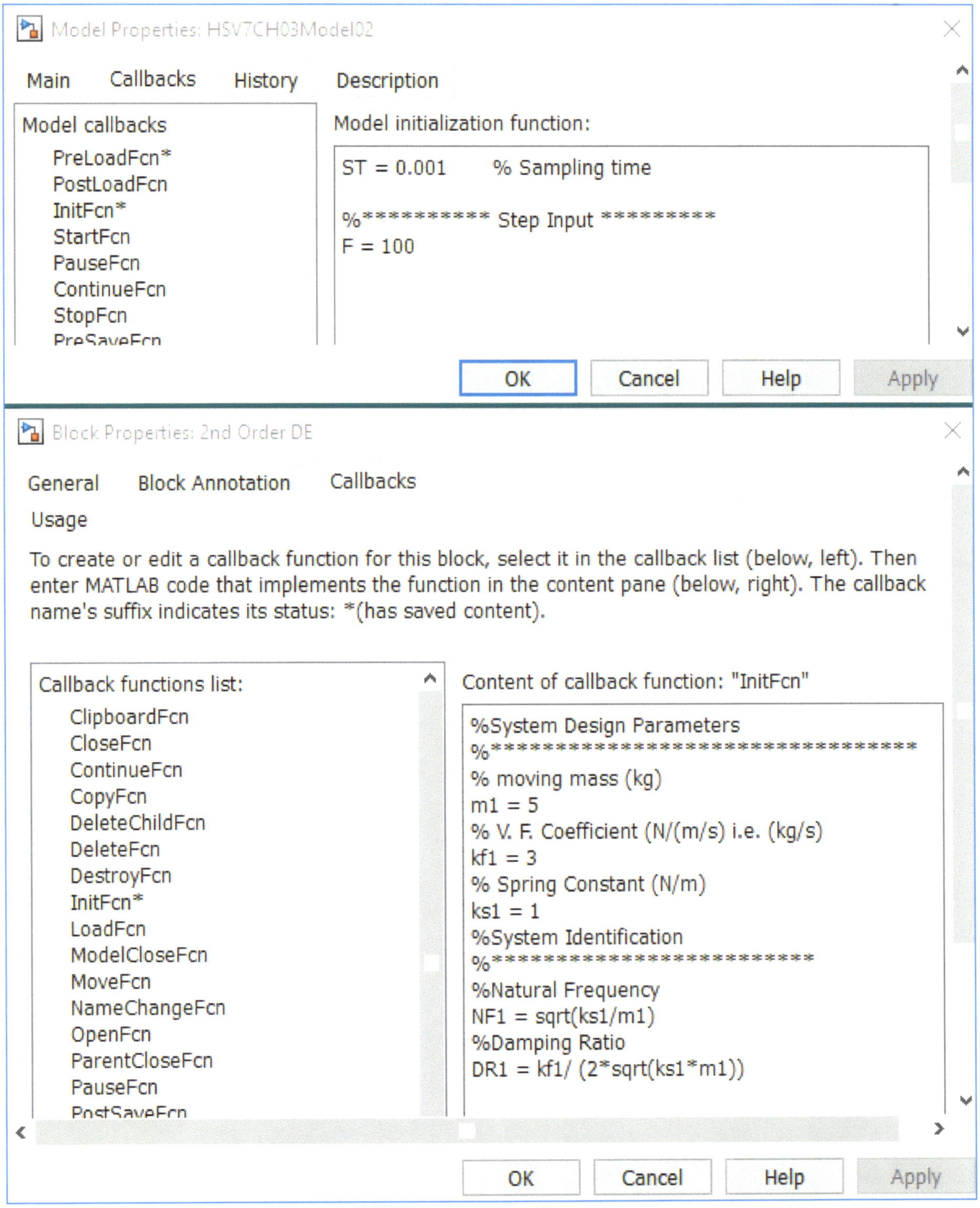

Fig. 3.4 – Design Parameters for the Previous Model

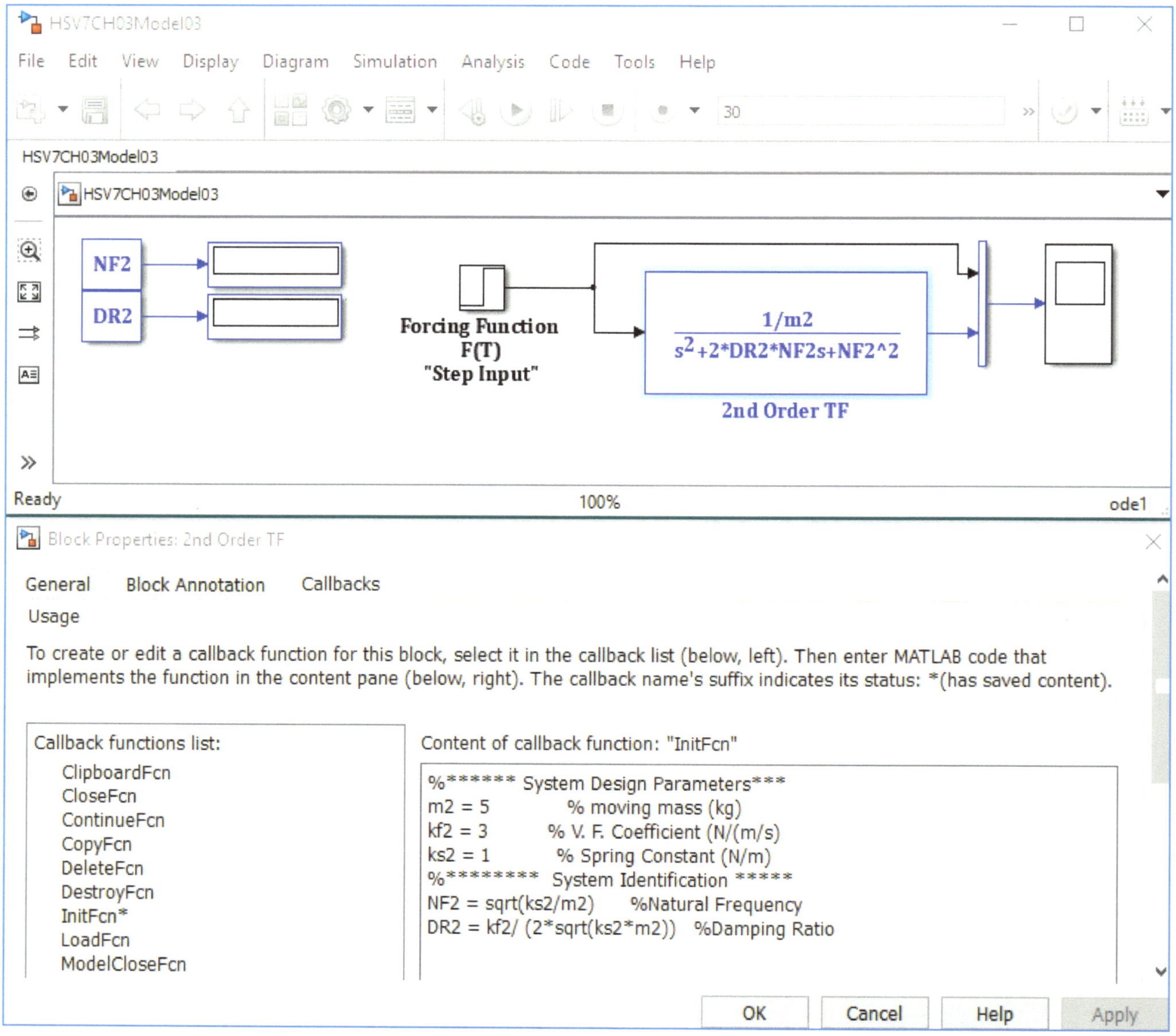

Fig. 3.5 – Simulation Model for Second-Order System Based on Transfer Function

3.4- Performance Simulation of Second-Order Systems

As it has been previously stated, no matter how a dynamic system is mathematically modeled and no matter what the input signal is, the simulation results should be the same if the same design parameters are used. Figure 3.6 shows a new simulation model (HSV7CH03Model04) that combines the model in t-domain and the model in s-domain. The figure shows a *Sampling Time* of 1 ms. has been assigned as a global variable in the model properties. A step input **F = 100** is also assigned in the model properties.

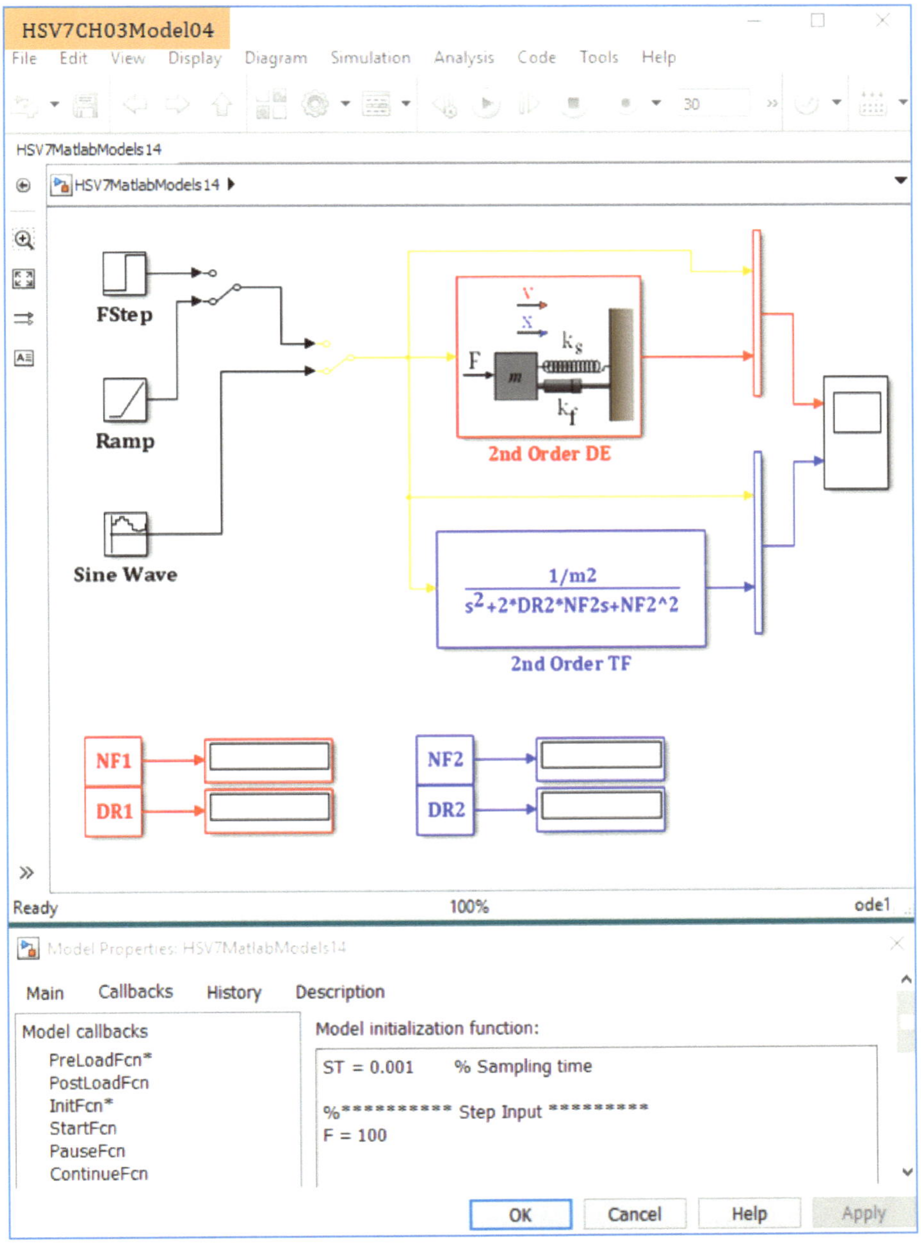

Fig. 3.6 – Simulation Models for Second-Order System

Figure 3.7 shows the response of a second-order system to a *Step Input*. The following observations can be seen from the simulation results:

Note 1: Dynamic response of two models (including the dynamic pattern, Natural Frequency, and Damping) are identical since same design parameters are assigned in both models.

Note 2: Since the design parameter **c = 1 (the spring constant k_s = 1)**, the steady state value in the step response equals the same magnitude as the input step function.

Note 3: Steady state value of the system response is when the external force is balanced with the spring force. This is seen whether the system is represented by a DE or a TF. To calculate the SS value when a TF is subjected to a step input, F(s) = 100/s, final value theorem shown in Eq. 3.11 is applied as follows.

$$x_{ss} = \lim_{s\to 0} s.\, x(s) = \lim_{s\to 0} \frac{sF(s)/m}{s^2 + 2\xi\omega_n s + \omega_n^2} = \lim_{s\to 0} \frac{s\,\frac{100}{s}/m}{s^2 + 2\xi\omega_n + \omega_n^2} = \frac{100}{m\omega_n^2}$$

$$\text{Eq. 3.8} \rightarrow \omega_n^2 = \frac{k_s}{m} = \frac{1}{m} \quad \textbf{Then} \quad x_{ss} = 100$$

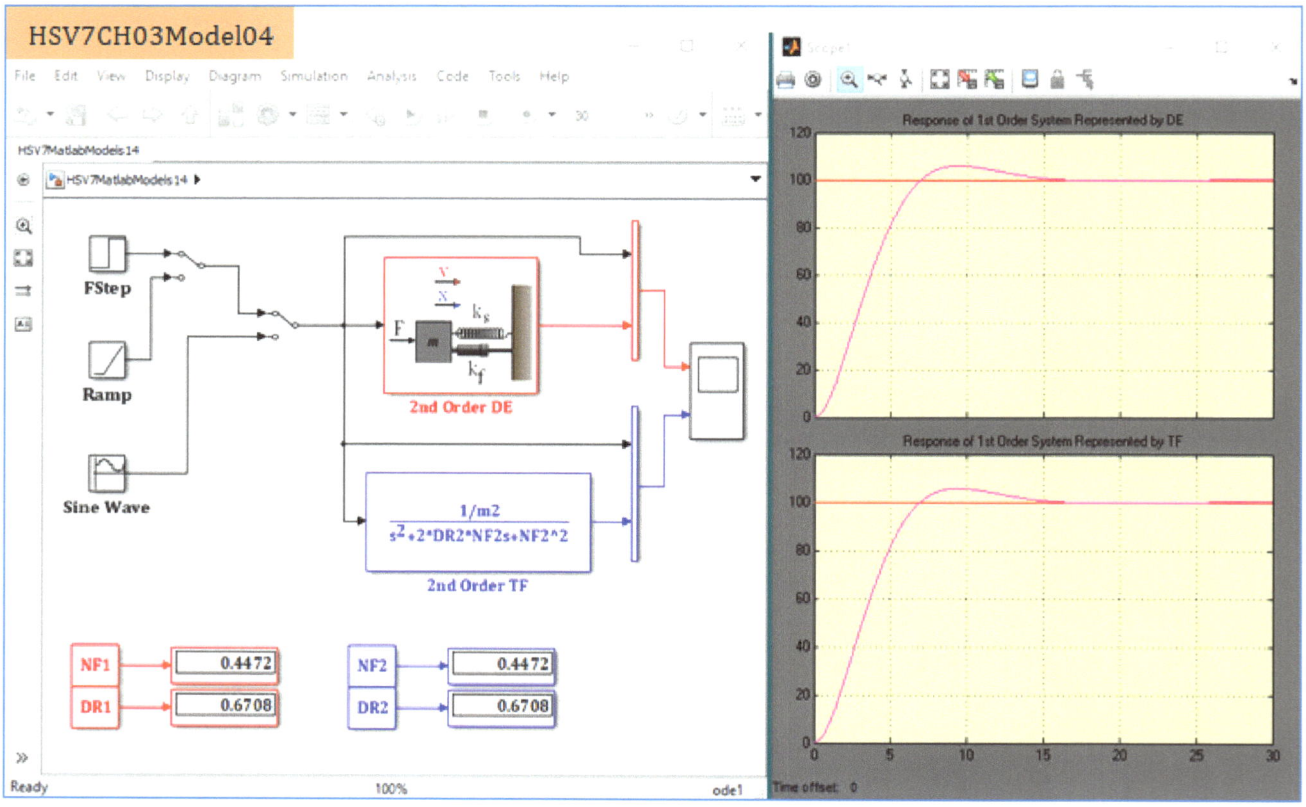

Fig. 3.7 – Step Response of a Second-Order System Based on two Models

Figure 3.8 shows the response of a second-order system to a *Ramp Input*, as it is subjected to gradual loading or commanding with constant rate. Again, simulation results show identical response despite using different model structures. Figure 3.9 shows the response of a second-order system to a *Sinusoidal Input*, as it is subjected to periodic loading or commanding. Results show that same response no matter the structure of the model.

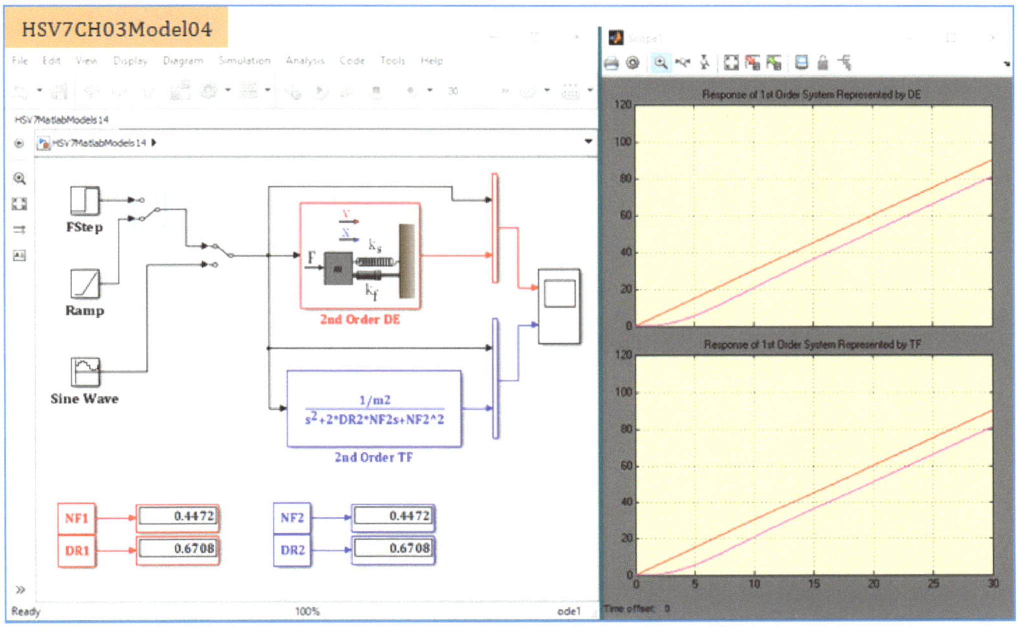

Fig. 3.8 – Ramp Response of a Second-Order System Based on two Models

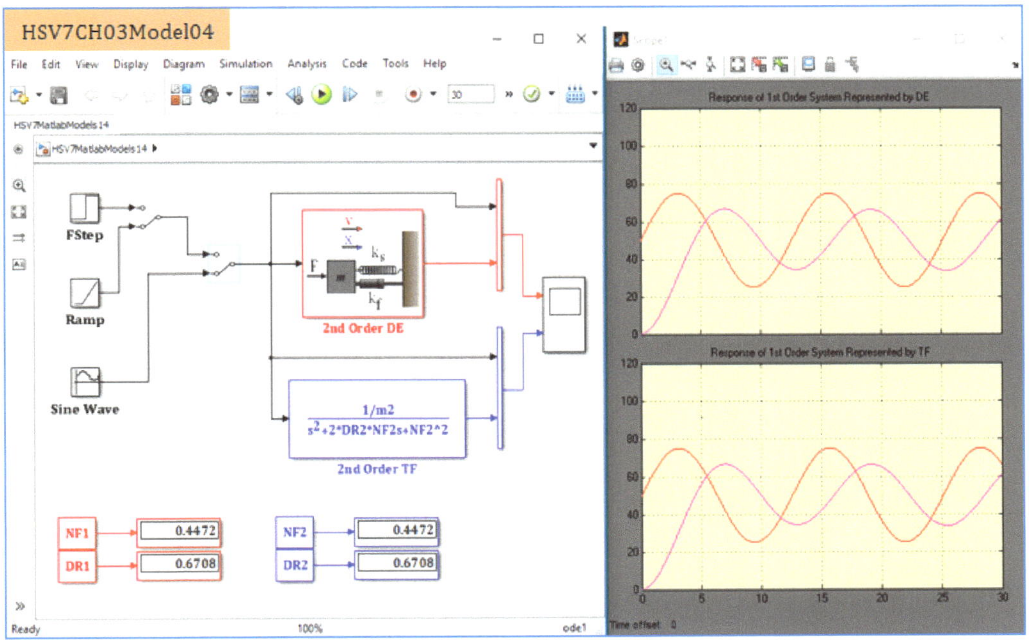

Fig. 3.9 – Harmonic Response of a Second-Order System Based on two Models

3.5- Step Response Analysis of Second-Order Systems

3.5.1- Identification of Second-Order System Based on Step Response

Second order systems are identified by two characterizing parameters: *Damping Ratio* **ξ** and *Natural Frequency* **ω_n**. Figure 3.10 shows a typical step response of a second-order system. Typically, 2nd order systems are designed to have such an underdamped **$(0 < \xi < 1)$** response in which the system experiences a few oscillations before it stabilizes around the steady state value. The following parameters are used as conventional standard measures to identify the response of the system.

- **SS = *Steady State Value***: Is the destination the system is targeting to reach.
- **SSE = Steady State Error**: is a defined % of deviation from the SS.
- **M_p = *Overshoot***: Is the value the system each on top of the SS.
- **PO = *Percentage Overshoot***: Is Mp as % of SS.
- **T = *Cycle Time***: Is the time the system takes to complete one oscillation cycle.
- **t_d = *Delay Time***: Is the time the system takes to reach to 50% of SS.
- **t_r = *Rise Time***: Is the time the system takes to reach to 90% of SS.
- **t_p = *Peak Time***: Is the time the system takes to reach to highest value (at first overshot).
- **t_s = *Settling Time***: Is the time the system takes to reach to the vicinity of SSE.

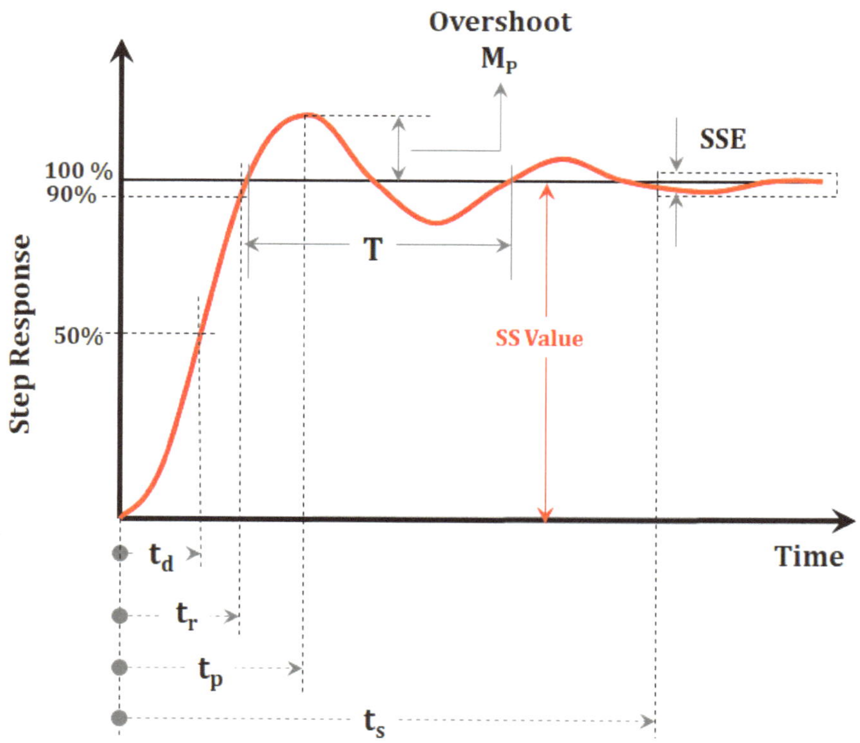

Fig. 3.10 – Step Response Analysis for Second-Order Systems

It was found analytically that, for such underdamped 2nd order systems **($0 < \xi < 1$)**, the following equations are applicable:

$$\textbf{Natural Frequeny [Hz]} = f_n = \frac{1}{T[s]} \qquad 3.13$$

$$\textbf{Natural Frequeny [rad/s] (defined in Eq. 3.8)} = \omega_n = 2\pi f_n \qquad 3.14$$

$$\textbf{Rise Time [s]} = t_r \approx 1.8/\omega_n \qquad 3.15$$

$$\textbf{Peak Time [s]} = t_P = \frac{\pi}{\omega_n\sqrt{1-\xi^2}} \qquad 3.16$$

$$\textbf{Settling Time [s]} = t_s = \frac{4.6}{\xi\omega_n} \text{ for SSE} = \pm 1\% \qquad 3.17A$$

$$\textbf{Settling Time [s]} = t_s = \frac{4}{\xi\omega_n} \text{ for SSE} = \pm 2\% \qquad 3.17B$$

$$\textbf{Settling Time [s]} = t_s = \frac{3}{\xi\omega_n} \text{ for SSE} = \pm 5\% \qquad 3.17C$$

$$\textbf{Overshoot} = M_P = SS\ e^{\left[\frac{-\pi\xi}{\sqrt{1-\xi^2}}\right]} \qquad 3.18$$

$$\textbf{Percentage Overshoot} = M_P(\%) = 100(M_P/SS) = 100\ e^{\left[\frac{-\pi\xi}{\sqrt{1-\xi^2}}\right]} \qquad 3.19$$

Figure 3.11 shows a plot of M_P (%) versus ξ. This plot will be used later to generate transfer function of a 2nd order system based on existing dynamics. The figure shows that

- at ξ = 1, no overshoot will occur.
- at ξ = 0.5, M_P (%) = 16%.
- at ξ = 0.7, M_P (%) = 5%.

That verify Eq. 3.19.

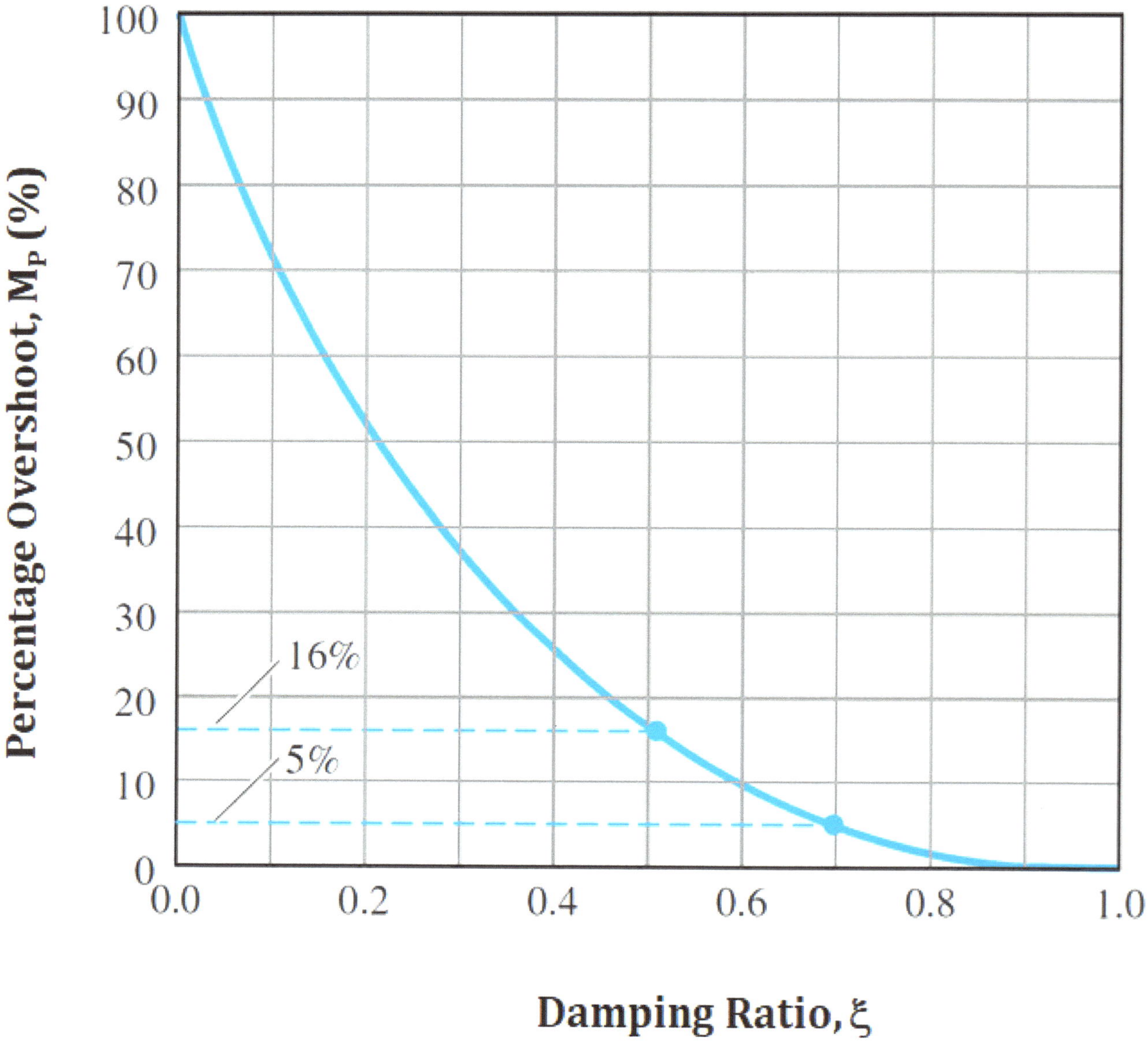

Fig. 3.11 – Percentage Overshoot versus Damping Ratio for Underdamped 2nd Order Systems (Courtesy of Prentice Hall ISBN 0-13-032393-4)

To verify the set of previous equations, as shown in Fig. 3.12, they are uploaded to be solved by Matlab-Simulink assuming that the mass = 1, viscous friction coefficient = 0.5, and spring constant = 1. Figure 3.13 shows the simulation model (HSV7CH03Model05). Simulation results verify the previous analysis as follows:

- Eq. 3.8 $\rightarrow = \ \omega_n[\text{rad/s}] = \sqrt{k_s/m}\ = 1$
- Simulation Results (Fig. 3.13) $\rightarrow$ T = 6.28 s.
- Eq. 3.13 $\rightarrow$ f_n [Hz] = 1/T = 1/6.28.
- Eq. 3,14 $\rightarrow$ ω_n [rad/s] = 2 πf_n = 1 that verifies Eq. 3.8.
- Eq. 3.9 $\rightarrow = \ \xi = \ [k_f/2\sqrt{k_s m}] = \ 0.5/2 \ = \ 0.25$
- Since spring constant = 1, then steady state value = F = 100 (verified by Fig. 3.13).
- Eq. 3.15 $\rightarrow$ Rise Time = 1.8/1 = 1.8 s (verified by simulation results in Fig. 3.13).
- Eq. 3.16 $\rightarrow$ Peak Time = 3.24 s (verified by simulation results in Fig. 3.13).
- Eq. 3.17B $\rightarrow$ Settling Time = 16 s (verified by simulation results in Fig. 3.13).
- Eq. 3.18 $\rightarrow$ Overshoot = 44.45 (verified by simulation results in Fig. 3.13).
- Eq. 3.19 $\rightarrow$ % Overshoot = 44.45% (verified by simulation results in Fig. 3.13).
- Fig. 3.11$\rightarrow$ at ξ = 0.25. % Overshoot = 44.45 that verifies Eq.3.19.

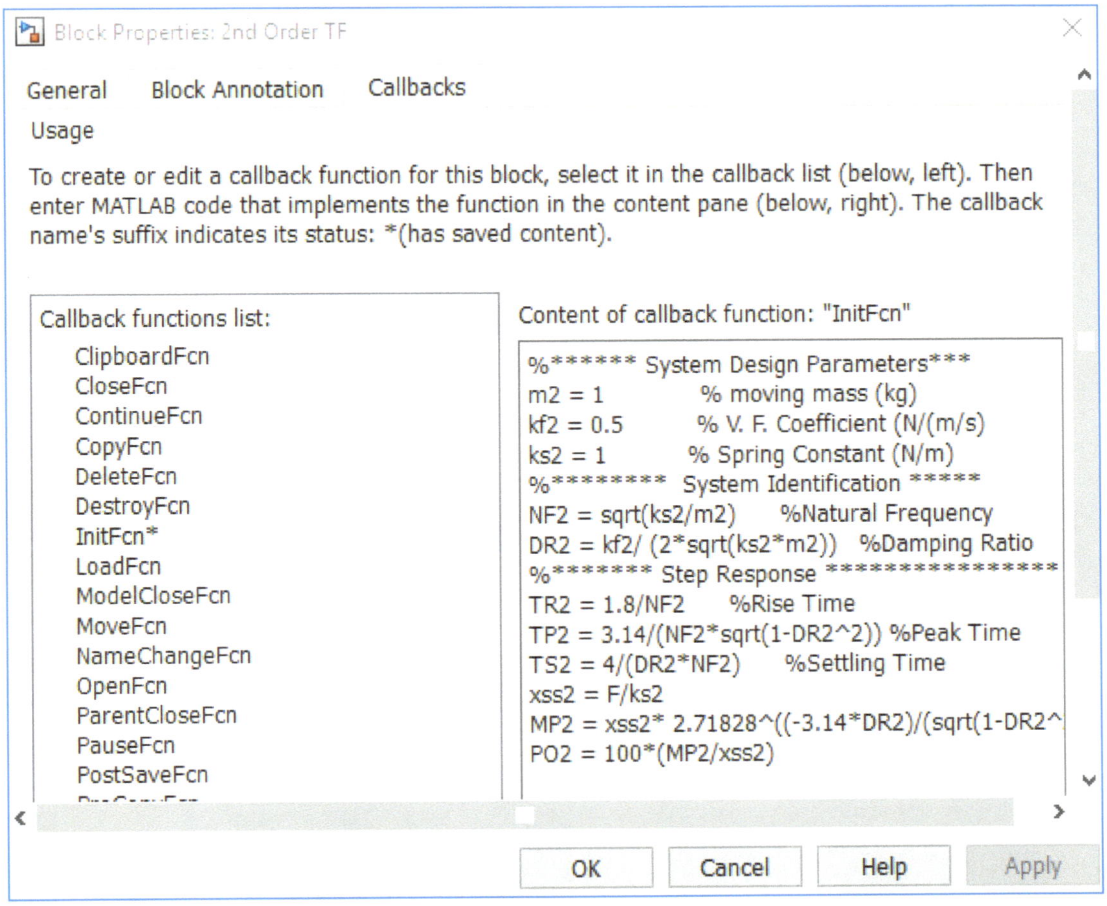

Fig. 3.12 – Block Properties for Model HSV7CH03Model05

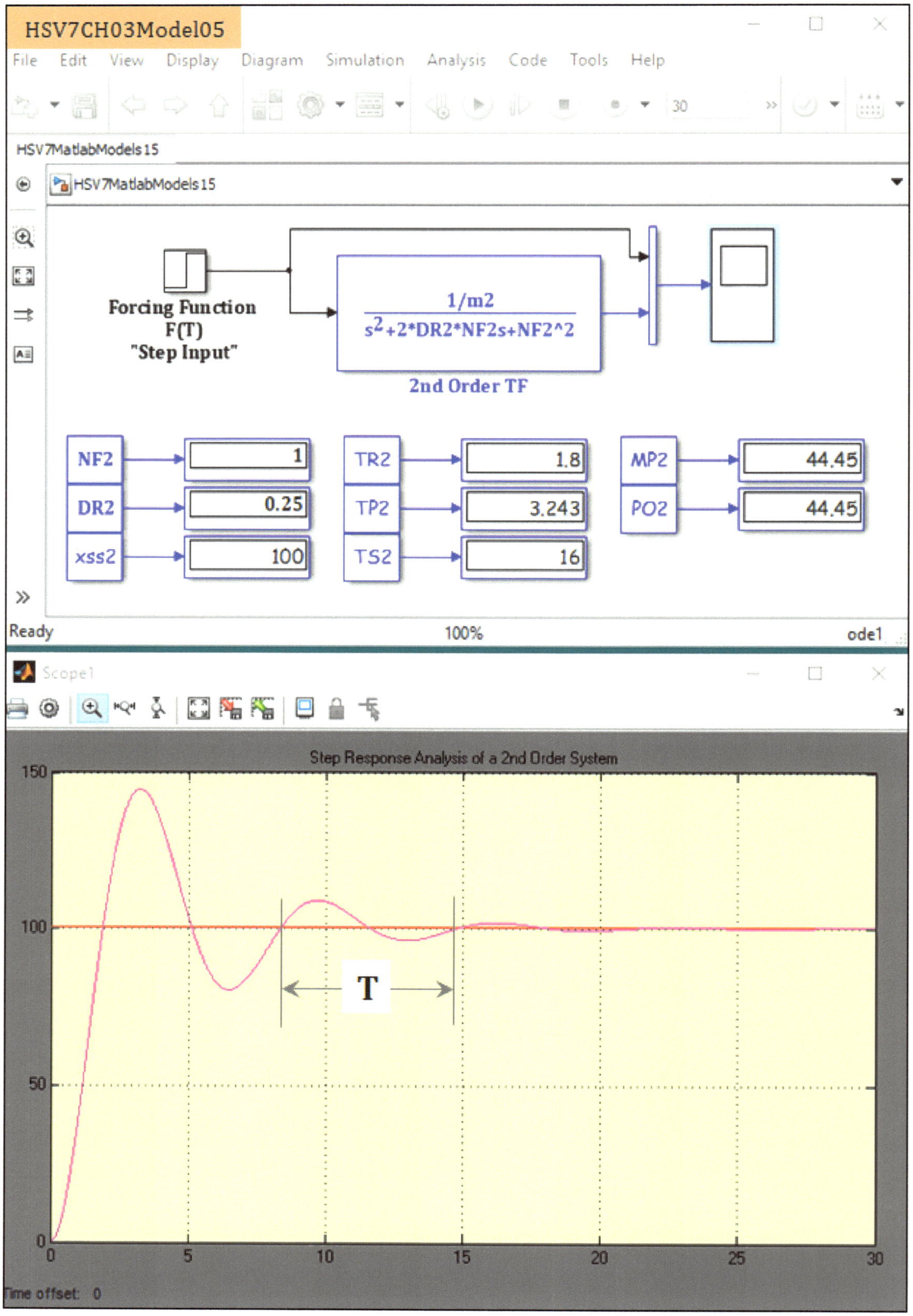

Fig. 3.13 – Verification of a Step Response of a Second Order System

3.5.2- Effect of Design Parameters on Second-Order System Step Response

Simulation model (HSV7CH03Model06) is used to investigate the effect of the physical design parameters on response of second order systems. The following subsections present the results.

3.5.2.1- Effect of Moving Mass

Figures 3.14A and 3.14B show that, by increasing the moving mass incrementally by 1, it was found that:

$$\mathbf{m} \uparrow \underline{(\mathbf{k}_f = 1 \text{ and } \mathbf{k}_s = 1)} \rightarrow \omega_n \downarrow \text{ and } \xi \downarrow \rightarrow \mathbf{M_p(\%)} \uparrow \text{ \& Settling Time} \uparrow$$

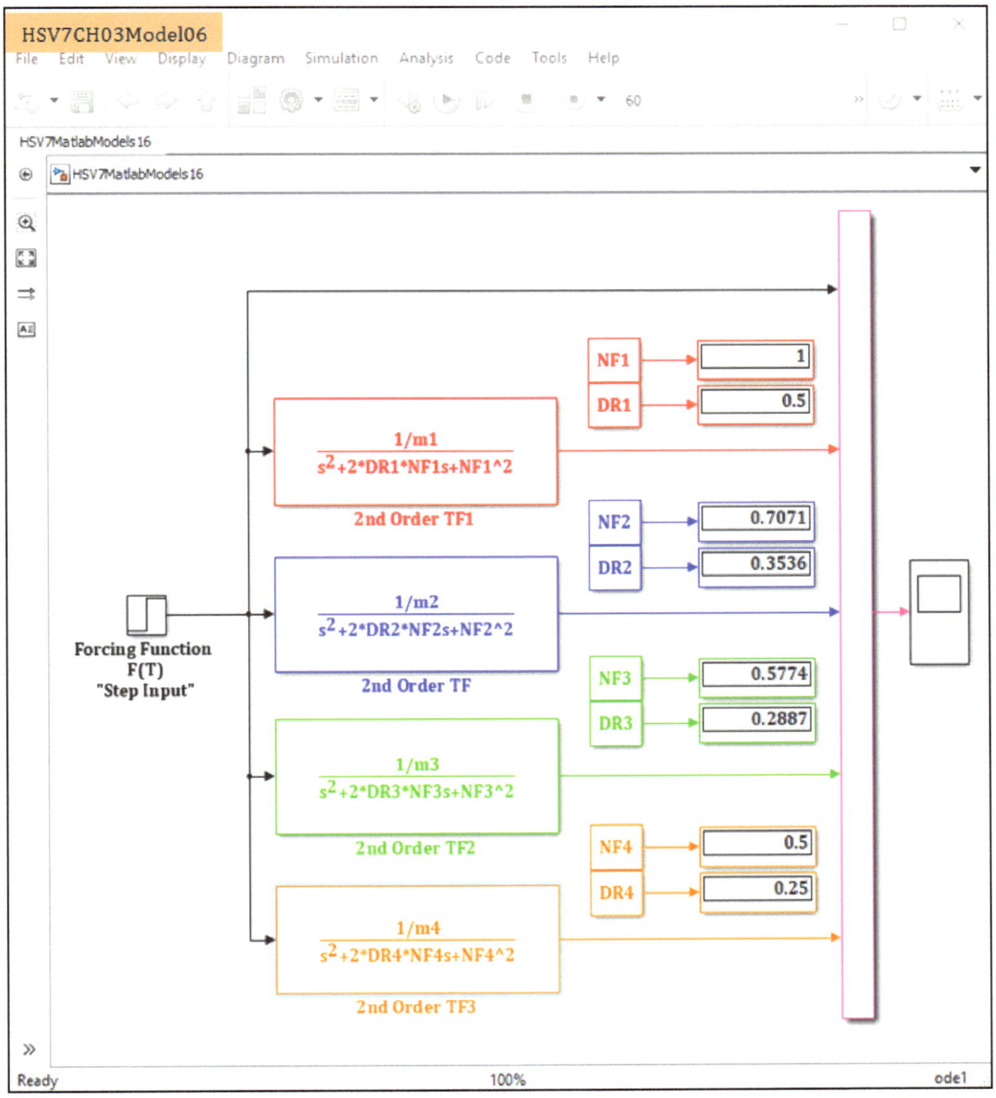

Fig. 3.14A – Model to Investigate Effect of Moving Mass on Step Response of a 2nd Order System

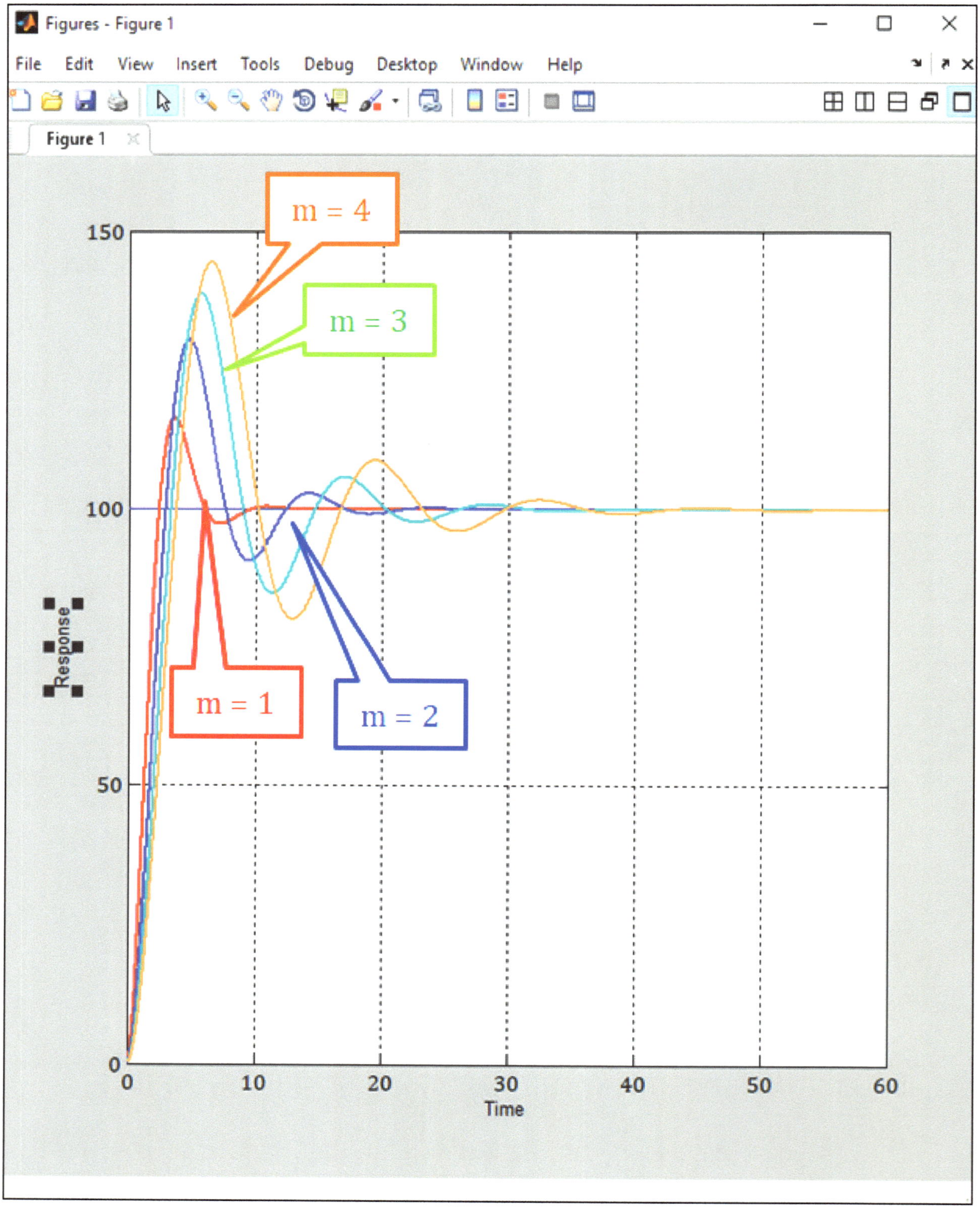

Fig. 3.14B – Effect of Moving Mass on Step Response of a 2nd Order System

3.5.2.2- Effect of Friction Coefficient

Figures 3.15A and 3.15B show that, by increasing the friction coefficient incrementally by 0.2, it was found that:

$\mathbf{k_f}$ ↑ (m = 1 and $\mathbf{k_s}$ = 1) → $\boldsymbol{\omega_n}$ (no change) and ξ ↑ → $\mathbf{M_p(\%)}$ ↓ & Settling Time ↓

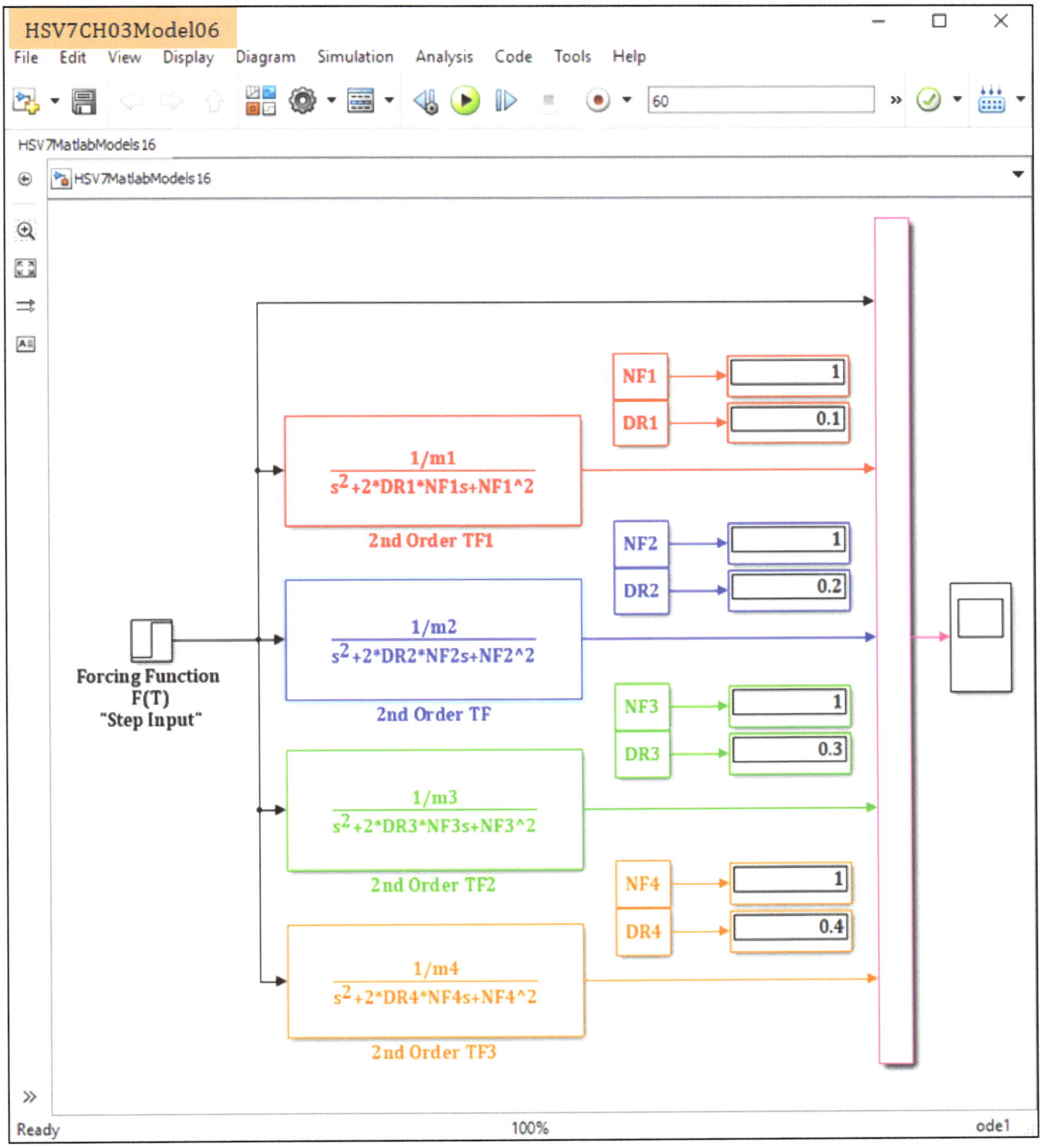

Fig. 3.15A – Model to Investigate Effect of Friction Coefficient on Step Response of a 2nd Order System

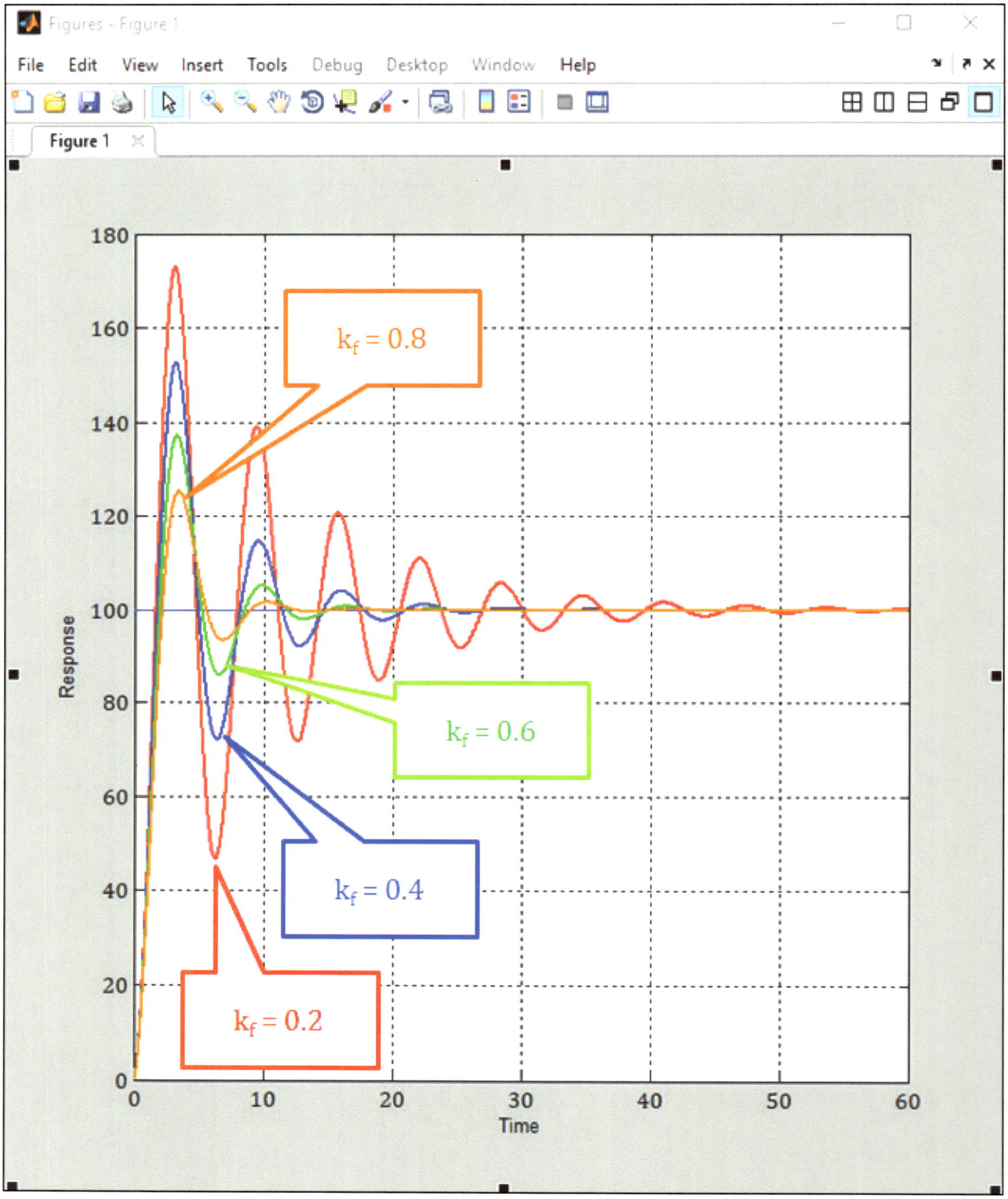

Fig. 3.15B – Effect of Friction Coefficient on Step Response of a 2nd Order System

3.5.2.3- Effect of Spring Constant

Figures 3.16A and 3.16B show that, by increasing the spring constant incrementally by 0.25, it was found that: System may become more oscillatory to the limit of being unstable. System settles at a magnitude different from magnitude of the forcing step function.

$$\mathbf{k_s} \uparrow \underline{(\mathbf{m} = 1 \text{ and } \mathbf{k_f} = 0.5)} \rightarrow \boldsymbol{\omega_n} \uparrow \text{ and } \xi \downarrow \rightarrow \mathbf{x_{ss}} \downarrow$$

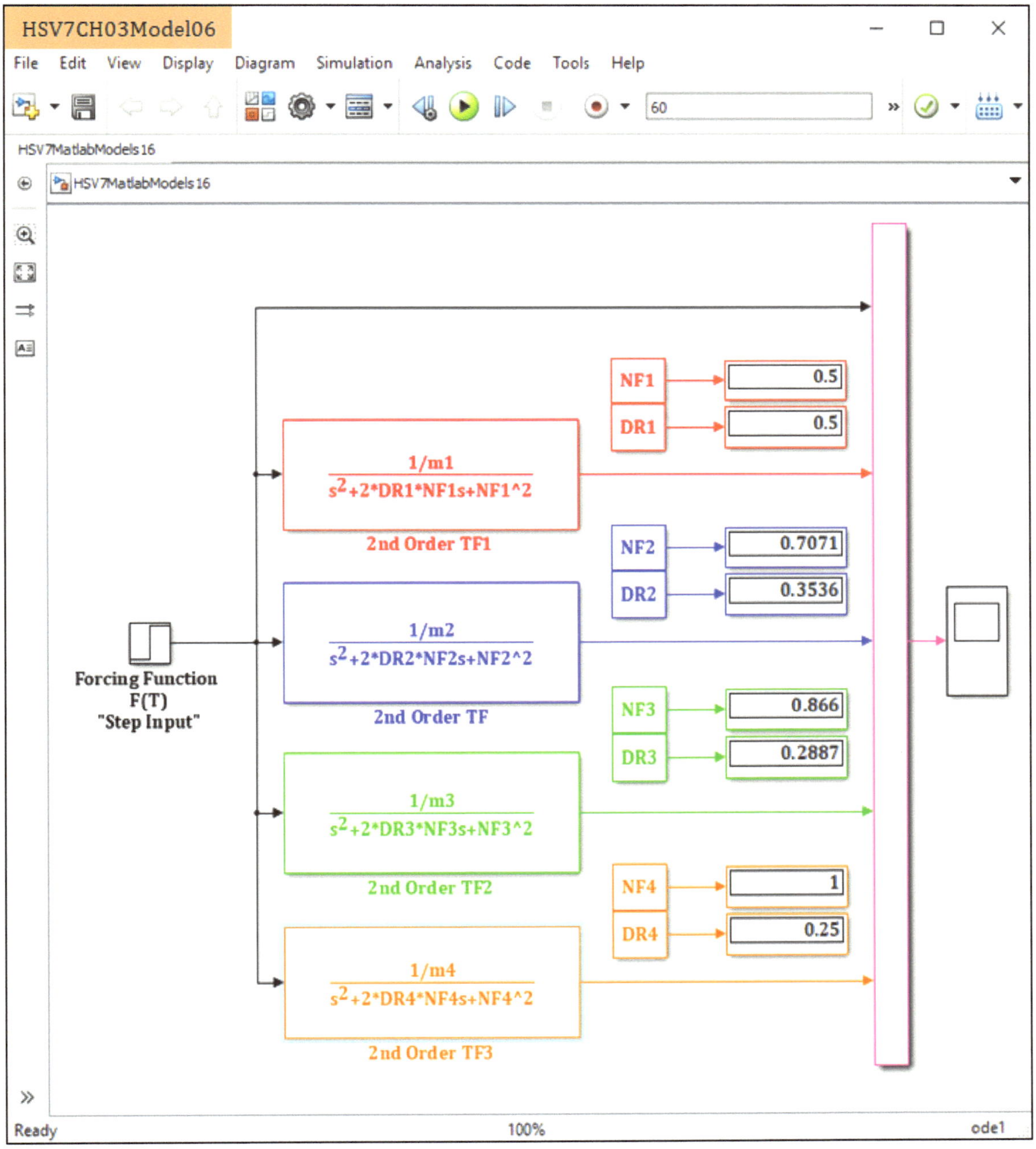

Fig. 3.16A – Model to Investigate Effect of Spring Constant on Step Response of a 2nd Order System

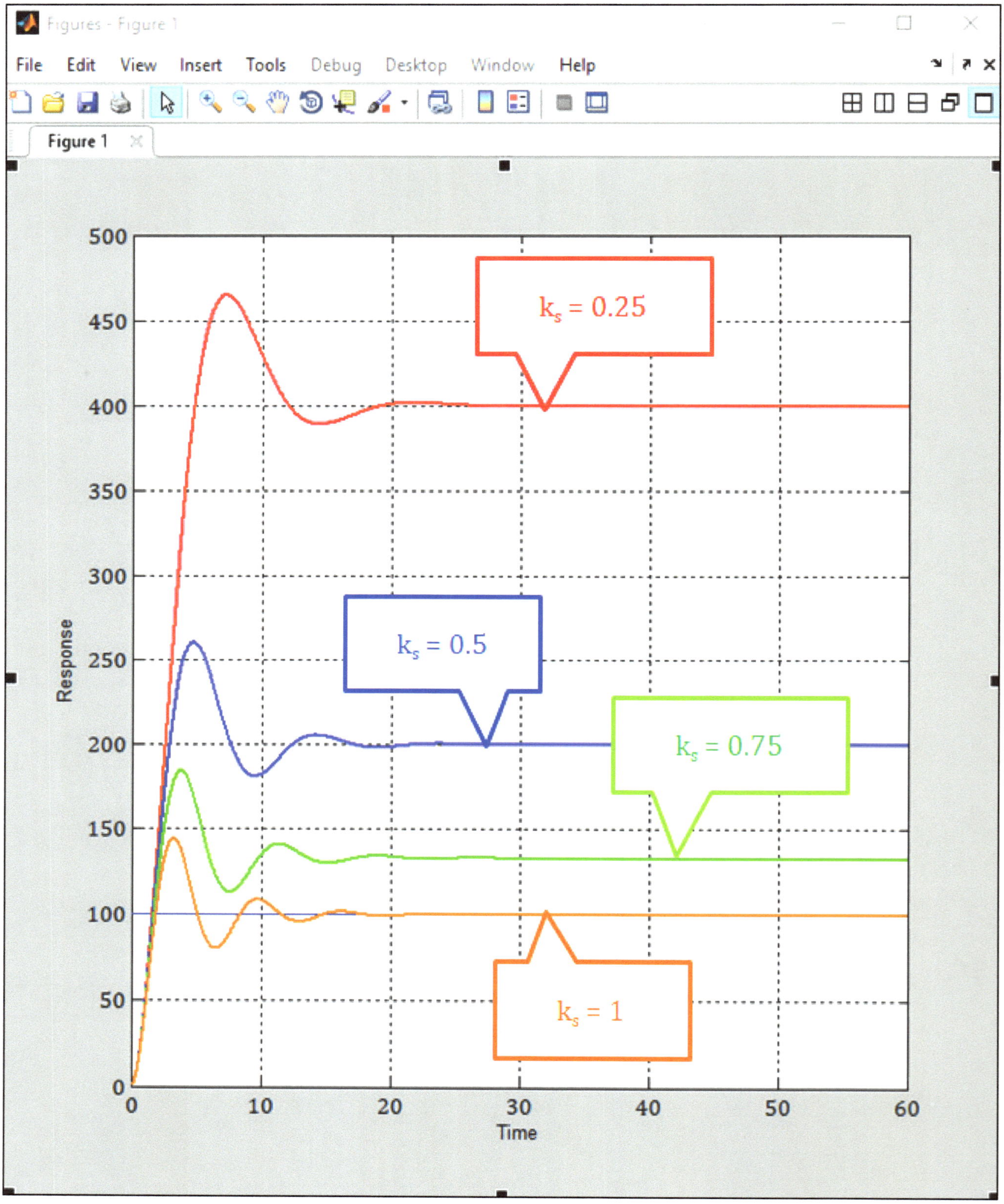

Fig. 3.16B – Effect of Spring Constant on Step Response of a 2nd Order System

3.5.2.4- Effect of Natural Frequency

Figures 3.17A and 3.17B show that, by increasing the natural frequency incrementally by 1, it was found that:

$\omega_n \uparrow$ (ξ is constants) → System speed of response ↑.

The following four sets of data are used to generate the results shown in the figure (m = 1, k_f = 1, and k_s = 1), (m = 0.25, k_f = 0.5, and k_s = 1), (m = 0.1111, k_f = 0.3333, and k_s = 1), and (m = 0.0625, k_f = 0.25, and k_s = 1).

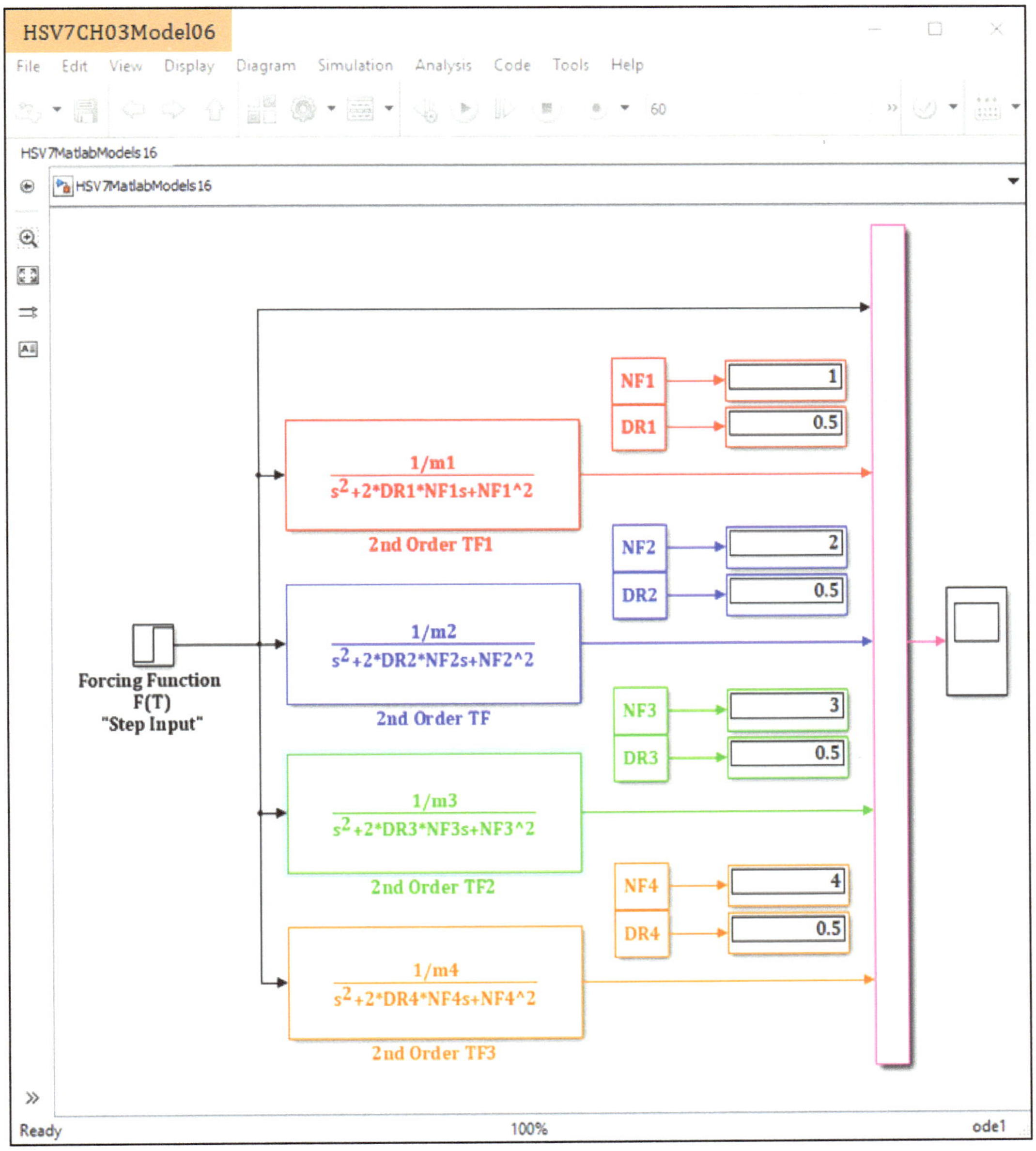

Fig. 3.17A – Model to Investigate Effect of Natural Frequency on Step Response of a 2nd Order System

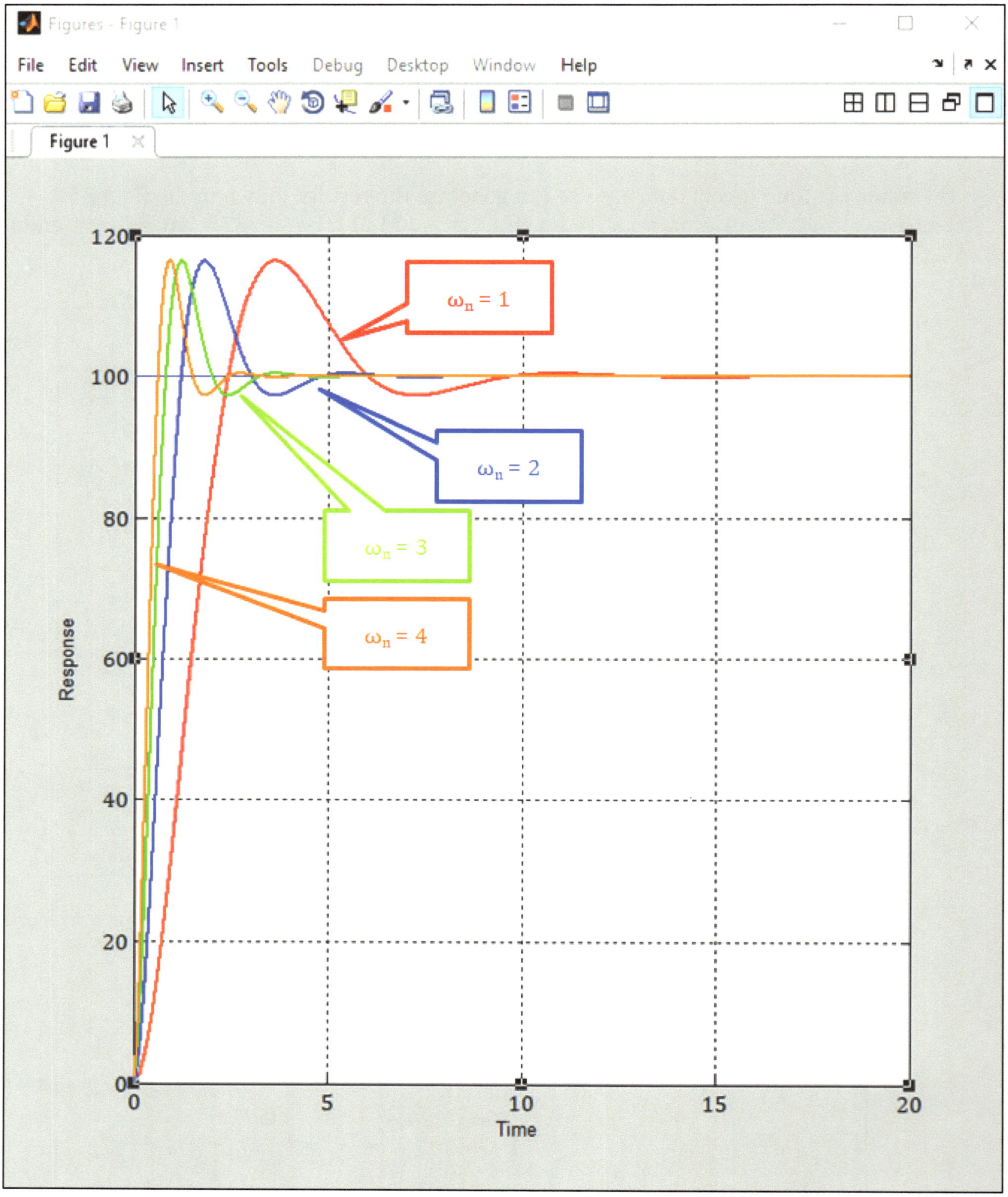

Fig. 3.17B – Effect of Natural Frequency on Step Response of a 2nd Order System

3.5.2.5- Effect of Damping Ratio

As shown in Fig 3.18A, the following four sets of data are used to generate the results shown in the figure ($m = 1$, $k_f = 0$, and $k_s = 1$), ($m = 0.25$, $k_f = 1$, and $k_s = 1$), ($m = 1$, $k_f = 2$, and $k_s = 1$), and ($m = 1$, $k_f = 3$, and $k_s = 1$). As a result, damping ratio is incrementally increased by 0.5.

Figure 3.18B shows that, by increasing the it was found that:

$\xi \uparrow$ (ω_n is constants) $\rightarrow$ System speed of response $\downarrow$ i.e. sluggish system

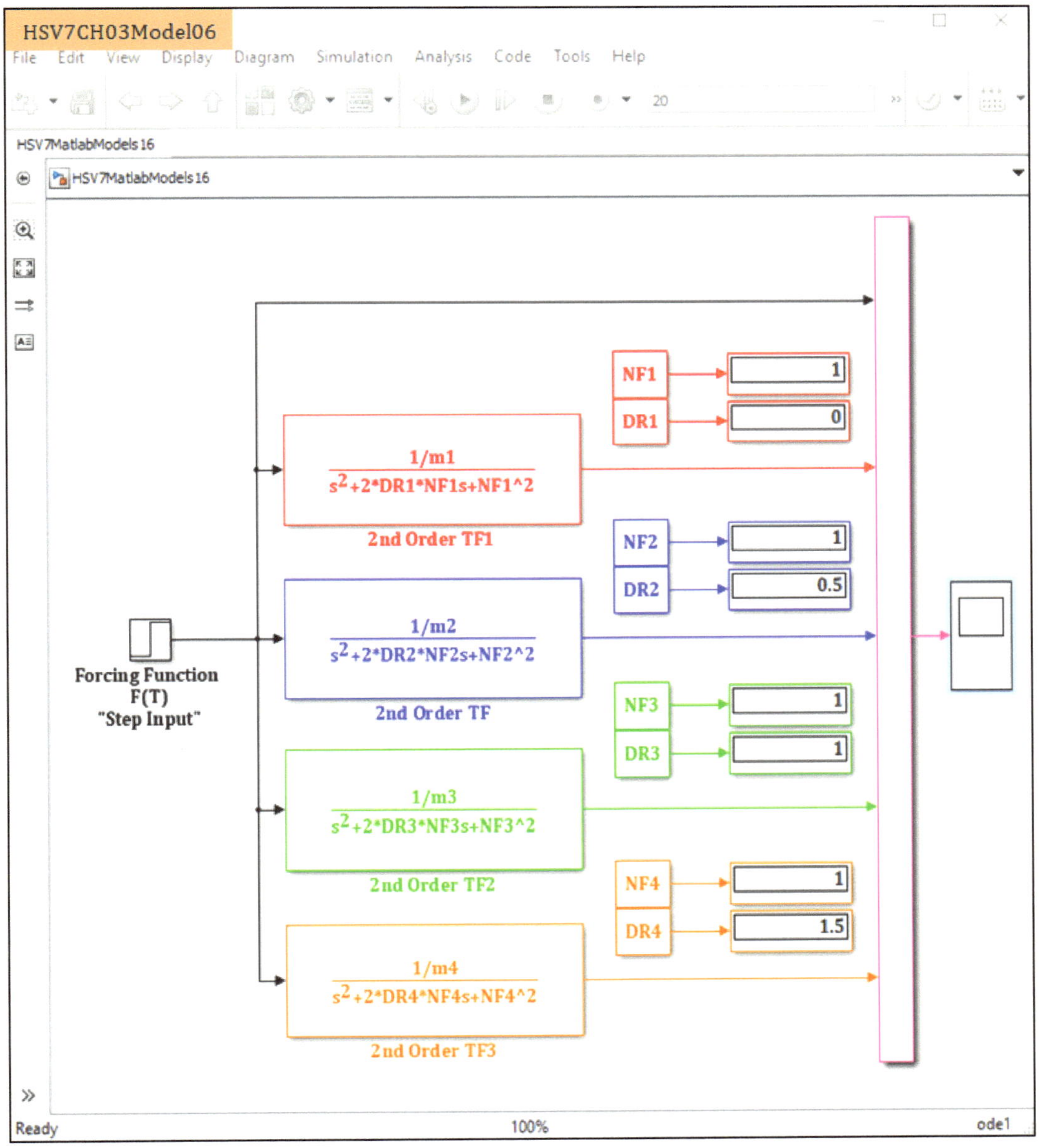

Fig. 3.18A – Model to Investigate Effect of Damping Ratio on Step Response of a 2nd Order System

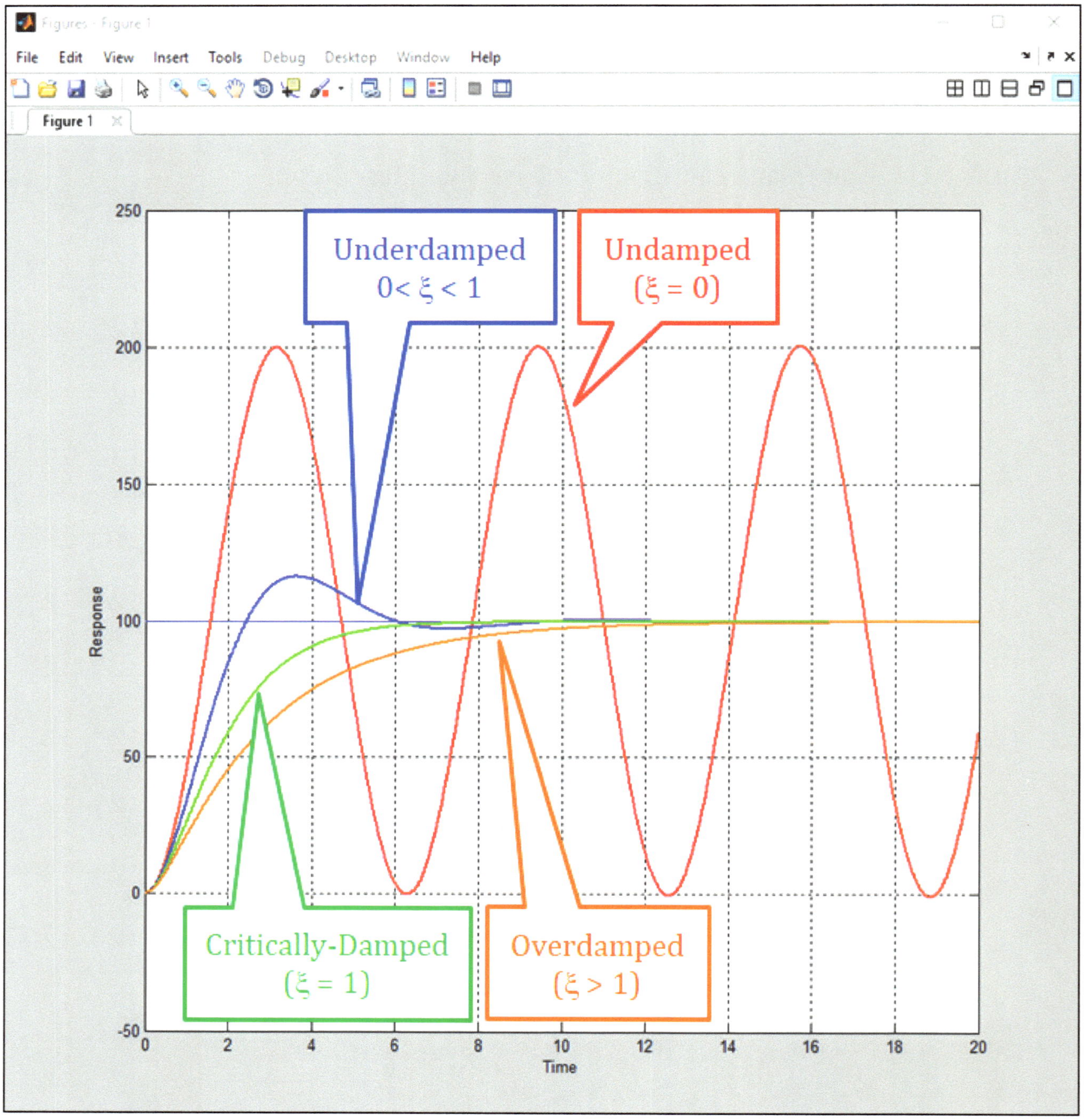

Fig. 3.18B – Effect of Damping Ratio on Step Response of a 2nd Order System

3.5.2.6- Difference between 1st Order and Overdamped 2nd Order Step Response

As shown in Fig. 3.19, simulation model (HSV7CH03Model07) contains TF for a 1st order system and a TF for an overdamped 2nd order system. The figure shows that the response of the two systems looks similar. However, the initial slope for a 2nd order system is different so that the overall response has s-shape.

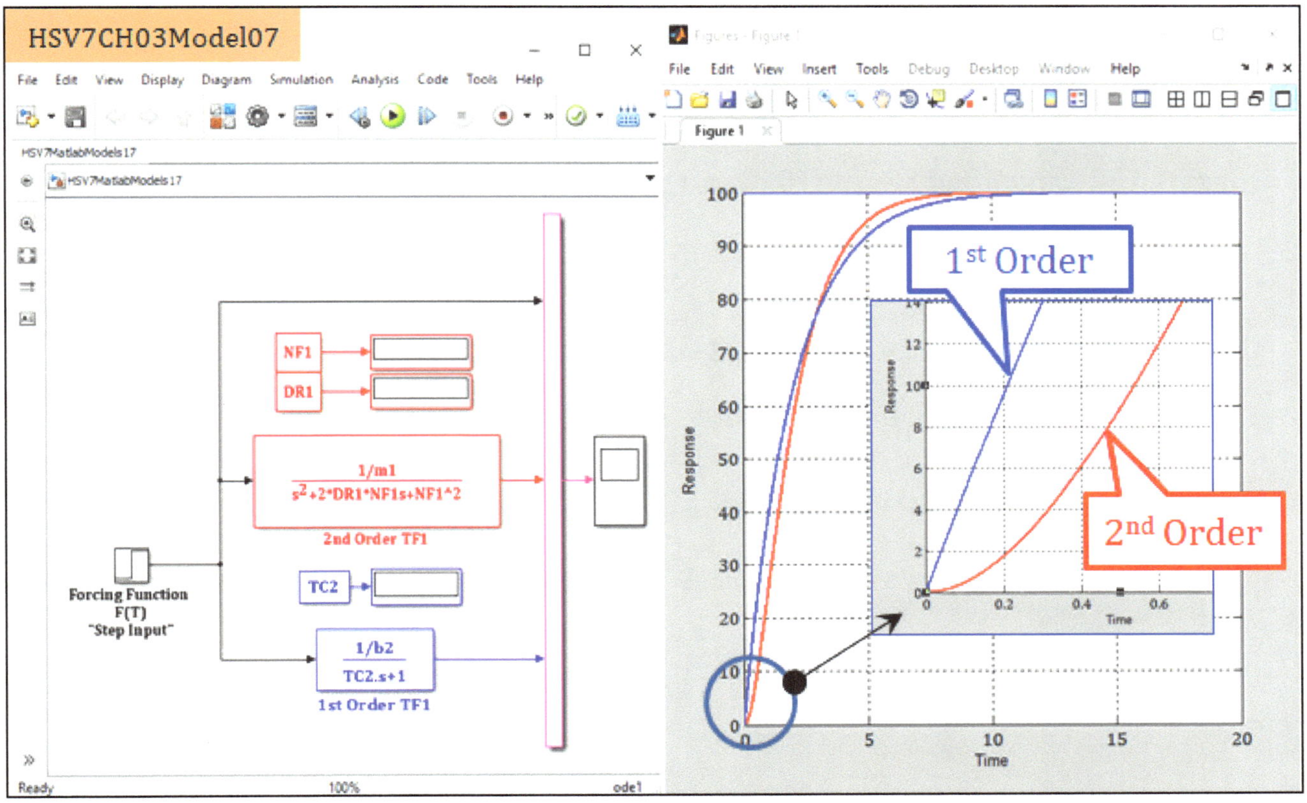

Fig. 3.19 – Effect of Damping Ratio

3.5.2.7- Effect of Damping Ratio as Presented on s-Plane

Equation 3.20 is known as the Characteristic Equation of a 2nd order system where **r** is called the roots of the system.

$$\mathbf{r^2 + 2\xi\omega_n r + \omega_n^2 = 0} \qquad 3.20$$

The general solution of Eq. 3.20 results in

$$\mathbf{r_{1,2} = -\xi\omega_n \pm \omega_n\sqrt{(\xi^2 - 1)} = -\xi\omega_n \pm j\omega_n\sqrt{(1 - \xi^2)}} \qquad 3.21$$

Where **j** is an imaginary number $= \sqrt{-1}$

As shown in Fig. 3.20, based on damping ratio value, the following system responses are defined.

Case 1: Unstable System (ξ < 0):

$$\xi < 0 \rightarrow r_{1,2} = -\xi\omega_n \pm j\omega_n\sqrt{(1-\xi^2)} = \text{ +ve Real Part (no matter the imagenry par is)}$$

Case 2: Undamped System (ξ = 0) is Marginally Stable:

$$\xi = 0 \rightarrow r_{1,2} = \pm j\omega_n = \pm \text{ Imaginary Part}$$

Case 3: Underdamped System (0 < ξ < 1) is Stable:

$$0 < \xi < 1 \rightarrow r_{1,2} = -\xi\omega_n \pm j\omega_n\sqrt{(1-\xi^2)} = -\text{ve Real Part} \pm \text{ Imagenary Part}$$

Case 4: Critically Damped System (ξ = 1) is Stable:

$$\xi = 1 \rightarrow r_1 = r_2 = -\xi\omega_n = -\text{ve Real Part (Coincided Roots)}$$

Case 5: Overdamped System (ξ > 1) is Stable:

$$\xi > 1 \rightarrow r_{1,2} = -\xi\omega_n \pm \omega_n\sqrt{(\xi^2-1)} = -\text{ve Real Part (seprated Roots)}$$

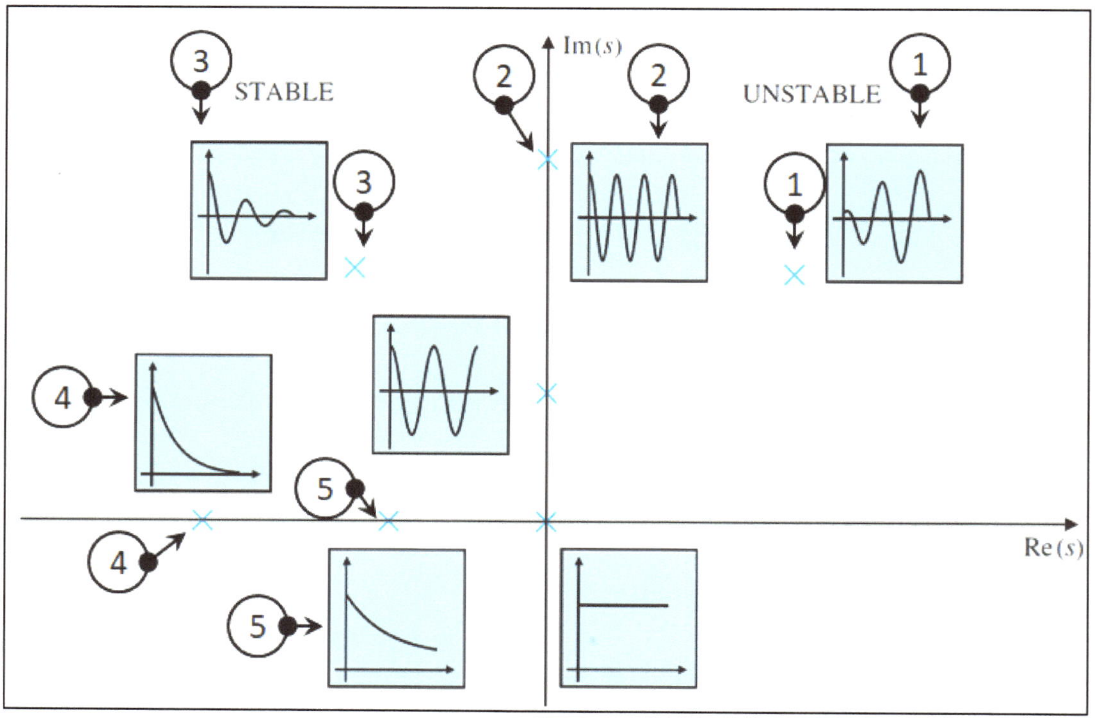

Fig. 3.20 – Effect of Damping Ratio as Presented on s-Plan

Figure 3.21 helps to identify a second-order system based on the location of its roots.

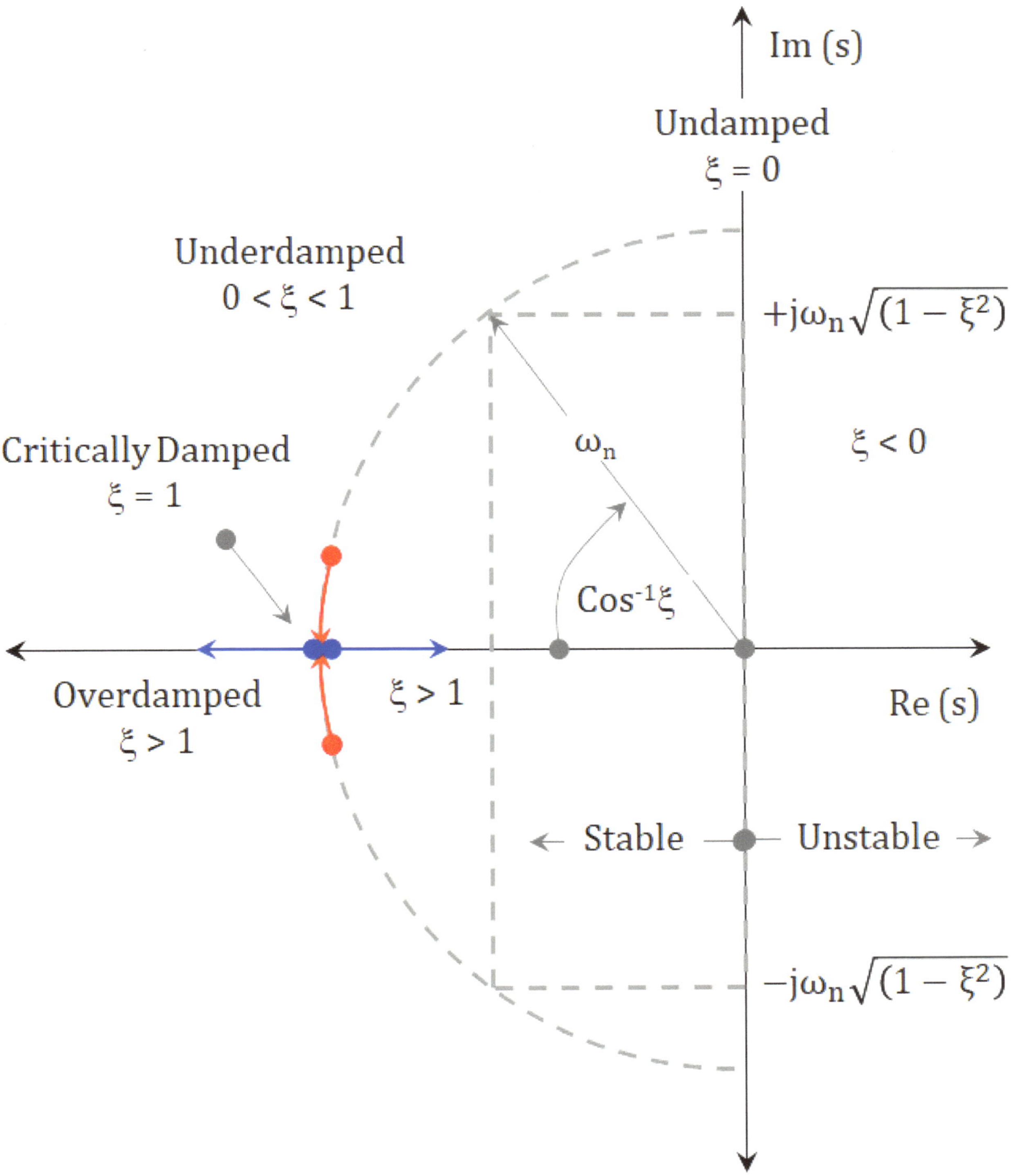

Fig. 3.21 – Location of 2nd Order System Roots on s-Plan based on Damping Ratio

3.6- Second-Order System Modeling Based on Step Response

The question now is, like what has been done for a first-order system, can a second-order system model be developed based on existing dynamic response. The answer is "certainly yes" following the road map shown in Fig. 3.22. As, the time constant is the main characterizing parameter for a 1st order system, a 2nd order system has the **natural frequency** and the **damping ratio** are the main charactering parameters.

❖ **For Underdamped Systems:**

If the step response of a 2nd order system shows an underdamped response, follow the left side of the road map shown in Fig. 3.22 as follows:

Step 1- Natural Frequency ω_n:

- **Complete Cycle Time T**: step response → Define T.
- T & Eq. 3.13 & Eq. 3.14 → ω_n
- **OR Rise Time t_r**: step response → Define t_r.
- t_r & Eq. 3.15 → ω_n.

Step 2- Damping Ratio ξ:

- **Peak Time t_p**: step response → Define t_p. t_p & Eq. 3.16 → ξ..
- **OR Settling Time t_s**: step response → Define t_s. t_s & Eq. 3.17 → ξ.
- **OR Overshoot Mp**: step response → Define M_p. M_p & Eq. 3.19 & Curve → ξ.

❖ **For Critically Damped and Overdamped Systems:**

If the step response of a 2nd order system shows critically or overdamped response, follow the right side of the road map shown in Fig. 3.22 as follows:

Step 1: Settling Time t_s: step response → Define t_s.

Step 2:

- **Damping Ratio ξ:** Assume a starting value of Damping Ratio $\xi = 1$.
- **Natural Frequency ω_n: t_s** & Eq. 3.17 → Calculate a starting value of ω_n.

❖ **Developing a Model:**

Once ξ and ω_n are calculated, based on the available physical parameters, follow the appropriate step to continue building the model:

Step 3A: Develop the NTF (in cases where physical design parameters are not known):

- **Develop Normalized Transfer Function:** ξ & ω_n, Eq. 3.12 → NTF.
- **Model Tuning:** It is to be remembered that Eq. 3.17 is used for underdamped system, so it has been used here just to get a starting value for ω_n. Model response still needs to be tuned by incremental increases of ξ and/or ω_n.

Step 3B: Develop the TF (in cases where physical design parameters are known):

- **Develop Transfer Function:** Mass **m**, Eq. 3.10 → TF.

OR Step 3C: Developing DE (in cases where physical design parameters are known):

- **Spring Constant k_s:** m, Eq. 3.8 → k_s.
- **Friction Coefficient k_f:** m and k_s, Eq. 3.9 → k_f = 2
- **Develop DE:** m, k_f and k_s, Eq. 3.5 → DE.

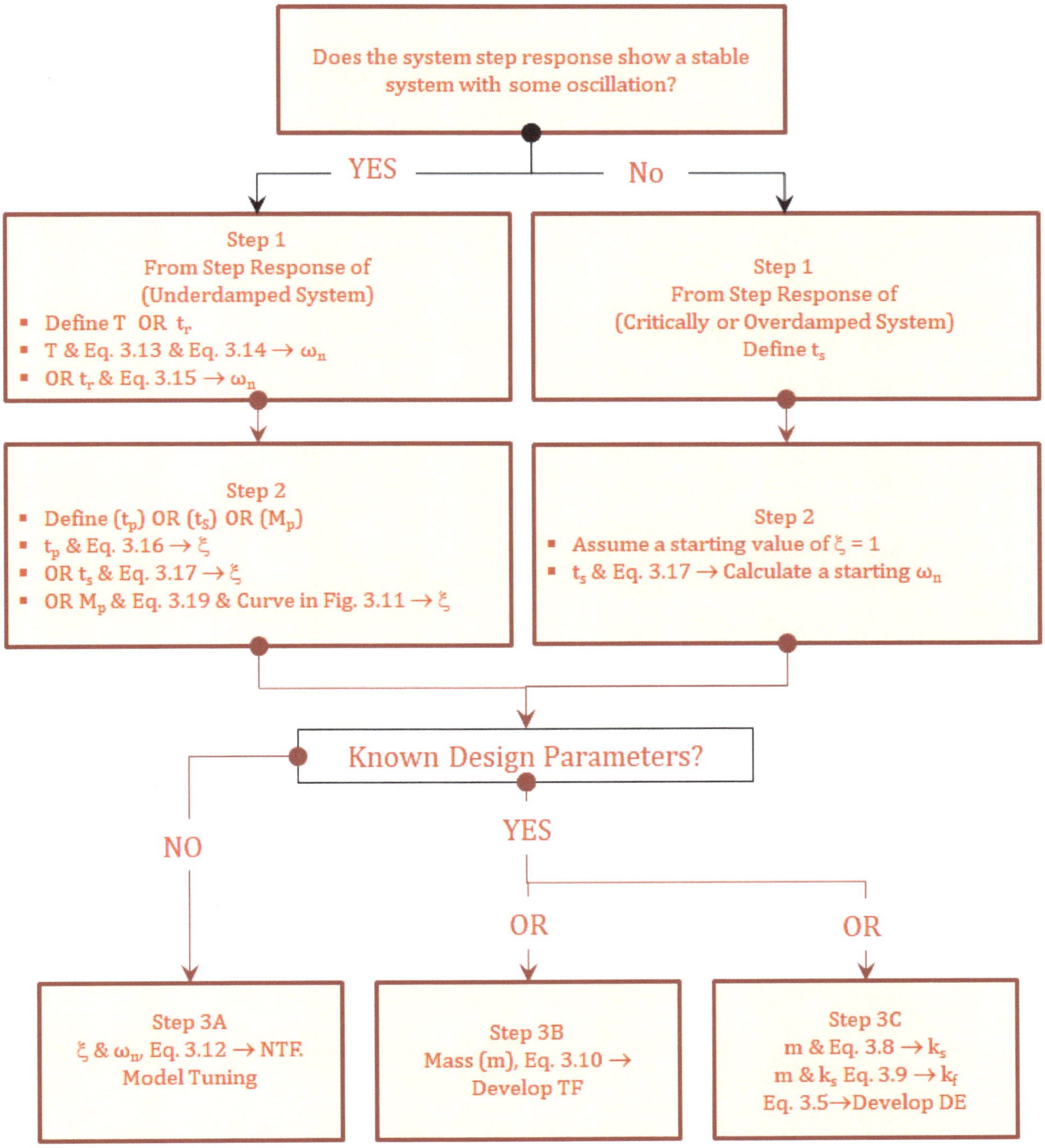

Fig. 3.22 – Road Map to Develop a Second-Order Model based on Step Response

Case Study 1 (Critically or Overdamped 2nd Order System ξ >=1):

Given: A step response of a stroke-controlled proportional directional valve shows no oscillation and settles at 100% stroke in 18 ms. Spool stroke = 10 mm, spool mass = 100 grams, maximum electromagnetic force from the proportional solenoid **F** = 122.5 N.

It is to be noted that this case study only includes valve dynamics. The consequence of this valve dynamics (e.g. flow and leakage condition through the valve) will be discussed in the relevant chapter during building the complete model for the valve.

This spool type proportional directional valve has a spool moving against a return spring. Therefore, the second-order system shown in Fig. 3.1 represents the valve. Since the valve response shows no oscillation, the valve is either critically or overdamped. Then, the following steps show the procedure of developing the model.

Step 1: Settling Time t_s: step response → Settling time (given) t_s = 18 ms.

Step 2:

- **Damping Ratio ξ:** Assume a starting value of Damping Ratio ξ = 1.
- **Natural Frequency ω_n:**
- Assuming SSE = ± 1%, Eq. 3.17A → starting ω_n (rad/s) = 4.6 / 0.018 = 255.

Step 3A: Develop the NTF (in cases where physical design parameters are not known):
As shown in Fig. 3.23, simulation model (HSV7CH03Model08) is constructed containing the NTF that is repeated to check the valve response at various commands. Damping ratio is kept equal to 1, but the natural frequency is increased to **350 rad/s** to meet the required response. Because NTF is used, input must be spool stroke so that the output will be the valve dynamic response that settles at the input value.

Step 3B: Develop the TF (in cases where physical design parameters are known):
As shown in Fig. 3.24, simulation model (HSV7CH03Model09) is constructed to represent the valve response using TF defined by Eq. 3.10. Spool mass **m** = 0.1 kg is added to the model properties. Valve model is subjected to various force commands so that the model calculates % of spool stroke considering the maximum stroke = 10 mm.

Results shows that, weather step 3A or 3B are used, results are the same. Results shows that spool moves proportional to the input force.

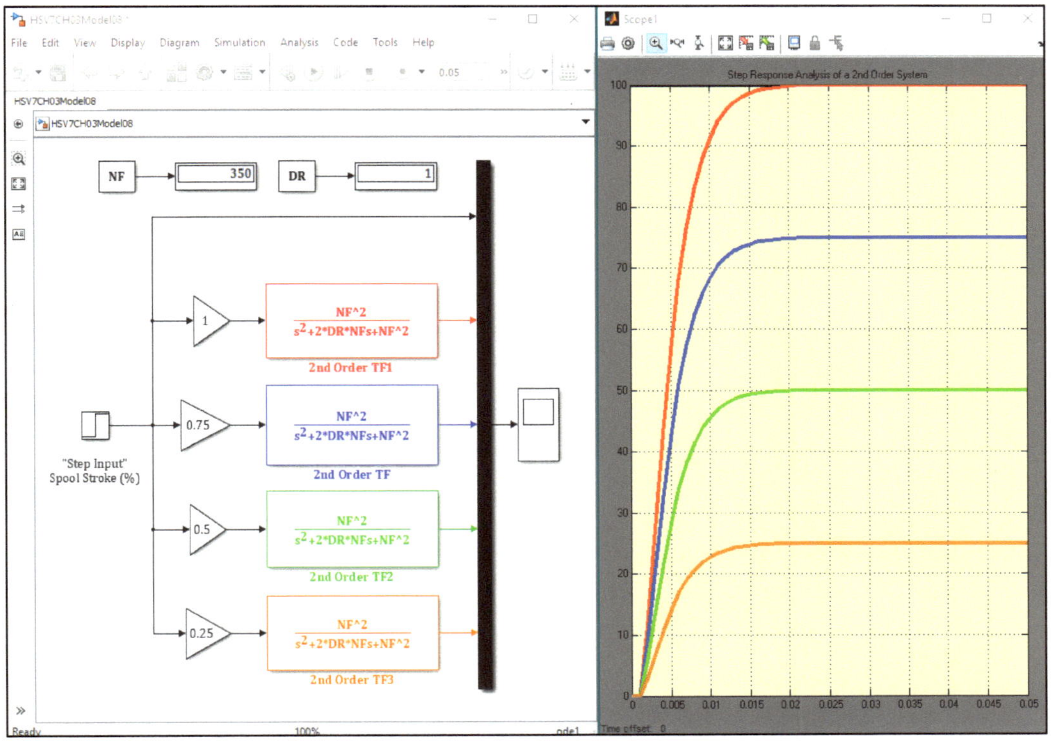

Fig. 3.23 – Second-Order Valve Dynamics Based on Normalized Transfer Function

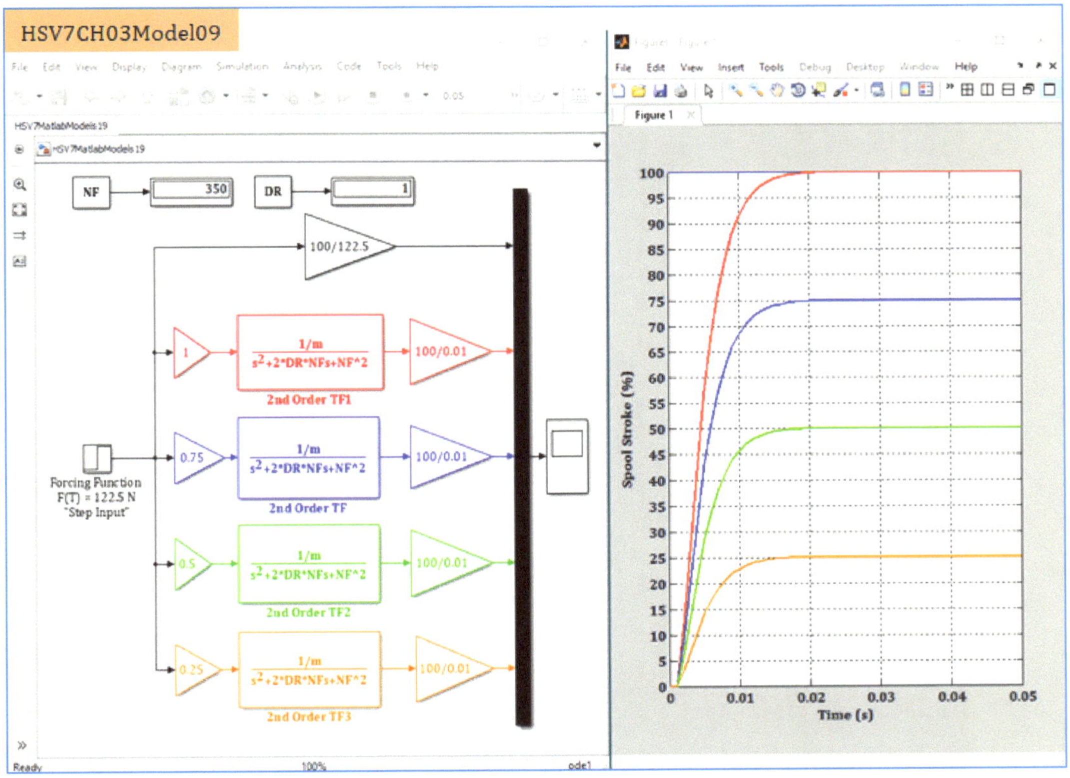

Fig. 3.24 – Second-Order Valve Dynamics Based on Transfer Function

OR Step 3C: Developing DE (in cases where physical design parameters are known):

- **Spring Constant k_s:** spool mass m, Eq. 3.8 → $k_s = m\,(\omega_n)^2 = 0.1\,(350)^2 = 12250$ N/m.
- **Friction Coefficient k_f:** Eq. 3.9 → $k_f = 2\xi\,(mk_s)^{0.5} = 2 \times 1\,(0.1 \times 12250)^{0.5} = 70$ kg/s.
- **Develop DE: m, k_f and k_s**, Eq. 3.5 → DE as follows.

$$\mathbf{m[kg]} \times \mathbf{a}\left[\frac{\mathbf{m}}{\mathbf{s^2}}\right] + \mathbf{k_f}\left[\frac{\mathbf{kg}}{\mathbf{s}}\right] \times \mathbf{v}\left[\frac{\mathbf{m}}{\mathbf{s}}\right] + \mathbf{k_s}\left[\frac{\mathbf{N}}{\mathbf{m}}\right] \times \mathbf{x\,[m]} = \mathbf{F(t)\,[N]} \qquad 3.22$$

$$\mathbf{0.1} \times \mathbf{a}\left[\frac{\mathbf{m}}{\mathbf{s^2}}\right] + \mathbf{70} \times \mathbf{v}\left[\frac{\mathbf{m}}{\mathbf{s}}\right] + \mathbf{12250} \times \mathbf{x\,[m]} = \mathbf{F(t)\,[N]} \qquad 3.23$$

As shown in Fig. 3.25, simulation model (HSV7CH03Model10) is constructed to represent the valve response using DE defined by Eq. 3.23. Spool mass, friction coefficient, and spring constant are added to the model properties. Valve model is subjected to various force commands so that the model calculates % of spool stroke considering the maximum stroke = 10 mm. Simulation results shows identical response as in Fig. 3.24 when TF is used.

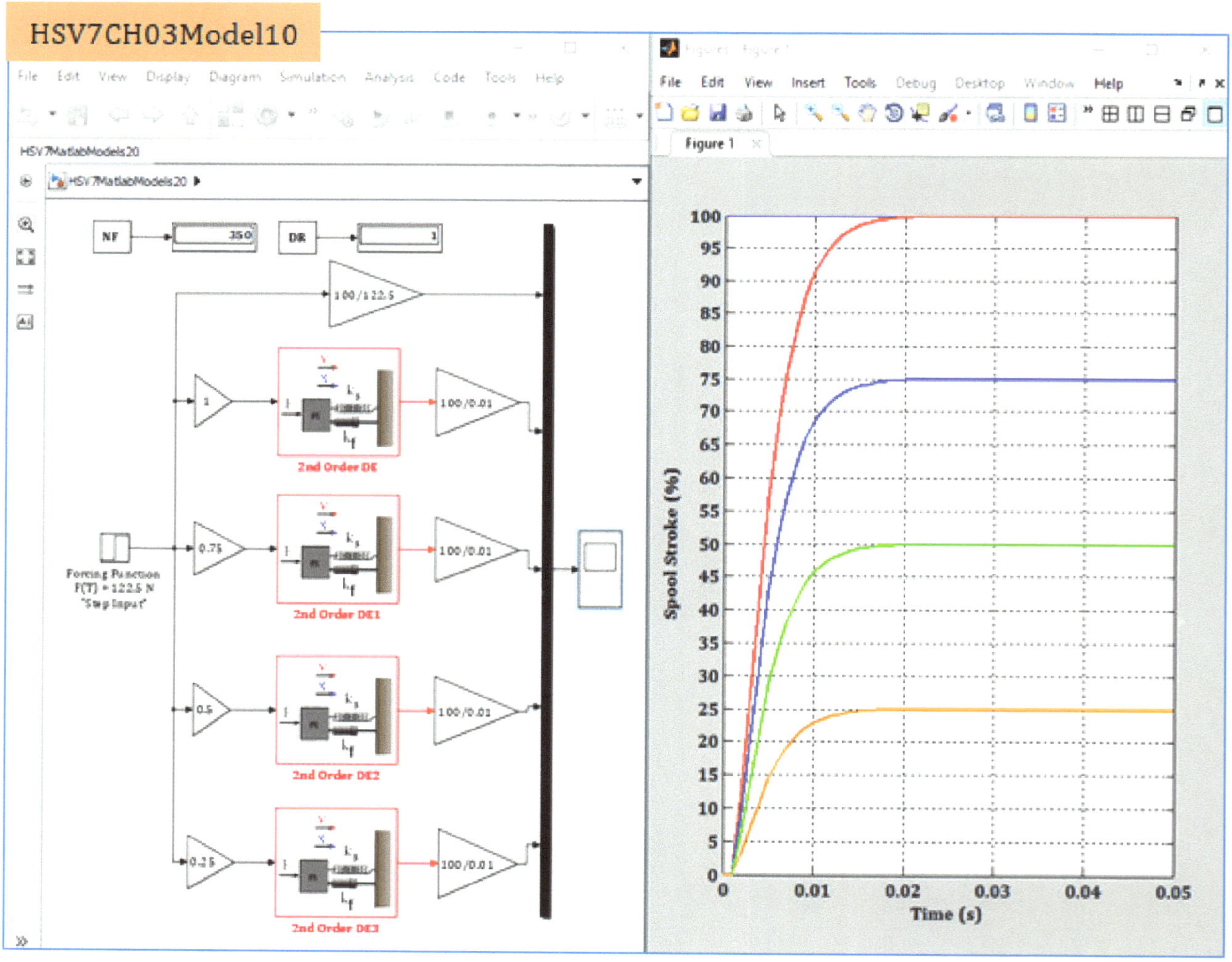

Fig. 3.25 – Second-Order Valve Dynamics Based on Differential Equation

Case Study 2 (Underdamped 2nd Order System $0 < \xi < 1$):

Given: A step response of a hydraulic actuator is shown in Fig. 3.26. The following information is extracted from the step response when the cylinder is extended by **F = 40 N** force:

- Rise Time t_r = 0.287 s.
- Peak Time t_p = 0.51 s.
- Settling Time t_s = 3.7.
- Overshoot M_p (%) = 53.
- Equivalent Moving Mass **m** = 10 kg.
- Targeted Actuator Stroke = 100 mm.

Since the step response shows that the system is underdamped, the following steps show the procedure for developing the model.

Step 1- Natural Frequency ω_n:

- **Complete Cycle Time T**: step response → T (from peak-to-peak) = 1 s.
- Eq. 3.13 → f_n [Hz] = 1/T = 1/1= 1.
- Eq. 3,14 → ω_n [rad/s] = 2 πf_n = 6.28
- **OR Rise Time t_r**: step response → $t_r \approx 0.287$
- Eq. 3.15 → ω_n [rad/s] = 1.8/ $t_r \approx 6.28$

Step 2- Damping Ratio ξ:

- **Peak Time t_p**: step response → t_p = 0.51. Eq. 3.16 → $\xi = 0.2$
- **OR Settling Time t_s**: step response → t_s = 3.7 s. Eq. 3.17 → $\xi \approx 0.2$
- **OR Overshoot Mp (%)** 53. Eq. 3.19 → M_p % = 53 & Curve (Fig. 3.11) → $\xi \approx 0.2$

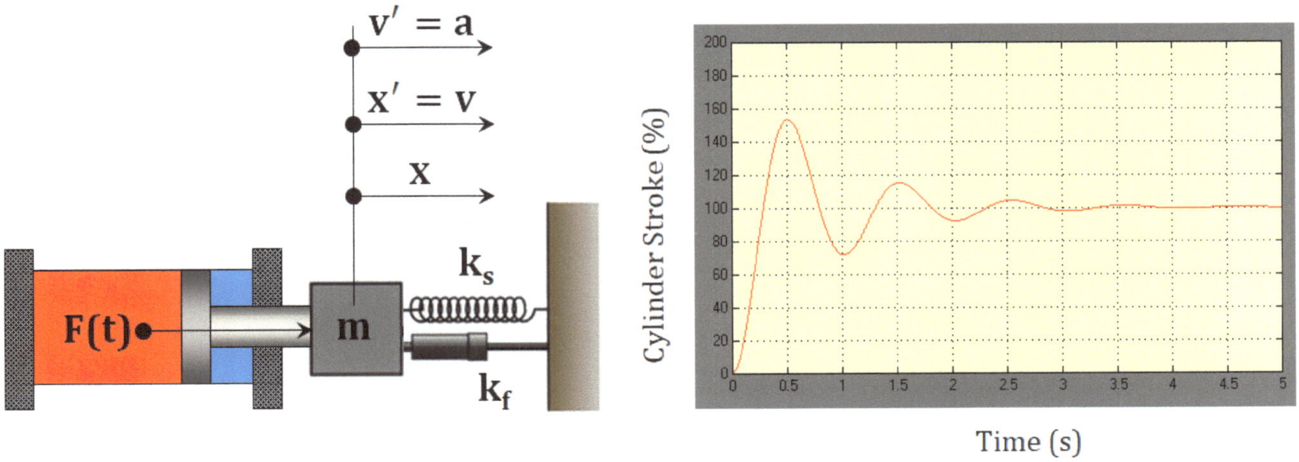

Fig. 3.26 – Experimental Step Response of a Hydraulic Actuator

Step 3A: Develop the NTF (in cases where physical design parameters are not known):
As shown in Fig. 3.27, simulation model (HSV7CH03Model11) is constructed containing the NTF represents the pattern of the actuator's dynamics. Simulation results show that the calculated values from the model are very close to the experimental measurements.

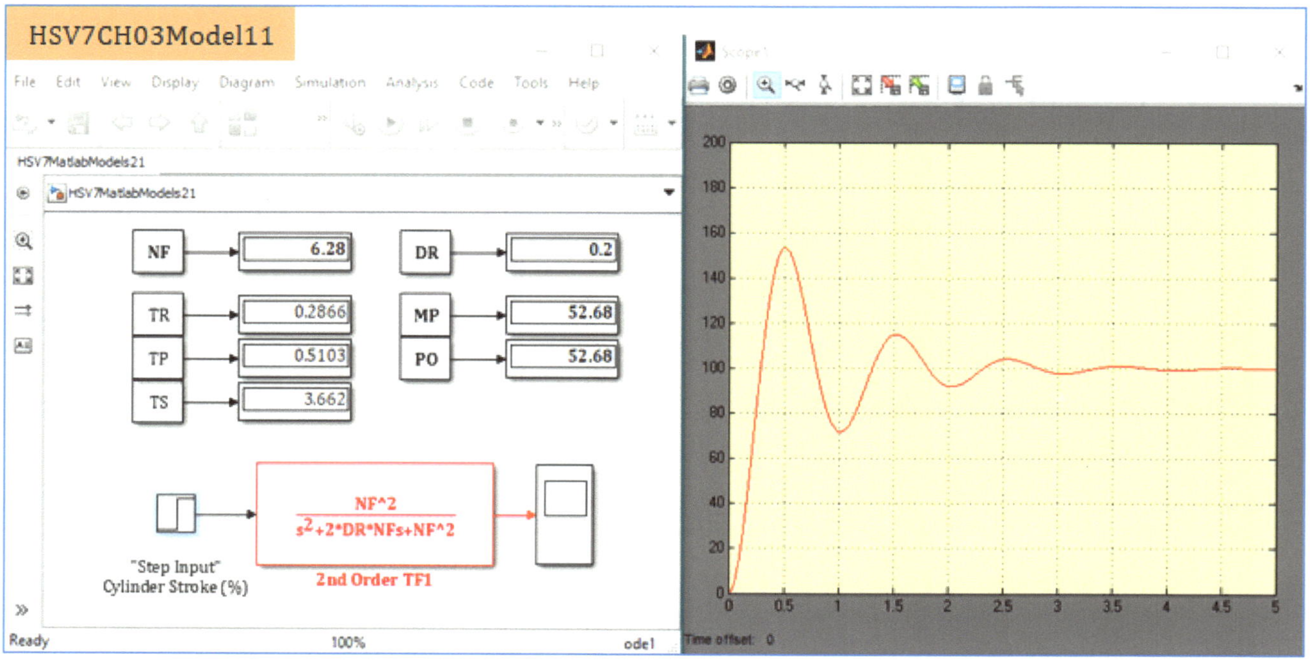

Fig. 3.27 – Simulated Step Response of a Hydraulic Actuator based on NTF

Step 3B: Develop the TF (in cases where physical design parameters are known):
As shown in Fig. 3.28, simulation model (HSV7CH03Model12) is constructed to represent the actuator response using TF defined by Eq. 3.10. Equivalent moving mass **m** =10 kg is added to the model properties. Simulation results show that, in order to drive the actuator to a targeted stroke of 100 mm, the driving force should be **F** = 40 N.

OR Step 3C: Developing DE (in cases where physical design parameters are known):

- **Spring Constant k_s:** spool mass m, Eq. 3.8 → $k_s = m\,(\omega_n)^2 = 10\,(6.28)^2 = 394.4$ N/m.
- **Friction Coefficient k_f:** Eq. 3.9 → $k_f = 2\xi\,(mk_s)^{0.5} = 2 \times 0.2\,(10 \times 394.4)^{0.5} = 25.12$ kg/s.
- **Develop DE: knowing m, k_f** and **k_s**, Eq. 3.5 → DE as follows.

$$\mathbf{m[kg]} \times \mathbf{a}\left[\frac{\mathbf{m}}{\mathbf{s^2}}\right] + \mathbf{k_f}\left[\frac{\mathbf{kg}}{\mathbf{s}}\right] \times \mathbf{v}\left[\frac{\mathbf{m}}{\mathbf{s}}\right] + \mathbf{k_s}\left[\frac{\mathbf{N}}{\mathbf{m}}\right] \times \mathbf{x\,[m]} = \mathbf{F(t)\,[N]} \qquad 3.22$$

$$\mathbf{10} \times \mathbf{a}\left[\frac{\mathbf{m}}{\mathbf{s^2}}\right] + \mathbf{25.12} \times \mathbf{v}\left[\frac{\mathbf{m}}{\mathbf{s}}\right] + \mathbf{394.4} \times \mathbf{x\,[m]} = \mathbf{F(t)\,[N]} \qquad 3.23$$

As shown in Fig. 3.29, simulation model (HSV7CH03Model13) is constructed to represent the actuator response using DE defined by Eq. 3.23. Identical Results were found.

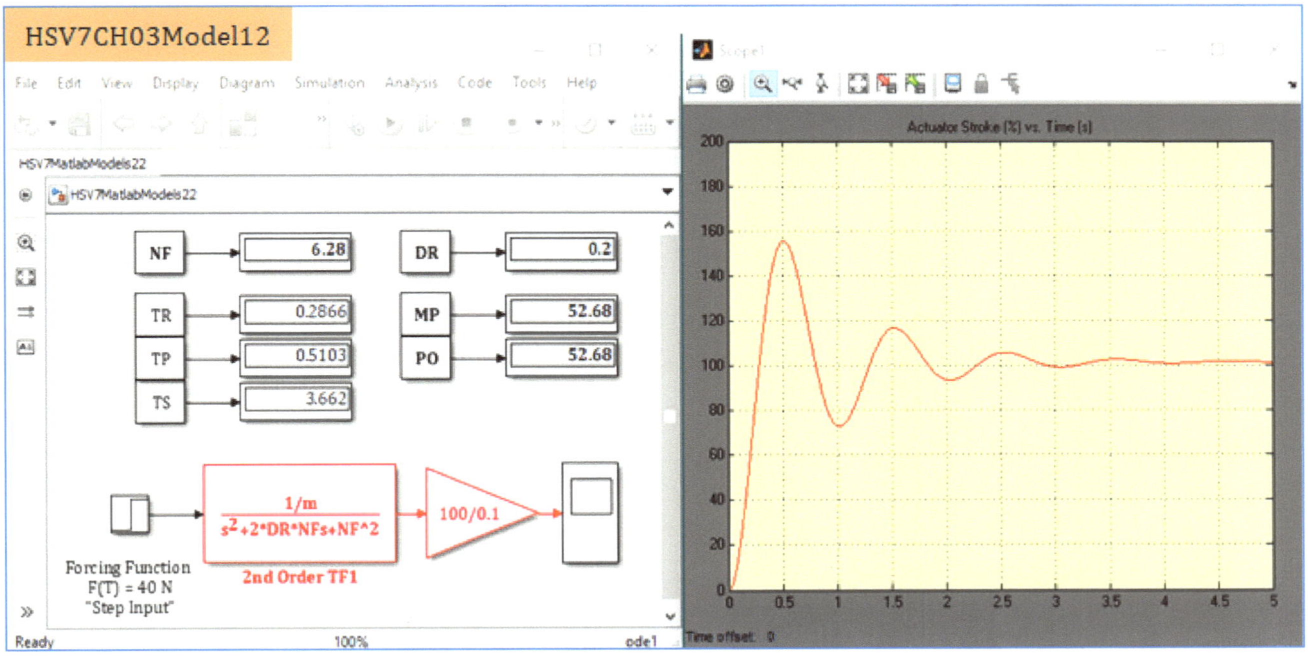

Fig. 3.28 – Simulated Step Response of a Hydraulic Actuator based on TF

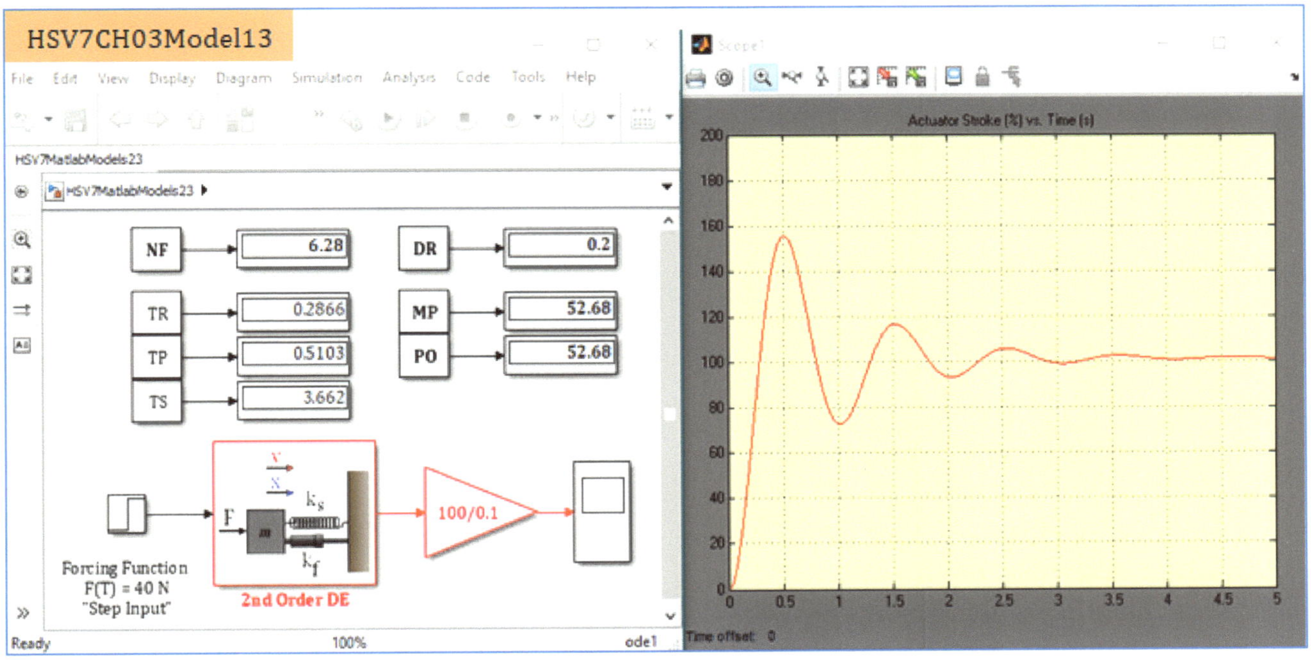

Fig. 3.29 – Simulated Step Response of a Hydraulic Actuator based on Differential Equation

3.7- Frequency Response Analysis of Second-Order Systems

3.7.1- Identification of Second-Order Systems Based on Frequency Response

Mathematical solutions for second-order frequency response are as follows:

$$\mathbf{x(t) = B\sin(\omega t + \varphi)} \qquad 3.24$$

$$\mathbf{B = A/\sqrt{\left[\frac{2\xi\omega}{\omega_n}\right]^2 + \left[1 - \frac{\omega^2}{\omega_n^2}\right]^2}} \qquad 3.25$$

$$\mathbf{\varphi = -\tan^{-1}\left[\left(\frac{2\xi\omega}{\omega_n}\right) / \left(1 - \frac{\omega^2}{\omega_n^2}\right)\right]} \qquad 3.26$$

$$\mathbf{AR\ (dB) = 20\log\frac{B}{A} = 20\log\left[1/\sqrt{\left[\frac{2\xi\omega}{\omega_n}\right]^2 + \left[1 - \frac{\omega^2}{\omega_n^2}\right]^2}\right]} \qquad 3.27$$

Where:
A = Amplitude of the harmonic input (exciting) signal.
ω (rad/s) = Frequency of the harmonic input (exciting) signal.
ω_n (rad/s) = Frequency of the harmonic input (exciting) signal.
B = Amplitude of the system response.
AR = Amplitude Ratio of the system response.
B = Amplitude of the system response.
φ (deg) = Phase lag of the system response.

As shown in Fig. 3.30, *Bode Plot* for a second-order system can be drawn as follows.

- $\omega < 0.1\,\omega_n \rightarrow$ AR $\approx$ 0 dB and $\varphi = 0$ deg.
- ω = Bandwidth $\rightarrow$ AR = -3 dB.
- $\omega = \omega_n \rightarrow \varphi = -90$ deg.
- $\omega > \omega_n \rightarrow$ AR decays 40 dB/decade of exciting frequency.
- $\omega > 10\,\omega_n \rightarrow \varphi = -180$ deg.

Straight lines show the approximate frequency response. Asymptotic line represents the accurate frequency response.

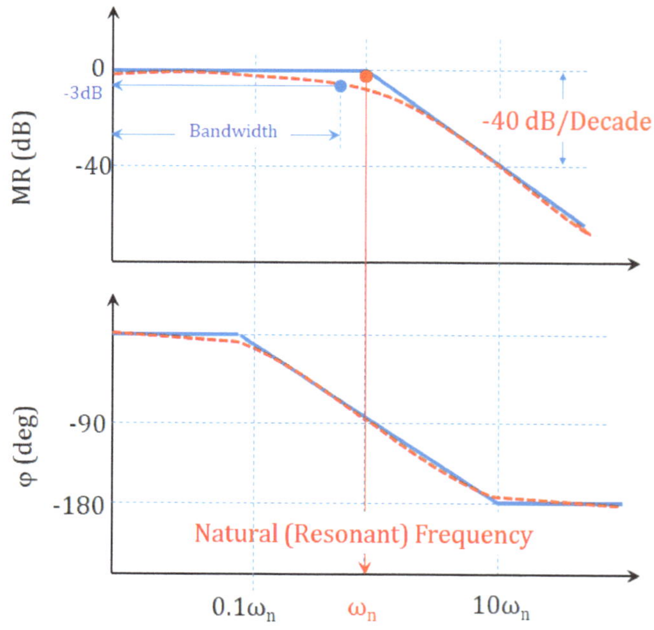

Fig. 3.30 – Bode Plot of a 2nd Order System

3.7.2- Effect of Design Parameters on Second-Order System Frequency Response

3.7.2.1- Effect of Damping Ratio

Figure 3.31 shows that reducing the damping ratio increases the overshoot of the system and obviously the system becomes sluggish. Figure 3.32 shows the same effect but on a Bode Plot.

3.7.2.2- Effect of Exciting Frequency

Exciting frequency on a 2nd Order system has the same effect as on a 1st Order system previously discussed in Fig. 2.24.

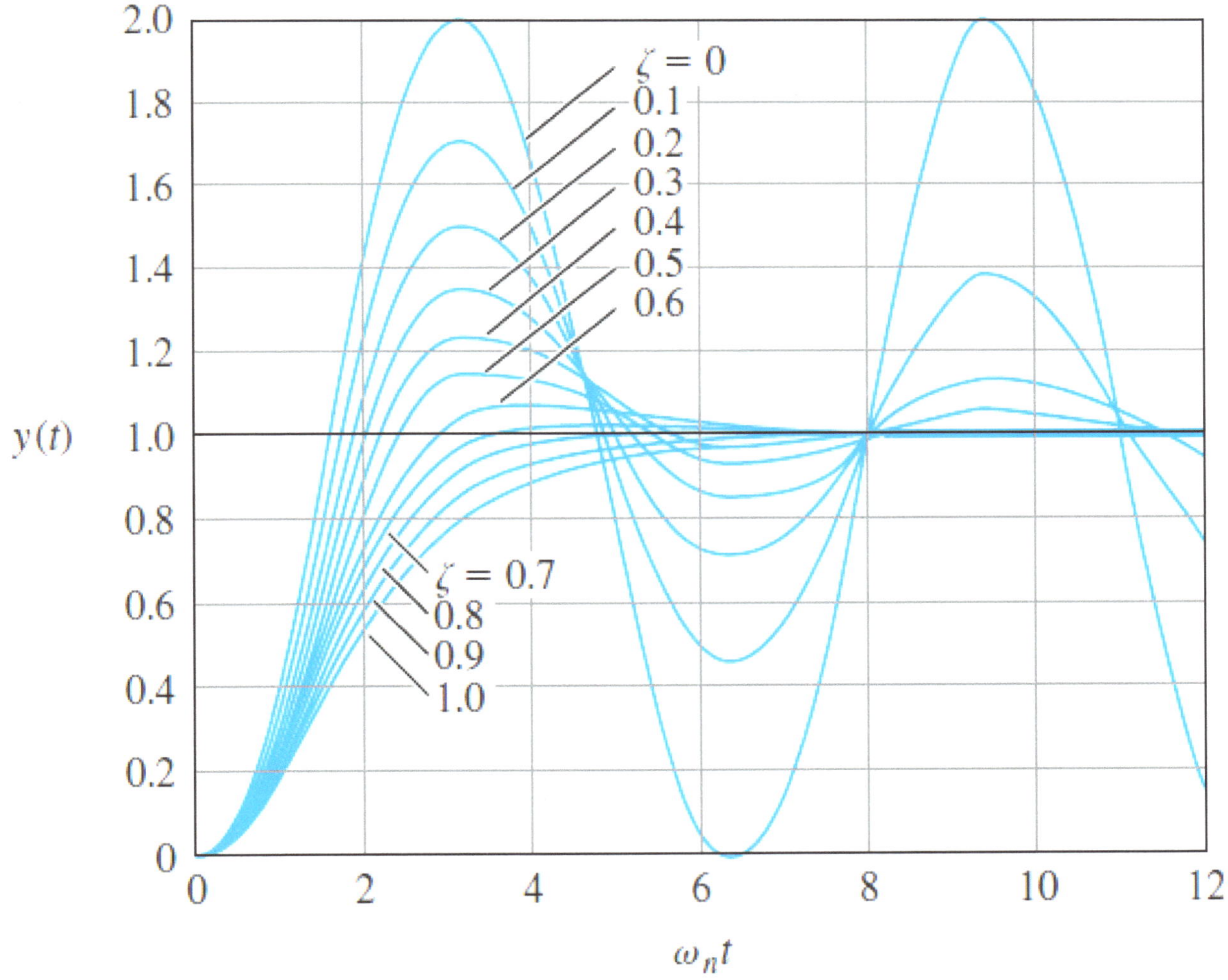

Fig. 3.31 – Effect of Damping Ratio on Harmonic Response of a 2nd Order System

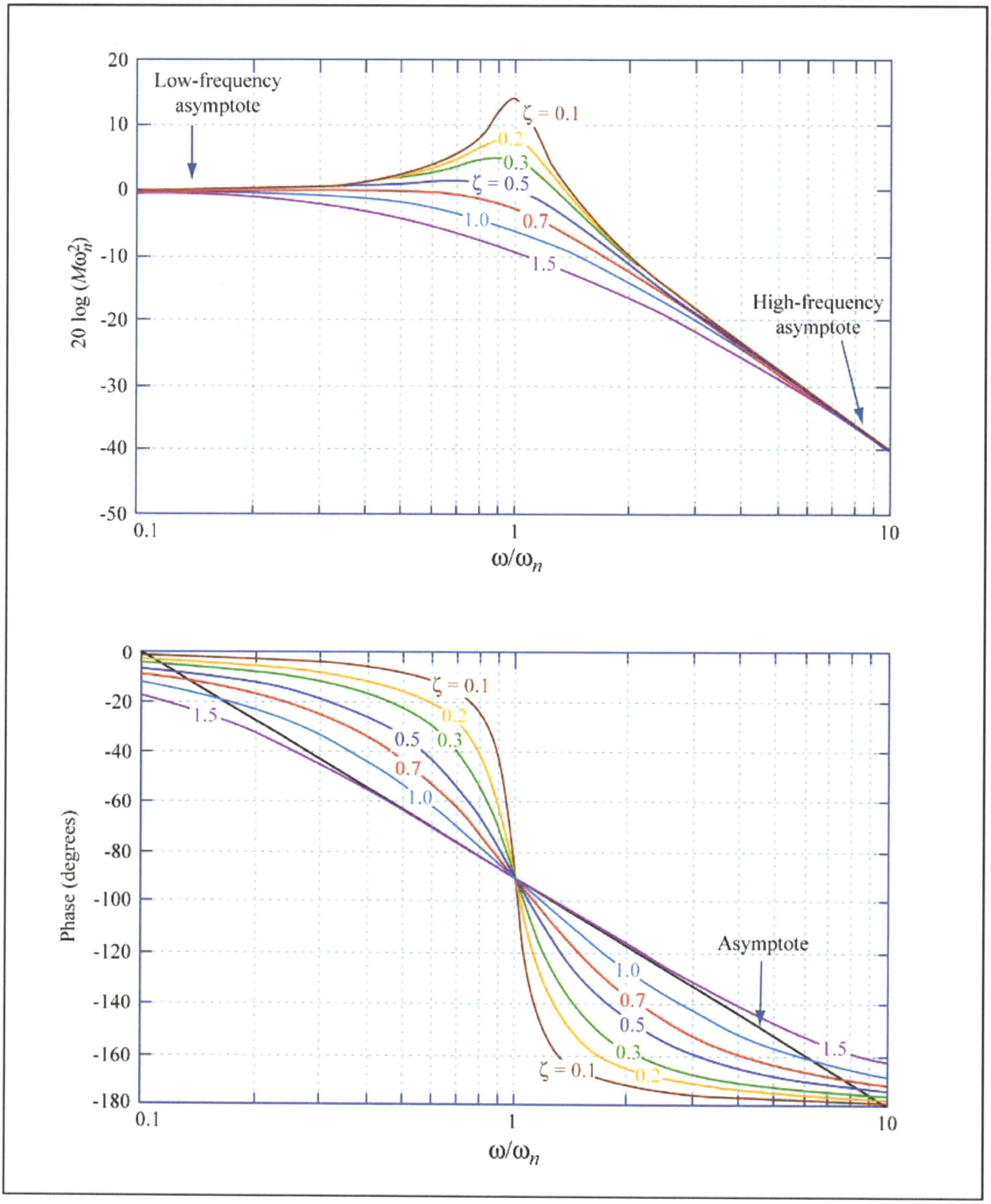

Fig. 3.32 – Effect of Damping Ratio on Frequency Response of a 2nd Order System

3.7.3- Stability Analysis Based on Bode Plot

In order to investigate stability of a dynamic system using a Bode Plot, the following definitions are assigned:

- **Gain Crossover Frequency (ω_{cg}):** Is the frequency at which **AR** starts to descend for first time below unity (equivalent to 0 dB).
- **Phase Crossover Frequency (ω_{cp}):** Is the frequency at which the system lags by 180 deg. i.e. phase angle **φ** first reaches (-180).
- **Gain Margin (GM):** Is the value of **AR** (at ω_{cp}).
- **Phase Margin (PM):** Is the value of **φ** (at ω_{cg}).

As shown in Fig. 3.33, for a dynamic system to be stable:

- **Stable System:**
  - GM = +ve value by which the gain curve would be increased (lifted) to reach unity in % (0 in db).
  - PM = +ve value by which the phase curve would be decreased (lowered) to reach -180.
  - Both gain margin and phase margin must be positive, which means the system is at least underdamped and lagging.
  - Large +ve GM and large +ve PM means highly damped and sluggish system.
  - Small +ve GM and Small +ve PM means underdamped and responsive system.
  - Dynamic systems are typically designed with GM = 10 dB and PM = 45°. that makes the system stable and lightly damped.
- **Marginally Stable System:** GM = 0 and PM = 0.
- **Unstable System:** GM = -ve and PM = -ve.

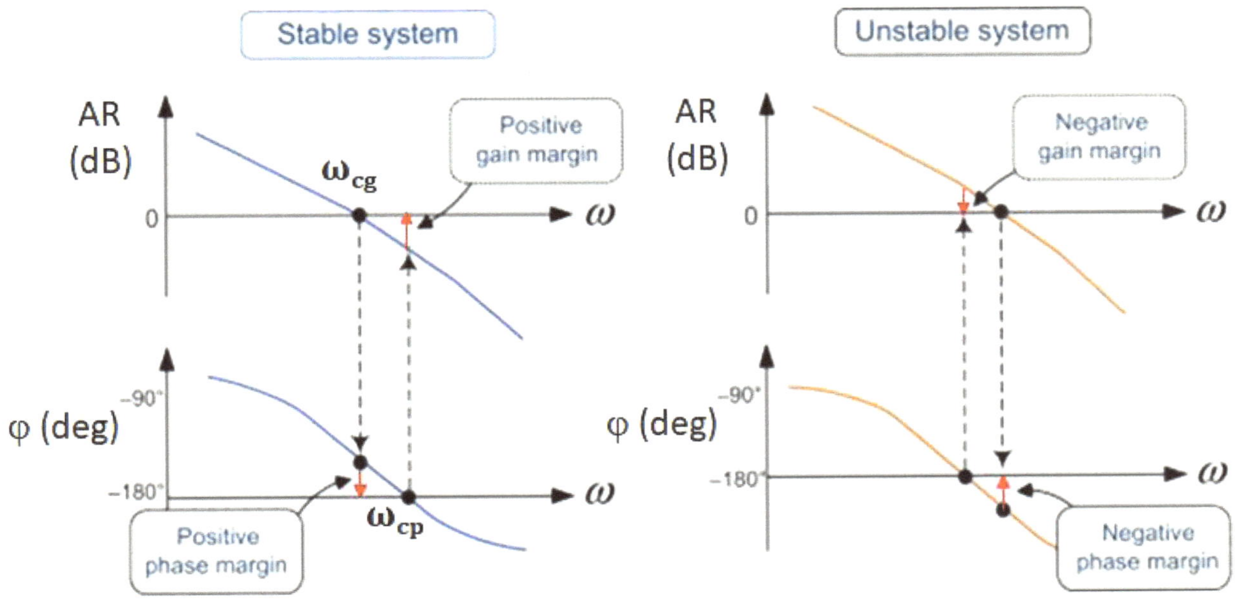

Fig. 3.33 – Stability Analysis Based on Bode Plot

Figure 3.34 shows how to calculate phase margin and gain margin. Figure 3.35 shows an example of two systems. System (A) is stable and at least critically damped if not overdamped. System (B) is an underdamped stable system.

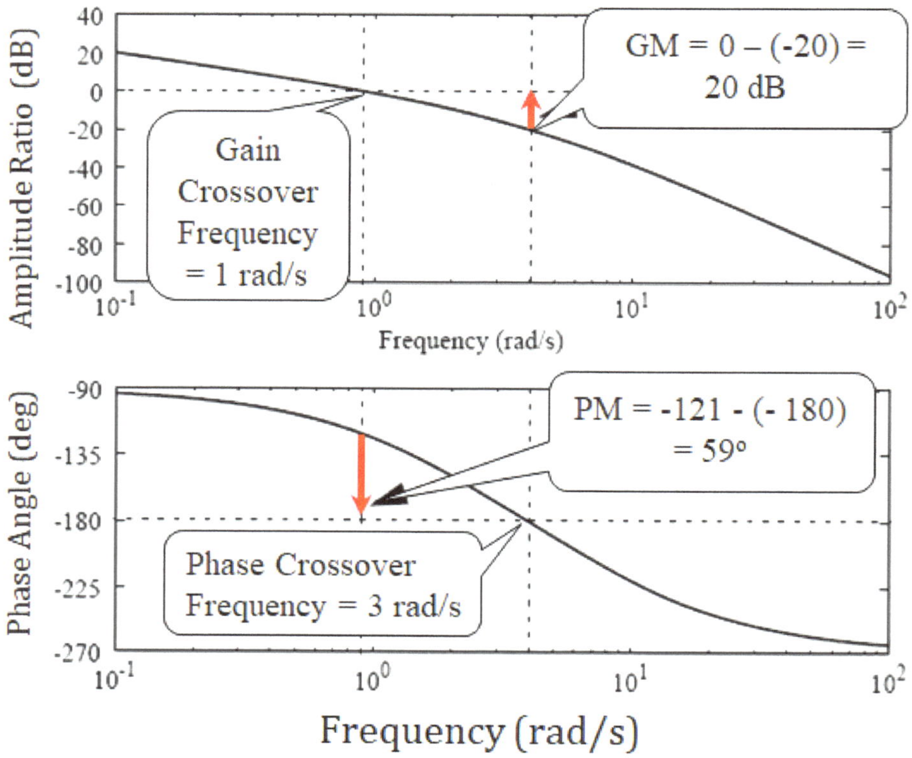

Fig. 3.34 – Stability Analysis Based on Bode Plot, Example 1

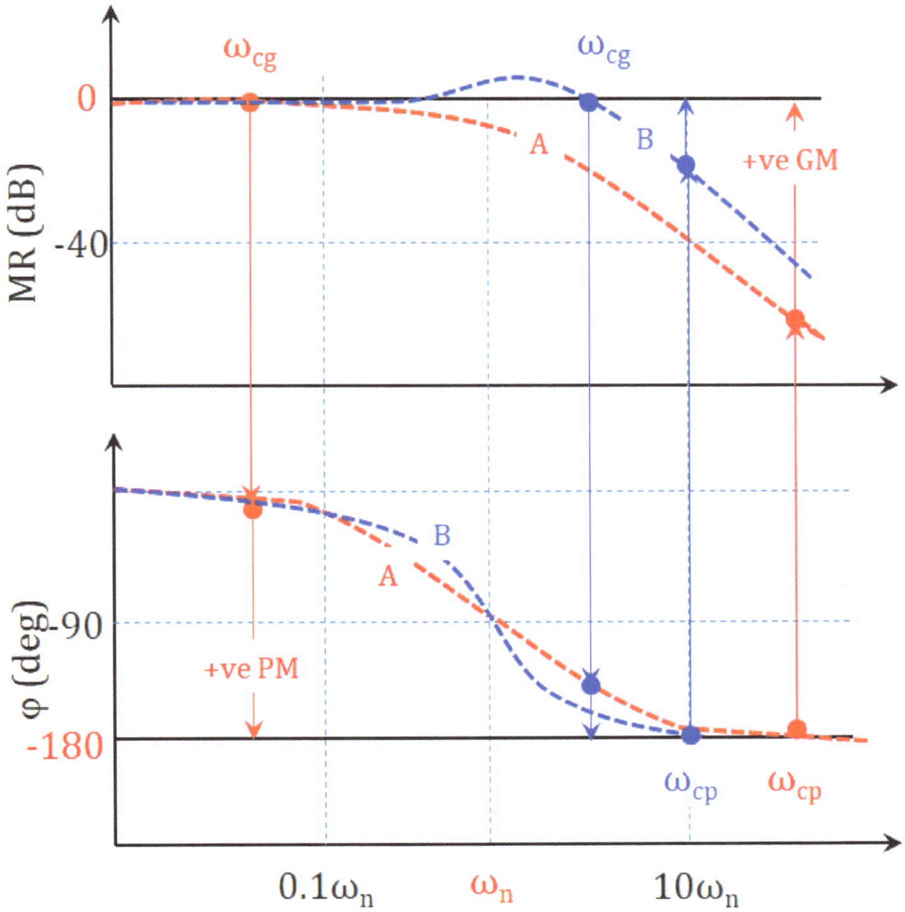

Fig. 3.35 – Stability Analysis Based on Bode Plot, Example 2

3.8- Second-Order System Modeling Based on Frequency Response

Figure 3.36 shows the road map for developing a model of a second order system based on frequency response.

Step 1- Natural Frequency ω_n: Find the natural frequency at phase angle = -90 deg.

Step 2- Damping Ratio ξ:

❖ **For Underdamped Systems:**

If the frequency response of a 2nd order system shows underdamped response, calculate the % overshoot then use the Curve in Fig 13 $\rightarrow \xi$.

❖ **For Critically Damped and Overdamped Systems:**

Assume a starting value of Damping Ratio $\xi = 1$ that can be tuned later to get better results.

❖ **Developing a Model:**

Once ξ and ω_n are calculated, based on the available physical parameters, follow the appropriate step to continue building the model as it has been discussed previously.

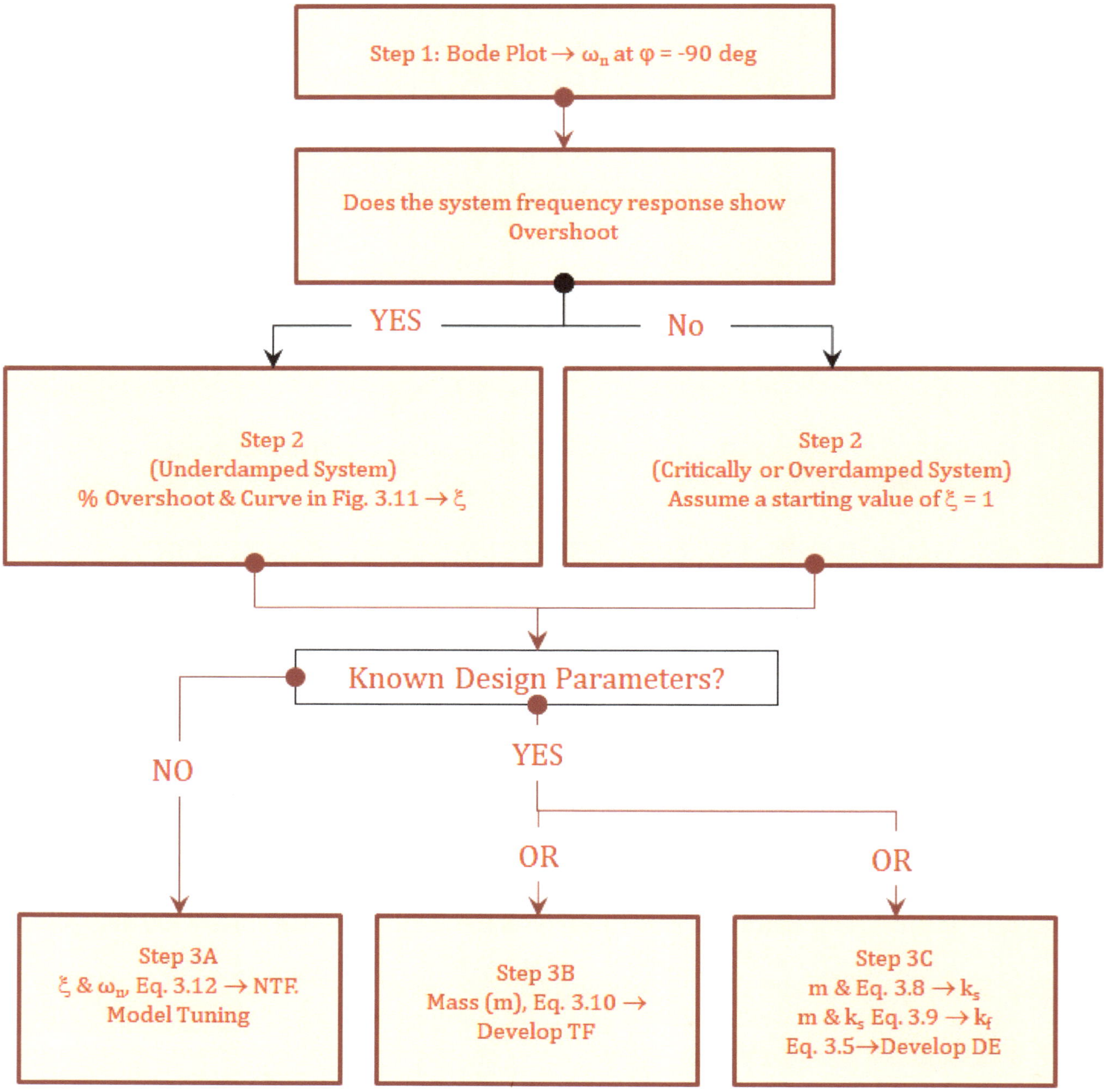

Fig. 3.36 – Road Map to Develop a Second-Order Model based on Frequency Response

Chapter 4
Modeling Approaches for Hydraulic Components and Systems

Objectives

This chapter explores the different approaches when modeling a hydraulic component versus modeling a hydraulic system at large. The chapter presents the basic idea and the structure of lumped modeling, an adopted modeling approach for application engineers.

Brief Contents

Chapter 4: Modeling Approaches for Hydraulic Components and Systems

4.1- Modeling Approaches for Component Developers

4.1.1- Basic Purposes of Modeling for Component Developers

In order for the hydraulic **component manufacturers** to compete in the very tough market, their R&D divisions are working hard to improve their components characteristics. Therefore, hydraulic component developers usually adopt detailed modeling approach for the following purposes:

- Model kinematics, static and dynamic performance of the component.
- Investigate the effect of the operating conditions on the component performance.
- Optimize a component design to makes it more competitive.

4.1.2- Features of Modeling for Component Developers

Such detailed models for component developers have the following features:

- Requires deep understanding to the construction and principle of operation of the component.
- Requires access to classified information such as constructional dimensions, tolerances, material used, mass, friction coefficients, spring constants, etc.
- Complex models that require lots of computational effort.
- Multiple models may be needed, one for kinematic and dynamic analysis, one for stress analysis, one for thermal analysis, one for cavitation analysis, one for CFD analysis, etc.
- Models built mostly using customized software.
- Models are expensive, owned by component manufacturers, and not available publicly because it is considered part of the component manufacturer's know-how.
- Models are dedicated for a specific component/size/brand and are not applicable for other components or used universally.

4.2- Modeling Approaches for Application Engineers

4.2.1- Basic Purposes of Modeling for Application Engineers

Machine builders and application engineers build machine models for the following purposes:

- Study the interaction between various subsystems in the machine, such as engine, transmission, hydraulic functions, etc.
- Investigate the kinematic and dynamic performance of the machine.
- Investigate the machine operation such as: duty cycles, energy efficiency, reliability, maintenance protocols, effect of hydraulic fluid, etc.

4.2.2- Features of Modeling for Application Engineers

Machine builders are the end users of hydraulic components. They have the challenge of selecting the proper component from a wide range of manufacturers. Machine builders are not concerned much about the internal design and construction of hydraulic components. Depending on the application, machine builders may have other interests in hydraulic components such as the static and dynamic characteristics, cost, efficiency, lifetime, operational conditions, sensitivity to contaminations, noise, etc.

An example of that is if a machine designer has decided to use a pressure-compensated pump in an excavator, does it mean anything for the machine design whether the pump is a swashplate or a vane pump? The machine designer may be more concerned with the characteristics of the pump rather than its internal design.

Therefore, machine builders and application engineers usually adopt simplified modeling approaches for the components that form the system for the following reasons:

- Avoid containing large number of equations.
- Reduce the model complexity and the required computational effort.
- Avoid asking for construction details and classified information on every component.
- Focus more on overall machine performance.
- Components should be modeled for ease of connecting to other upstream and downstream components.

4.3- Lumped Modeling Approach for Application Engineers

4.3.1- Introduction to Lumped Modeling Approach

System designers can mathematically model simple systems by a set of equations to be solved instantaneously. Part of the problem to follow such a conventional approach is that if the system layout is changed, the whole set of equations will be changed too.

There are several software packages that offer generic models for hydraulic components such as Matlab-Simhydraulics, Automation Studio, Easy 5, etc. These software packages give access to recharacterize the components but do not give access to change the mathematics under the mask of the model.

For all that has been discussed previously, an alternative *Lumped* modeling approach is proposed for modeling hydraulic components within a machine model. Lumped modeling approach, as discussed in this textbook, has the following features:

- System model composed of lumped component models hooked together.
- Component lumped models are based on static and dynamic characteristics published by the component's manufacturers.
- Component lumped models represent generic components that can be re-characterized and used repeatedly.
- Structured with predefined input/output to make it easy for connecting with downstream and upstream models.

Lumped models of hydraulic components can be classified based on the energy as follows:

- Energy Source: e.g. pump.
- Energy Sink: e.g. reservoir.
- Energy Storage: e.g. accumulator.
- Energy Transmission: e.g. conductor.
- Energy Consumers: e.g. actuators and valves

Lumped models can be classified based on the direction of flow as follows:

- *Irreversible* Flow Components: e.g. pressure relief valves.
- *Reversible* Flow Components: e.g. double acting cylinder. This model can be a combination of two irreversible flow components.

4.3.2- Lumped Model Structure for Hydraulic Components

Figure 4.1 shows structure of a lumped model for a generic hydraulic component. The figure shows that the model contains the following elements:

- **Flow Path:** Flow calculation moves forward. Model receives inlet flow from the upstream model, calculates the output flow, sends it forward to the downstream model, and calculates the differential pressure across the component.

- **Pressure Path:** In order to estimate the accumulated pressure at the pump, pressure calculation should start at the load side. Therefore, inlet flow is used to calculate the differential pressure. Then differential pressure is added to the outlet pressure received from the downstream component model to calculate the inlet pressure, and finally sends it backward to the upstream model.

- **Static Characteristics:** All the required calculations inside the model use static characteristics that are load to the model.

- **Dynamic Characteristics:** Calculated values pass through the dynamic characteristics to carry the dynamic pattern of the modeled component. Such dynamic characteristics is presented by normalized transfer function.

- **Inputs/Outputs:** Additional predefined inputs and outputs are used for interactive condition monitoring, performance investigations, and model debugging.

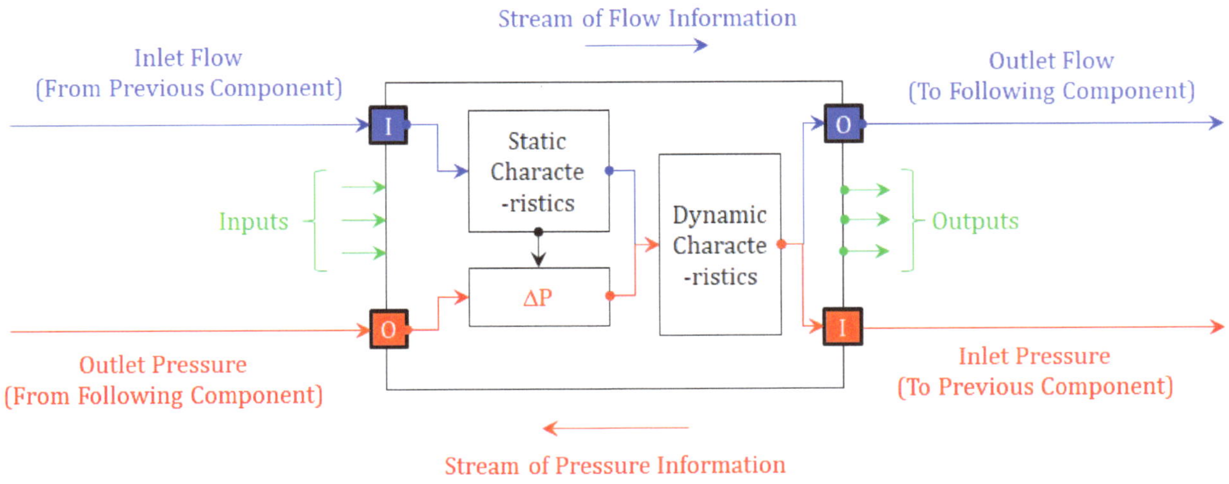

Fig. 4.1- Structure of a Lumped Model for a Hydraulic Component

4.3.3- Lumped Model Structure for Open Circuits

Figure 4.2 shows structure of a lumped model for an open hydraulic circuit. The figure shows that the model contains the following elements:

- **Middle Components:** They are components between pump model and reservoir model. Such component models are structured as described in Fig. 4.1.
- **Pump Model:** Pump is an energy source and appear as the first component model that generates flow based on outlet pressure with no need for inlet flow information. A pump for an open circuit is a unidirectional pump that has dedicated suction and discharge ports.
- **Intake line Model:** This model is needed only if cavitation analysis is required. Intake line model receives flow withdrawn by the pump based on which it calculates the line losses. Then based on tank pressure, from tank model, pump inlet pressure is calculated.
- **Reservoir Model:** Reservoir is the last component model that receives both the flow withdrawn by the pump and flow return by actuators, based on which fluid level and tank pressure are calculated. If it is a closed reservoir, flow withdrawn by the pump and return by actuator are used to calculate the variable tank pressure $\mathbf{P_T}$. If the reservoir is open, then tank pressure has constant atmospheric pressure.

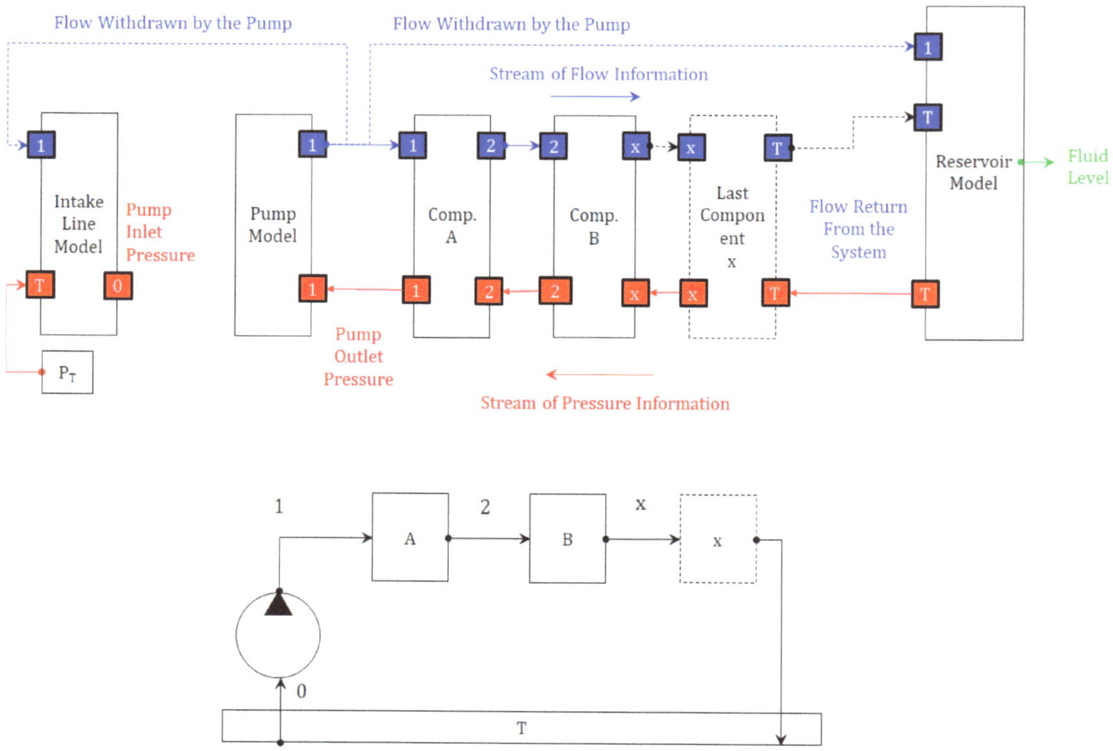

Fig. 4.2- Structure of a Lumped Model for an Open Hydraulic Circuit

Figure 4.3 shows an example of a simple hydraulic system that contains a tank, a unidirectional pump, a throttle valve, and a unidirectional hydraulic motor. As it has been described previously in Fig. 4.2:

- Flow information is moving forward. Outlet flow from the pump (at point 1) is forwarded to the throttle model. Outlet flow from the throttle model (at point 2) is forwarded to the motor model. Outlet flow from the motor model (at point 0) is forwarded to the tank model. If the tank is a closed tank, then tank model uses the information of the flow returns by the motor and withdrawn by the pump to calculate the fluid level in the tank and tank pressure. Atmospheric pressure is assumed for open tanks.

- Pressure information is moving backward. Reservoir pressure (at point 0) and motor loading conditions are used to calculate the motor inlet pressure (at point 2). Throttle outlet pressure (at point 2) is used by the throttle model to calculate the throttle inlet pressure (at point 1). Pump outlet pressure (at point 1) is used by the pump model to calculate the pump outlet flow (at point 1).

- Intake line model here is skipped for no need for cavitation analysis.

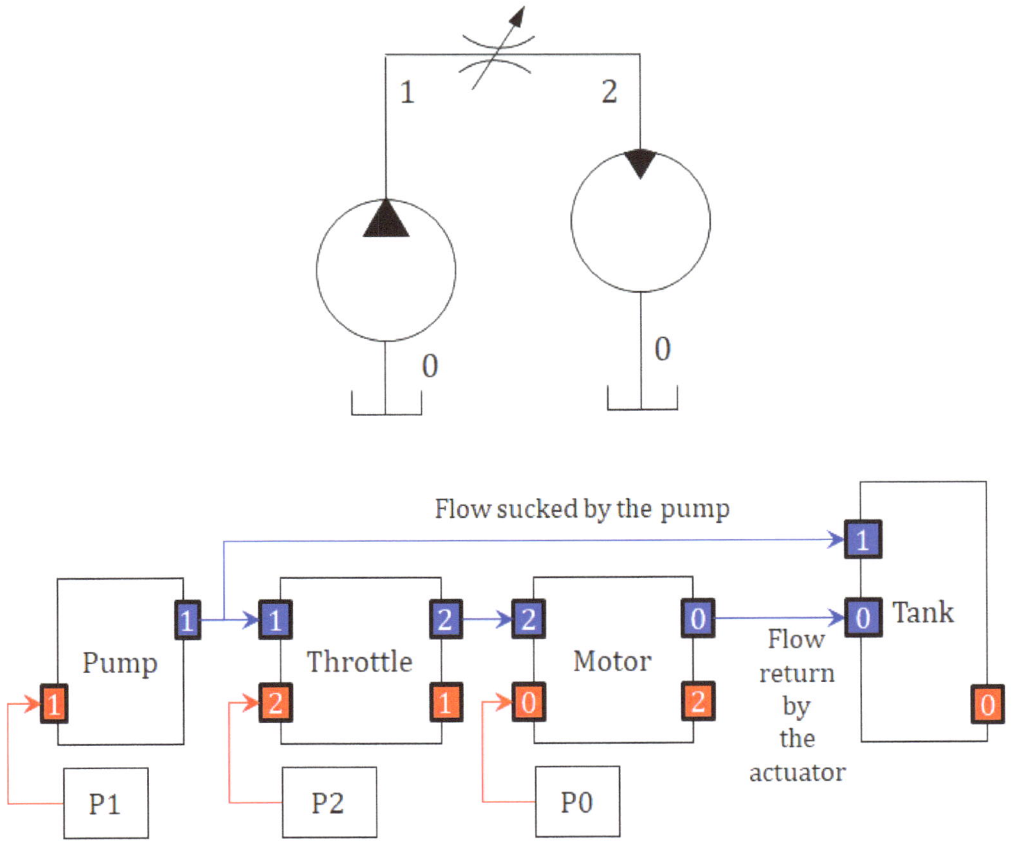

Fig. 4.3- Example of Hydraulic System Lumped Model

4.3.4- Lumped Model Structure for Closed Circuits

Figure 4.4 shows structure of a lumped model for a closed hydraulic circuit. The figure shows that the model contains the following elements:

- **Pump Model:** A pump for a closed circuit is either a bidirectional pump (usually in industrial applications) or an over-center pump (usually for mobile applications). Such pumps have no dedicated suction and discharge ports. Each of the pump ports could be a suction, or a discharge depends on the pump direction of rotation (for bidirectional pumps) or direction of flow (for over-center pumps). As shown in the figure, pump model assumes generic inlet and outlet side of the pump and the motor. High pressure side is at the outlet side of the pump that is the inlet side of the motor. Low pressure side is at the return side of the motor that is the inlet side of the pump. Model assigns number 1 and 2 for physical ports on the pump and the motor. A sign identifier is used to indicate whether port 1 is inlet or outlet and so for port 2. Like a unidirectional pump, outlet flow is calculated based on outlet pressure.

- **Motor Model (High-Pressure Side):** Motor Model receives torque, based on which motor inlet (pump outlet) high pressure side is calculated. Motor model receives pump outlet (motor inlet) flow, based on which motor RPM is calculated. No information about motor outlet (pump inlet) flow is needed.

- **Boosting Pump Model (Low-Pressure Side):** This model is needed only if cavitation analysis is required. Otherwise, a constant low-pressure side can be assigned.

- **Reservoir Model:** Reservoir model receives case drain from the pump and motor. Reservoir model also receives information about flow withdrawn by the boosting pump. If it is a closed reservoir, variable tank pressure is calculated based on flow information. If the reservoir is open, then tank pressure has constant atmospheric pressure.

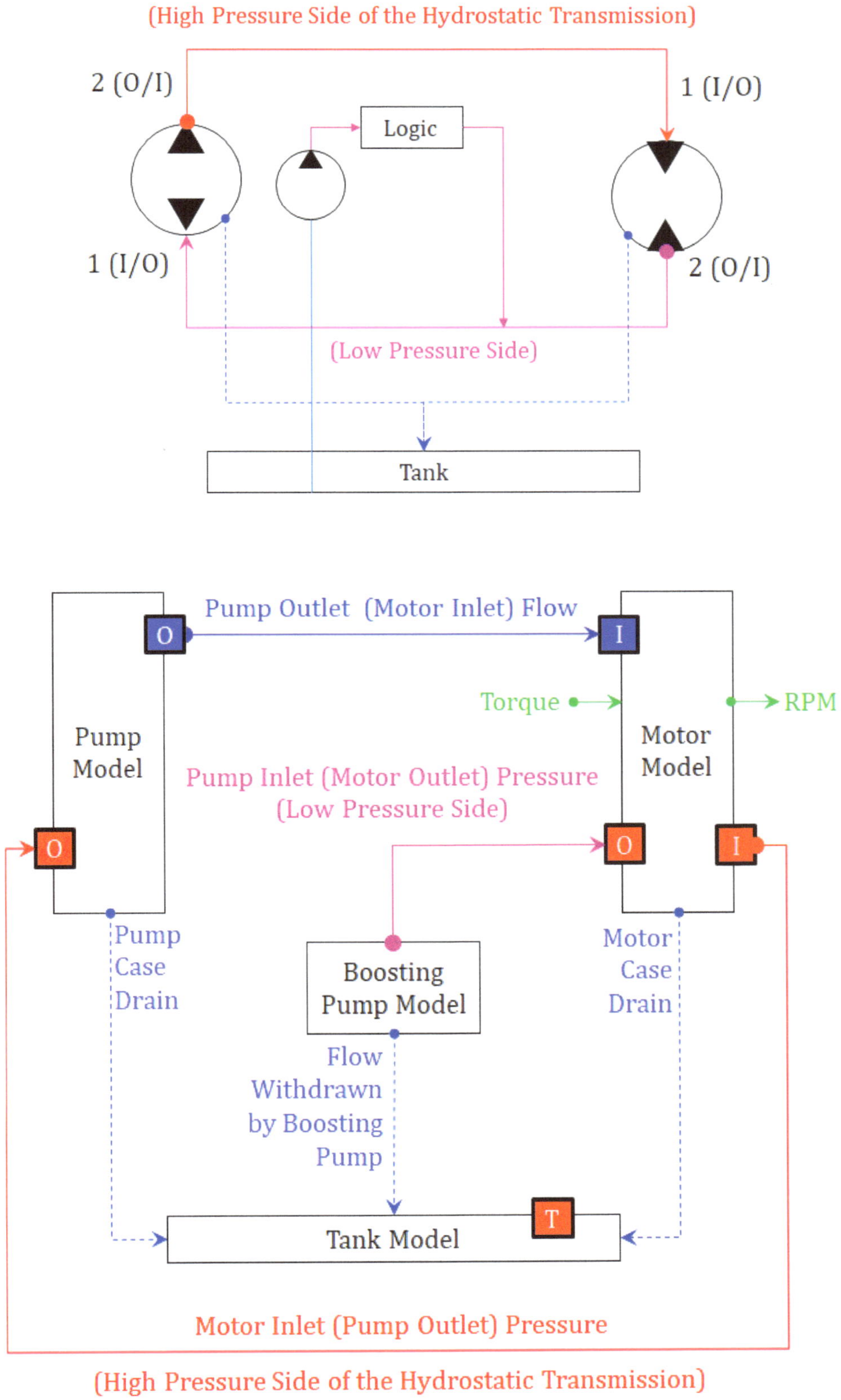

Fig. 4.4- Structure of a Lumped Model for a Closed Hydraulic Circuit

4.3.5- Simulating the Static Characteristics of a Hydraulic Component in a Lumped Model

The previous section discussed the structure and the communication between lumped models. The other core part of a lumped model is the static characteristics. Static characteristics of a hydraulic component must be loaded into the lumped model considering the following rules:

Complexity: Avoid unnecessary information to reduce the complexity of the model. The amount of information loaded to the model should be enough to get the predefined outputs. Depending on the nature of the hydraulic circuit operating conditions, various levels of the model can be built.

Source: As it has been mentioned previously, static characteristics of hydraulic components are either reported by the manufacturer in tabulated or graphical forms. Alternatively, it developed experimentally.

Linear vs. Nonlinear Characteristics: As shown in Fig. 4.5, it has been mentioned previously, static characteristics of hydraulic components are linear or nonlinear. Easiest method to model a linear characteristic are to develop the straight-line equation that represent it. If the static characteristics is nonlinear, depending on the degree of nonlinearity, the modeler has the following choices:

- **Lightly Nonlinear**: linearize it and develop an approximate straight-line equation.
- **Fairly Nonlinear:** Develop curve fit equation.
- **Highly Nonlinear and Contains Discontinuity:** convert it into a lookup table.

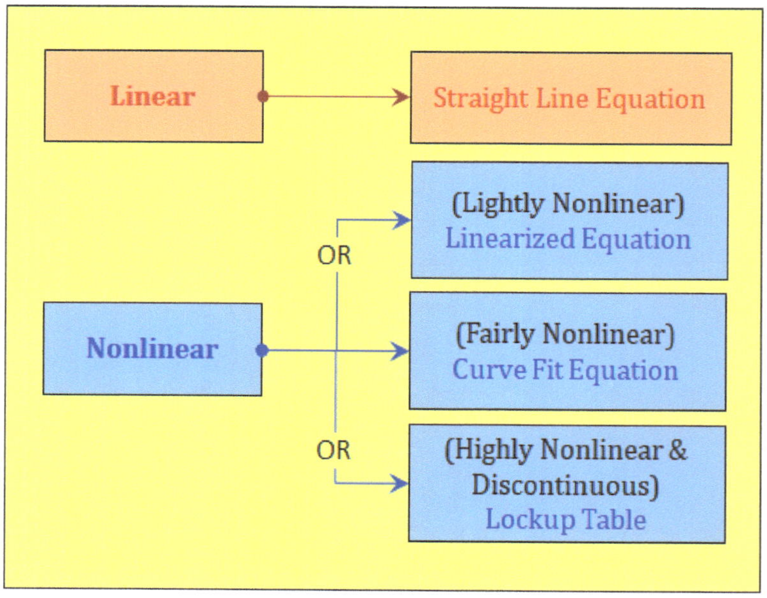

Fig. 4.5- Modeling Static Characteristics

Boundaries: Usually manufacturers test their components within the allowable ranges of the operating conditions. However, the available static characteristics may not cover the component performance under application operating conditions. Therefore, in some cases extrapolation may be needed to estimate the component performance beyond the scope of the reported characteristics. This extrapolation must always kept be within the allowable boundaries of the operating conditions for this component. The following examples are good practice for this rule:

Example 1: Pump flow is measured versus a range of 100 bar of pump pressure. Maximum operating pressure reported by the manufacturer is 200 bar. Q-P characteristics of the pump are loaded as a lookup table. Any extrapolation process should not exceed 200 bar pump pressure. In other words, the modeler shouldn't assume that the pump will follow the same performance at 300 bar.

Example 2: An equation $\mathbf{Q = f\ (\Delta P)^{0.5}}$ is used to model a directional valve. A modeler shouldn't assume that this equation is valid for all ranges of differential pressure. The reason is that every valve has power limit beyond which this equation isn't valid.

Multidimensional Static Characteristics: In some cases, a static characteristic is a *Multidimensional* characteristic, i.e. it depends on multiple variables. It can be approximated as a unidimensional characteristic by considering the variable that has the dominant effect and ignore the effect of other variables. Otherwise, the following modeling techniques are useful.

Multidimensional Lookup Table: Useful for nonlinear multidimensional characteristics. For example, pump flow as function of pressure and rpm can be loaded to the model in the form of a two-dimension lookup table.

Taylor's Series: Useful for linear or lightly nonlinear multidimensional characteristics.

$$\mathbf{z\ (x\ \&\ y) = z_o + \left[\frac{\delta z}{\delta x}\right]_y (x - x_o) + \left[\frac{\delta z}{\delta y}\right]_x (y - y_o) + Nonlinear\ Terms} \qquad 4.1$$

So, linearized Taylor's Series is as follows:

$$\mathbf{z\ (x\ \&\ y) = z_o + \left[\frac{\delta z}{\delta x}\right]_y (x - x_o) + \left[\frac{\delta z}{\delta y}\right]_x (y - y_o)} \qquad 4.2$$

Where:

- $\mathbf{z_o}$ is value of **z** function as defined at specific conditions; namely $\mathbf{x_o}$ and $\mathbf{y_o}$.
- **z** is value of **z** function as the conditions of **x** and **y** conditions change.
- $\mathbf{(\delta z/\delta x)_y}$ = rate of change of **z** with respect to **x** at constant **y**.
- $\mathbf{(\delta z/\delta y)_x}$ = rate of change of **z** with respect to **y** at constant **x**.

An example of this is fluid viscosity is function of pressure and temperature.

4.3.6- Simulating the Dynamic Characteristics of a Hydraulic Component in a Lumped Model

To avoid modeling a component based on its internal design, a normalized transfer function is used to represent the components dynamic behavior. Developing a normalized transfer function is based on published or experimentally developed step or frequency response.

4.3.7- Example of Lumped Modeling

Figure 4.6 shows a servo-controlled constant power operated swash plate pump used in a hydraulic press. In order to model such a pump based on its internal design, lots of dimensional and physical design parameters are needed in addition to a lot of effort is required to build the model. Obviously, this isn't needed in view of building a model for the whole press operation. Therefore, based on static and dynamic characteristic that are reported by the pump manufacturer or developed experimentally, a pump model can be developed as follows.

The figure shows that the pump Q-P static characteristic is a unidimensional where the pump is driven at a constant rpm. However, when the pump pressure changes, flow will not change at no time! It supposed to follow the dynamic pattern of the pump shown by the given step response. Such a dynamic pattern can be modeled by a Normalized transfer Function. Then the model will be constructed as shown in Fig. 4.7. The steady state characteristics are loaded into the model as lookup table and multiplied by the NTF. So, any change of the pump pressure will be associated with a change in flow that settles at the corresponding steady state value after certain dynamic behavior is assigned by the NTF.

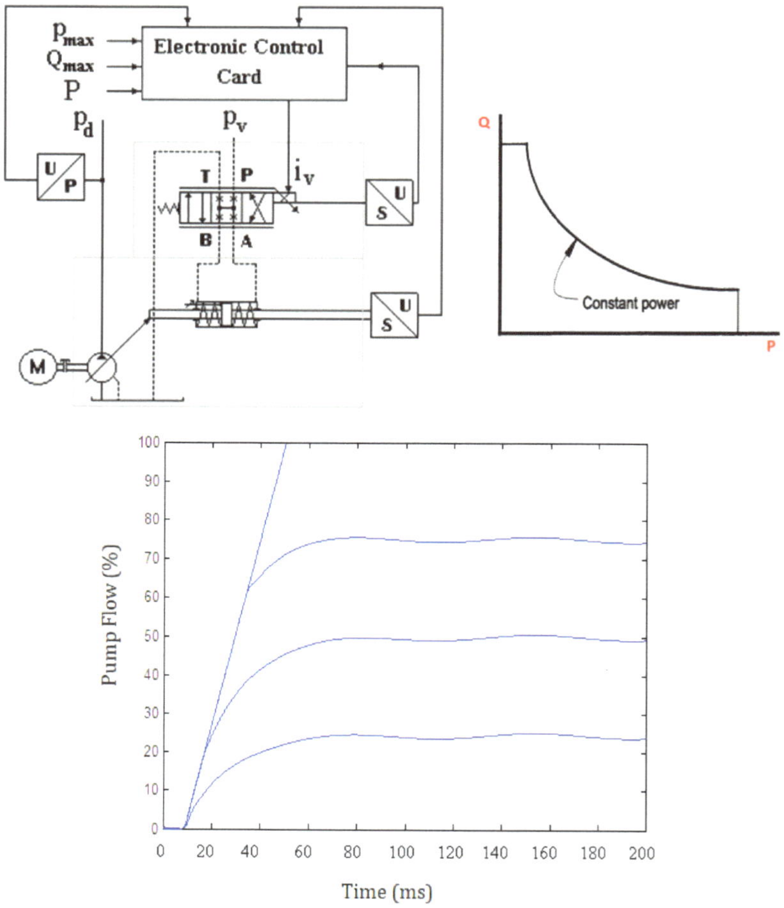

Fig. 4.6- Servo-Controlled Constant Power Swashplate Pump

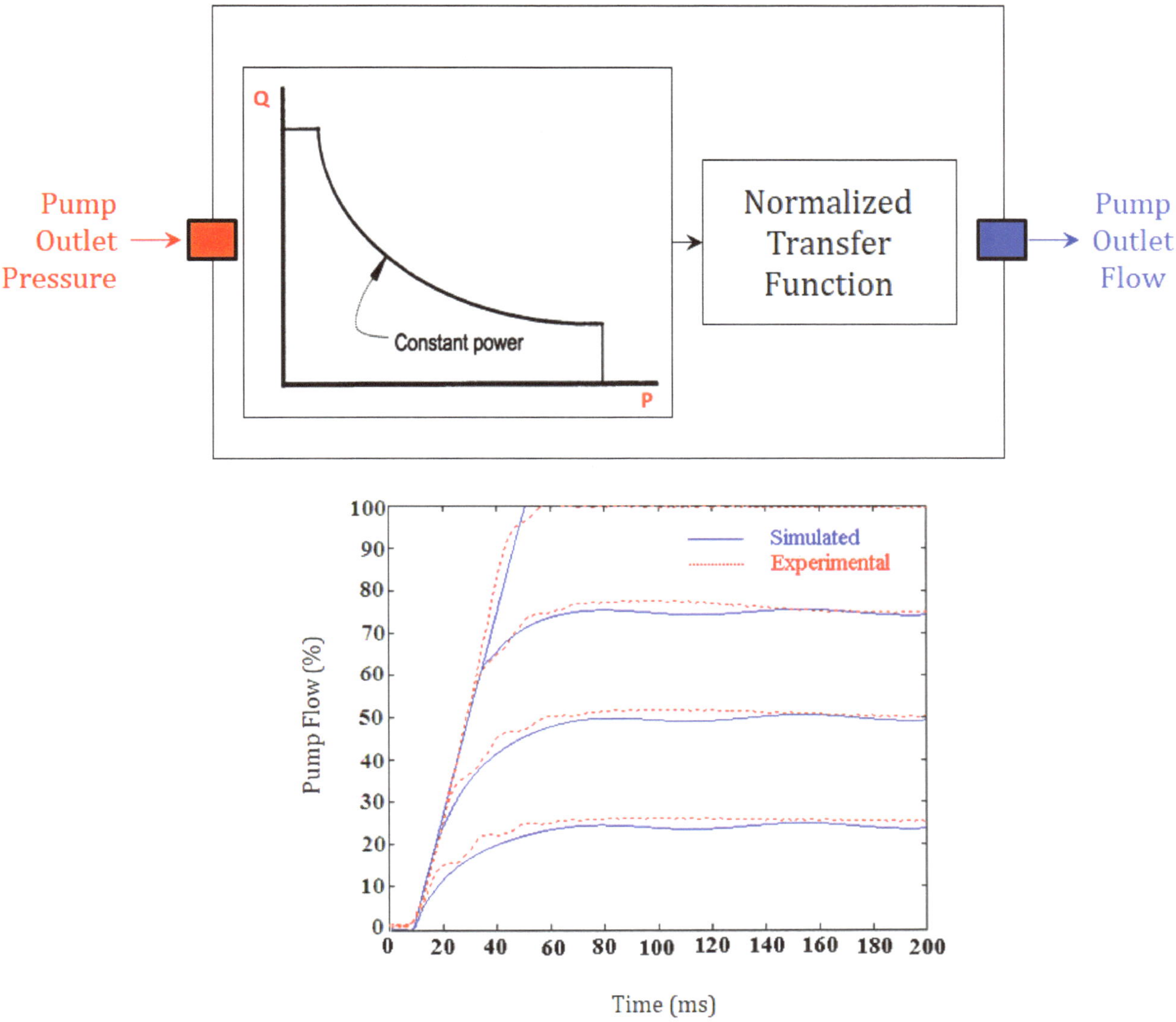

Fig. 4.7- Generic Lumped model for Servo-Controlled Constant Power Swashplate Pump

This approach, conceptually, can be used to build models for large systems. For example, as shown in Fig. 4.8, several self-contained electro-hydraulic actuators can be used in large facilities such as power plants, oil & gas industry, etc. To model these large plants, lumped model approach is more convenient where these actuators as whole will be capsulated in a lumped model based on its static and dynamic characteristics.

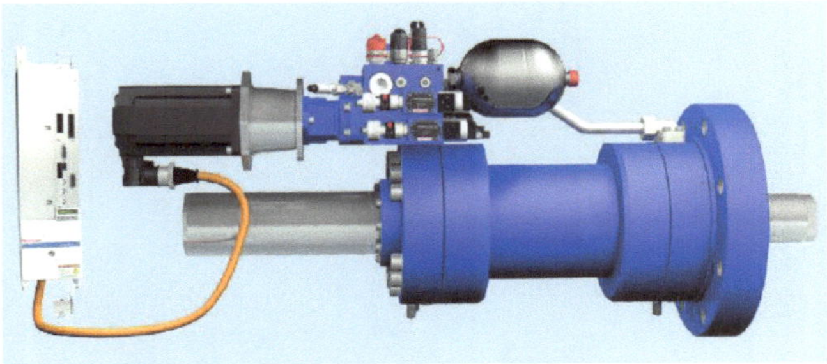

Fig. 4.8- Self-Contained Electro-Hydraulic Actuator (Courtesy of Bosch Rexroth)

Chapter 5
Modeling of Fluid Properties

Objectives

This chapter presents different techniques to model hydraulic fluid properties based on available information. Properties considered in this chapter are bulk modulus, density, specific gravity and viscosity. In modeling such properties, effects of working temperature and pressure are considered. Case studies are presented, and Matlab-Simulink models were built and validated based on given information.

Brief Contents

Chapter 5: Modeling of Fluid Properties

5.1- Introduction to Fluid Properties Modeling

Fluid properties modeling is a crucial step in modeling a hydraulic system. Many of the equations used in modeling hydraulic components include fluid properties. The challenge is these fluid properties are never maintained constant during machine operation. Most of the fluid properties change with the working conditions. Therefore, fluid properties must be modeled and updated on continuous basis. Unlike modeling a hydraulic component, fluid properties model is not connected to upstream and downstream components. However, since fluid properties are contained in most of component models, each fluid property must be assigned to a global variable in a “GoTo” block so that it can be recalled by a “From” block by any other component model as presented previously in Fig. 2.4.

Figure 5.1 shows the generic structure of a fluid properties lumped model. The figure shows that, since the model is not hooked to other models. it contains only the working temperature and pressure as the only interactive inputs. Other fluid properties are loaded inside model properties page. They also can be taken out of the model properties page and placed outside as interactive inputs. The model calculates some important outputs.

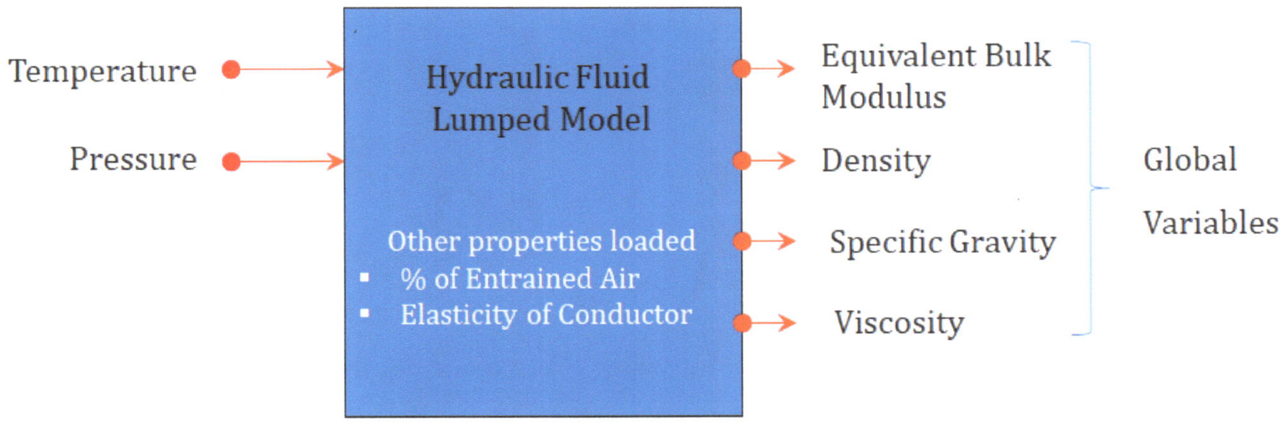

5.1- Generic Lumped Model of Fluid Properties

As shown in the figure, working pressure and temperature are inputs required by the model. Obviously, pressure and temperature changes throughout the system. Then, if only one block of fluid properties modeling is used, this means that fluid properties are modeled for a specific location in the system. A number of blocks can be used to model the fluid properties at several locations. Global variables can be names as Viscosity1, Viscosity2, etc.

5.2- Modeling of Hydraulic Fluid Bulk Modulus

5.2.1- Definition and Mathematical Expression of Bulk Modulus

Definition: The *Bulk Modulus* of a fluid is a property that indicates how stiff a fluid is. Bulk Modulus is the reciprocal of compressibility. The higher the bulk modulus, the stiffer and less compressible the fluid is. In other words, a fluid with high bulk modulus can carry large loads or be exposed to high pressure with less change in its volume.

Figure 5.2 shows a practical representation of fluid bulk modulus. A volume of oil is confined in a cylinder that has an effective area A. if the oil volume is subjected to a pressure increase from p1 to p2, oil volume decreases and load drifts down a distance x. If oil has a lower bulk modulus, its compressibility increases, and the drifting distance increases too.

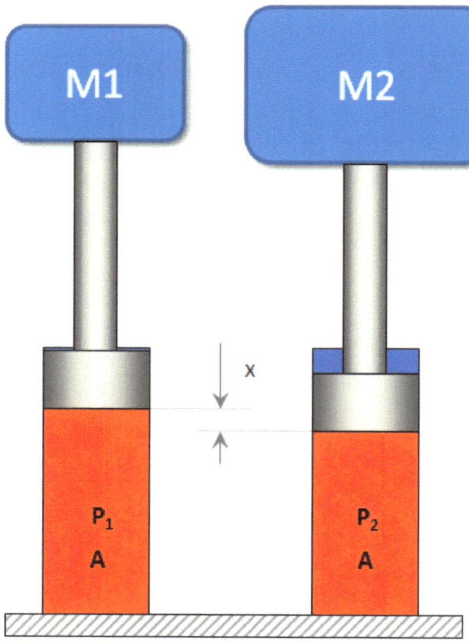

Fig. 5.2- Hydraulic Fluid Bulk Modulus vs. Compressibility

Mathematical Expression: Bulk Modulus is a fluid property that is equivalent to modulus of elasticity of a piece of metal and spring constant of a spring. All these properties are expressed in form of the ratio between the effort variable and the reactional change. *Modulus of Elasticity* for a specimen of a metal is determined by the ratio between the compressive or tensile stress and the longitudinal change in the specimen. *Spring Constant* of a spring is determined by the ratio between the compressive or tensile force and the longitudinal change in the spring.

Similarly, Bulk Modulus is determined by the ratio of the pressure applied on a confined volume of oil and the volumetric change of the oil volume.

Equation 5.1 shows the mathematical expression of hydraulic fluid bulk modulus. The minus sign indicates the inverse proportionality between the change of the pressure and the resulting volumetric change of the fluid. The unit of bulk modulus is a unit of pressure (bar or psi) because the denominator of the equation is dimensionless.

$$\beta = -\frac{\Delta P}{(\frac{\Delta V}{V_0})} \tag{5.1}$$

As shown in Table 5.1, in order of magnitude, water has a bulk modulus that is approximately 30% higher than mineral oil. Table 5.2 shows the volumetric strain for basic types of hydraulic fluids as a response to 100 bar (1450 psi) increase in pressure. It is obvious that the volumetric strain decreased with increasing water content in the fluid.

Fluid Type	Bulk Modulus at 20 °C and 10,000 psi
ISO 32 Mineral Oil	260,000 psi
Water-in Oil Emulsion	333,000 psi
Phosphate Ester	440,000 psi
Water-Glycol	500,000 psi

Table 5.1- Bulk Modulus of Typical Hydraulic Fluids

COMPRESSIBILITY-VOLUME CHANGE (at 100 bar = 1450 psi)	
Fluid Type	**% ΔV Reduction**
Mineral Oil	0.7%
Vegetable-based Oil	0.5%
Water and Emulsified Water-Oil	0.4%
Water-Glycol and Synthetic Fluids (Polymers)	0.35%

Table 5.2- Compressibility of Typical Hydraulic Fluids

5.2.2- Case Study 1 for Modeling of Bulk Modulus

5.2.2.1- Effect of Temperature and Pressure on Bulk Modulus

It is known that bulk modulus decreases somewhat with increase of working temperature and slightly increases with increase of working pressure. In order to model such effects, information should be known for a specific fluid so that a modeler can select the adequate method of modeling static characteristics as presented in Chapter 4

Fig. 5.3 shows the effect of pressure and temperature on MIL-H-83282 hydrocarbon fluids. Reference bulk modulus of this fluid is 300,000 psi stated at reference temperature of 100°F, and reference pressure of 2500 psi and no entrained air. The figure shows that bulk modulus increases linearly $\approx$10% for every 2000 psi and decreases linearly $\approx$10% for every 40% ^{0}F.

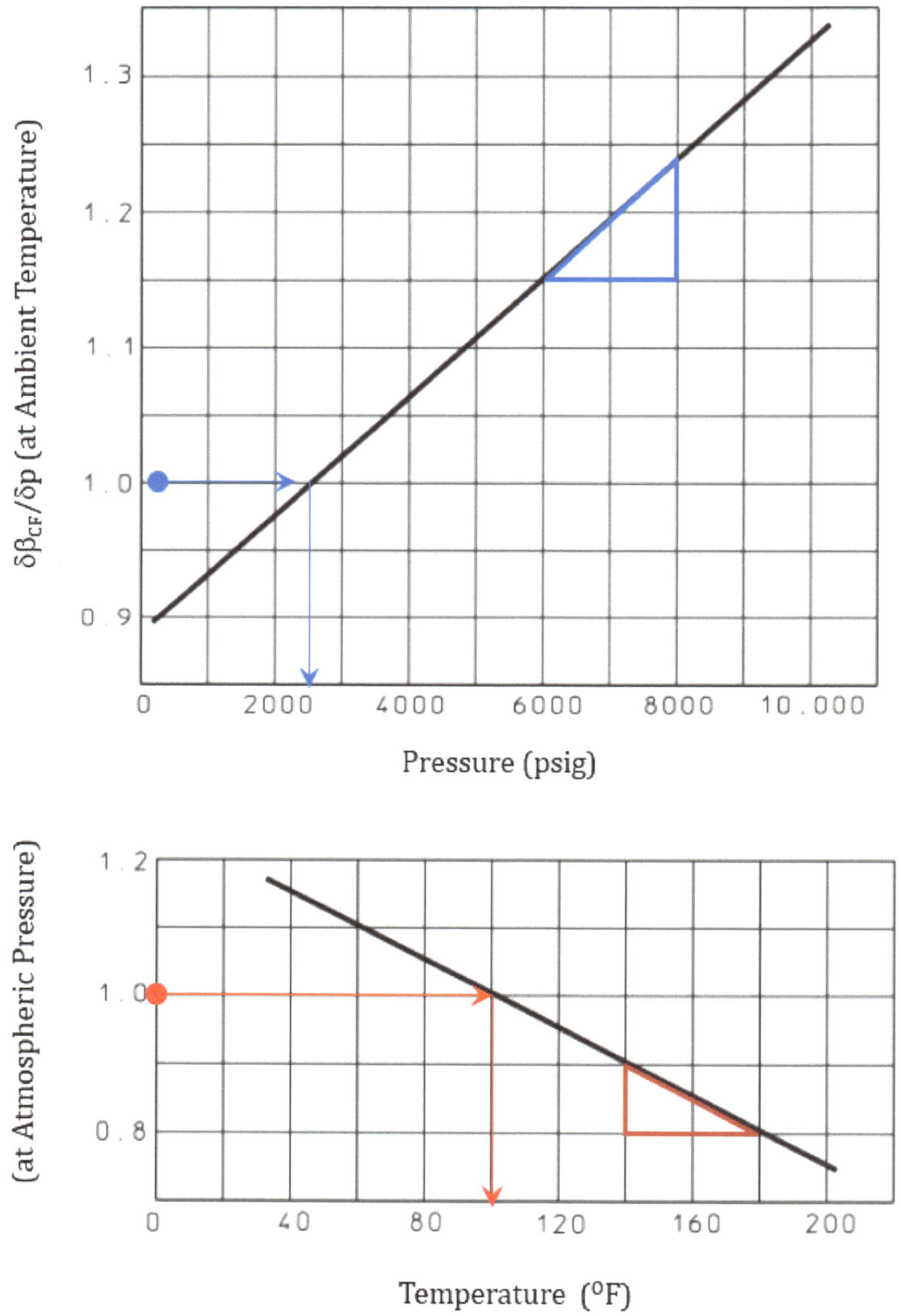

Fig. 5.3- Pressure and Temperature Correction Factors for MIL-H-83282 Hydraulic Fluid Bulk Modulus (Courtesy of the Lee Company)

Based on what has been presented in Chapter 4 this is a multidimensional linear static characteristic. Hence, by applying to linearized Taylor's series shown in Eq. 4.2, bulk modulus can be modeled as follows:

$$\beta_{TP} = \beta_o + \left[\frac{\delta\beta}{\delta T}\right]_P (T - T_o) + \left[\frac{\delta\beta}{\delta p}\right]_T (P - P_o) \qquad 5.2$$

But the characteristics given in Fig. 5.3 doesn't show the rate of change of bulk modulus with temperature and pressure, it rather shows the *Correction Factors*. Then equation 5.2 must be rewritten in the following form.

$$\beta_{TP} = \beta_o + \beta_o\left[\frac{\delta\beta_{CF}}{\delta T}\right]_P (T - T_o) + \beta_o\left[\frac{\delta\beta_{CF}}{\delta P}\right]_T (P - P_o) \Rightarrow$$

$$\beta_{TP} = \beta_o[1 + \left[\frac{\delta\beta_{CF}}{\delta T}\right]_P (T - T_o) + \left[\frac{\delta\beta_{CF}}{\delta P}\right]_T (P - P_o)] \qquad 5.3$$

Where:

- β_{TP} = Bulk modulus corrected based on current temperature **T** and current pressure **P**.
- β_o = 300k psi = Reference bulk modulus at reference T_o [100 ^{0}F] and P_o. [= 2500 psig].
- $(\delta\beta/\delta T)_P$ = Rate of change of **β** with respect to **T** at constant **P.**
- $(\delta\beta/\delta P)_T$ = Rate of change of **β** with respect to **P** at constant **T.**
- Given Characteristics → $(\delta\beta_{CF}/\delta T)_P$ = (-0.1/40) = 0.0025 [1/^{0}F].
- Given Characteristics → $(\delta\beta_{CF}/\delta P)_T$ = (0.1/2000) = 0.00005 [1/psi].

5.2.2.2- Effect of Entrained Air on Bulk Modulus

While bulk modulus is drastically decreased with increasing entrained air at low pressure, it is slightly decreased at high pressure. Lots of research work including experimental tests have done to determine the effect of the entrained air. Figure 5.4 shows a result of one of these studies. To model such an effect, using these curves, multidimensional lookup table is needed.

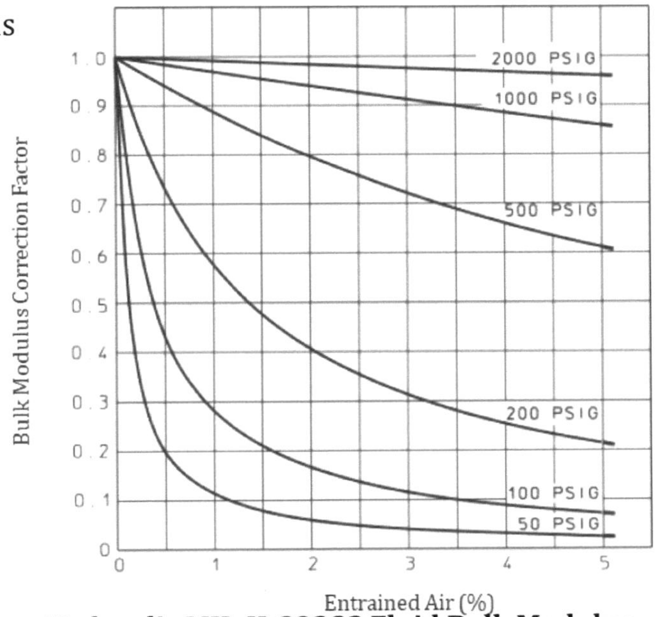

Fig. 5.4- Effect of Entrained Air on Hydrocarbon Hydraulic MIL-H-83282 Fluid Bulk Modulus (Courtesy of the Lee Company)

Alternatively, Eq. 5.4 calculates the Equivalent Bulk Modulus based on air content and elasticity of the transmission line walls.

$$\frac{1}{\beta_E} = \frac{1}{\beta_C} + \frac{V_F}{V_T}\frac{1}{\beta_{TP}} + \frac{V_A}{V_T}\frac{1}{\beta_A} = \frac{1}{\beta_C} + \frac{(V_T - V_A)}{V_T}\frac{1}{\beta_{TP}} + \frac{V_A}{V_T}\frac{1}{\beta_A} \quad 5.4$$

Where:

- β_E = Effective bulk modulus based on entrained air and conductor elasticity.
- β_C = *Modulus of Elasticity* of the transmission line (conductor).
- β_A = Air bulk modulus.
- V_F / V_T = Ratio of fluid volume to the total volume.
- V_A / V_T = Ratio of air volume to the total volume.

By defining V_{Air} as the % of air in the total volume, the Eq. 5.4 can be rewritten as follows.

$$\frac{1}{\beta_E} = \frac{1}{\beta_C} + \frac{(100 - V_{Air})/100}{\beta_{TP}} + \frac{V_{Air}/100}{\beta_A} \quad 5.5$$

Example:

- Amount of entrained air = 1%
- Oil bulk modulus = 300 x 10^3 psi (corrected based of temperature and pressure)
- Conductor bulk modulus = 200 x 10^6 psi
- Air bulk modulus = 2800 psi

Then,

$$\frac{1}{\beta_E} = \frac{1}{\beta_C} + \frac{(100 - V_{Air})/100}{\beta_{TP}} + \frac{V_{Air}/100}{\beta_A} = \frac{1}{200 \times 10^6} + \frac{0.99}{3 \times 10^3} + \frac{0.1}{2800} = \frac{1}{145.5 \times 10^3}$$

This example shows that, increasing air content by 1% drops the oil bulk modulus from 300 x 10^3 to 145.5 x 10^3 that is equivalent to 51.5% reduction of the fluid's bulk modulus with no air content.

5.2.2.3- Structure of Lumped Model for Case Study 1

As shown in Fig. 5.5, simulation model (HSV7CH05Model01) is built to solve Eq. 5.3 and 5.5. By running the simulation model, it validates the case given in Case Study 1. The model results in the reference bulk modulus (300,000 psi) at the given reference temperature (100 ^{0}F) and pressure (2500 psi) at no air content (VAIR = 0). Conductor bulk modulus was considered very high on purpose to have negligible effect.

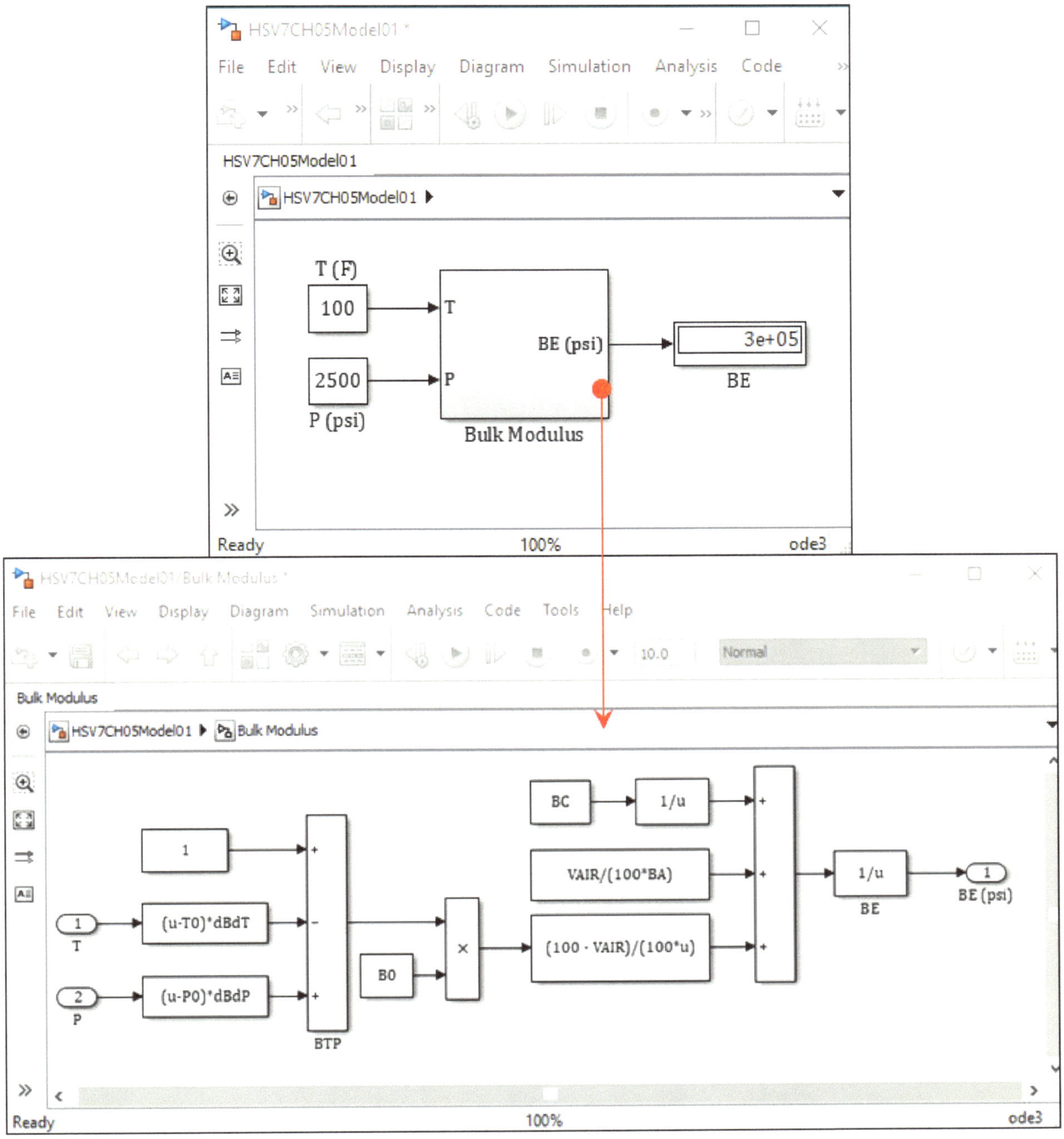

Fig. 5.5- Structure of Bulk Modulus Model

Figure 5.6 shows the properties added to the model.

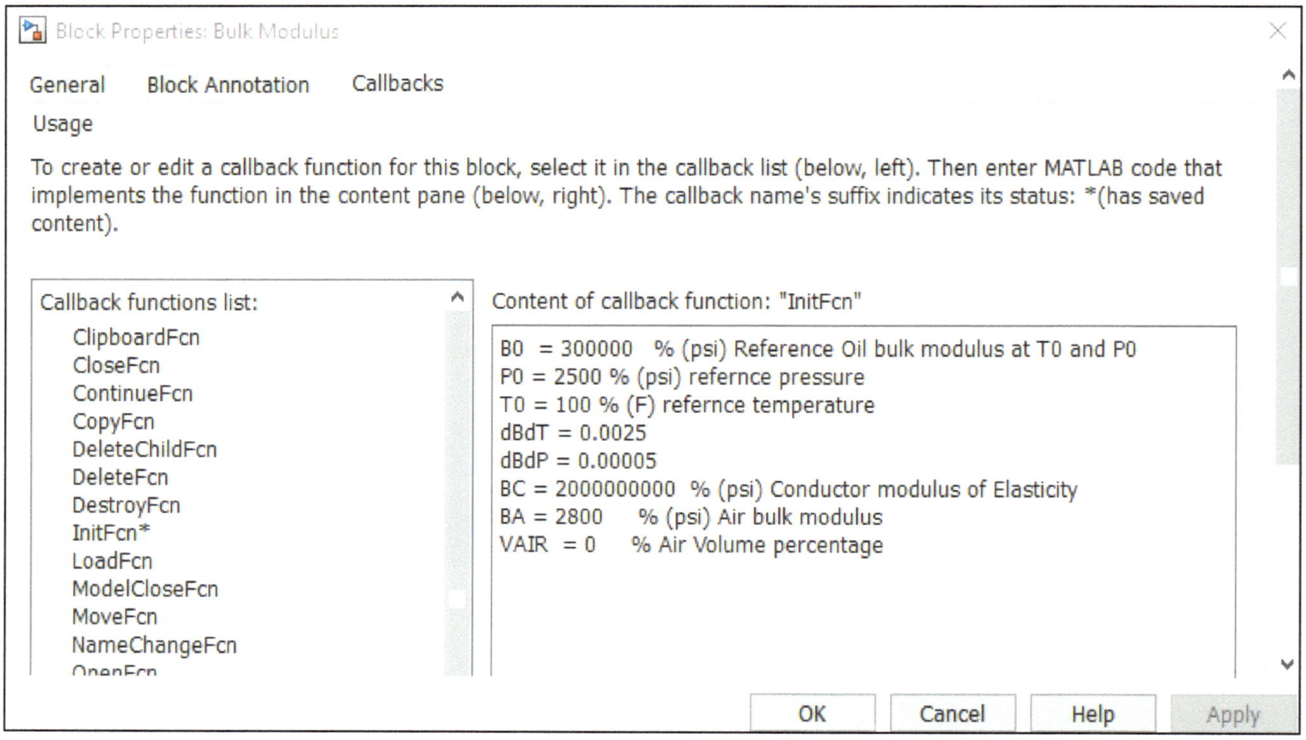

Fig. 5.6- Properties of the Previous Model

By considering 1% entrained air, equivalent bulk modulus decreased from 300,000 psi to 145,500 psi which validates the given values in in Case Study 1.

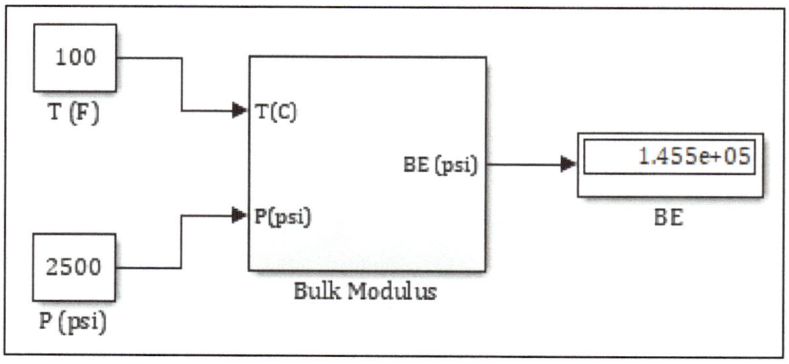

Fig. 5.7- Validating Case Study 1

5.3- Modeling of Hydraulic Fluid Density and Specific Gravity

5.3.1- Definitions and Mathematical Expression of Density

Hydraulic Fluid *Density* is defined as the amount of mass in a unit volume of the fluid. Equations 5.6 and 5.7 show the mathematical quantification of fluid density in Metric and English system of units; respectively.

$$\rho_f \left[\frac{kg_m}{m^3}\right] = \frac{\text{Mass }(kg_m)}{\text{Volume }(m^3)} \qquad 5.6$$

Where density of water at ambient conditions $\rho_w = 1000\ kg_m/m^3$

$$\rho_f \left[\frac{lb_m}{ft^3}\right] = \frac{\text{Mass }(lb_m)}{\text{Volume }(ft^3)} \qquad 5.7$$

Where density of water at ambient conditions $\rho_w = 62.4\ lb_m/ft^3$

5.3.2- Definitions and Mathematical Expression of Specific Weight

Hydraulic Fluid *Specific Weight* is defined as the weight of a unit volume of the fluid. Equations 5.8 and 5.9 shows the mathematical quantification of fluid specific weight in Metric and English system of units; respectively.

$$\gamma_f \left[\frac{kg_f}{m^3}\right] = \frac{\text{Weight }(kg_f)}{\text{Volume }(m^3)} = \frac{\text{Mass }(kg_m) \times g\ (m/s^2)}{\text{Volume }(m^3)} = \rho_f \times g \qquad 5.8$$

Then specific weight of water $\gamma_w = \rho_w \times g = 1000\ [kg_m / m^3] \times 9.81\ [m/s2]$
$= 1000 \times 9.81\ [kg_m .m/ s^2]\ [1/m^3]$
$= 1000 \times 9.81\ [N/m^3] = 1000\ [kg_f /m^3]$

$$\gamma_f \left[\frac{lb_f}{ft^3}\right] = \frac{\text{Weight }(lb_f)}{\text{Volume }(ft^3)} = \frac{\text{Mass }(lb_m) \times g\ (ft/s^2)}{\text{Volume }(ft^3)} = \rho_f \times g \qquad 5.9$$

Then specific weight of water $\gamma_w = \rho_w \times g = 62.4\ [lb_m / ft^3] \times 32.2\ [ft/s2]$
$= 62.4 \times 32.2\ [lb_m. ft/s^2]\ [1/ft^3]$
$= 62.4 \times 32.2\ [\text{Poundal}/ft^3] = 62.4\ [lb_f /ft^3]$

Where **g** is the gravitational acceleration.

5.3.3- Definitions and Mathematical Expression of Specific Gravity

Hydraulic Fluid *Specific Gravity* is defined as the ratio of the specific weight of the fluid and the specific weight of a very well-known fluid, which is water. It is a dimensionless value. Equation 5.10 shows the mathematical quantification of fluid specific gravity. So, a hydraulic fluid that has a SG < 1 means it is lighter than water and vice versa.

$$\mathbf{SG_f} = \frac{\gamma_f}{\gamma_w} = \frac{\rho_f}{\rho_w} \qquad \mathbf{5.10}$$

5.3.4- Case Studies for Modeling of Density and Specific Gravity

5.3.4.1- Case Study 2 based on Effect of Temperature and Pressure on Density

Like bulk modulus, it is known that density decreases with increasing temperature and increases with decreasing pressure. In order to model such effects, information should be known for a specific fluid so that a modeler can select the adequate method of modeling the static characteristics as presented in Chapter 4.

Figure 5.8 shows the effects of pressure and temperature on density of a hydraulic fluid. Based on what has been presented in Chapter 4, this is a multidimensional linear static characteristic. In order to model such an effect, multidimensional lookup table is needed.

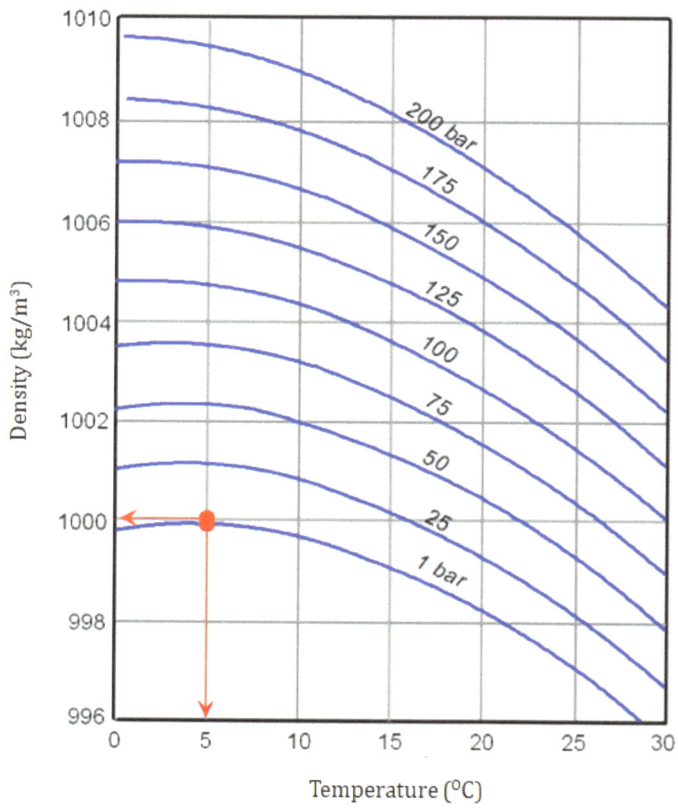

Fig. 5.8- Effects of Pressure and Temperature on Fluid Density (www.engineeringtoolbox.com)

To avoid building multidimensional lookup table, the following alternative method is found useful. Reference density of this fluid is 1000 kg/m3 stated at reference temperature of 5°C, and reference pressure of 1 bar gauge pressure. By looking at the figure, the rate of change of the density with respect to pressure (at constant temperature) seems approximately linear and equal to (1 kg/m^3 per 25 bar). Eq. 5.11 shows correcting fluid density based on pressure.

$$\rho_P = \rho_o + \left[\frac{\delta\rho}{\delta P}\right]_T (P - P_o) \tag{5.11}$$

- ρ_P = Density corrected based on current working pressure **P**.
- ρ_o = 1000 kg/m^3 = Reference density at reference values T_o [5 ^{0}C = 41 ^{0}F] and P_o [14.5 psig = 1 barg].
- $(\delta\rho/\delta P)_T$ = Rate of change of **ρ** with respect to **P** at constant **T**
- Given Characteristics → $(\delta\rho/\delta P)_T \approx (1/25) = 0.04$ [(kg/m^3)/bar].

Rate of change of density with respect to temperature is clearly nonlinear, but it behaves similar at various pressures. So, an approximate correction factor can be developed. One easy method is to develop the correction factor 1-D lookup table as shown in Table 5.3. For more determination of the correction factor, the lookup table can be used to generate the curve fit equation as shown in Fig. 5.9. However, in this case study, just a lookup table is used.

T [C]	0	5	10	15	20	25	30
DT	1	1	1	0.999	0.998	0.997	0.995

Table 5.3- Temperature-Density Correction Factor Lockup Table

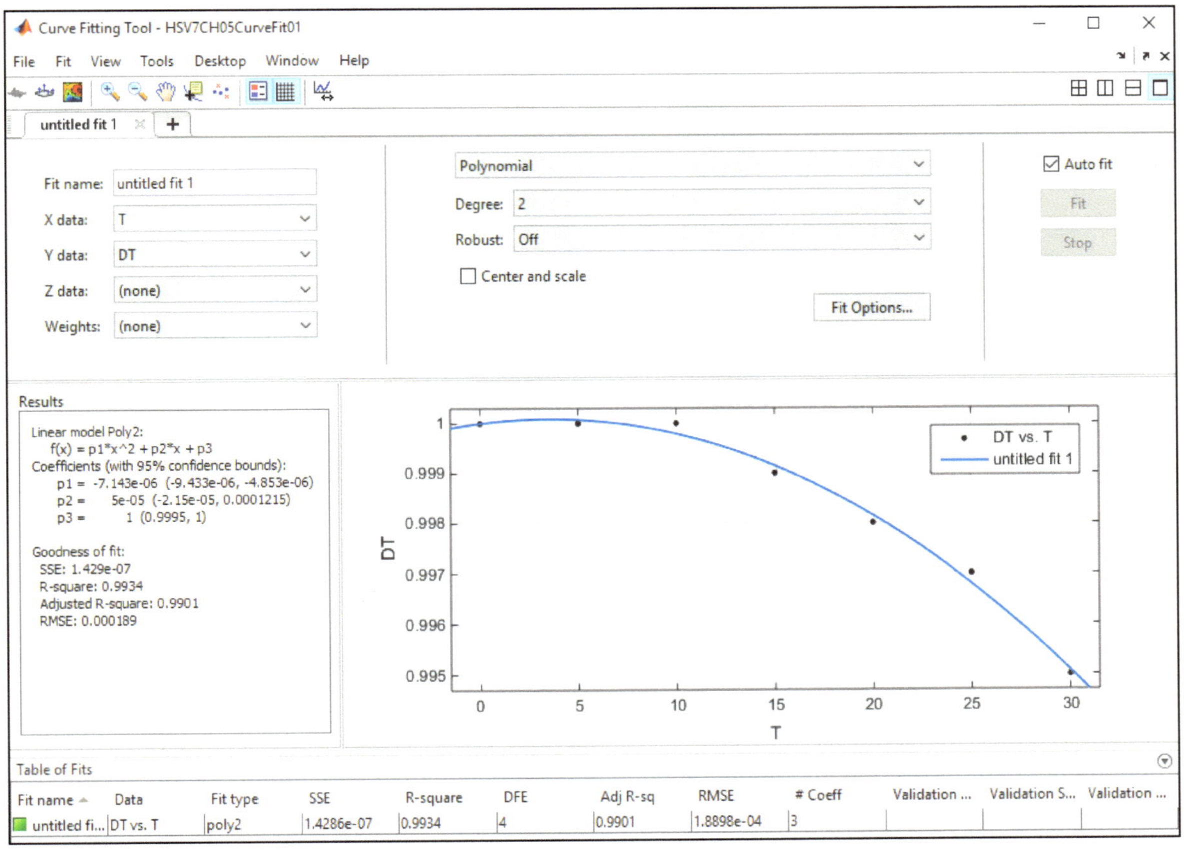

Fig. 5.9- Temperature-Density Correction Factor Curve Fit Equation

5.3.4.2- Structure of Lumped Model for Case Study 2

Figure 5.10 shows a simulation model (HSV7CH05Model02) built based on Case Study 2 and validates the values given in this case study. The model is structured to calculate the density based on pressure first, then the result is multiplied by a correction factor based on temperature. Specific gravity is then calculated based on the resulting density divided by density of water (1000 kg/m³).

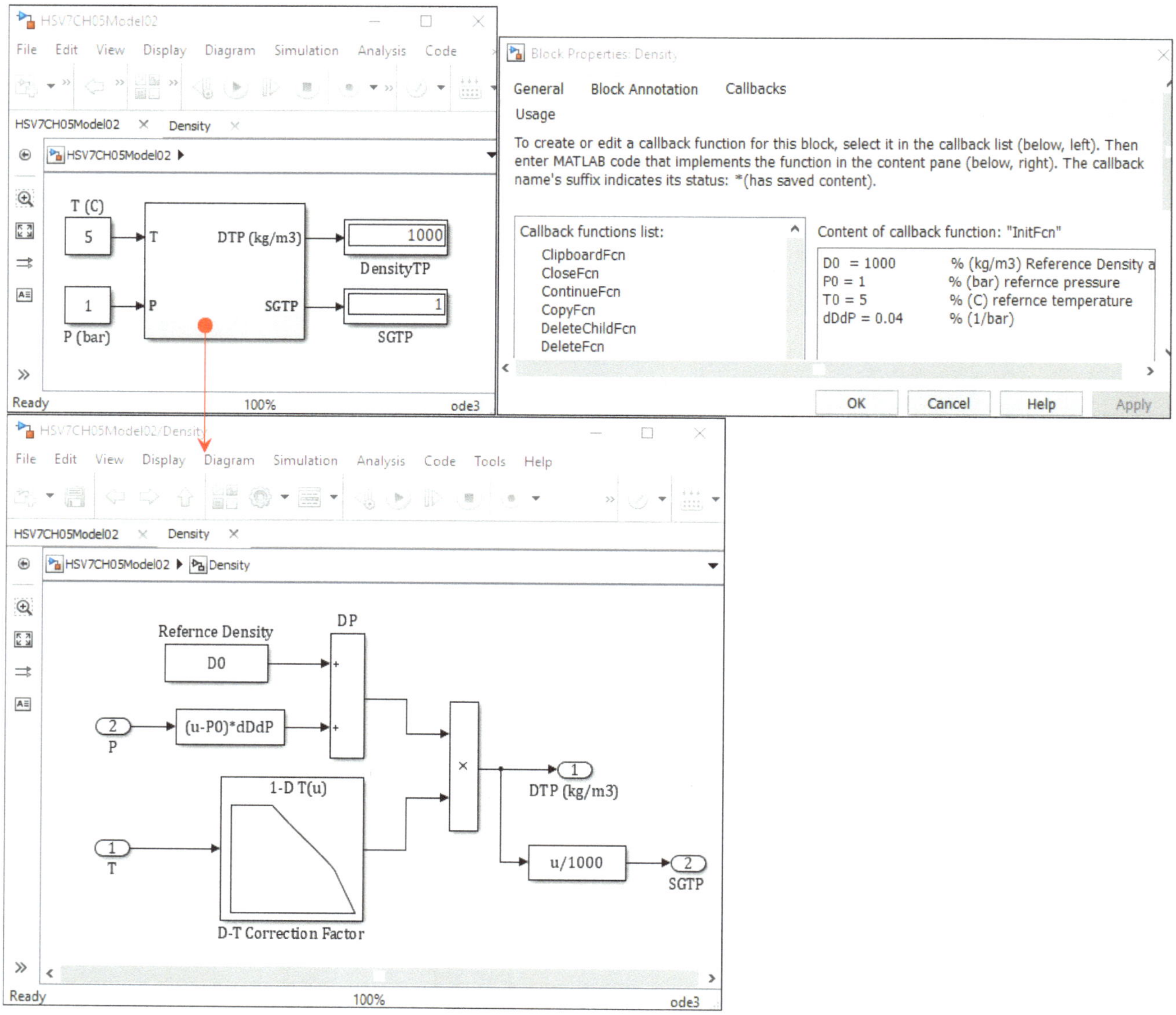

Fig. 5.10- Simulation Model for Case Study 2

5.3.4.3- Case Study 3 based on Density as Function of other Properties.

This is an alternative modeling method for the density if curves for density versus pressure and temperature were not found. Equation 5.12 is the linearized Taylor's Series shown in Eq. 4.2 as applied to density.

$$\rho_{TP} = \rho_o + \left[\frac{\delta\rho}{\delta P}\right]_T (P - P_o) + \left[\frac{\delta\rho}{\delta T}\right]_p (T - T_o) \qquad 5.12$$

- ρ_{TP} = Density corrected based on current temperature **T** and current pressure **P**.
- ρ_o = Reference density as defined at reference values T_o and P_o.
- $(\delta\rho/\delta T)_P$ = Rate of change of ρ with respect to **T** at constant **P.**
- $(\delta\rho/\delta P)_T$ = Rate of change of ρ with respect to **P** at constant **T**

This equation can be mathematically manipulated to calculate the density as function of the bulk modulus and the *Cubical Thermal Expansion Coefficient* (α) of the fluid as follows.

$$\text{Eq. 5.1} \rightarrow \beta = -V_o \left[\frac{\delta P}{\delta V}\right]_T \text{ \& since } V\uparrow \rightarrow \rho\downarrow \text{ then } \rightarrow \beta = \rho_o \left[\frac{\delta P}{\delta \rho}\right]_T \rightarrow \left[\frac{\delta P}{\delta \rho}\right]_T = \frac{\beta}{\rho_o}$$

$$\text{Hence,} \quad \left[\frac{\delta\rho}{\delta P}\right]_T = \frac{\rho_o}{\beta} \qquad 5.13$$

$$\alpha = \frac{1}{V_o}\left[\frac{\delta V}{\delta T}\right]_p \text{ \& since } V\uparrow \rightarrow \rho\downarrow \text{ then } \rightarrow \alpha = -\frac{1}{\rho_o}\left[\frac{\delta \rho}{\delta T}\right]_p$$

$$\text{Hence,} \quad \left[\frac{\delta\rho}{\delta T}\right]_p = -\alpha\,\rho_o \qquad 5.14$$

Substitute Equations 5.12 and 5.13 in 5.11 results in

$$\rho_{TP} = \rho_o \left[1 + \frac{(P - P_o)}{\beta_{TP}}\right] - \alpha_{TP}(T - T_o) \qquad 5.15$$

Then based on the knowledge of fluid bulk modulus as corrected based on temperature and pressure (β_{TP}) and the Cubical thermal expansion coefficient of the fluid as corrected base on temperature and pressure (α_{TP}), fluid density as corrected based on temperature and pressure (ρ_{TP})can be calculated.

5.3.4.4- Structure of Lumped Model for Case Study 3

Figure 5.11 shows a generic structure for a simulation model for case study 3. In order to consider the dependency of the thermal expansion coefficient on pressure and temperature, information shown in Fig. 5.12 is required. Proper modeling methods for this information should be applied similar to what has been discussed previously.

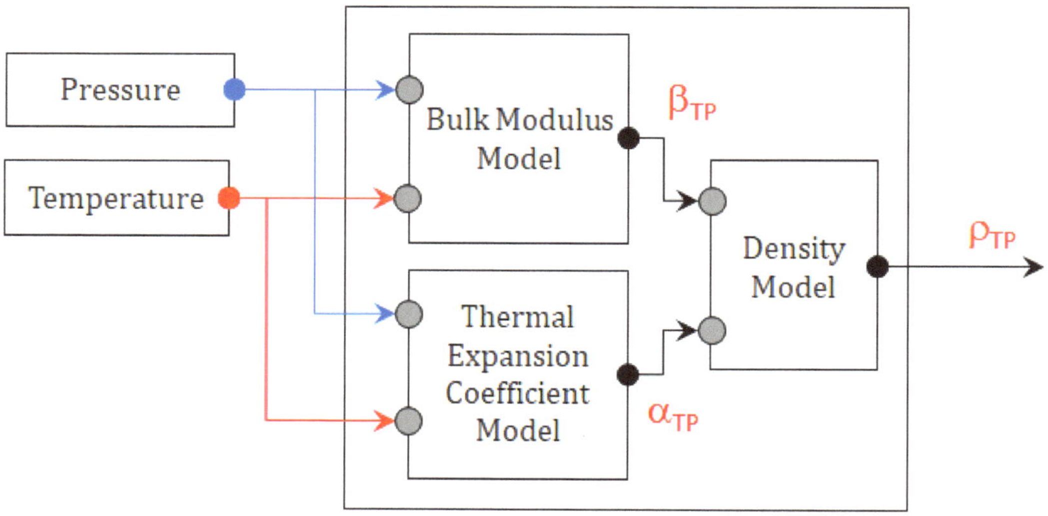

Fig. 5.11- Simulation Model for Case Study 3

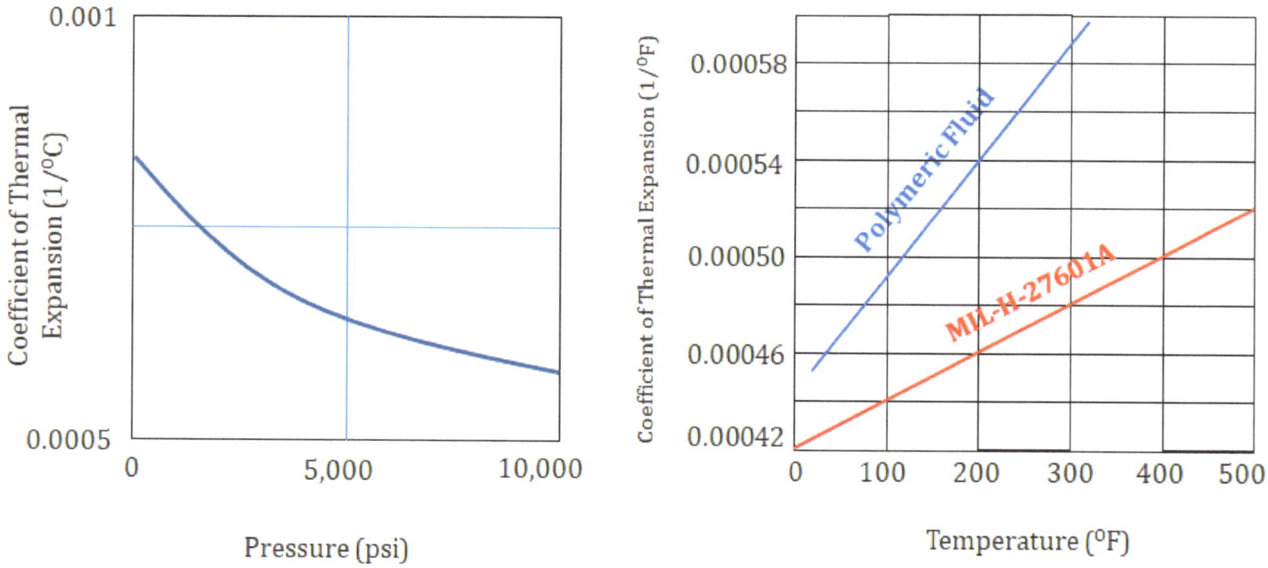

Fig. 5.12- Example of Effects of Temperature and Pressure on Thermal Expansion Coefficient

5.3.4.5- Case Study 4 based on Effect of Temperature and Pressure on Specific Gravity

In case studies 2 and 3, models are structured based on information available on density, based on which the specific gravity is calculated. As shown in Fig. 5.13, information may be available for specific gravity. In this case model will solve for the specific gravity and then density is calculated based on it. Similar modeling method as in case study 2 is applied here. Reference specific gravity of this fluid is selected to be 1 at reference temperature of 100°F, and reference pressure of 0 psi gauge pressure. By looking at the figure, the rate of change of the specific gravity with respect to. the pressure (at constant temperature) is linear and equal (≈ 0.004 per 1000 bar). Then Eq. 5.16 shows correcting fluid density based on pressure.

$$\mathbf{SG_P = SG_o + \left[\frac{\delta SG}{\delta P}\right]_T (P - P_o)} \qquad \mathbf{5.16}$$

- $\mathbf{SG_P}$ = Density corrected based on current working pressure **p**.
- $\mathbf{SG_o}$ = 1 = Reference density at reference values $\mathbf{T_o}$ [= 100 ^{o}F] and $\mathbf{p_o}$ [= 0 psig].
- $\mathbf{(\delta SG/\delta P)_T}$ = Rate of change of **SG** with respect to **P** at constant **T.**
- Given Characteristics → $\mathbf{(\delta SG/\delta P)_T}$ ≈ (0.004/1000) = 4x10^{-6} [(1/psi].

The rate of change of specific gravity w.r.t pressure at constant temperature is the slope of the line. The figure shows that the slope decreases consistently approximately 0.2% per every 100 ^{o}F as temperature increases. So, the resulting specific gravity based on pressure, must be corrected based on temperature.

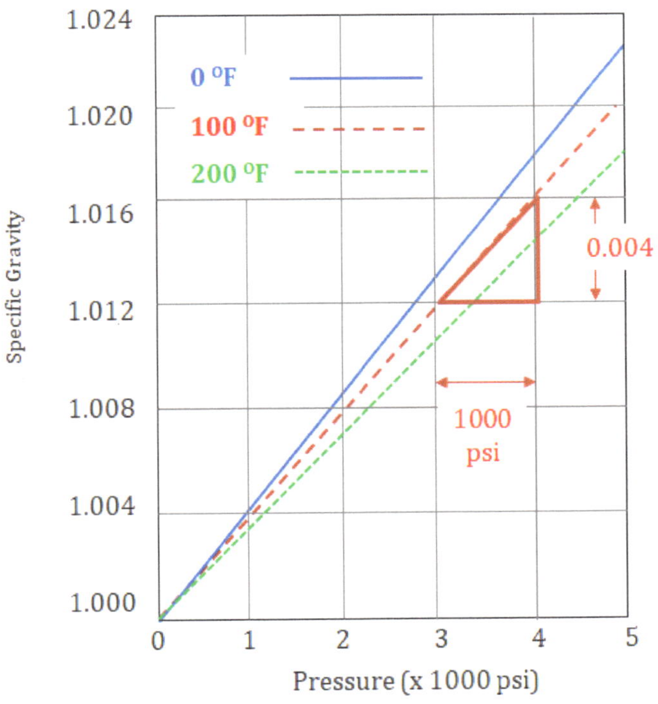

Fig. 5.13- Specific Gravity versus Pressure

5.3.4.6- Structure of Lumped Model for Case Study 4

Figure 5.14 shows a simulation model (HSV7CH05Model03) built based on Case Study 4 and validates the values given in this case study. The model is structured to calculate the specific gravity based on pressure first, then the result is multiplied by a correction factor based on temperature. Density is then calculated based on the resulting specific gravity multiplied by density of water (1000 kg/m^3).

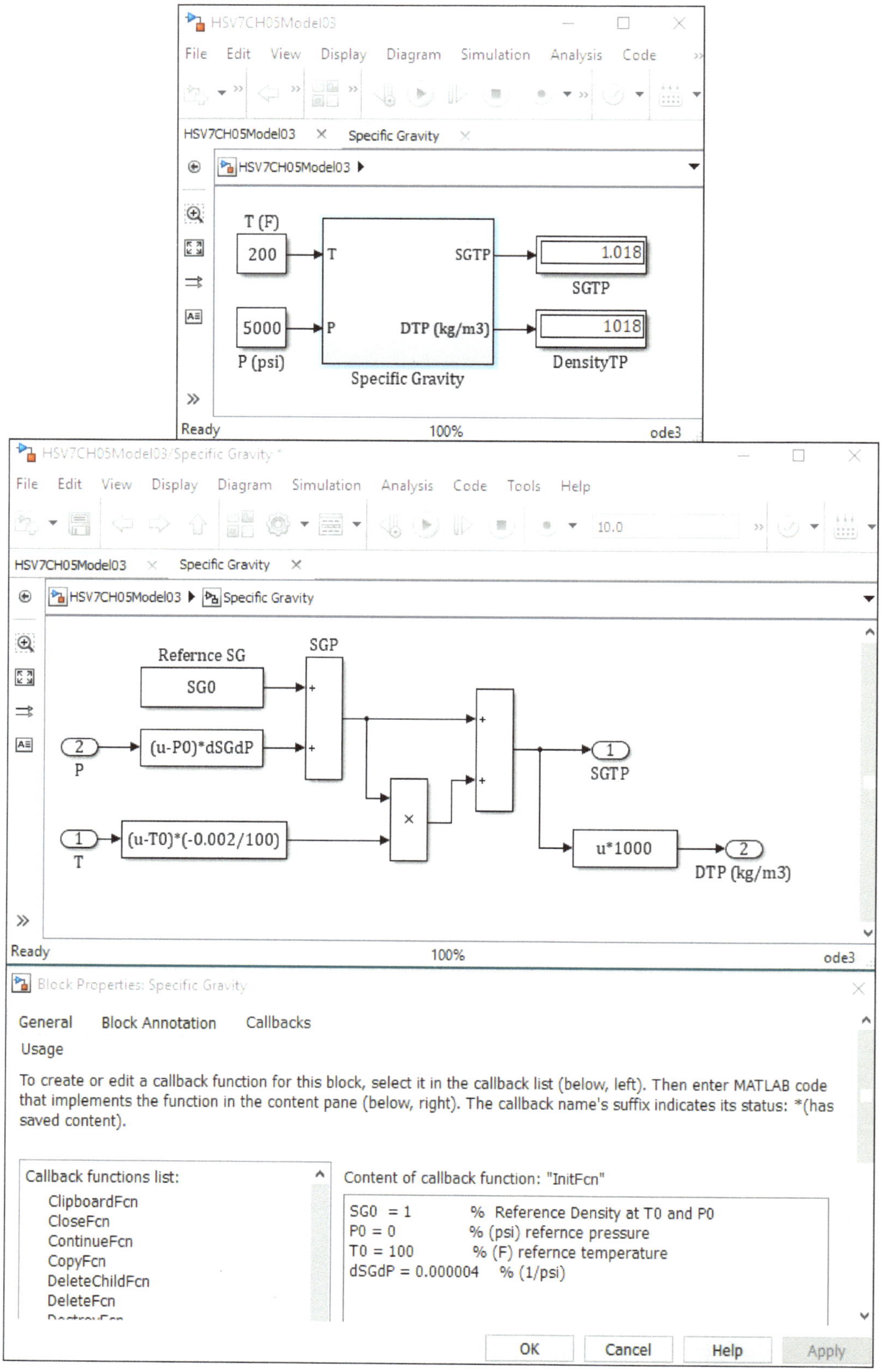

Fig. 5.14- Simulation Model for Case Study 4

5.4- Modeling of Hydraulic Fluid Viscosity

5.4.1- Definitions and Mathematical Expression of Viscosity

Viscosity is a property that reflects the resistance of the fluid to flow. The piston-cylinder physical model shown in Fig. 5.15 is used to drive the mathematical equation of fluid *Viscosity*. In this physical model, the piston is pushed into the cylinder with a force **F** and a speed **v**. The clearance between the piston and the cylinder is **y**. The contact area between the piston and the cylinder is **A**. The two-plate model, shown on the right side of the figure, shows that the fluid film between the moving surface and the fixed surface is laminar. While the boundary layer at the fixed surface is stationary, the boundary layer at the moving surface has the same velocity **v**. In between, the layers of the fluid move relative to each other with a velocity gradient (**v/y**).

As shown in Equation 5.17, **F** is directly proportional to **v** and **A**; and inversely proportional to **y**. The proportionality symbol (alpha) can be replaced by an equality operator (=) by multiplying the right-hand side of the equation by a constant. The constant **μ** (named "mu") is defined as the *Dynamic (Absolute) Viscosity*. Then, dynamic viscosity can be defined as the ratio between the *Shear Stress* (**F/A**), that causes the fluid layers to move relative to each other, and the rate of fluid distortion or *Shear Rate* (**v/y**) in the clearance between the moving surfaces. Equation 5.18 defines the *Kinematic (Relative) Viscosity* **ν** (named "new") as the ratio of the dynamic viscosity **μ** to the fluid density **ρ**. Kinematic (Relative) Viscosity is more commonly used than the dynamic viscosity.

$$\mathbf{F} \propto \frac{\mathbf{vA}}{\mathbf{y}} \rightarrow \mathbf{F} = \mu\frac{\mathbf{vA}}{\mathbf{y}} \rightarrow \textbf{Dynamic (Absolute) Viscosity } \mu = \frac{\mathbf{F/A}}{\mathbf{v/y}} \qquad 5.17$$

$$\textbf{Kinematic (Relative) Viscosity } \nu = \frac{\mu}{\rho} \qquad 5.18$$

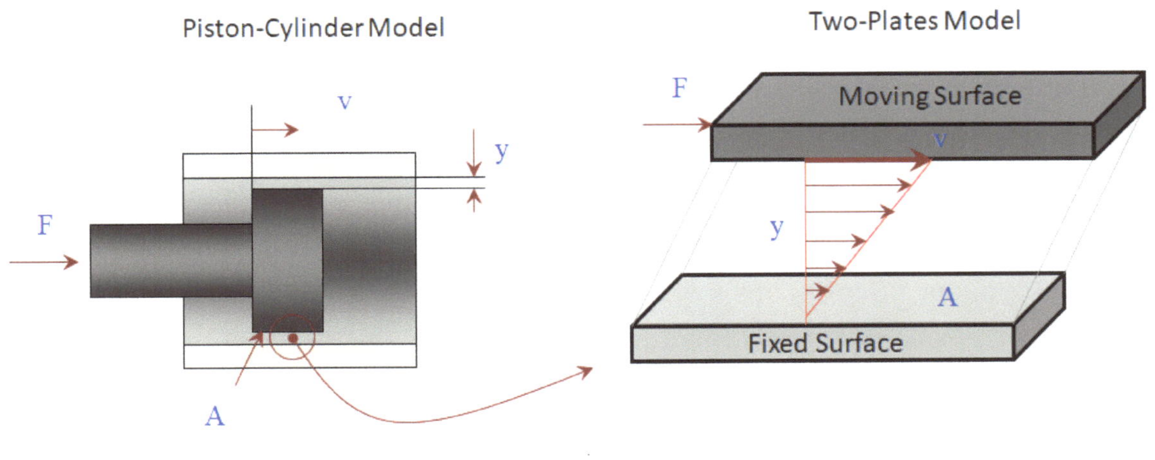

Fig. 5.15- Developing Mathematical Expression for Fluid Viscosity

Units of Fluid Viscosity are as follows:

Metric Units of Dynamic (Absolute) Viscosity $\mu = \dfrac{\text{Shear Stress}}{\text{Shear Rate}}$ is $\left[\dfrac{\text{N.s}}{\text{m}^2}\right]$.

Industrial Unit of Dynamic Viscosity is Centipoise, $1\text{cP} = 10^{-3}\left[\dfrac{\text{N.s}}{\text{m}^2}\right]$.

Metric Units of Kinematic (Relative) Viscosity $\nu = \dfrac{\mu}{\rho} =$ is $\left[\dfrac{\text{m}^2}{\text{s}}\right]$.

Industrial Unit of Kinematic Viscosity is Centistoke, $1\text{cSt} = 10^{-6}\left[\dfrac{\text{m}^2}{\text{s}}\right] = \left[\dfrac{\text{mm}^2}{\text{s}}\right]$.

For balancing the units, Eq. 5.18 →

Kinematic (Relative) Viscosity $\nu\left[\dfrac{\text{m}^2}{\text{s}}\right] = \dfrac{\mu\left[\dfrac{\text{N.s}}{\text{m}^2}\right]}{\rho\left(\dfrac{\text{kg}}{\text{m}^3}\right)}$

$$\rightarrow \mathbf{10^{-6}\nu\,(cSt)} = \frac{\mathbf{\mu\,(cP)x\,10^{-3}}}{\boldsymbol{\rho}\left(\mathbf{\frac{kg}{m^3}}\right)} \longrightarrow \mathbf{\nu\,(cSt)} = \frac{\mathbf{1000\,x\,\mu\,(cP)}}{\boldsymbol{\rho}\left(\mathbf{\frac{kg}{m^3}}\right)} = \frac{\mathbf{1000\,x\,\mu\,(cP)}}{\mathbf{SG} \times \boldsymbol{\rho}_{\mathbf{W}}\left(\mathbf{\frac{kg}{m^3}}\right)} \qquad 5.19$$

Knowing that the water density is equal to 1000 kg/m3, equation 5.20 can be used to solve for the dynamic viscosity as follows:

$$\mu\,(\text{cP}) = \frac{\left[\mathbf{\nu\,(cSt)} \times \mathbf{SG} \times \boldsymbol{\rho}_{\mathbf{W}}\left(\mathbf{\frac{kg}{m^3}}\right)\right]}{\mathbf{1000}} = \mathbf{\nu\,(cSt)} \times \mathbf{SG} \times \boldsymbol{\rho}_{\mathbf{W}}\left(\frac{\mathbf{g}}{\boldsymbol{cc}}\right) \qquad 5.20$$

Where SG is the fluid specific gravity.

Water has dynamic viscosity of 1 cP at 25 ^{0}C. and kinematic viscosity of approximately 1 cSt.

5.4.2- Case Study 5 for Modeling of Viscosity Based on Effect of Temperature and Pressure

Like every other hydraulic fluid property, viscosity decreases with increasing working temperature and increases with increasing working pressure. In order to model such effects, information should be known for a specific fluid so that a modeler can select the adequate method for modeling static characteristics as presented in Chapter 4.

As shown in Fig. 5.16, for ISO VG 100 hydraulic fluid, reference temperature is 40 ^{0}C and reference viscosity is 100 cSt. The figure reveals that viscosity-temperature relationship is linear. This is a deceiving impression and a modeler should be careful. Since the viscosity is presented on a logarithmic scale, this isn't a linear characteristic and needs a lookup table to be modeled.

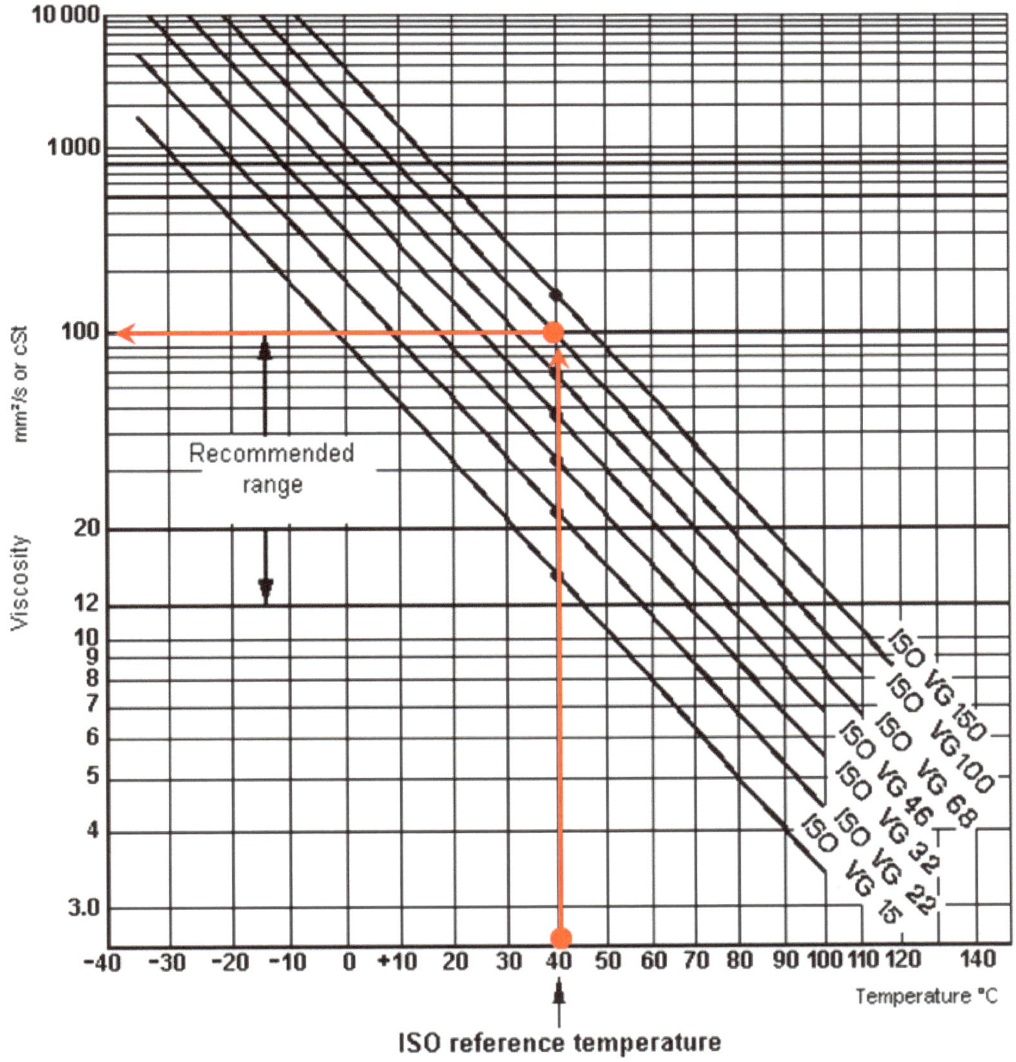

Fig. 5.16- Effect of Temperature on Viscosity on Mineral Oil.

Then, the model must correct the viscosity based on the rate of change of viscosity versus pressure at constant temperature. Assuming viscosity increases approximately 1% per every 10 bar, then the corrected viscosity based on temperature must be multiplied by a correction factor. Figure 5.17 shows a simulation model (HSV7CH05Model04) built based on Case Study 5 and validates the values given in this case study. In this model, all constants are placed inside the model rather than in the model properties as alternative method. It is to be noted that the boundary of this model is temperature range of 0-70 ^{0}C. This means that this model is valid or more accurate within this range of working temperature. Out of this range, the lookup table will extrapolate linearly and may result in inaccurate results.

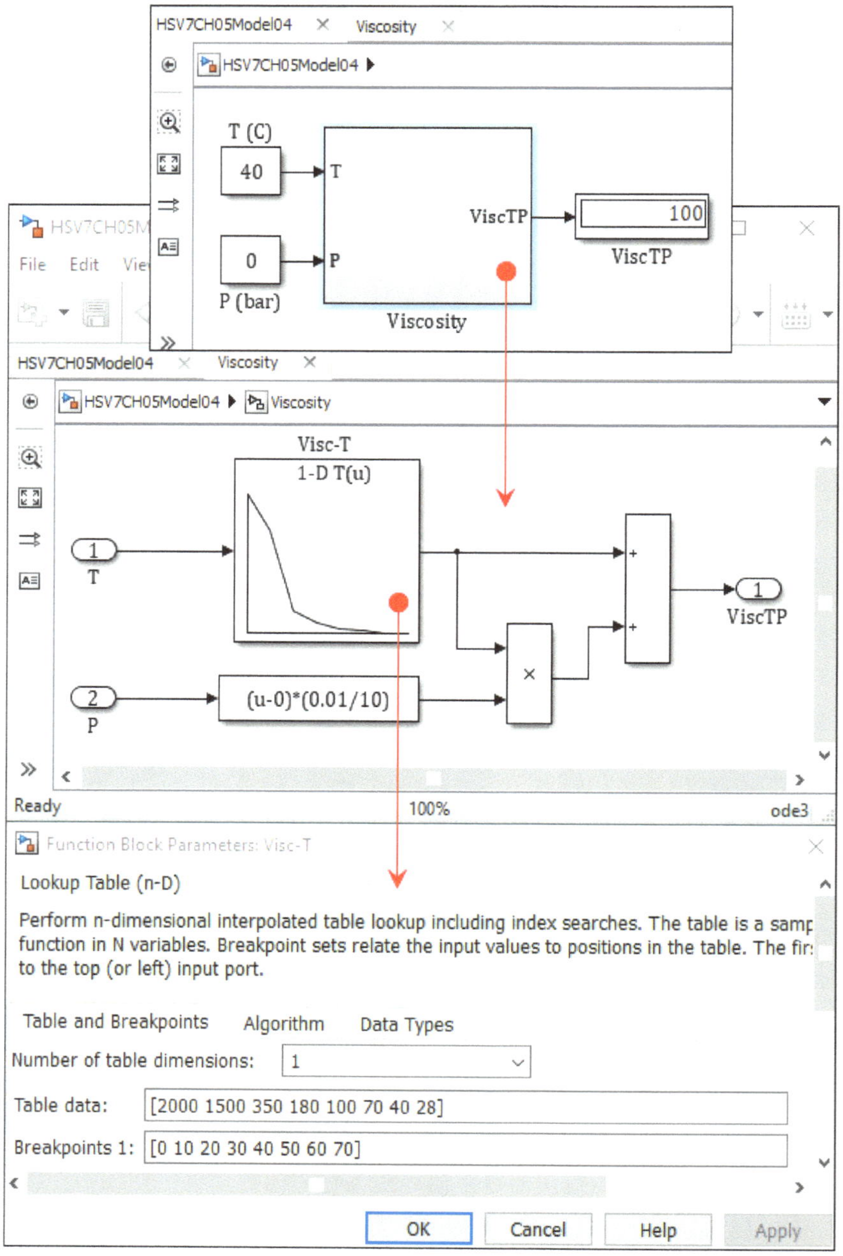

Fig. 5.17- Simulation Model for Case Study 5

5.5- Lumped Model for Hydraulic Fluid Properties

Figure 5.18 shows simulation model (HSV7CH05Model05) that combines the main fluid properties. The model contains the ability to choose the units of temperature and pressure. Outputs of the model are assigned to global variables so that it can be accessed from anywhere in the model. Ramping temperature and pressure are used to check the model. Figure 5.19 shows the result.

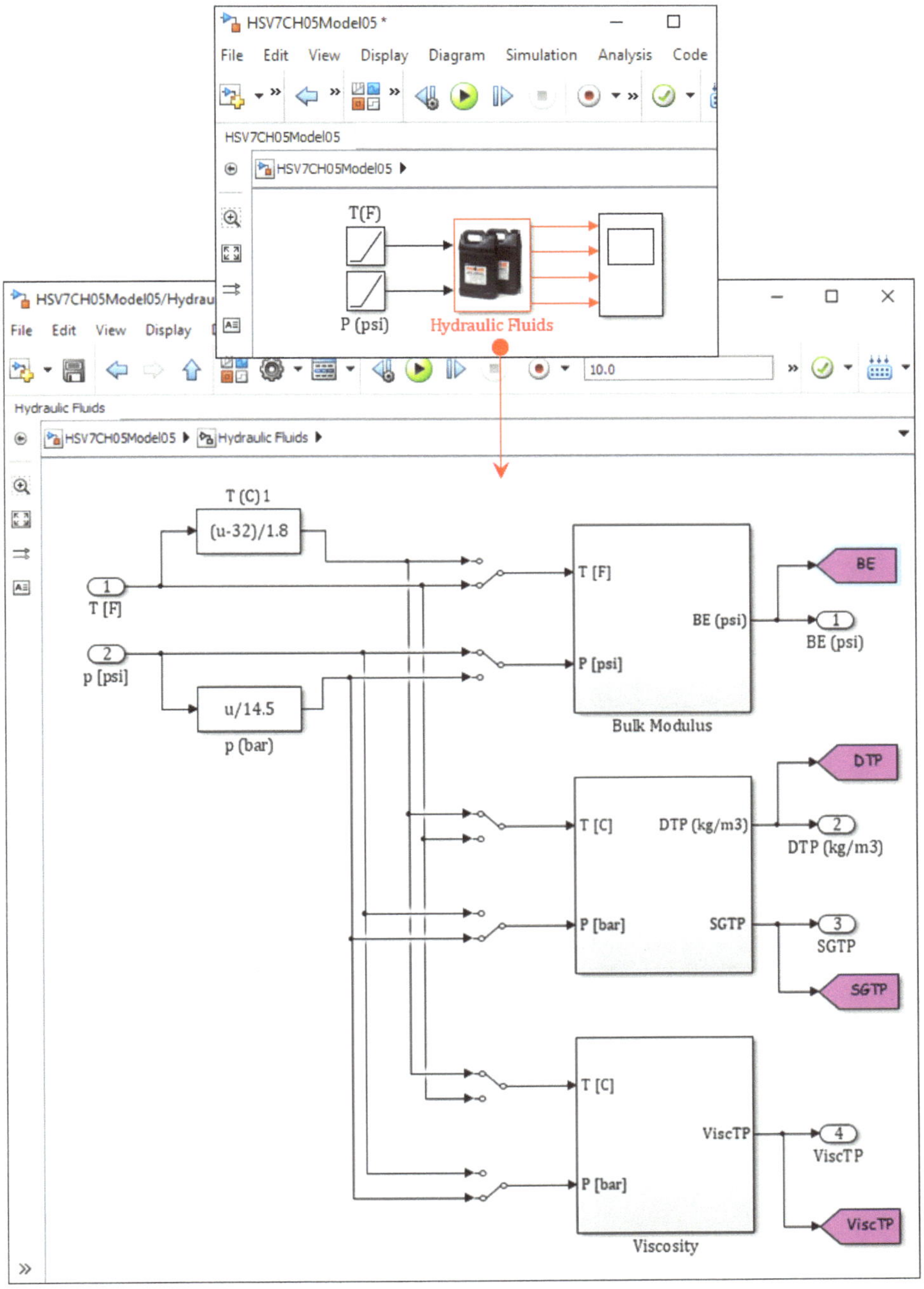

Fig. 5.18- Lumped Model for Hydraulic Fluid Properties

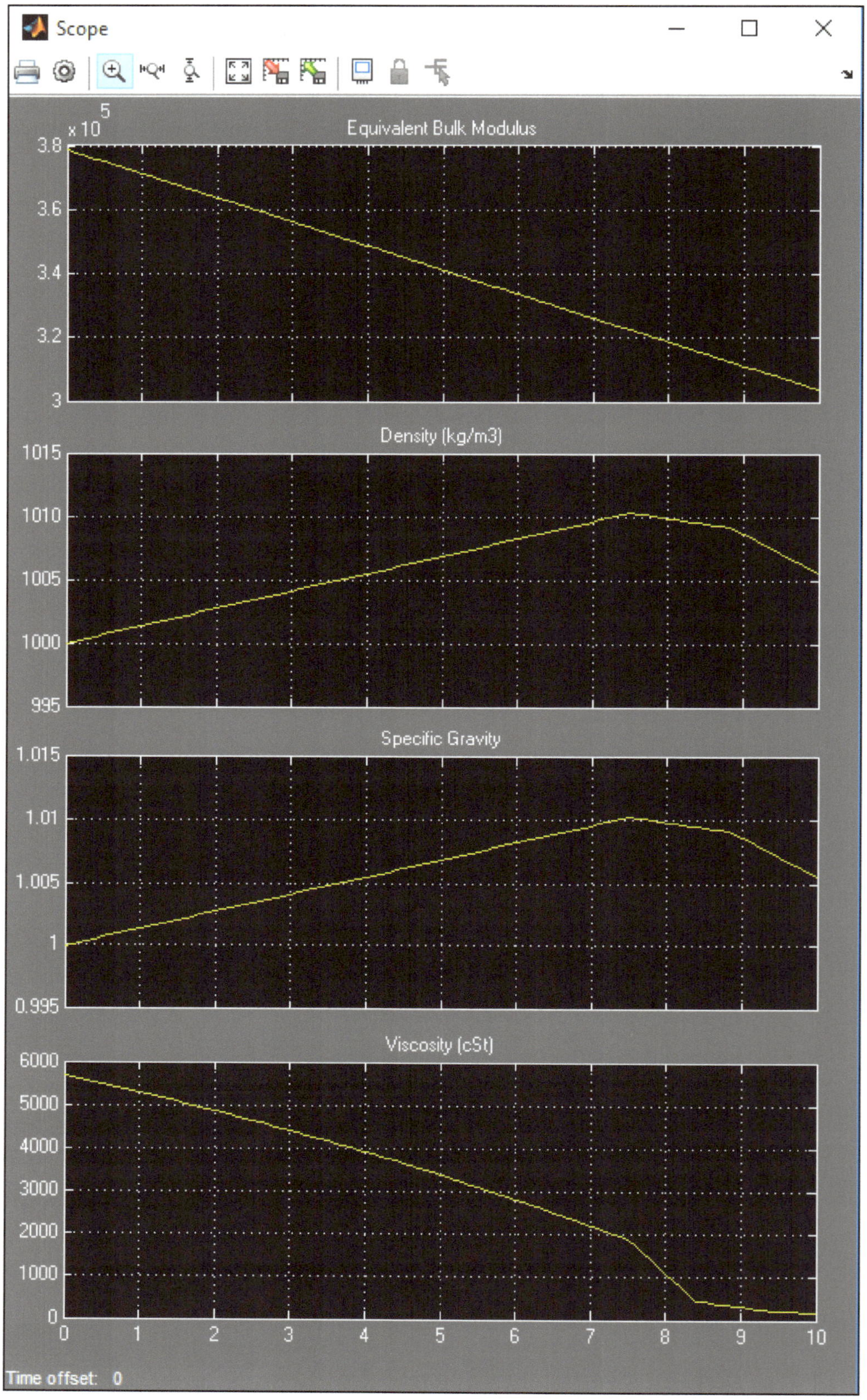

Fig. 5.19- Results of Simulating Hydraulic Fluid Properties

Chapter 6
Modeling of Hydraulic Transmission Lines

Objectives

This chapter presents concepts of modeling transmission lines, fittings and orifices. Model for a transmission line considers compressible fluid so that effect of line capacitance can be investigated. Developed models were validated based on other software.

Brief Contents

Chapter 6: Modeling of Hydraulic Transmission Lines

6.1- Modeling of Seamless Hydraulic Transmission Lines

As it has been presented in Chapter 1, seamless hydraulic transmission line shown in Fig. 6.1 has resistive effect, inductive effect and capacitive effect. Therefore, modeling of transmission line can be built on concept of finite elements as follows.

1. **Lumped Resistance**: In this portion of the line, pressure drops due to fluid-fluid and fluid-wall friction, but flow will be the same as Qin.
2. **Lumped Inductance**: In this portion of the line, pressure drops due to fluid inertia, but flow will be the same as Qin.
3. **Lumped Capacitance:** In this portion of the line, outlet flow changes based on the fluid compressibility and the rate of change of the differential pressure across the line.

Figure 6.2 shows the stream of calculations in a generic transmission line model structure. The model uses the inlet flow to calculate the line pressure drop due to line resistance and inductance. This pressure drop is added to the outlet pressure to get the inlet pressure. Outlet pressure and inlet flow are used with the line capacitance to calculate the outlet flow.

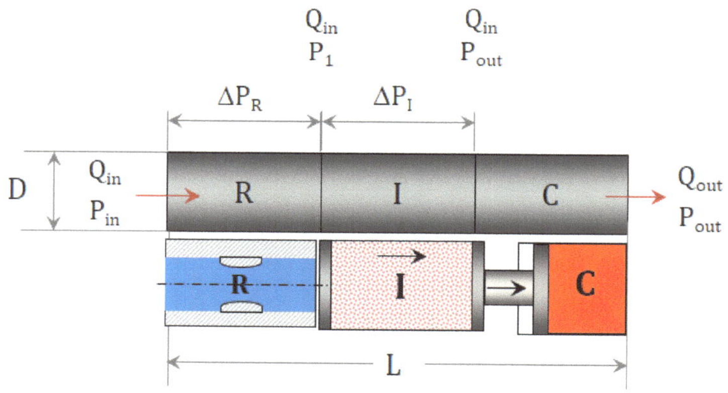

Fig. 6.1 – Modeling Pressure Drop in Hydraulic Transmission Lines

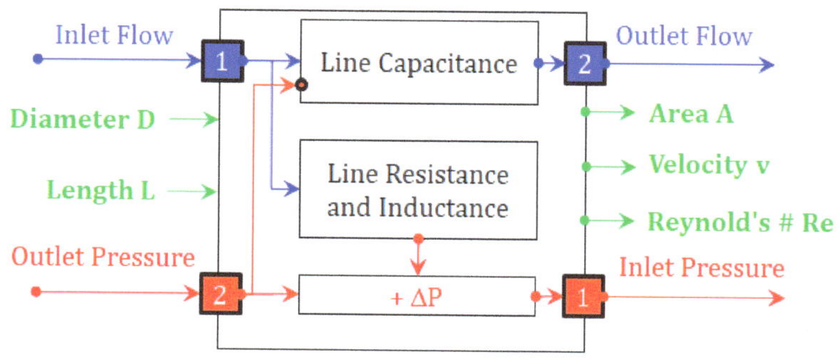

Fig. 6.2- Structure of a Generic Model for a Hydraulic Transmission Line

The challenge to build a Simulink model is to make sure units are balanced. Therefore, the following set of variables with the assigned units are used to build the model.

- **ρ (kg/m³)** = Fluid Density. Fluid properties must be interactively corrected based on working temperature and pressure. Therefore, it will be delivered to this model through Global Variables from hydraulic fluid model.
- **v (m/s)** = Fluid Velocity.
- **D (mm)** = Line Diameter.
- **A [cm²) =** Line Area.
- **L (m)** = Line Length.
- **R_e (-) = Reynold's Number**
- **λ (-) =** Coefficient of Fluid Friction.
- **ν (cSt = mm²/s)** = Kinematic Viscosity.
- **Q_{in} (lit/min) =** Inlet Flow.
- **Q_{out} (lit/min) =** Outlet Flow.
- **P_{in} (bar) =** Inlet Pressure.
- **P_{out}(bar) =** Outlet Pressure.

6.1.1- Modeling Pressure Losses due to Line Resistivity

The following set of equations shows the sequence based on which the model is structured.

$$A\,[cm^2] = [\pi \times D^2(mm)]\,/(4 \times 100) \qquad 6.1$$

$$v\,[m/s] = \frac{Q_{in}(lit/min) \times \frac{1}{60 \times 1000}}{A\,(cm^2) \times 10^{-4}} = \frac{Q_{in}(lit/min)}{A\,(cm^2) \times 6} \qquad 6.2$$

$$R_e\,[-] = \frac{v\,[m/s] \times 1000 \times D\,[mm]}{\nu\,(Cst = mm^2/s)} \qquad 6.3$$

The *Coefficient of Fluid Friction* **λ** depends on the flow pattern (Laminar or Turbulent) inside the transmission line that is determined based on the *Reynolds's Number* as follows:

For Laminar and Transitional Flow (Reynolds's Number <= 3500):

$$\lambda = \frac{64}{R_e} \qquad 6.4$$

For Turbulent Flow (Reynolds's Number > 3500):

Equation 6.5 calculates the coefficient of friction based on an assumption that the inside surface of the hydraulic line is smooth. This assumption is feasible unless otherwise stated.

$$\lambda = \frac{0.3164}{R_e^{0.25}} \qquad 6.5$$

An alternative way to determine the coefficient of fluid friction **λ** is to use the *Moody Diagram* shown in Fig. 6.3. The first step is to locate the Reynolds's Number on the logarithmic x-axis then draw a vertical line.

If the flow is laminar, the vertical line intersects with the solid line, and then go left to find the coefficient of friction. For example, if $\mathbf{R_e}$ = 1,000 then coefficient of friction = 0.065.

If the flow is in transition, the vertical line intersects with the dashed line, and then go left to find the coefficient of friction. For example, if $\mathbf{R_e}$ = 3,000 then coefficient of friction = 0.0215.

If the flow is turbulent, the vertical line will intersect with one of the Relative Roughness curves. Relative roughness is defined as the ratio between the inside surface roughness **ε** of the fluid line and its nominal diameter **D**. The lowest relative roughness curve represents the smoothest fluid conductor. For example, $\mathbf{R_e}$ = 100,000 then coefficient of friction for a smooth pipe = 0.0185 approximately. This curve of lowest relative roughness is used to develop Eq. 6.5. Moody diagram can be converted into a lookup table to avoid using Eqs. 6.4 and 6.5 in the model.

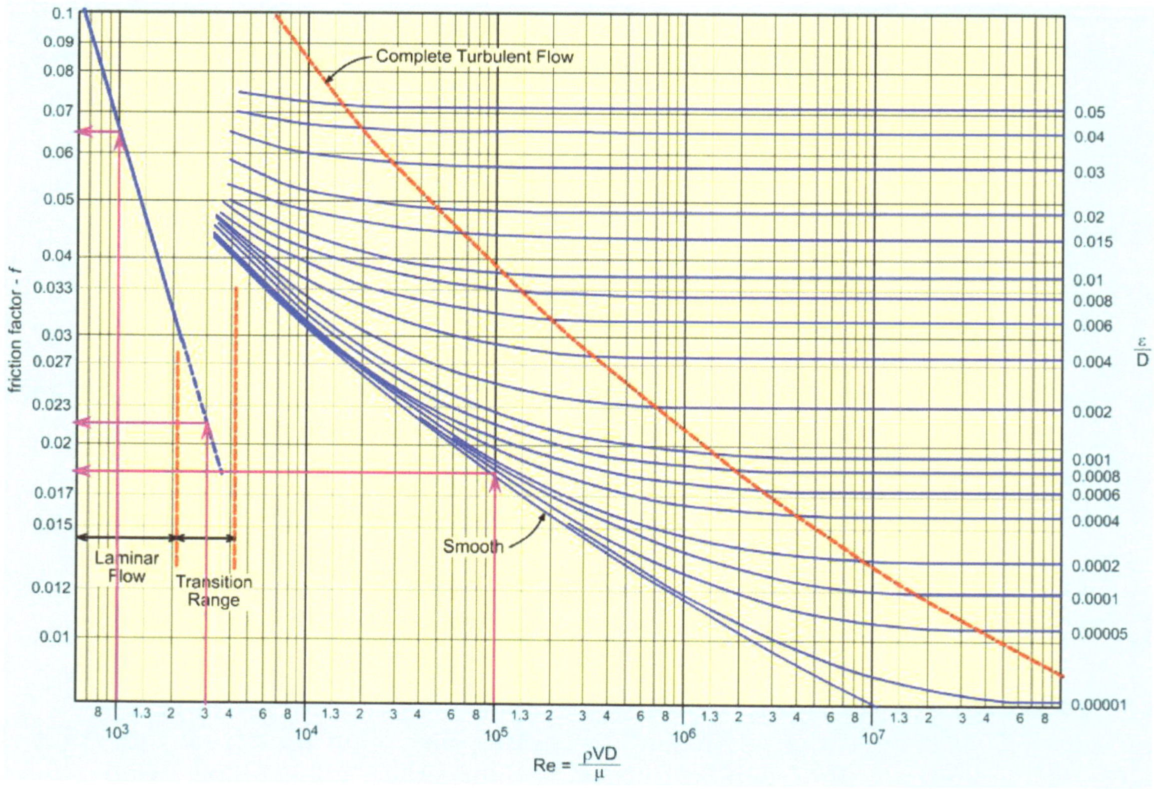

Fig. 6.3- Moody Diagram

Finally, Eq. 6.6 calculates the pressure drop across a hydraulic transmission line due to its resistivity. This pressure drop is conventionally known as "*Frictional Losses*".

$$\Delta P_R = \lambda \frac{L}{D} \frac{\rho v^2}{2} = \lambda \frac{L}{D} \frac{SG \times \rho_W \times v^2}{2} \quad 6.6$$

Equation 6.7 presents Eq. 6.6 with the units being balanced as follows:

$$\Delta P_R(\text{Pascal}) = \frac{\lambda \times L(m)}{D(mm) \times 10^{-3}} \frac{SG \times 1000\ (kg/m^3) \times v^2 (m/s)^2}{2}$$

$$\Longrightarrow \Delta P_R(\text{Pascal}) = \frac{10^6 \times \lambda \times SG \times L \times v^2}{2 \times D} \left[\frac{kg.m}{s^2} \frac{1}{m^2} = \frac{N}{m^2}\right]$$

$$\Longrightarrow \Delta P_R(\text{bar}) = \frac{\Delta p_R(\text{Pascal})}{10^5} = \frac{5 \times \lambda \times SG \times L(m) \times [v(m/s)]^2}{D(mm)} \quad 6.7$$

Figure 6.4 shows the structure of a Simulink model that solves Eq. 6.1 through 6.6. For better understanding of the structure, colors on the figure are made to match the colors on the text.

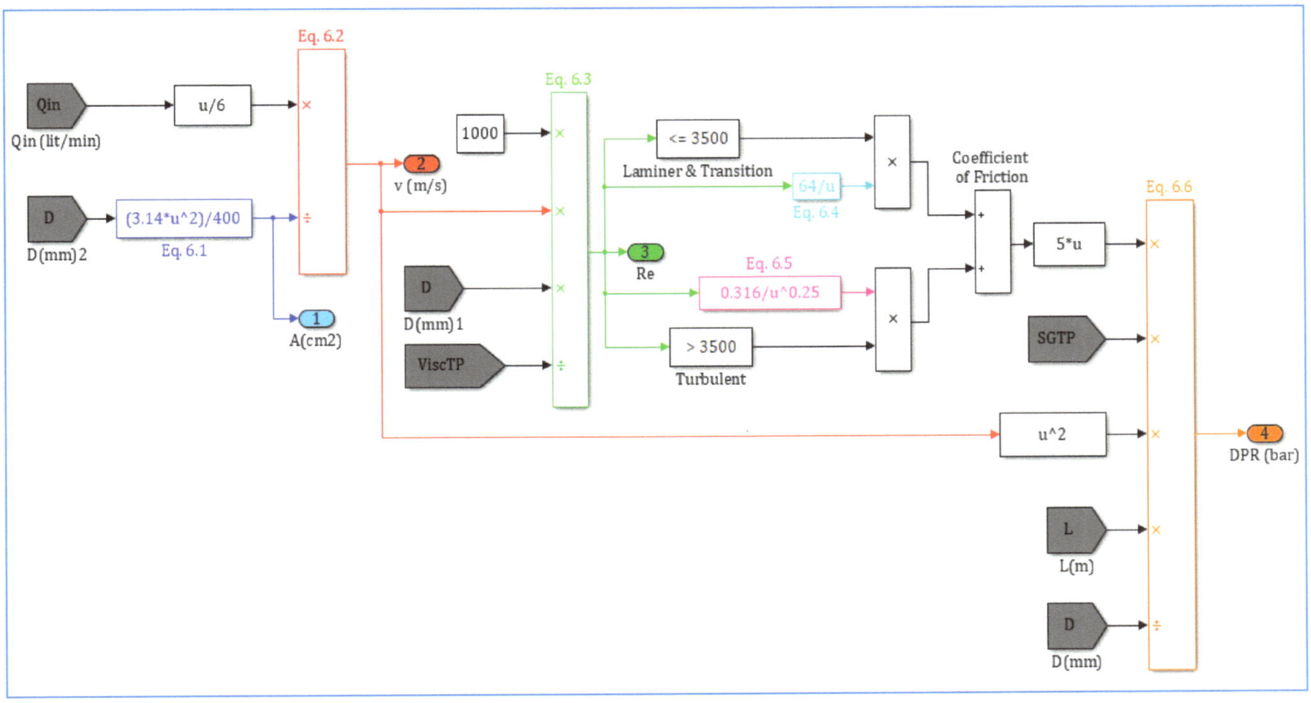

Fig. 6.4- Modeling Pressure Losses due to Line Resistivity

6.1.2- Modeling Pressure Losses due to Line Inductance

Referring to Eq. 1.6D and Eq. 1.7 in Chapter 1, Eq. 6.8A is used to calculate he pressure drop across a hydraulic transmission line due to its inductance. The equation is developed based on the concept that a pressure drop is required to accelerate a mass of fluid in motion inside a transmission line of known length and diameter.

$$\Delta P_I = \frac{\rho L}{A} \times \frac{dQ_{in}}{dt} \tag{6.8A}$$

Equation 6.8B presents Eq. 6.8A with the units being balanced as follows:

$$\Delta P_I\ (\text{Pascal}) = \frac{\rho\left[\frac{kg}{m^3}\right] L[m]}{A[cm^2] \times 10^{-4}} \times \frac{dQ_{in}\left[\frac{lit}{min}\right] \times \frac{1}{60 \times 1000}}{dt[s]}$$

$$\Rightarrow \Delta P_I\ (\text{Pascal}) = \frac{\rho L}{A \times 10^{-4}} \times \frac{dQ_{in} \times \frac{1}{60 \times 1000}}{dt} \left[\frac{kg \times m \times m^3}{m^3 \times m^2 \times s^2}\right]$$

$$\Rightarrow \Delta P_I\ (\text{Pascal}) = \frac{\rho L}{A \times 6} \times \frac{dQ_{in}}{dt} \left[\frac{kg.m}{s^2}\frac{1}{m^2} = \frac{N}{m^2}\right]$$

$$\Rightarrow \Delta P_I\ (\text{bar}) = \frac{\rho\left[\frac{kg}{m^3}\right] L[m]}{A[cm^2] \times 6 \times 10^5} \times \frac{dQ_{in}(lit/min)}{dt} \tag{6.8B}$$

Figure 6.5 shows the structure of a Simulink model that solves Eq. 6.8.

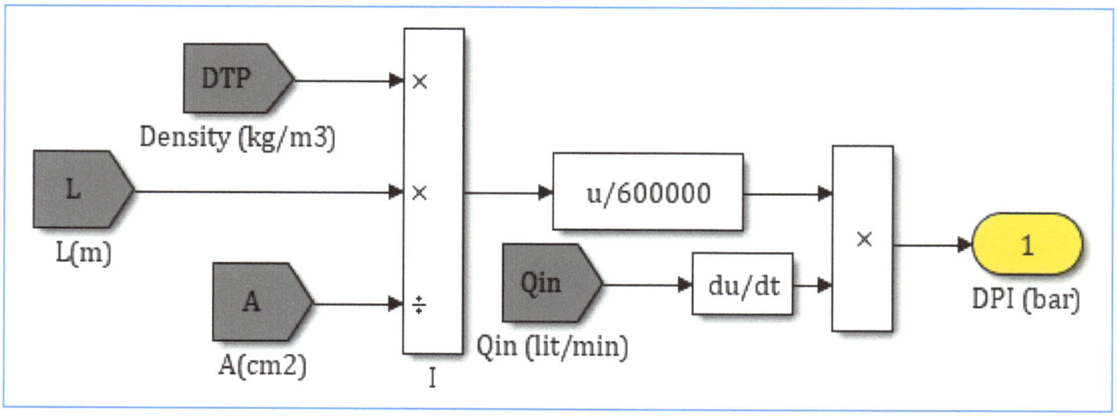

Fig. 6.5- Modeling Pressure Losses due to Line Inductance

6.1.3- Modeling Inlet Pressure of Transmission Line

$$\mathbf{P_{in}\,(bar) = P_{out}\,(bar) + \Delta P_R\,(bar) + \Delta P_I\,(bar)} \qquad 6.9$$

6.1.4- Modeling Outlet Flow due to Line Capacitance

Referring to Eq. 1.5D in Chapter 1,

$$\Delta P_{out} = \frac{\text{Equivalent Bulk Modulus } (\beta_E)}{\text{Container Volume } (V)} \int Q\,d \Rightarrow \frac{dP_{out}}{dt} = \frac{\beta_E}{V}[Q_{in} - Q_{out}]$$

$$\Rightarrow Q_{out} = Q_{in} - \frac{V}{\beta_E}\frac{dP_{out}}{dt} = Q_{in} - \frac{A \times L}{\beta_E}\frac{dP_{out}}{dt} \qquad 6.10A$$

Equation 6.10B presents Eq. 6.10A with the units being balanced as follows:

$$Q_{out}\left[\frac{lit}{min}\right] = Q_{in}\left[\frac{lit}{min}\right] - \frac{A\,[cm^2] \times \frac{1}{10000} \times L[m]}{\beta_E\,[bar]}\,\frac{dP_{out}[bar]}{dt\,[s] \times \frac{1}{60}} \times 1000$$

$$Q_{out}\left[\frac{lit}{min}\right] = Q_{in}\left[\frac{lit}{min}\right] - \frac{6 \times A\,[cm^2] \times L[m]}{\beta_E\,[bar]}\,\frac{dP_{out}[bar]}{dt\,[s]} \qquad 6.10B$$

Figure 6.6 shows the structure of a Simulink model that solves Eq. 6.10.

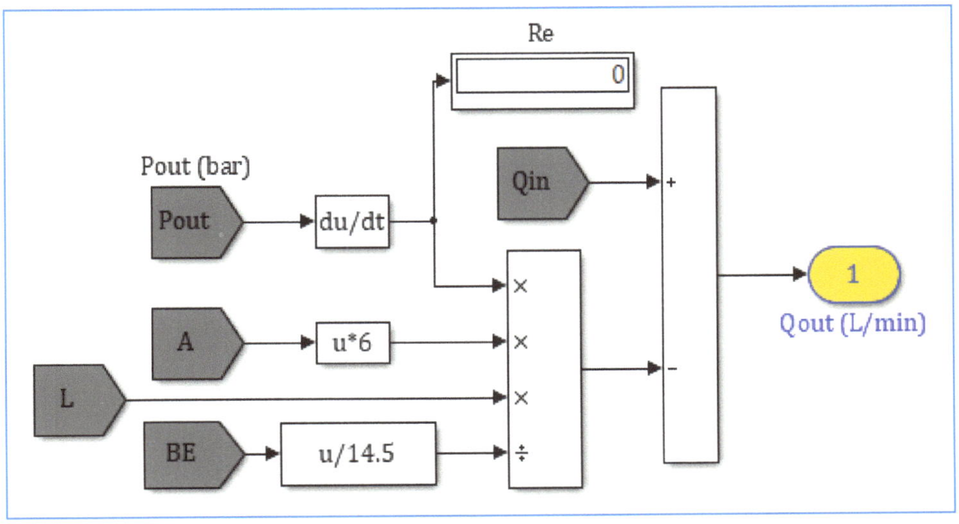

Fig. 6.6- Modeling Outlet Flow due to Line Capacitance

6.1.4- Case Studies for Modeling a Hydraulic Transmission Line

Figure 6.7 shows simulation model (HSV7CH06Model01) for a hydraulic transmission line. The following features were considered in the model to improve model capabilities:

- Model of hydraulic fluid is added to get an instantaneous update on hydraulic fluid properties used in the transmission line model.
- Most of the variables are assigned to global variables so that it can be accessed from anywhere in the model.
- Two Simulink blocks are extensively used ("GOTO" and "From") to reduce the crossing linking lines.
- In order to help a modeler to use the model to investigate various scenarios, inlet flow and outlet pressure can be selected as a constant or ramping value.

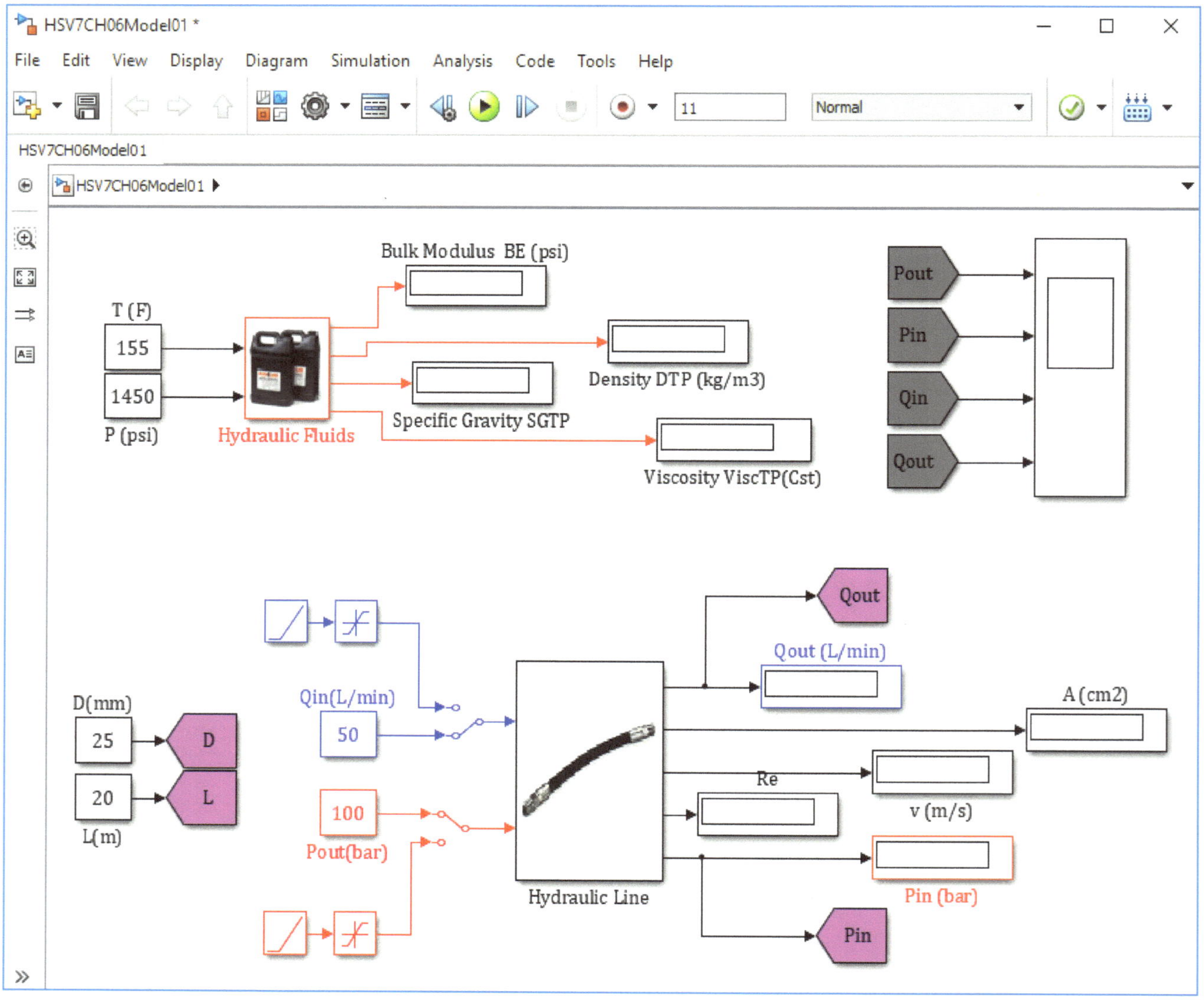

Fig. 6.7- Simulation Model for Hydraulic Transmission Line

Figure 6.8 shows the details of the model under the mask. The figure shows combining the three previous sub-models.

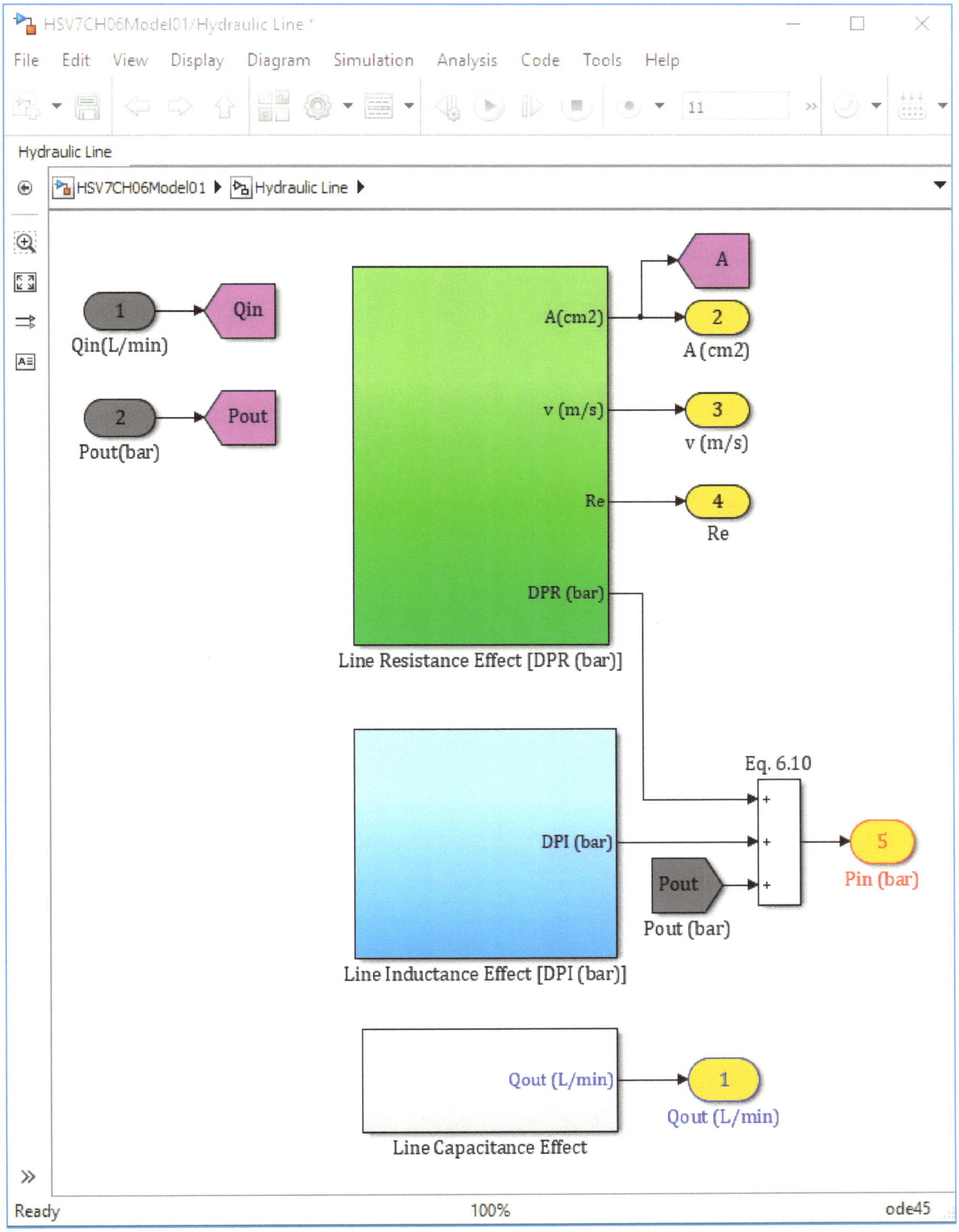

Fig. 6.8- Model Structure under the Mask of Transmission Line Model

6.1.4.1- Case Study 1: Steady State Condition

The upper part of Fig. 6.9 shows the results of simulating a hydraulic transmission line based on constant values of inlet flow and outlet pressure. The lower part is developed by running other software "Hydraulic Component Sizing Calculator - HCSC". The results from HCSC validates the results from the simulation model.

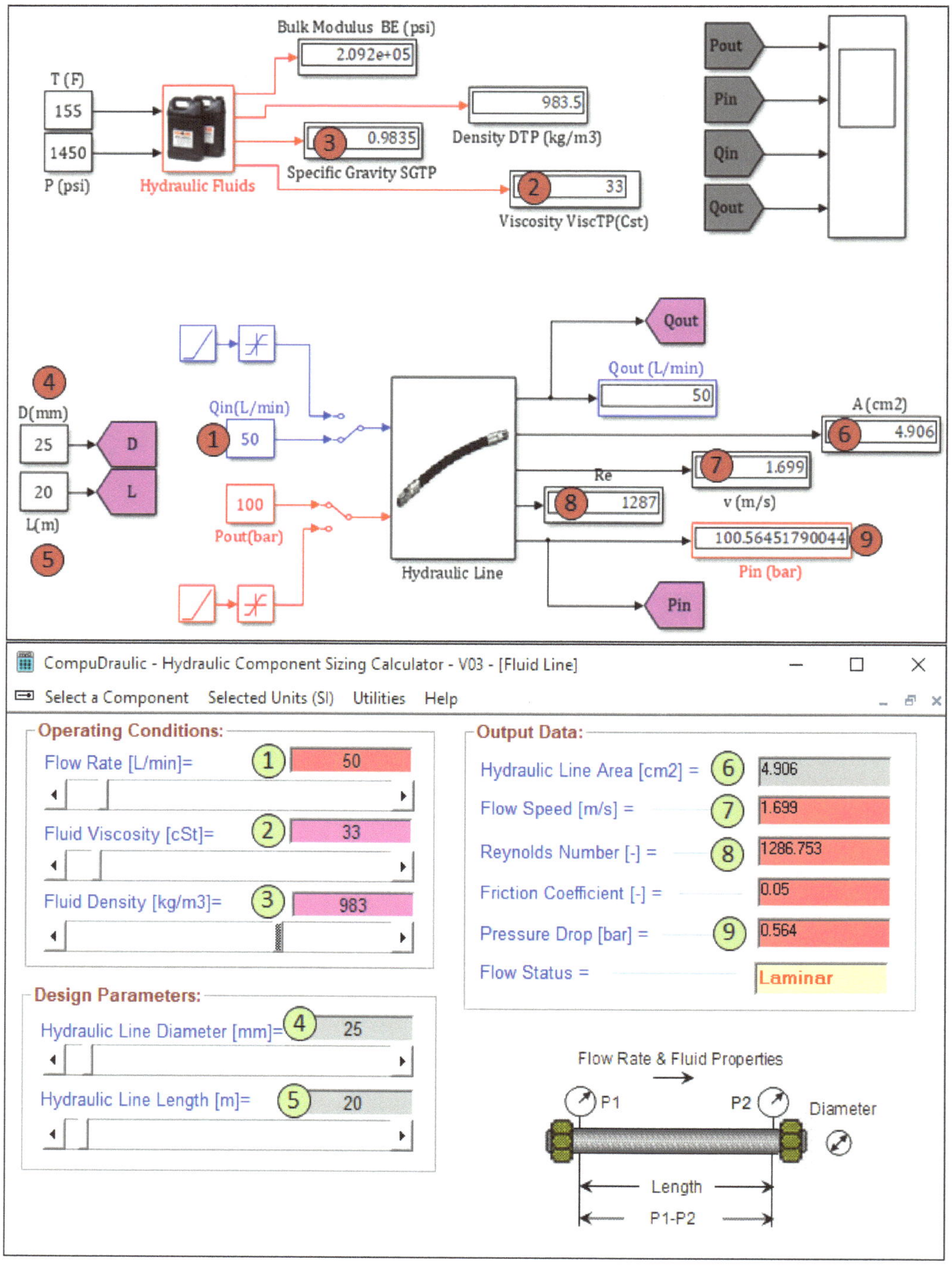

Fig. 6.9- Results of Case Study 1

6.1.4.2- Case Study 2: Surge Inlet Flow

During a machine operation, a hydraulic line may subject to a surge inlet flow. Figure 6.10 shows simulating this case by sudden increase of the inlet flow. The figure shows that, due to surge inlet flow, inlet pressure spikes up before returning to steady state condition.

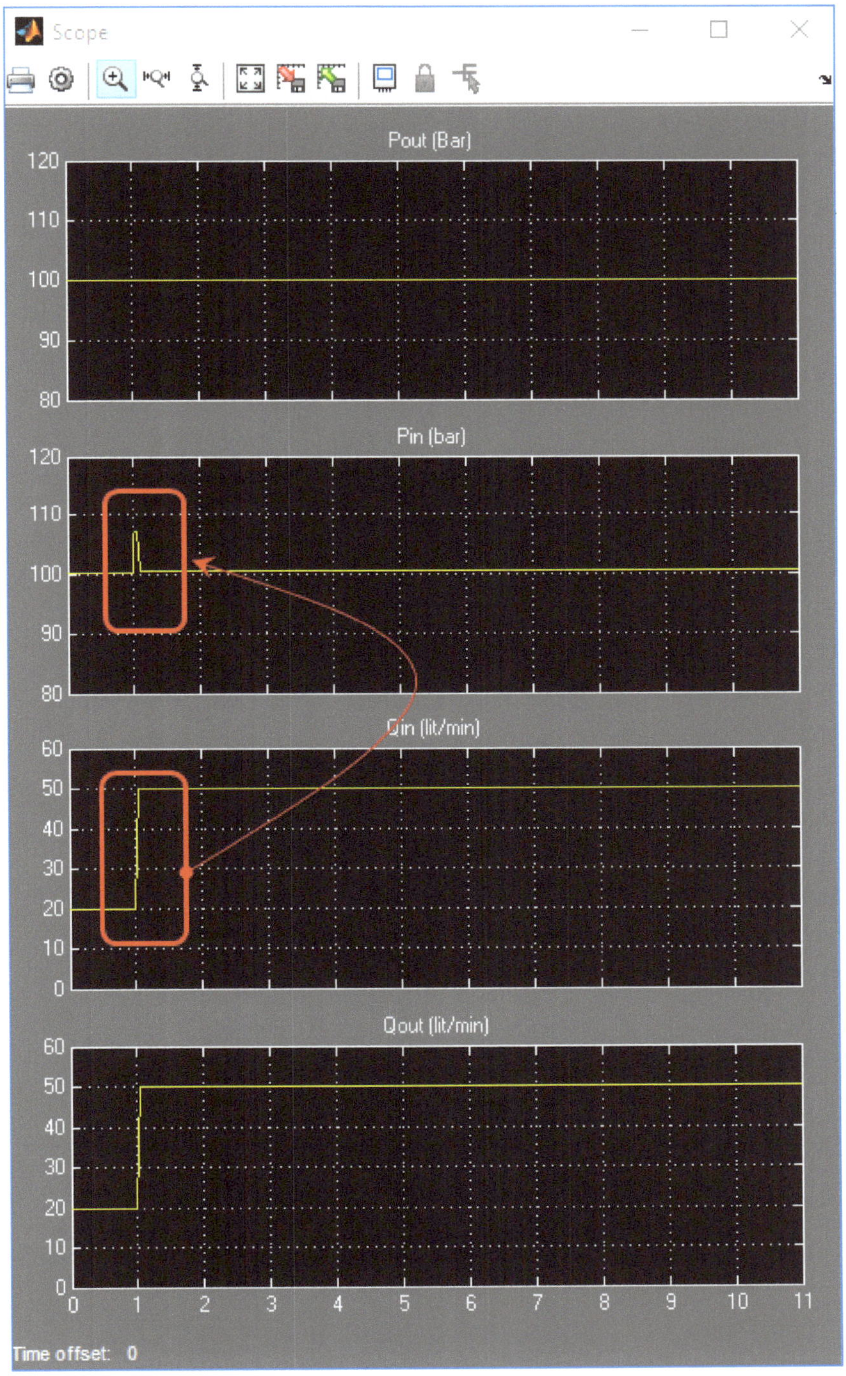

Fig. 6.10- Results of Case Study 2

6.1.4.3- Case Study 3: Surge Outlet Pressure

During a machine operation, a hydraulic line may subject to a surge outlet pressure, let's say during overrunning load or sudden closure of a valve. Figure 6.11 shows simulating this case by sudden increase in the outlet pressure. The figure shows that, due to surge outlet pressure, outlet flow surges backward before returning to steady state condition.

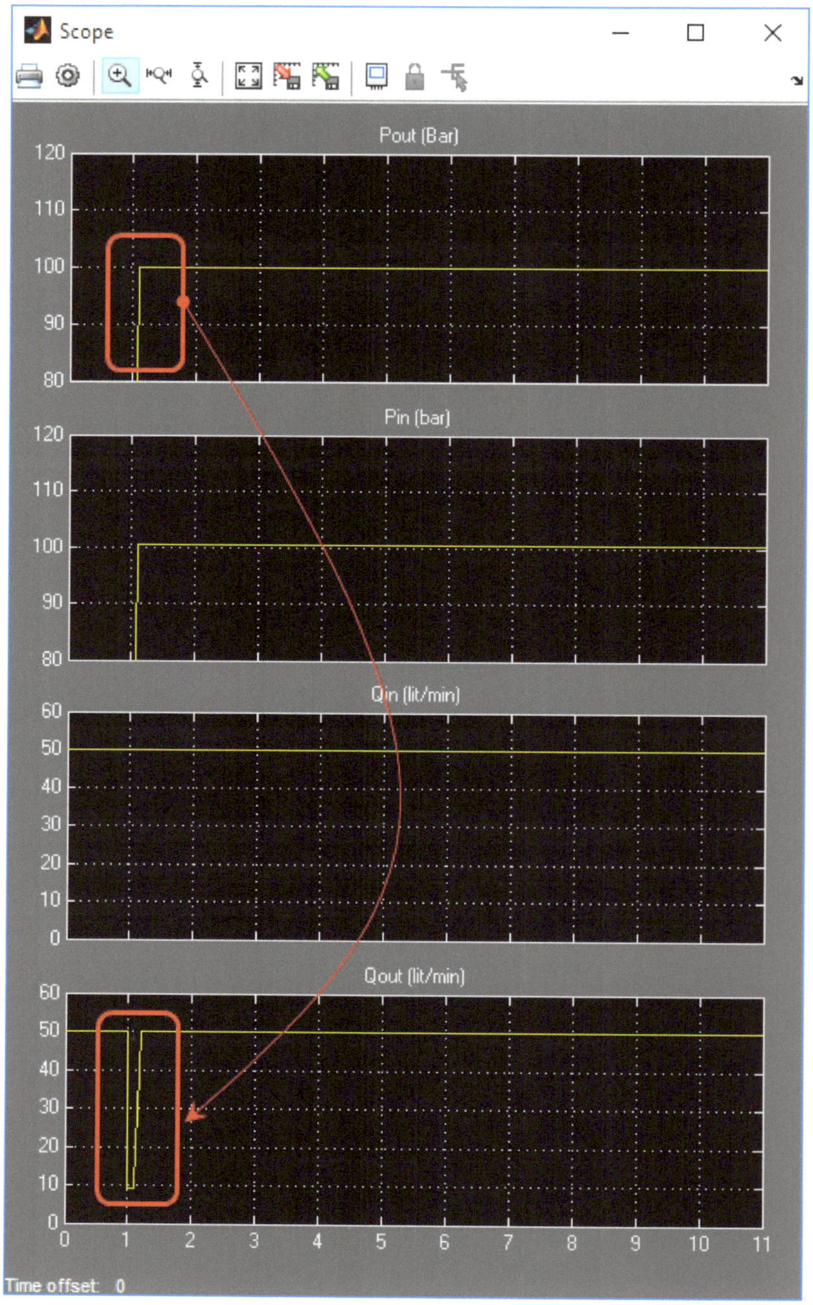

Fig. 6.11- Results of Case Study 3

6.2- Modeling of Hydraulic Fittings

Figure 6.12 shows the stream of calculations in a generic fitting model structure. The model uses the inlet flow to calculate the pressure drop across the fitting. This pressure drop is added to the outlet pressure to get the inlet pressure. Outlet flow will be considered equal to inlet flow because the piece of fitting has small dead volume, so effect of fluid capacitance can feasibly be ignored.

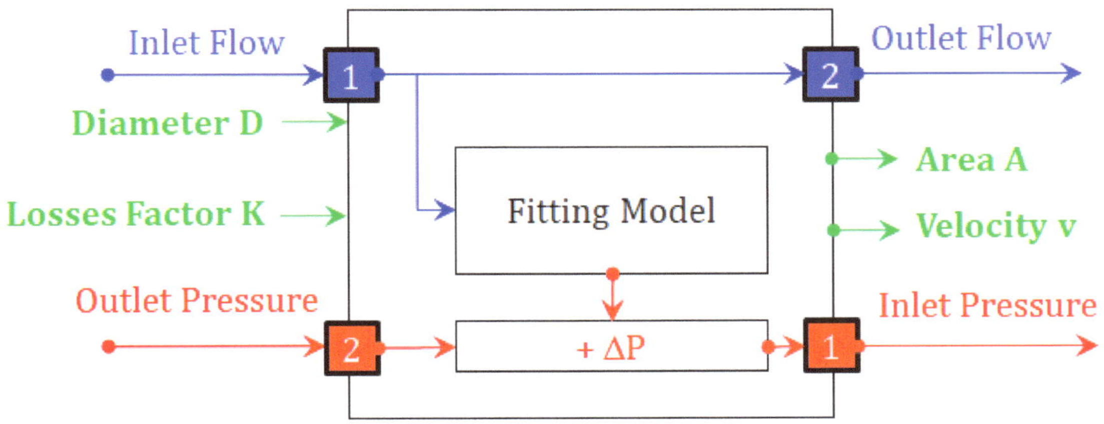

Fig. 6.12- Structure of a Generic Model for a Hydraulic Fitting

Modeling the Pressure Drop: Pressure drop in fittings is referred to as *Local Losses*. This term is used whenever the fluid velocity changes its magnitude (e.g. when a line diameter is changed as in fittings) or its direction (e.g. as in elbows). This pressure drop is mathematically expressed as follows:

$$\Delta p_L = K\,\frac{\rho v^2}{2} = K\,\frac{SG \times \rho_W \times v^2}{2} \qquad 6.11A$$

Equation 6.11B presents Eq. 6.11A with the units being balanced as follows:

$$\Delta p_L(\mathrm{bar}) = \frac{K \times SG \times 1000\,(\mathrm{kg/m^3}) \times v^2 (\mathrm{m/s})^2}{2 \times 10^5} = \frac{K \times SG \times v^2 (\mathrm{m/s})^2}{200} \qquad 6.11B$$

Where **K** is a lump sum resistivity factor and reported numerically for every specific shape of fitting as shown in Fig. 6.13. As shown in Fig. 6.14, sometimes the **K** factor is reported graphically. In such a case, a lookup table should be built to generate the proper **K** factor depending on the given data.

Modeling the Inlet Pressure: Inlet pressure of a fitting is calculated as follows:

$$p_{in}\,(\mathrm{bar}) = p_{out}\,(\mathrm{bar}) + \Delta p_L\,(\mathrm{bar}) \qquad 6.12$$

Other Modeling Method: An alternative easy method to model a hydraulic fitting is to convert the local losses in the fitting into a lookup table as function of the inlet flow.

Fitting	90⁰ bend	90⁰ angle	T-Piece	Double angle	Valve
	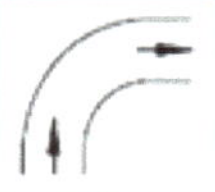	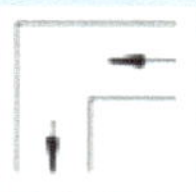	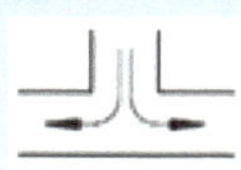		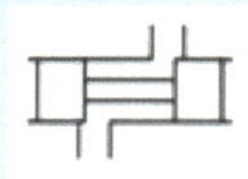
K	0.5 - 1	1.2	1.3	2	5 - 15
Sudden Contraction	Well-Rounded	Slightly Rounded	Sharp Edged	Shoulder	
	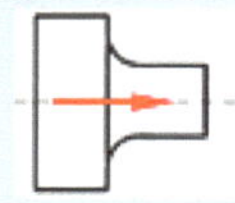	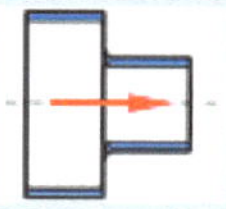	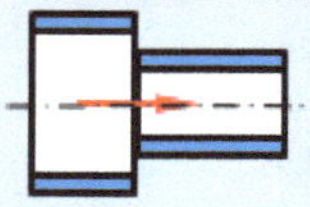	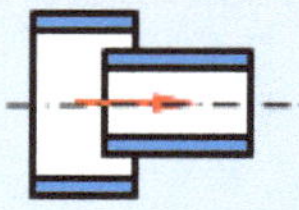	
K	0.04	0.2	0.5	0.8	
Sudden Expansion	Well-Rounded	Slightly Rounded	Sharp Edged	Shoulder	
	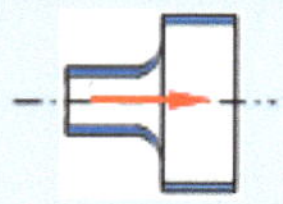	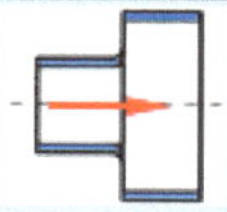	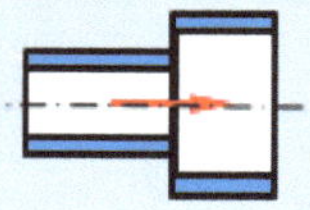	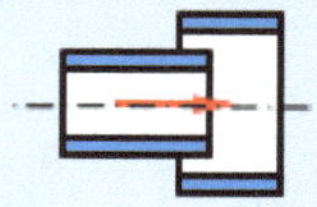	
K	1	1	1	1	

Fig. 6.13- Losses Factor in Fittings

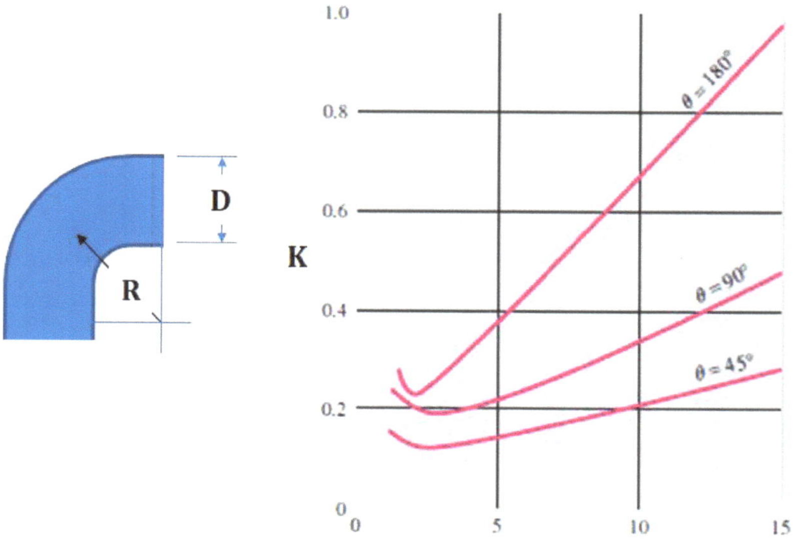

Fig. 6.14- Losses Factor in Elbows

Simulation Model: Figure 6.15 shows the structure of a simulation model (HSV7CH06Model02). The model is built to simulate the local losses with the choice of using a lookup table or the previously described set of equations.

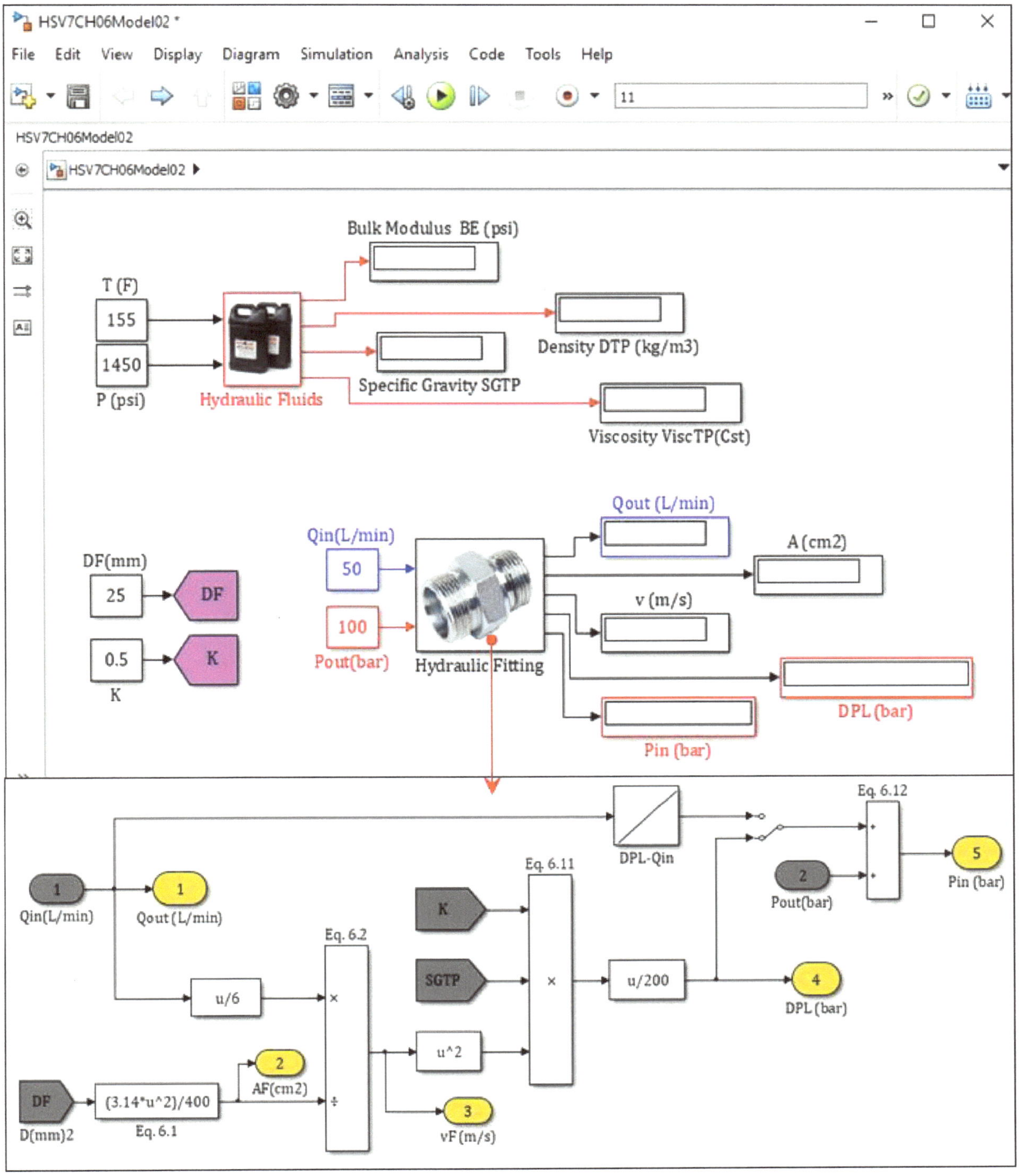

Fig. 6.15- Simulation Model for Hydraulic Fitting

Model Validation: As shown in Fig. 6.16, a sudden contraction is assumed in the hydraulic line shown in Fig. 6.9 from 25 mm diameter to 10 mm. Simulation results shows 0.38 bar for a losses factor of 0.7. Simulation results were validated based on other software "Automation Studio".

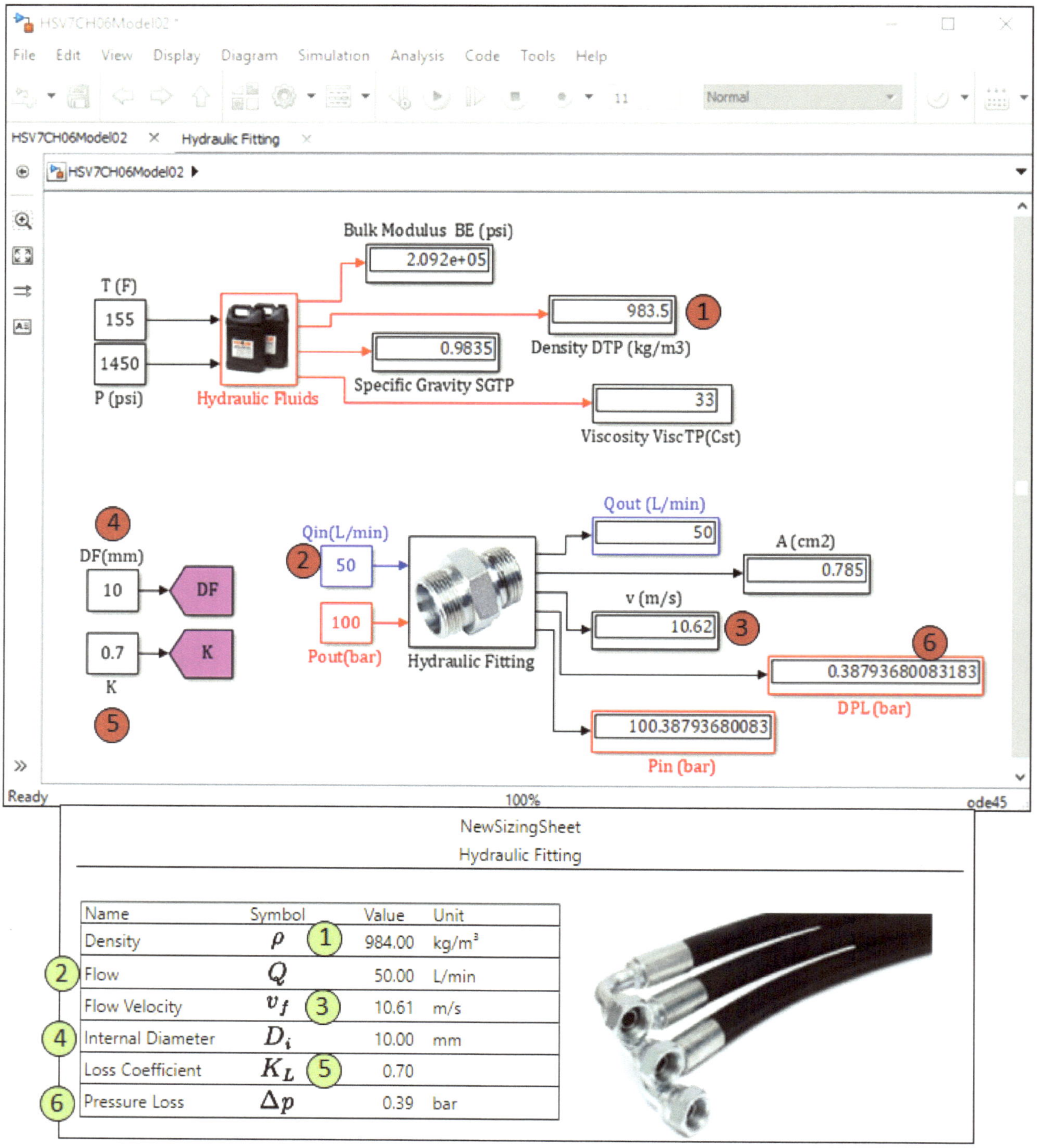

Name	Symbol	Value	Unit
Density	ρ	984.00	kg/m³
Flow	Q	50.00	L/min
Flow Velocity	v_f	10.61	m/s
Internal Diameter	D_i	10.00	mm
Loss Coefficient	K_L	0.70	
Pressure Loss	Δp	0.39	bar

Fig. 6.16- Validation of Hydraulic Fitting Model

6.3- Modeling of Hydraulic Orifices

Figure 6.17 shows the stream of calculations in an orifice generic model structure. The model uses the inlet flow to calculate the pressure drop across the orifice. This pressure drop is added to the outlet pressure to get the inlet pressure. Outlet flow will be considered equal inlet flow because the orifice has small dead volume, so effect of fluid capacitance can feasibly be ignored.

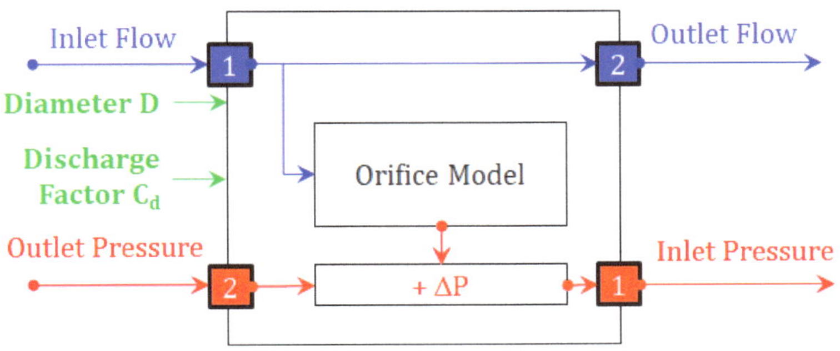

Fig. 6.17- Structure of a Generic Model for a Hydraulic Orifice

Modeling the Pressure Drop: Equation 6.13 shows that the flow through an orifice is a nonlinearly function of the pressure drop across the orifice as follows.

$$Q_{in} = C_d A \sqrt{\frac{2 \times \Delta P_O}{\rho}} \qquad 6.13$$

Where C_d is the *Discharge Coefficient* that depends on the orifice shape and equals 0.611 for sharp edged orifices. Equation 6.13 is rewritten in a reverse form to solve for the pressure as follows:

$$\Delta P_O = \frac{\rho}{2}\left[\frac{Q_{in}}{C_d A}\right]^2 = \frac{SG \times \rho_w}{2}\left[\frac{Q_{in}}{C_d A}\right]^2 \qquad 6.14A$$

Equation 6.14B presents Eq. 6.14A with the units being balanced as follows:

$$\Delta P_O(Pascal) = \frac{SG \times 1000\ (kg/m^3)}{2}\left[\frac{Q_{in}[lit/min] \times \frac{1}{60 \times 1000}}{C_d \times A\ [mm^2] \times \frac{1}{10^6}}\right]^2$$

$$\Delta P_O(Pascal) = \frac{SG \times 1000 \times [Q_{in}(lit/min)]^2}{2 \times C_d^2 \times A^2\ [mm^2]}\left[\frac{100}{6}\right]^2 \quad \left[\frac{kg.\,m^6}{m^3}\frac{1}{s^2}\frac{1}{m^4}\right]$$

$$\Delta P_O(Pascal) = \frac{10^7 \times SG \times [Q_{in}(lit/min)]^2}{72 \times C_d^2 \times A^2\ [mm^2]} \quad \left[\frac{kg.\,m}{s^2}\frac{1}{m^2} = \frac{N}{m^2}\right]$$

$$\Delta P_O(\mathbf{bar}) = \frac{\Delta P_O(\mathbf{Pascal})}{10^5} = \frac{1000 \times SG \times [Q_{in}(\mathbf{lit/min})]^2}{720 \times C_d^2 \times A^2\ [\mathbf{mm}^2]} \quad \mathbf{6.14B}$$

Modeling the Inlet Pressure: Inlet pressure of the orifice is calculated as follows:

$$P_{in}\ (\mathbf{bar}) = P_{out}\ (\mathbf{bar}) + \Delta P_O\ (\mathbf{bar}) \quad \mathbf{6.15}$$

Other Modeling Method: An alternative easy method to model a hydraulic orifice is to convert the local losses in the orifice into a lookup table as function of the inlet flow.

Simulation Model: Figure 6.18 shows the structure of a simulation model (HSV7CH06Model03). The model is built to simulate the local losses with the choice of using a lookup table or the previously descried set of equations.

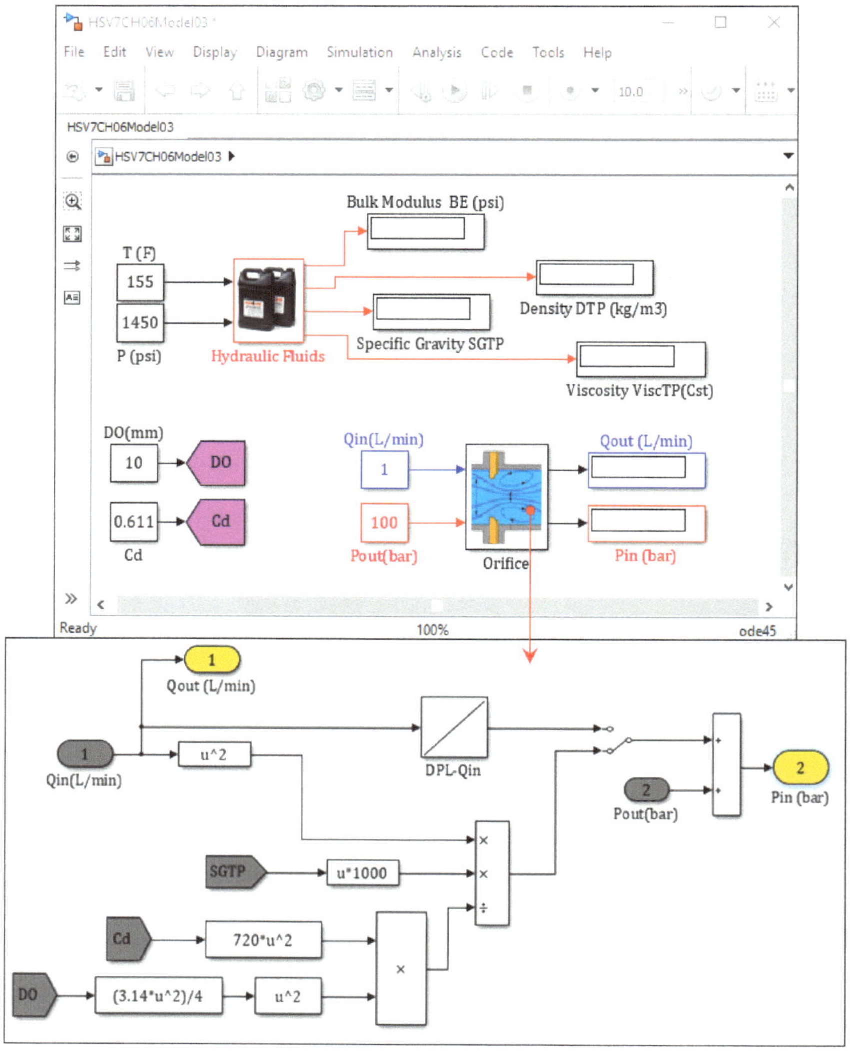

Fig. 6.18- Simulation Model for Hydraulic Orifice

Model Validation: As shown in Fig. 6.19, an orifice of 5 mm diameter is used. Simulation results shows 23.8 bar pressure drop across the orifice. Simulation results were validated based on other software "Automation Studio".

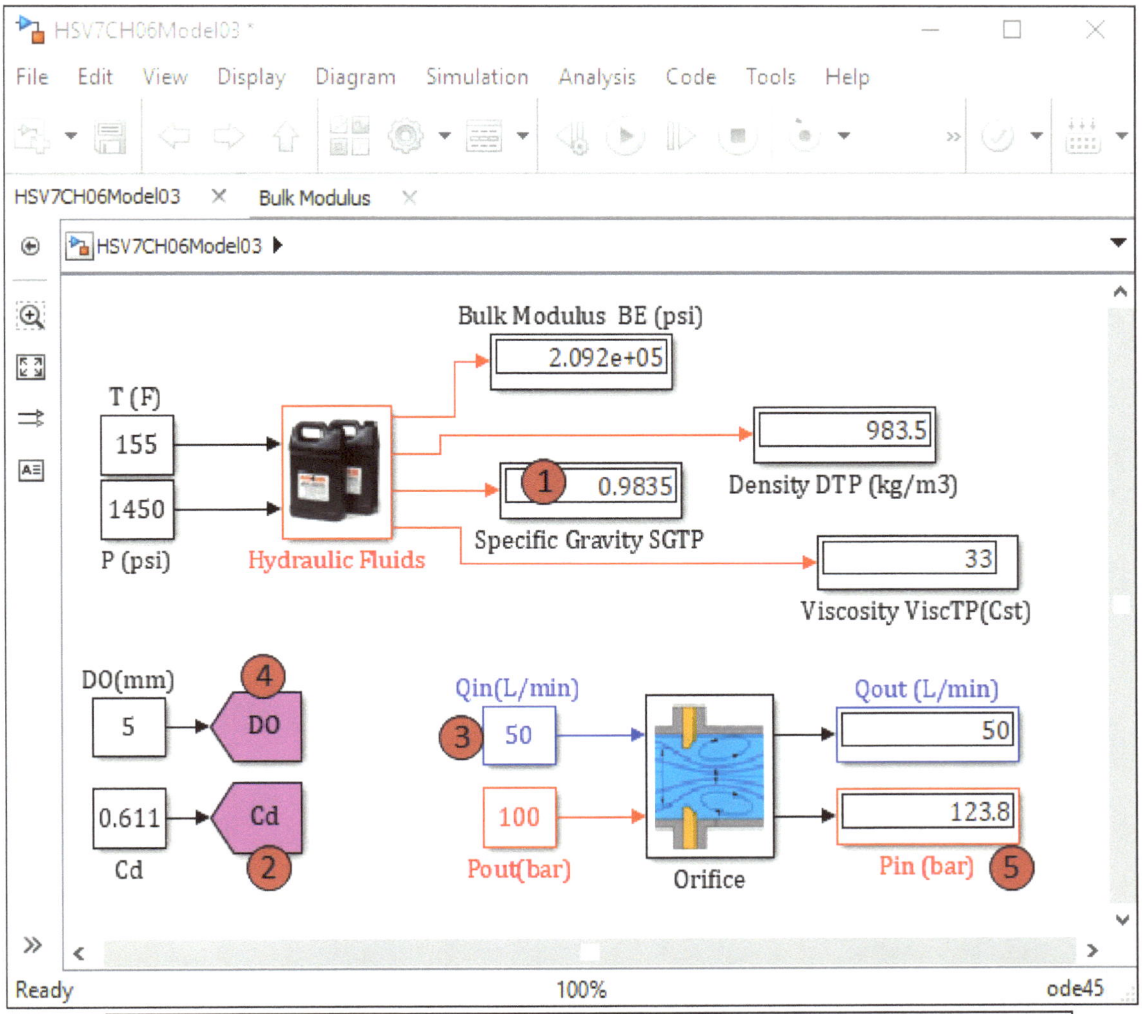

NewSizingSheet

Orifice

	Name	Symbol	Value	Unit
1	Density	ρ	983.00	kg/m³
	Diameter Ratio	β	0.50	
	Differential Pressure (5)	Δp	22.23	bar
2	Discharge Coefficient	C_d	0.61	
	Flow (3)	Q	50.00	L/min
4	Internal Diameter	d_o	5.00	mm
	Opening Area	A_o	19.63	mm²

Fig. 6.19- Validation of Hydraulic Fitting Model

6.4- Modeling Hydraulic Transmission Line Assembly

As shown in Fig. 6.20, in model (HSV7CH06Model04), a transmission line is assembled with a piece of fitting followed by an orifice. As shown in figure, for a 100-bar outlet pressure at the orifice, inlet pressure at the transmission line is 124.7 bar. An overall pressure drop is 24.7 bar including pressure drop in the three elements.

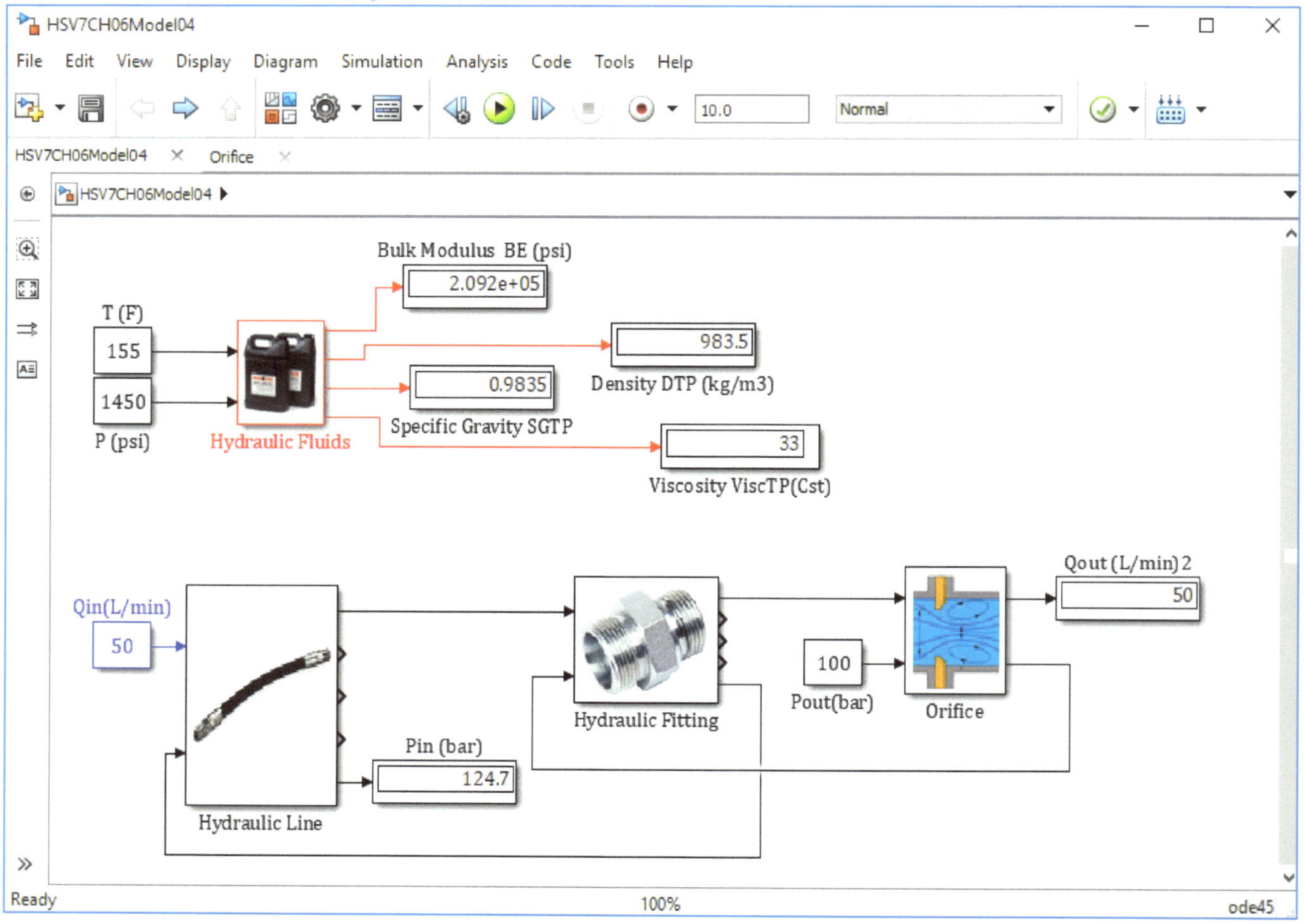

Fig. 6.20- Example of Modeling a Transmission Line Assembly

Chapter 7
Modeling of Hydraulic Pumps

Objectives

This chapter presents the lumped modeling concept as applied for fixed and variable displacement pumps. This chapter considers situations where a pump works under a constant or variable pressure and driving speed. Models for pressure-compensated, displacement-controlled, and torque-limited pumps are developed.

Brief Contents

Chapter 7: Modeling of Hydraulic Pumps

7.1- Lumped Model Structure of a Unidirectional Hydraulic Pump

Figure 7.1 shows a lumped model structure of a unidirectional hydraulic pump. The model contains the following features:

- Pump is the primary source of flow, so the pump model calculates the outlet flow based on outlet pressure with no need for inlet flow information.
- Pump lumped model takes different ascending levels based on, available information, assumptions, and the typical applications in which the pump runs. Inputs and outputs of the model depend on the model level.
- Pump static and dynamic characteristics are loaded to the model.

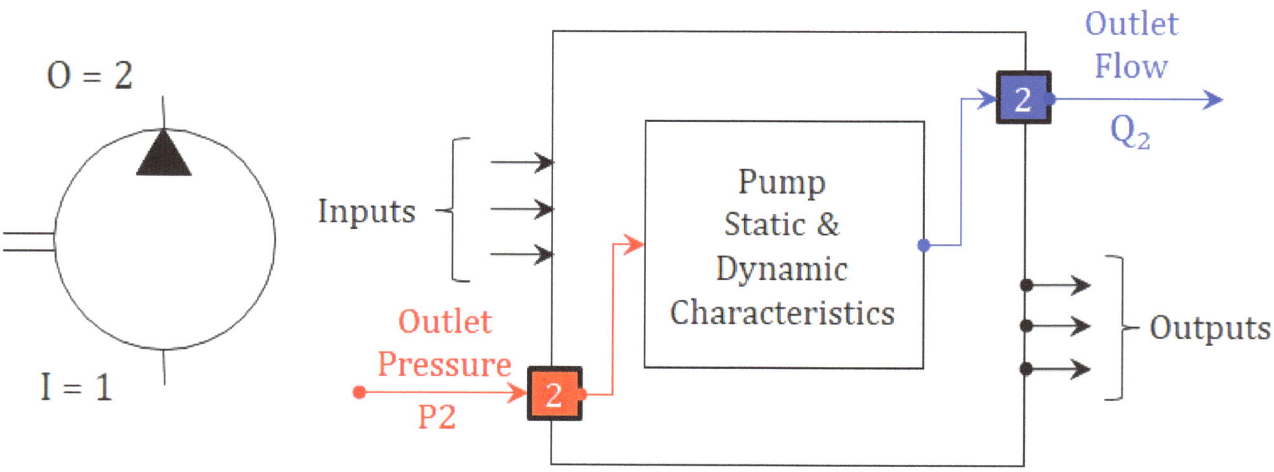

Fig. 7.1- Lumped Model Structure of a Unidirectional Hydraulic Pump

7.2- Lumped Model Structure of a Bidirectional Hydraulic Pump

Unlike a unidirectional pump, Fig. 7.2 shows that a model for a bidirectional pump has an identifier **U** is used to identify the pump direction of rotation, based on which flow and pressure on each physical port is assigned. If **U** is positive, then port 1 is the intake port and port 2 is the delivery port (**$P_1 = P_I$, $Q_1 = Q_I$, $P_2 = P_O$ and $Q_2 = Q_O$**) and vise-versa.

The logic for the identifier **U** can be performed in different way, such as using "If" statements. However, as shown in Fig. 7.3, the logic can be structured using Simulink logic blocks.

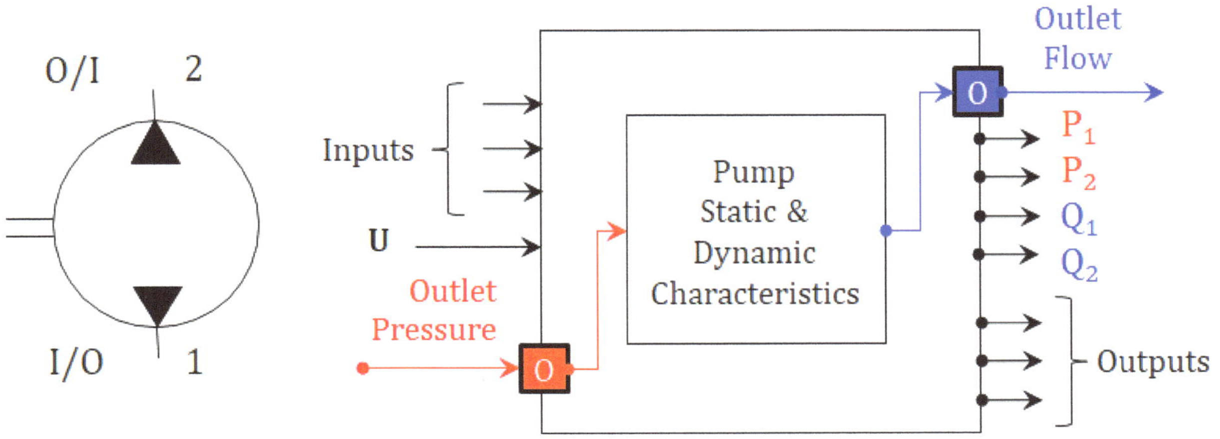

Fig. 7.2- Lumped Model Structure of a Bidirectional Hydraulic Pump

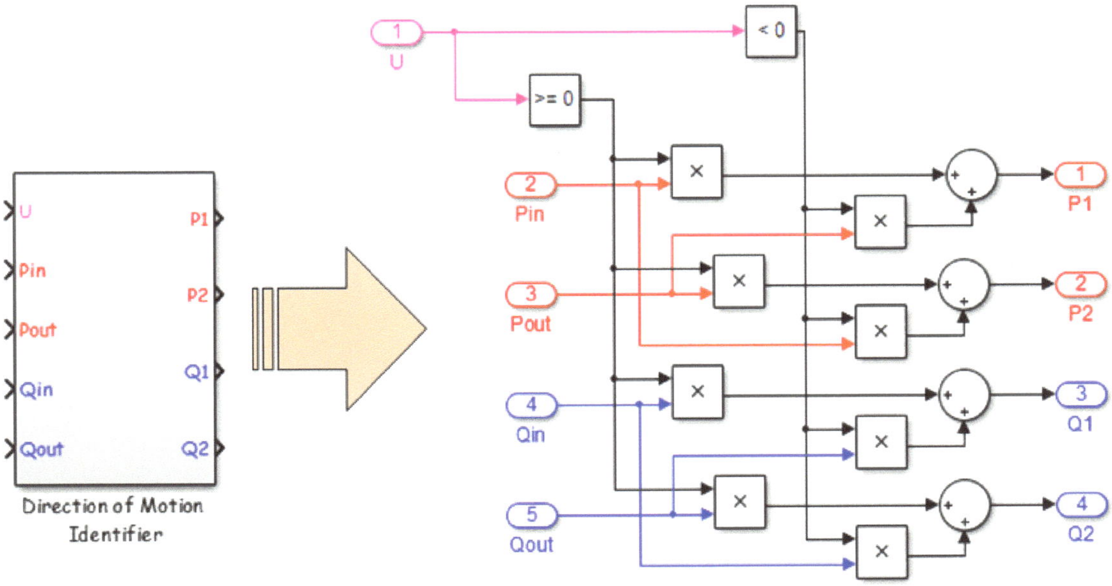

Fig. 7.3- - Logic of Identifying the Inlet and Outlet Ports of the Pump

7.3- Modeling Fixed Displacement Pumps

Part of the challenge in modeling hydraulic components in general is that these components work under various scenarios depending on the applications they are used for. For example, hydraulic pumps can work under fixed or variable driving speed and working pressure.

Ideally for modeling a pump, full range of test data should be available. Unfortunately, this is not the case all the time. Therefore, this chapter introduces various scenarios depending on the working conditions and the available information.

The following sections cover modeling static and dynamic characteristics for fixed displacement pumps first then variable displacement pumps. So, model level and complexity will be increased to cover various scenarios.

Fixed displacement pump static characteristics are reported in form of test data as shown in Fig. 7.4. This generic test data was developed for a 40 cc/rev pump size and will be considered in modeling the static characteristics of fixed pumps.

As shown in the figure, pumps are typically tested where pump actual flow and input power are reported versus allowable range of driving speeds and pressures.

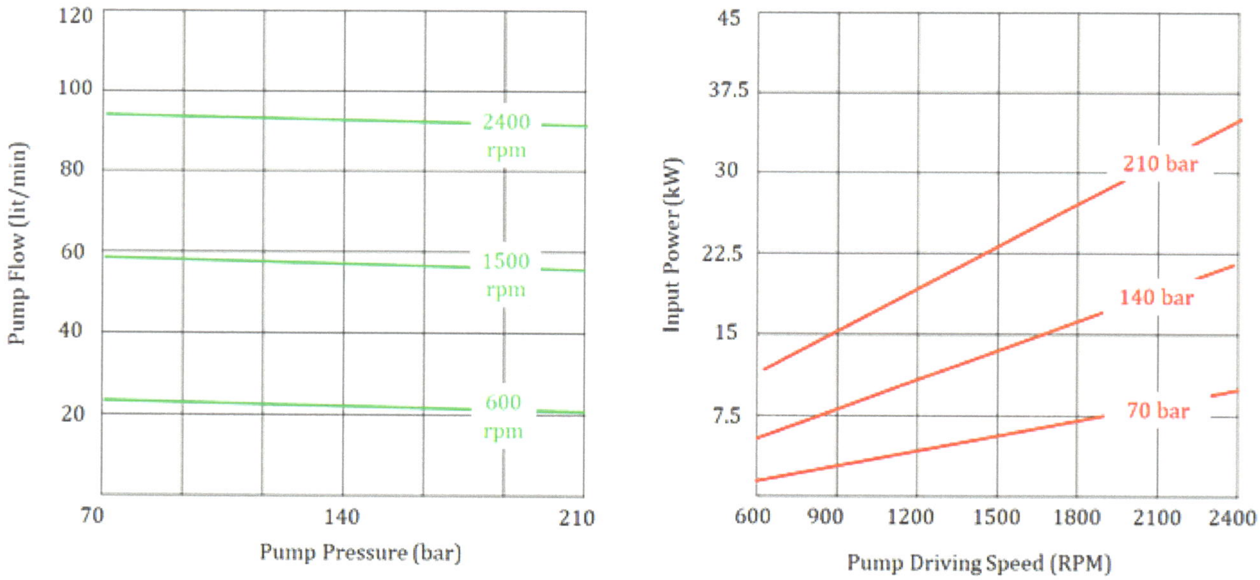

Fig. 7.4- Typical Static Characteristics of Fixed Displacement Pump in form of Test Data

Alternatively, as shown in Fig. 7.5., pump characteristics are reported in form of pump overall and volumetric efficiencies versus pump rpm and pressure. Practically, due to the difficulties of measuring the torque on the pump shaft, overall efficiency is reported instead of mechanical efficiency.

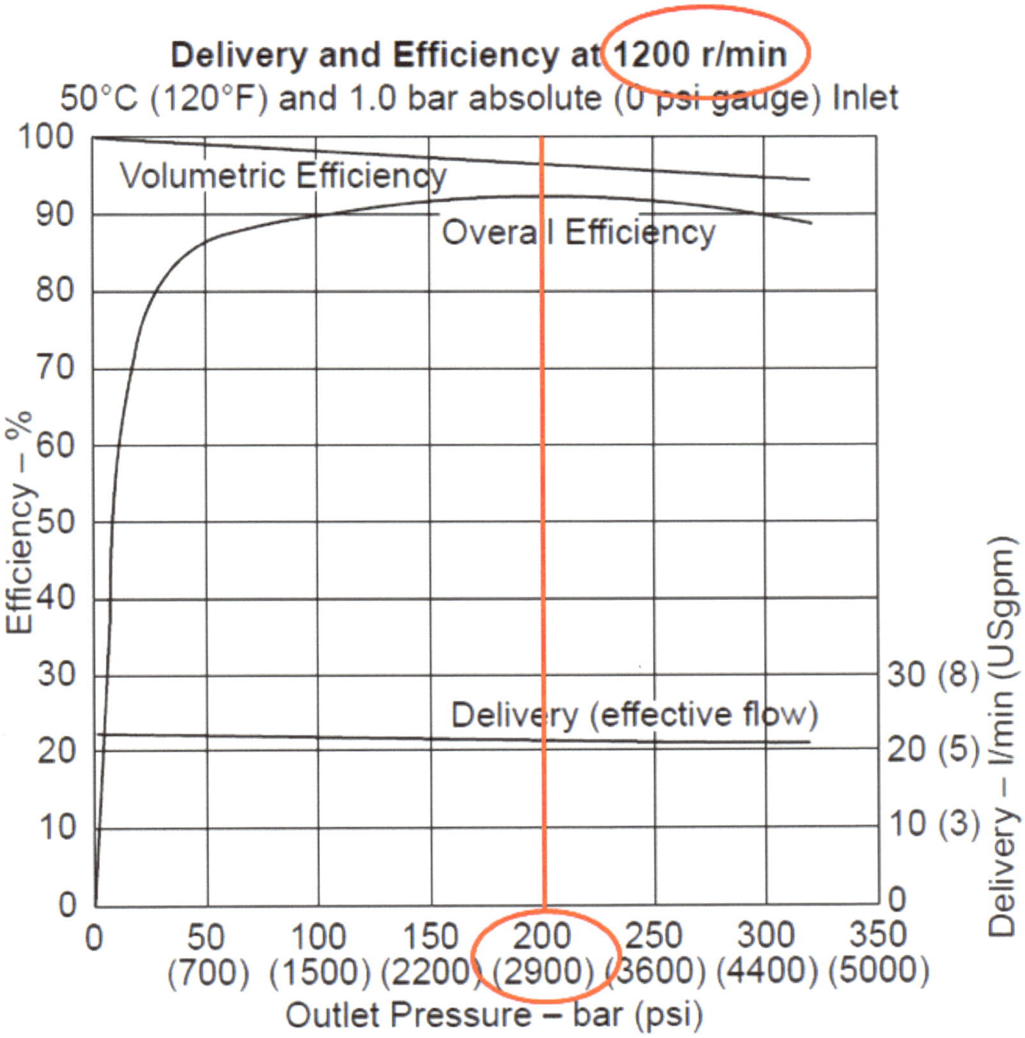

Fig. 7.5- Typical Static Characteristics of Fixed Displacement Pump in form of Efficiency Curves (Courtesy of Eaton)

7.4- Model #01 for an Ideal Fixed Displacement Pump

7.4.1- Model Features and Assumptions

Ideal Pump: In some applications, where pumps run continuously at constant rpm under low pressure, there is no reason to perform the efficiency calculations. So, the model represents just a source of constant flow as an ideal (100% efficient) pump. Then, no need for test data.

7.4.2- Pump Theoretical Flow Rate

Based on the 100% efficient pump assumption, Fig. 7.6 shows that theoretical pump flow $\mathbf{Q_{thp}}$ is function only of pump size $\mathbf{D_p}$ and driving speed $\mathbf{n_p}$ and is not affected by the pump outlet pressure $\mathbf{P_p}$.

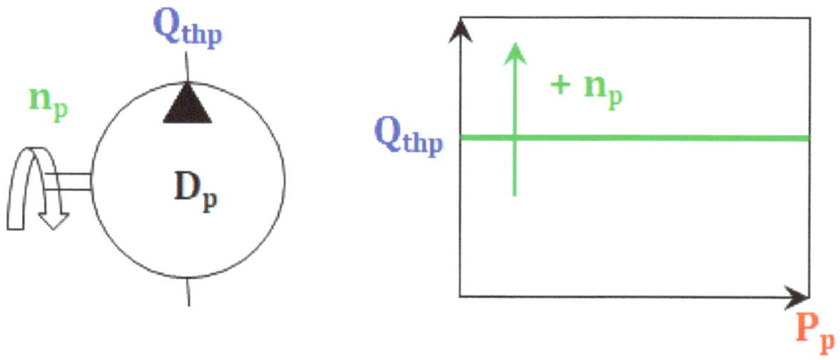

Fig. 7.6- Pump Theoretical Flow Characteristics

Therefore, Eq. 7.1 is used to calculate the pump theoretical flow as follows

$$\mathbf{Q_{thp} = D_p \times n_p} \qquad \mathbf{7.1}$$

Equations 7.1A and 7.1B represent Eq. 7.1 with the units being balanced in English and Metric systems of units; respectively.

$$\mathbf{Q_{thp}(gpm) = \frac{D_p\left(\frac{in^3}{rev}\right) \times n_p(rpm)}{231}} \qquad \mathbf{7.1A}$$

$$\mathbf{Q_{thp}(lit/min) = \frac{D_p\left(\frac{cc}{rev}\right) \times n_p(rpm)}{1000}} \qquad \mathbf{7.1B}$$

7.4.3- Theoretical Torque Acting on the Pump Drive Shaft

As shown in Fig. 7.7, a pump is a device that converts the input mechanical power into output hydraulic power. Therefore, starting with the power balance of a 100% efficient pump, Eq. 7.2 shows how to calculate the theoretical torque $\mathbf{T_{thp}}$ that acts on the pump shaft as function of the pump outlet pressure $\mathbf{P_p}$ and pump driving speed $\mathbf{n_p}$.

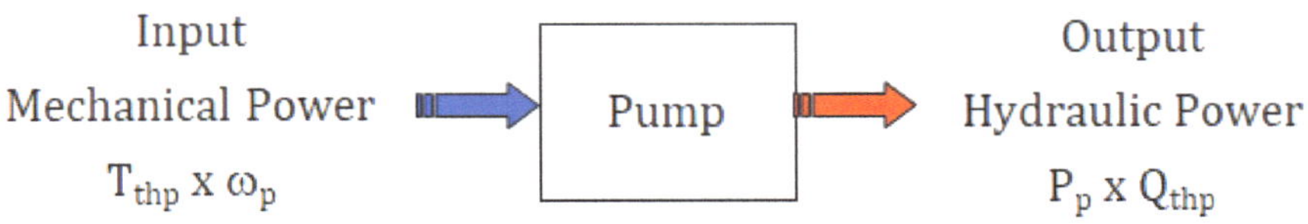

Fig. 7.7- Power Balance in an Ideal Pump

$$\mathbf{T_{thp} \times \omega_p = P_p \times Q_{thp} \rightarrow T_{thp} \times 2\pi n_p = P_p \times D_p \times n_p \rightarrow T_{thp} = \frac{P_p \times D_p}{2\pi}} \qquad \mathbf{7.2}$$

Equations 7.2A and 7.2B represent Eq. 7.2 with the units being balanced in English and Metric systems of units; respectively.

$$\mathbf{T_{thp}(lb.in) = \frac{P_p\,(psi) \times D_p(in^3/rev)}{2\pi}} \qquad \mathbf{7.2A}$$

$$\mathbf{T_{thp}(N.m) = \frac{P_p\,(bar) \times D_p(cc/rev)}{20\pi}} \qquad \mathbf{7.2B}$$

Figure 7.8 shows that the theoretical torque acting on the pump shaft is function only of the pump size and outlet pressure and is not affected by the pump driving speed.

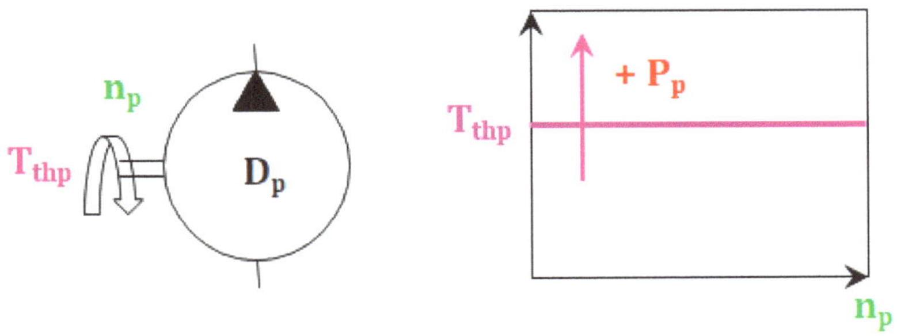

Fig. 7.8- Pump Theoretical Torque Characteristics

7.4.4- Model Structure

Figure 7.9 shows the structure of a pump lumped model 01.

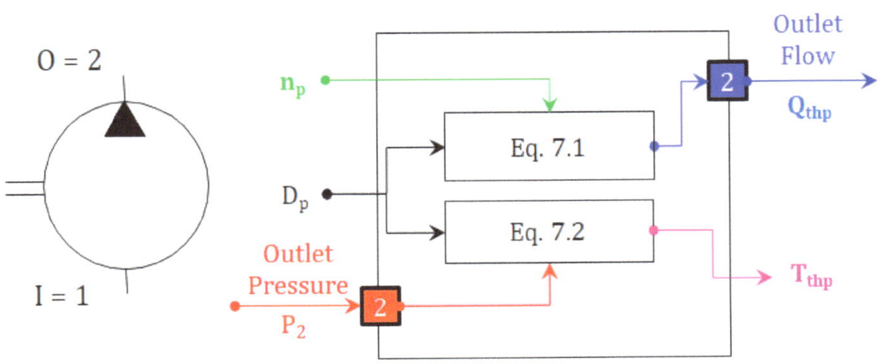

Fig. 7.9- Structure of Pump Model 01

7.4.5- Simulation Model

Figure 7.10 shows simulation model (HSV7CH07Model01) for ideal fixed displacement unidirectional pump built using Matlab-Simulink. As shown in the figure, metric system of units was used. The figure shows the upper mask and the interior details of the model.

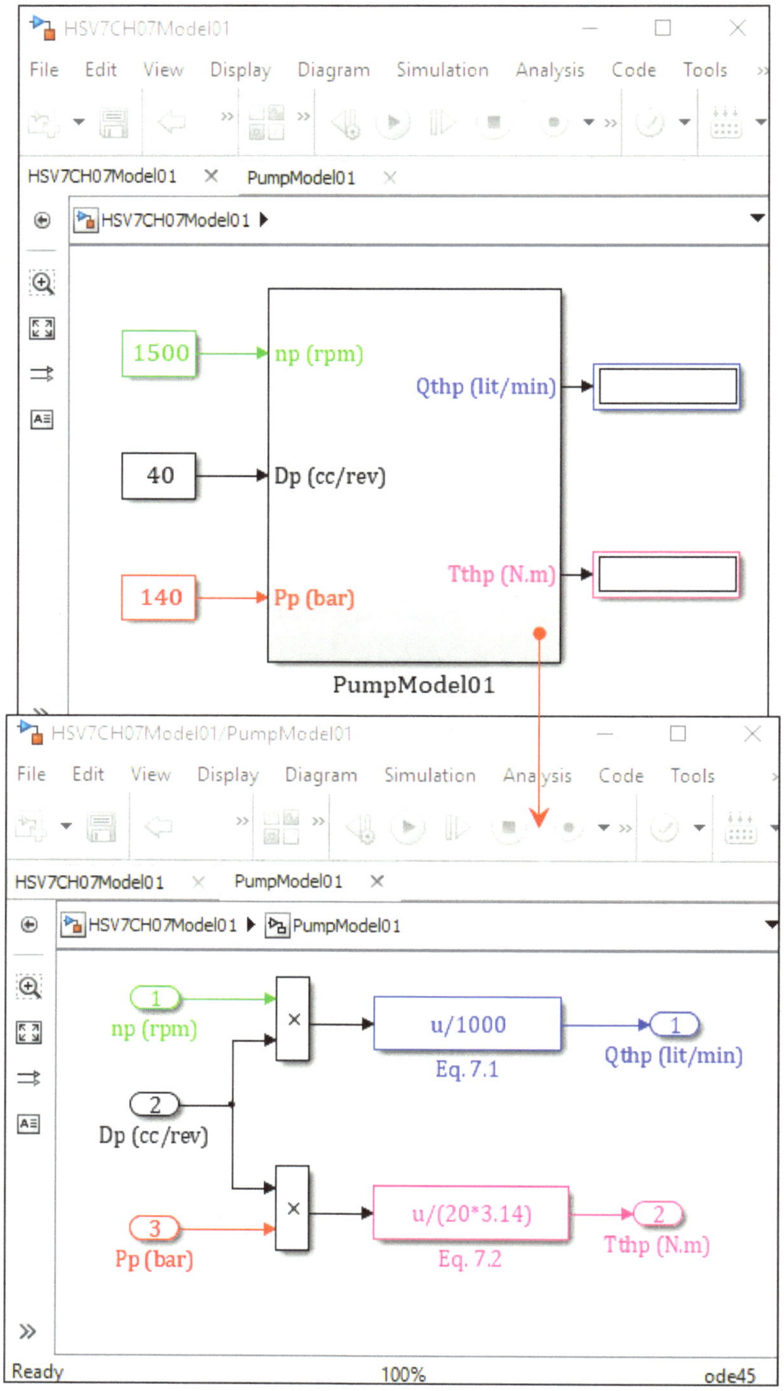

Fig. 7.10- Pump Simulation Model 01

7.5- Model #02A for a Fixed Displacement Pump Running at Constant Operating Conditions Based on Given Test Values

7.5.1- Model Features and Assumptions

Practical Pump Working at Constant Operating Conditions: In some applications, pumps run at constant rpm under constant operating pressure. This means that the pump works under constant efficiencies. In this case, only actual pump flow and input power (at this rpm and pressure) are needed to model the pump. Alternatively, two of the three pump efficiencies predefined at pump rpm and pressure are needed to model the pump.

7.5.2- Pump Actual Flow and Volumetric Efficiency

Referring to Fig. 7.11, Eq. 7.3 is used to solve for the pump actual flow $\mathbf{Q_p}$ or pump *Volumetric Efficiency* $\mathbf{e_{vp}}$ or leakage flow of the pump $\mathbf{q_{Lp}}$ depending on the available information.

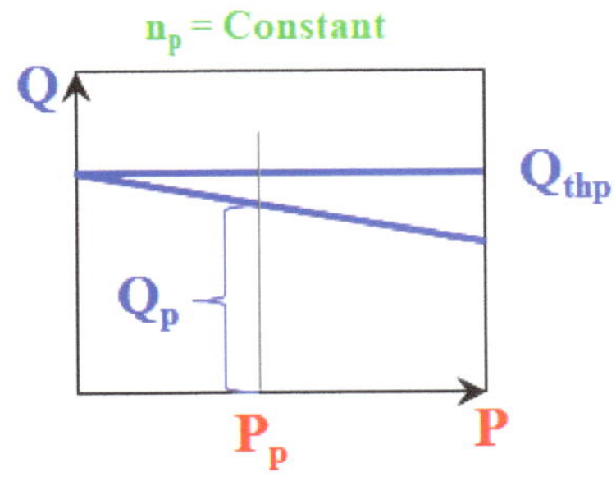

Fig. 7.11- Q-p Characteristics of a Fixed Pump

$$\mathbf{Q_p = e_{vp} \times Q_{thp} = Q_{thp} - q_L} \qquad 7.3$$

7.5.3- Pump Input Power and Overall Efficiency

Equations 7.4A and 7.4B shows calculation of pump *Overall Efficiency* and Power in English and Metric systems of units; respectively.

$$\mathbf{e_{op} = \frac{Pump\ Output\ (Fluid)Power\ [H_{out}]}{Pump\ Input\ (Shaft)Power[H_{in}]} = \frac{P_p(psi) \times Q_p(gpm)}{1714 \times H_{in}(HP)}} \qquad 7.4A$$

$$\mathbf{e_{op} = \frac{Pump\ Output\ (Fluid)Power\ [H_{out}]}{Pump\ Input\ (Shaft)Power[H_{in}]} = \frac{P_p(bar) \times Q_p(lit/min)}{600 \times H_{in}(kW)}} \qquad 7.4B$$

7.5.4- Pump Mechanical Efficiency

Equation 7.5 shows the calculation of pump *Mechanical Efficiency*.

$$\mathbf{e_{mp} = e_{op}/e_{vp}} \qquad 7.5$$

7.5.5- Pump Actual Torque

Referring to Fig. 7.12, Eq. 7.6 is used to calculate the actual torque $\mathbf{T_p}$ acting on the pump shaft.

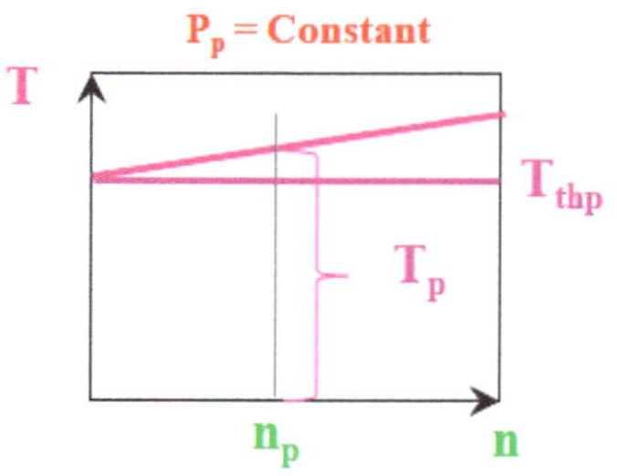

Fig. 7.12- T-n Characteristics of a Fixed Pump

$$\mathbf{T_p = \frac{T_{thp}}{e_{mp}}} \tag{7.6}$$

Equation 7.7 is used to calculate the power losses $\mathbf{H_L}$ that is converted into heat in the pump.

$$\mathbf{H_L = H_{in} - H_{out} = H_{in} - H_{in} \times e_{Op} = H_{in}(1 - e_{Op})} \tag{7.7}$$

7.5.6- Simulation Model

Figure 7.13 shows simulation model (HSV7CH07Model02A) for practical fixed displacement unidirectional pump built using Matlab-Simulink.

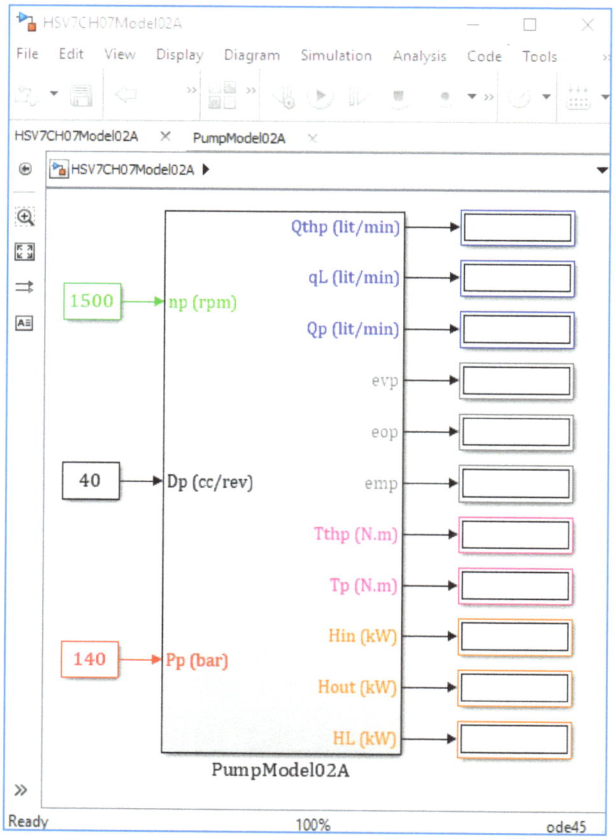

Fig. 7.13- Pump Simulation Model 02A

Figure 7.14 shows the details under the mask of the simulation model. In this model, "GOTO" and "From" Simulink blocks were used to reduce the linking line. Rather than loading the pump characteristics inside the model properties, it can be shown on the model page as constant for easier investigations and model debugging. The model solves Eq. 7.1 through 7.7 in the following sequence:

- Given Output Test Data (Q_p and H_{in}), at Input Operating Conditions (n_p and P_p), for pump size D_p:
- Eq. 7.1 $\rightarrow Q_{thp}$
- Eq. 7.2 $\rightarrow T_{thp}$
- Eq. 7.3 $\rightarrow e_{vp}$ and q_L
- Eq. 7.4 $\rightarrow H_{out}$ and e_{op}
- Eq. 7.5 $\rightarrow e_{mp}$
- Eq. 7.6 $\rightarrow T_p$
- Eq. 7.7 $\rightarrow H_L$

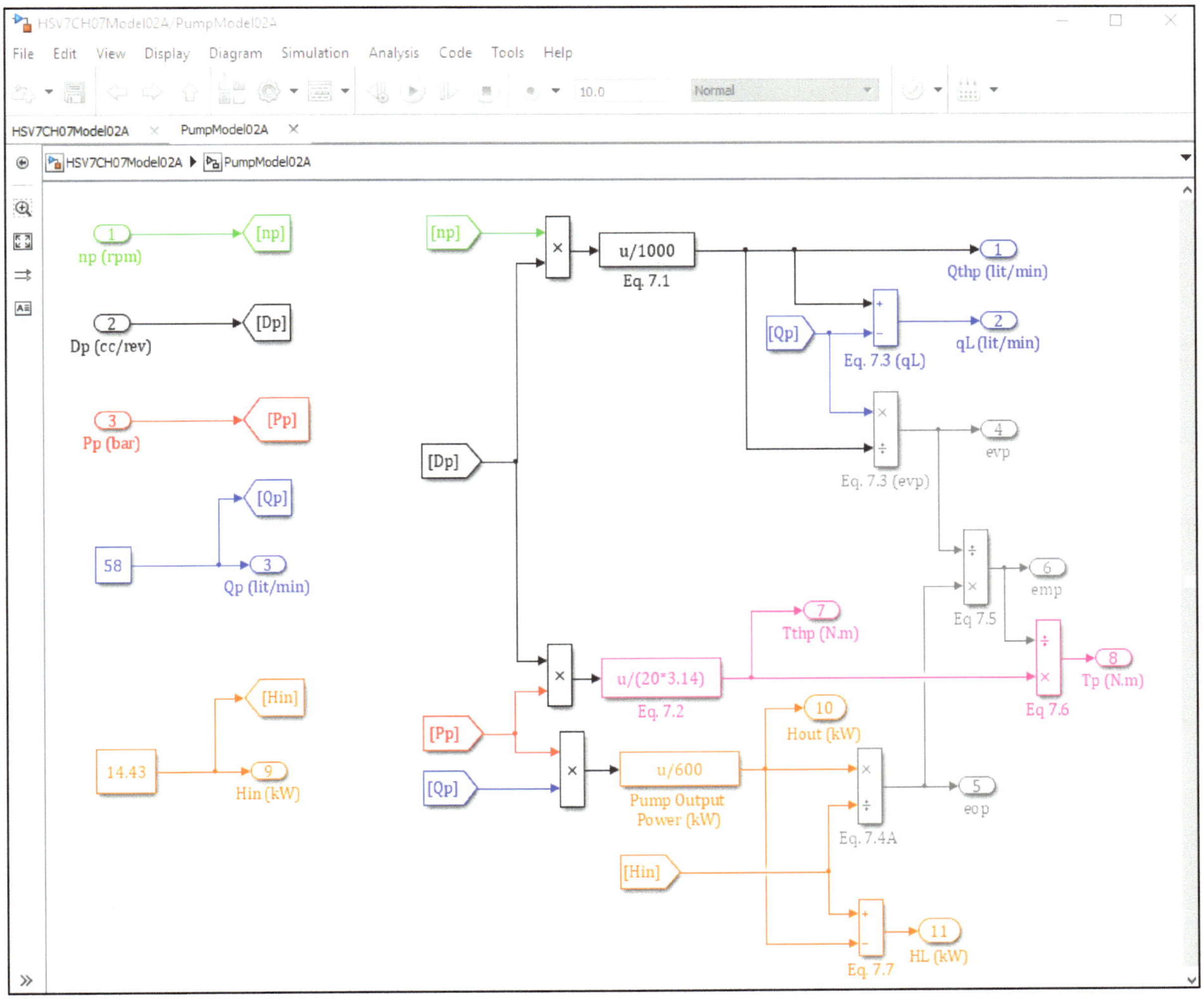

Fig. 7.14- Details under the Mask of Simulation Model 02A

7.5.7- Model Validation

Figure 7.15 shows locating the operating point (1500 rpm and 140 bar) of the pump on the test data to get pump flow and input power at the defined speed and pressure.

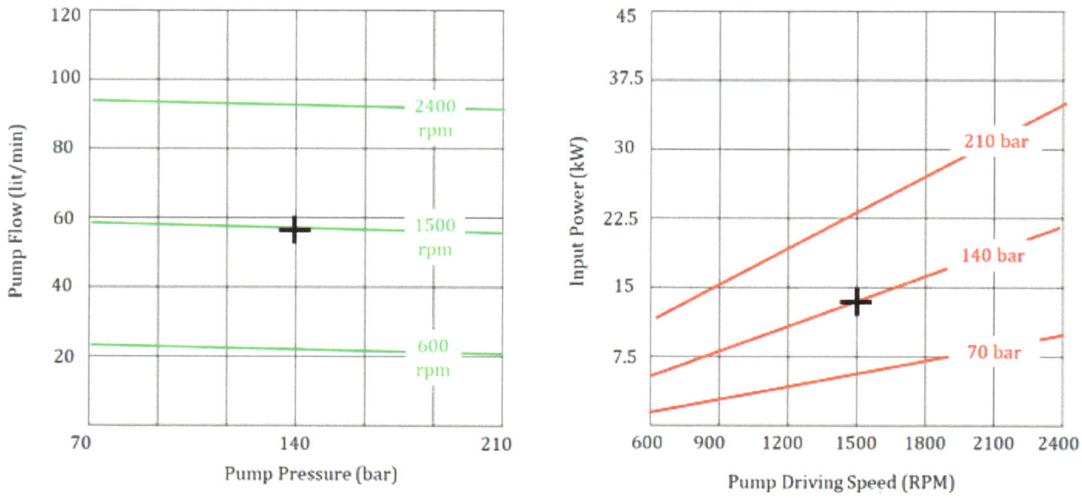

Fig. 7.15- Locating Operating Point of the Pump

Figure 7.16 shows that the model is validated based on other software (Hydraulic Components Sizing Calculator).

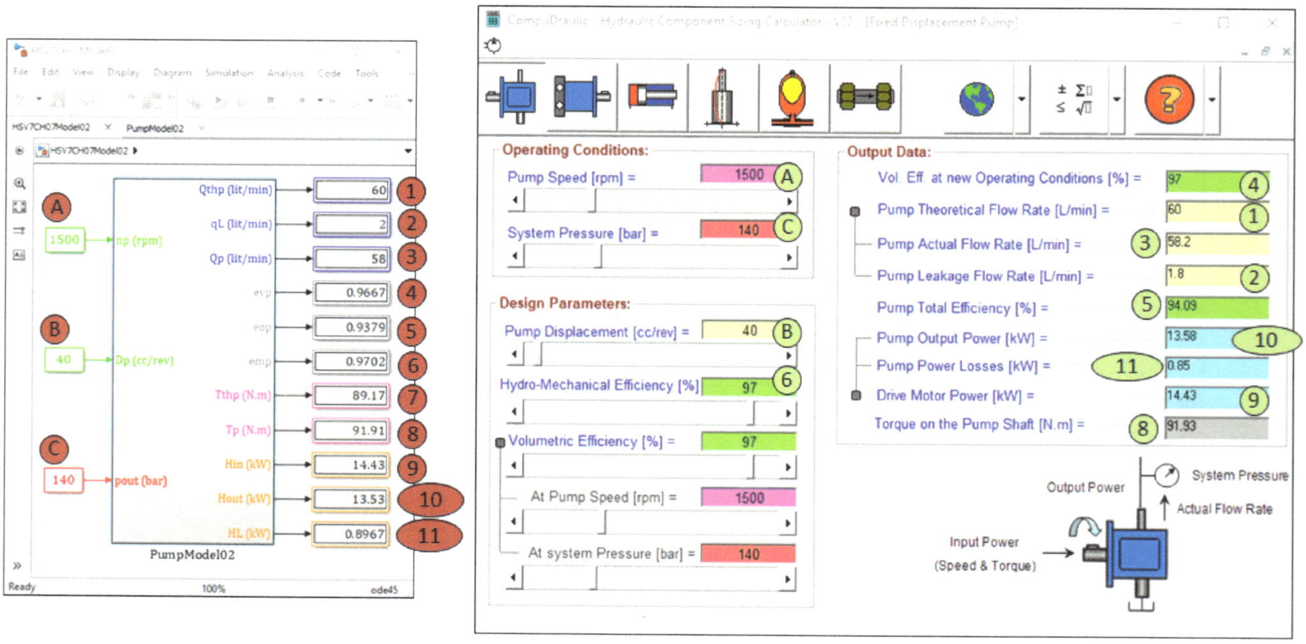

Fig. 7.16- Validation of Model 02A

7.6- Model #02B for a Fixed Displacement Pump Running at Constant Operating Conditions Based on Given Efficiency Values

7.6.1- Model Features and Assumptions

Same as in model 02A except that volumetric efficiency and overall efficiency are given instead of pump actual flow and input power.

7.6.2- Simulation Model

Figure 7.17 shows the upper mask of simulation model (HSV7CH07Model02B). When the model is fed by same efficiencies resulted in model 02A, it produces the same other results. This validates the model.

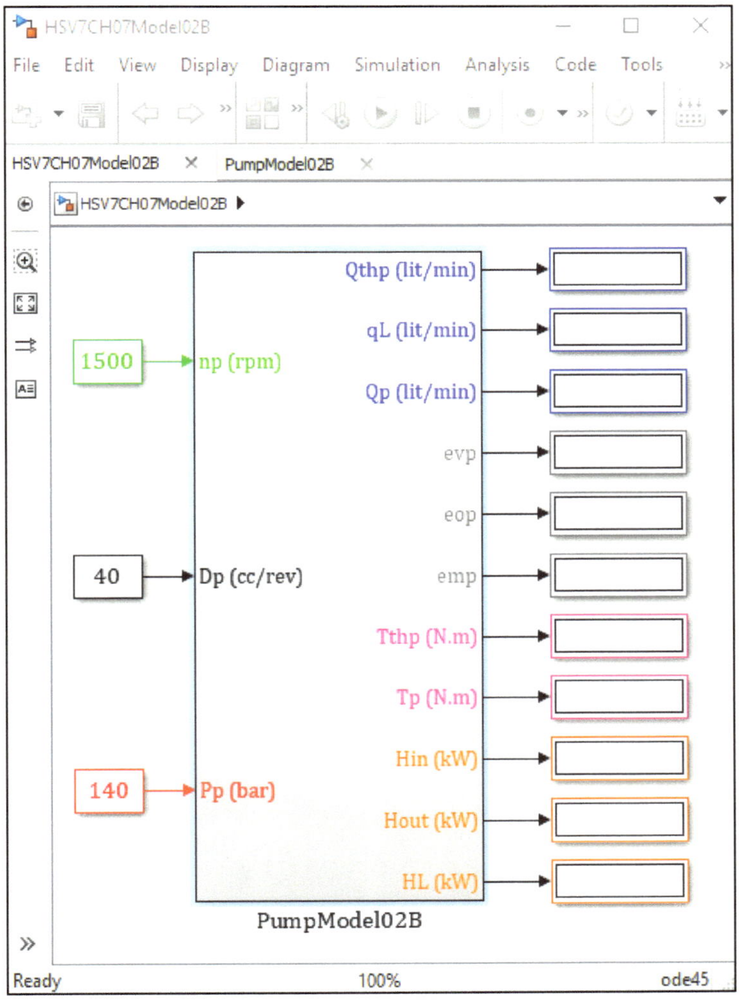

Fig. 7.17- Pump Simulation Model 02B

Figure 7.18 shows the details under the mask of simulation model 02B. The model solves the same set of Eq. 7.1 through 7.7 with the same sequence except that Eq. 7.3 solves for Q_p instead of e_{vp} and Eq. 7.4 solves for H_{in} instead of e_{op}. As follows:

- Given Efficiency Values (e_{vp} and e_{op}), Operating Conditions (n_p and P_p), and pump size D_p:
- Eq. 7.1 → Q_{thp}
- Eq. 7.2 → T_{thp}
- Eq. 7.3 → Q_p and q_L
- Eq. 7.4 → H_{out} and H_{in}
- Eq. 7.5 → e_{mp}
- Eq. 7.6 → T_p
- Eq. 7.7 → H_L

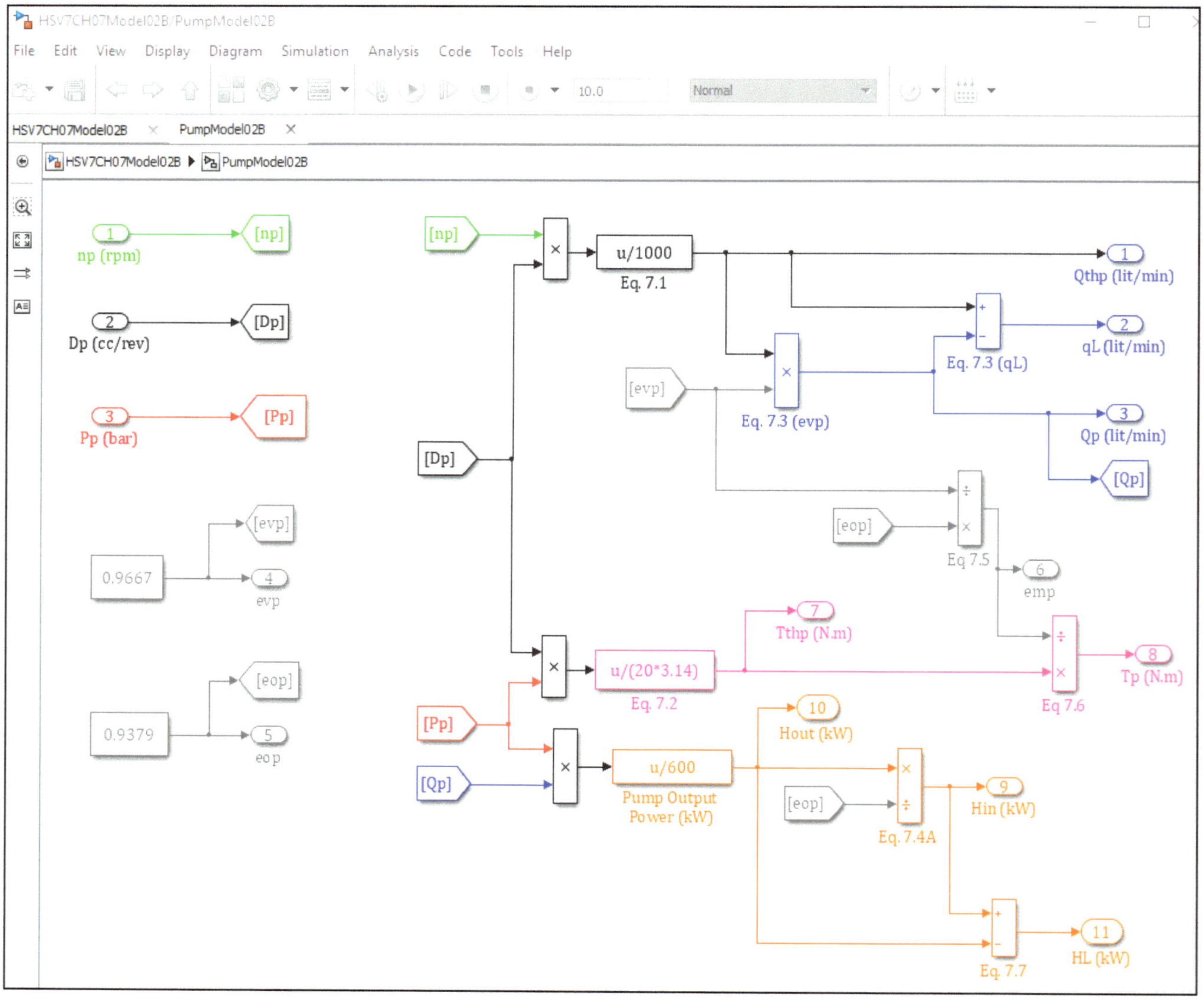

Fig. 7.18- Details under the Mask of Simulation Model 02B

7.7- Model #03A for a Fixed Displacement Pump Running at Variable Operating Conditions Based on Given Test Data

7.7.1- Model Features and Assumptions

Practical Pump Working at Variable Operating Conditions: In some applications, a pump may work under variable driving speed and operating pressure. This means that pump efficiencies are variable and model outputs must be updated interactively with changes in driving speed and working pressure.

7.7.2- Mathematical Model

Same set of equations 7.1 through 7.7 are still be used except that pump test data are loaded to the model in the form of two 2-D lookup tables instead of being given as constant values.

7.7.3- Simulation Model

Figure 7.19 shows the upper mask of the simulation model (HSV7CH07Model03A). The figure shows that the model has the following elements:

Driving Speed n_p: For a pump that works under variable operating speed, it receives the speed from the prime mover model. However, for investigating the pump performance under variable operating speed, a "Slider Gain" is temporarily used to replace the prime mover mode. Additionally, a "Ramp" block is used to develop the pump steady state characteristics versus the driving speed.

Pump Outlet Pressure P_p: Similarly, for a pump that works under variable operating pressure, pump outlet pressure should be received from the model of the component downstream the pump. However, for investigating the pump performance under variable operating pressure, a "Slider Gain" is temporarily used to feed the pump pressure. Additionally, a "Ramp" block is used to develop the pump steady state characteristics versus the operating pressure.

Pump Displacement D_p: It is very important to remember that the lookup tables are assigned for that specific pump size So, if the pump size is changed, then lookup table must be updated accordingly. That is why, rather than having the pump size as a variable input to the model, it has been loaded to the block properties as a pump constant.

Dynamic Effect of Operating Pressure Change (at D_p and n_p are Constants): Working pressure is used to calculate the driving torque and the internal leakage. So, due to modulus of elasticity of the driving shaft and how the internal leakage is developed within the pump clearances, it can't be assumed that these changes occur instantaneously with variations in pressure. Modeling this effect is very exhaustive and depends on specific pump design. However, in view of machine modeling, it can be feasibly represented by a fist order normalized transfer function with a reasonable time constant (e.g. 20 ms).

Dynamic Effect of Driving Speed Change (at D_p and P_p are Constants): Driving speed is used to calculate the pump flow and input power. Due to filling coefficient, fluid hydrodynamic effect, and inertial effect of pump rotating mass, it can't be assumed that the resulting pump flow and input power vary instantaneously with variation in pump driving speed. Therefore, similarly, dynamic effect of driving speed can be feasibly represented by a fist order normalized transfer function with a reasonable time constant (e.g. 20 ms).

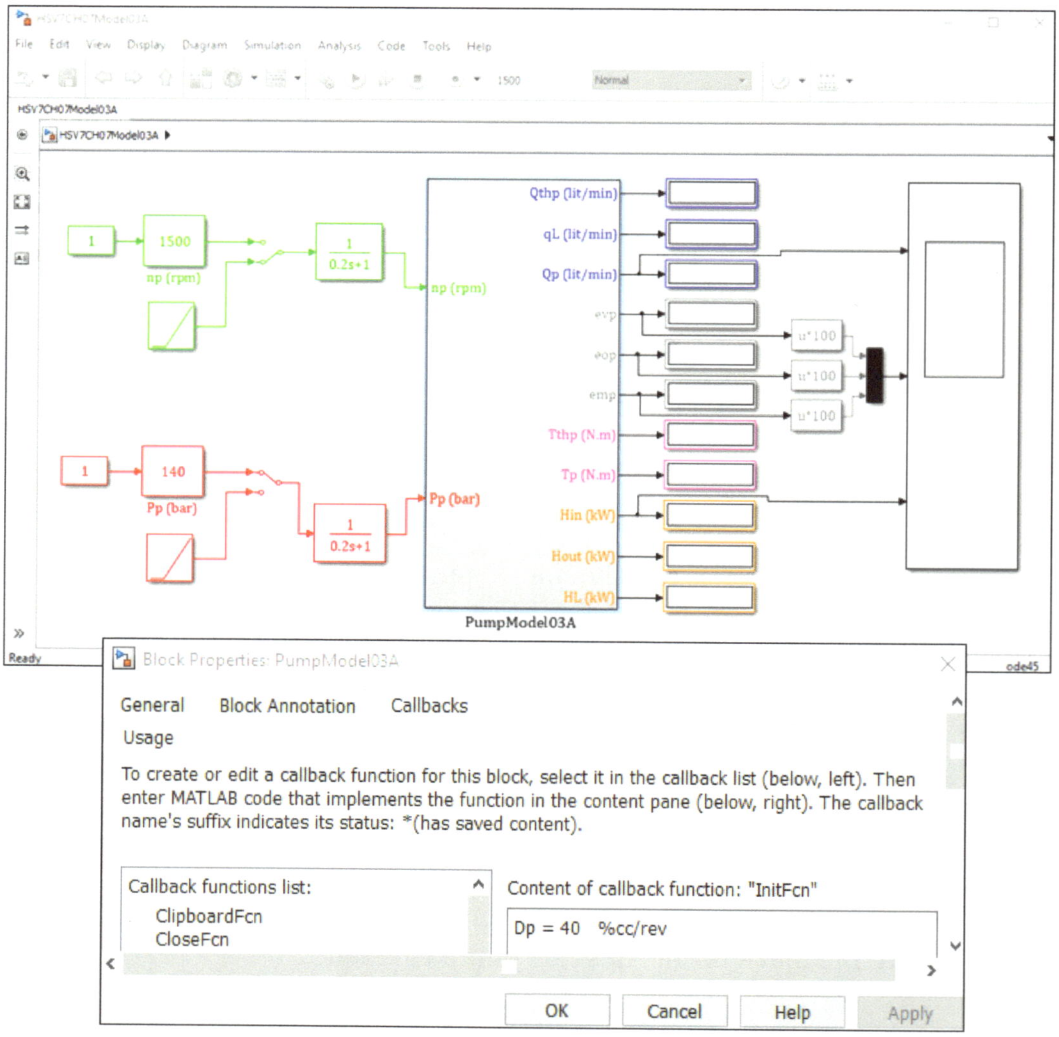

Fig. 7.19- Pump Simulation Model 03A

Figure 7.20 shows the details under the mask of simulation model 03A that is structured to override model 02A. Model 03A serves to simulate a pump that works under constant or variable speed and pressure. Two switches were added so that the model can select pump flow and input power as constants or variable from lookup table. However, if constant flow and input power are selected, this means these values are defined at whatever constant value selected at the pump speed and working pressure slider gains.

The figure shows how the model solves Eq 7.1 through 7.7. Equation numbers are listed accordingly on the model blocks.

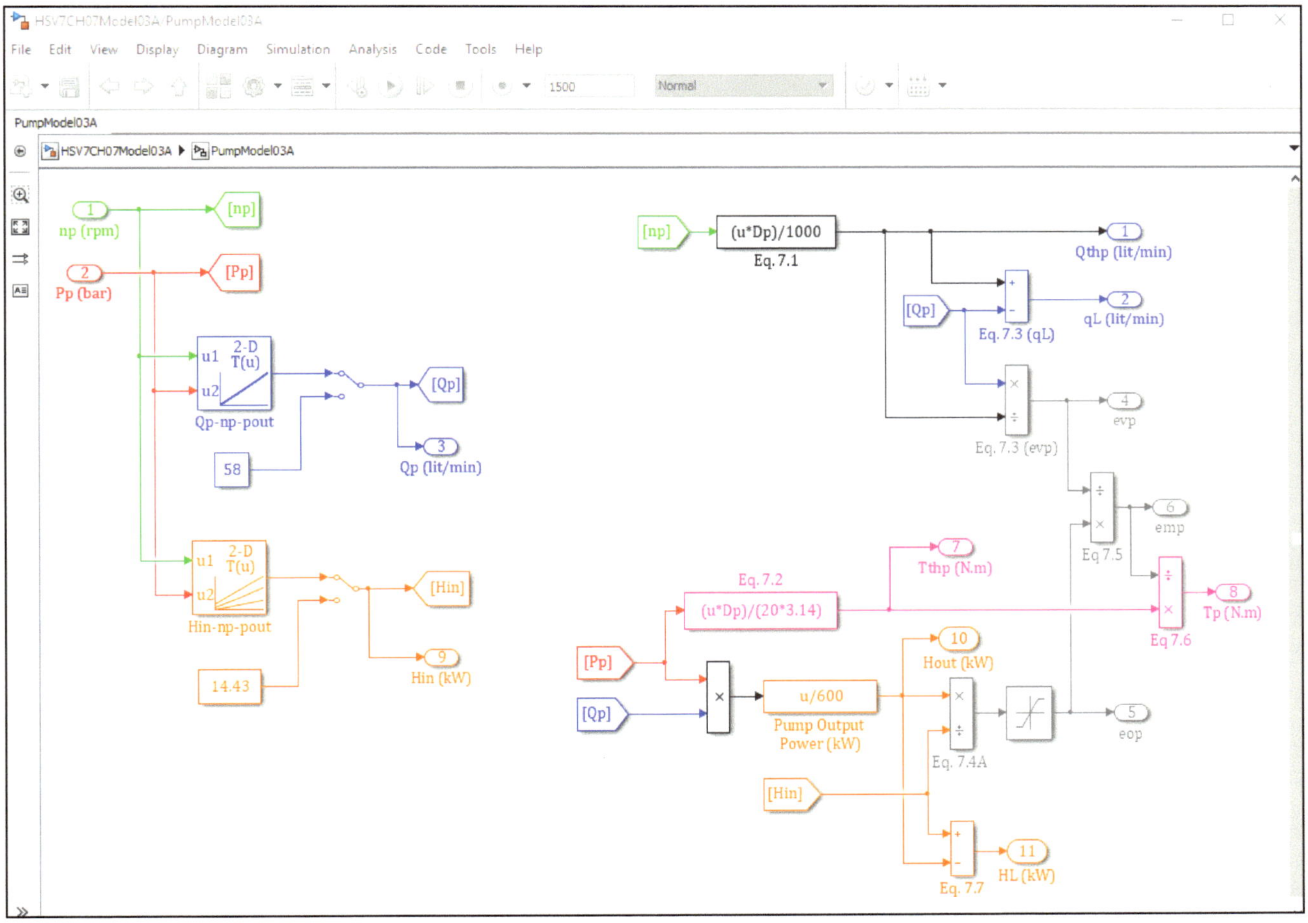

Fig. 7.20- Details under the Mask of Simulation Model 03A

Figures 7.21 and 7.22 show that the lookup tables were built based on the previously given generic characteristics. Since the pump characteristics are linear, then three points are just fine to build the lookup table. Algorithm of interpolation and extrapolation in the lookup tables are set to be "linear".

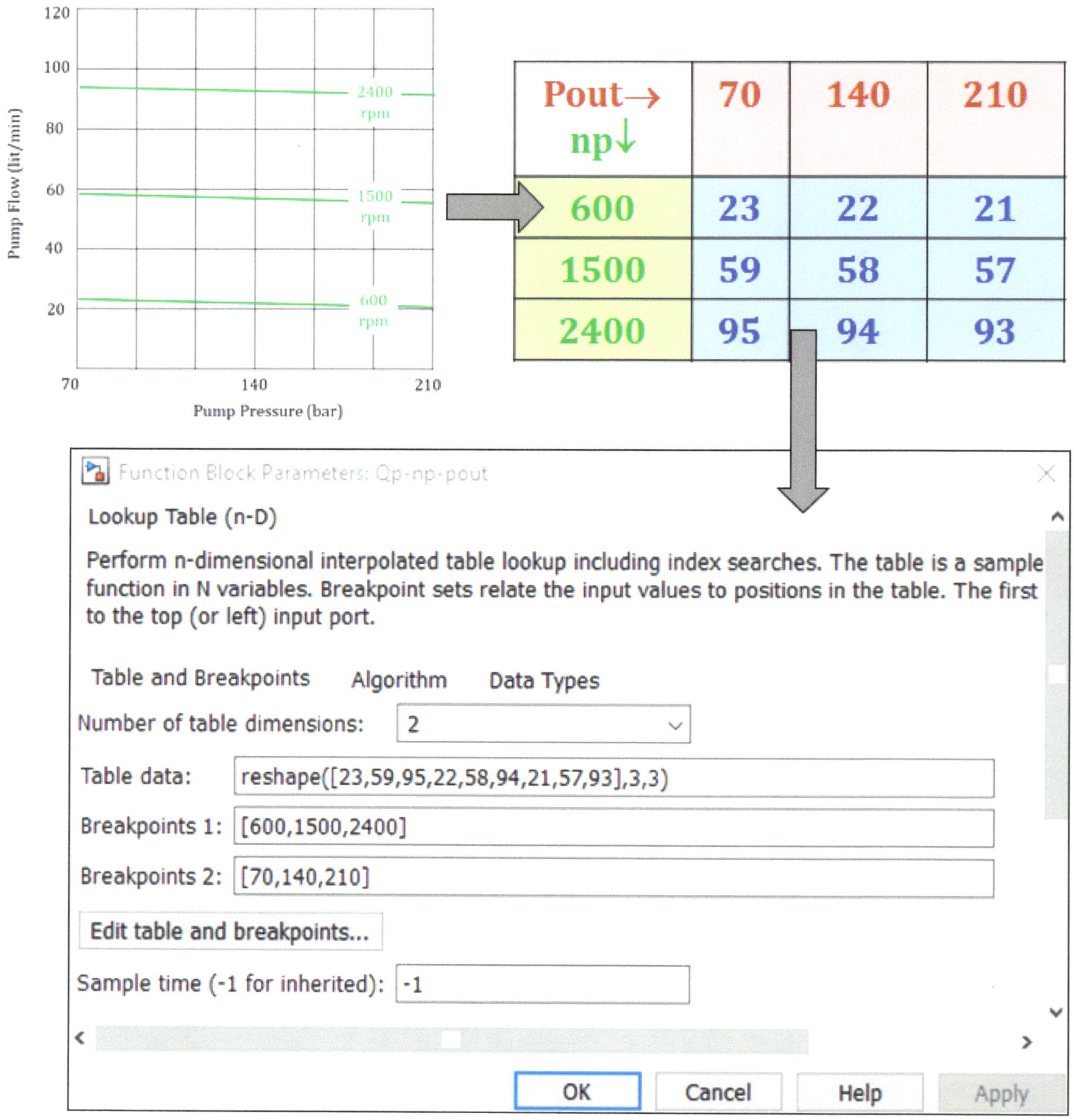

Fig. 7.21- Lookup Table for Pump Actual Outlet Flow versus Driving Speed and Pressure

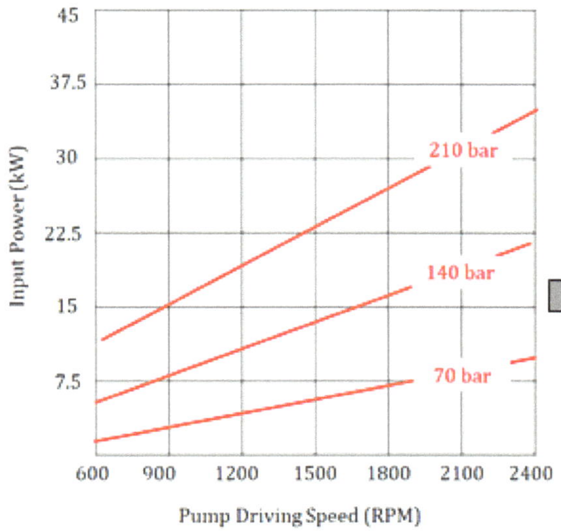

Pout→ np↓	70	140	210
600	2.89	5.71	8.45
1500	7.29	14.43	21.42
2400	11.78	23.33	34.63

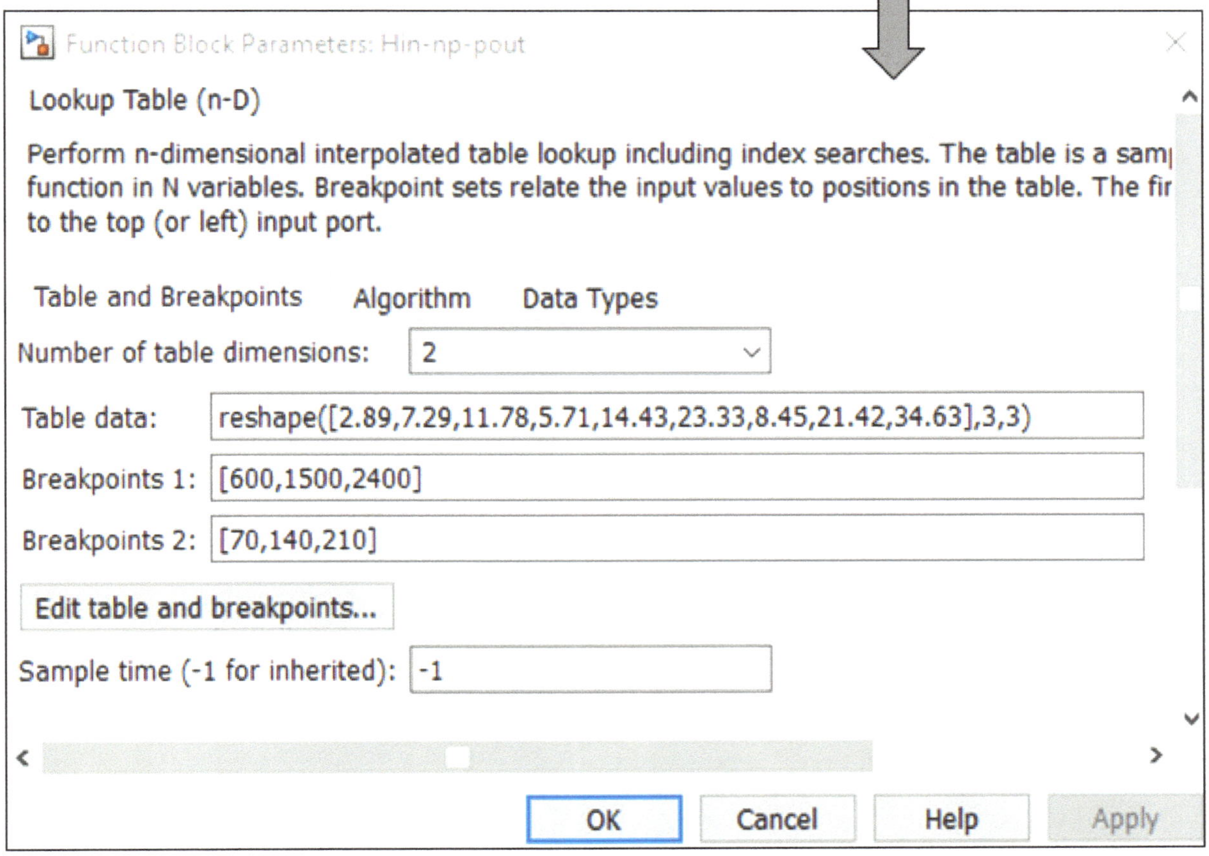

Fig. 7.22- Lookup Table for Pump Actual Input Power versus Driving Speed and Pressure

7.7.4- Model Validation

Figure 7.23 shows the pump static characteristics versus operating pressure at constant driving speed. As shown in the figure, for typical fixed displacement pumps, volumetric efficiency decreases, and mechanical efficiency increases with the increase of working pressure. These characteristics validate the generic characteristics loaded to the lookup tables.

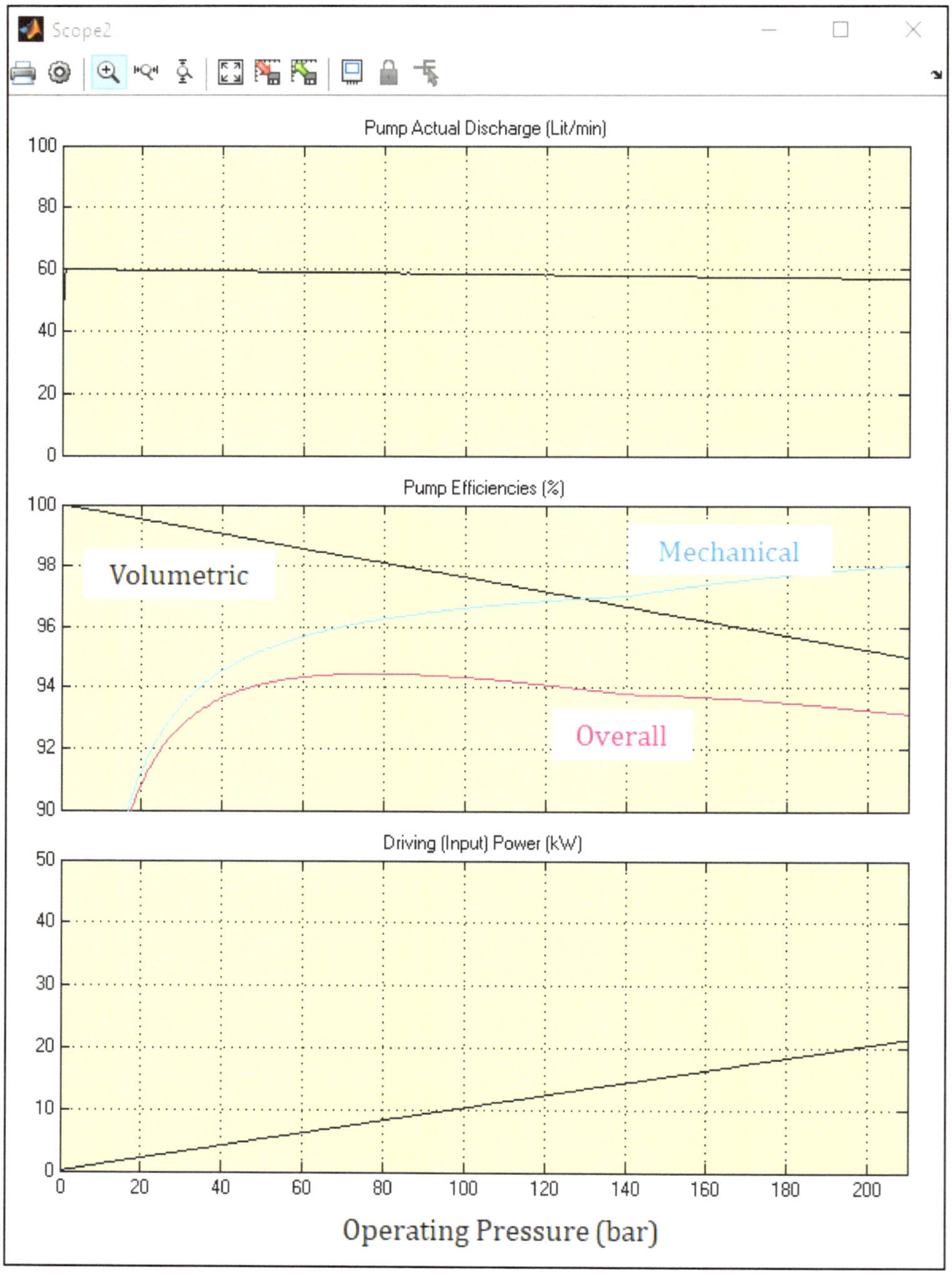

Fig. 7.23- Pump Static Characteristics versus Operating Pressure (at n_p = 1500 rpm)

Figure 7.24 shows the pump static characteristics versus driving speed at constant operating pressure. As shown in the figure, in typical fixed displacement pumps, volumetric efficiency increases, and mechanical efficiency decreases with the increase of driving speed. These characteristics validate the generic characteristics loaded to the lookup tables.

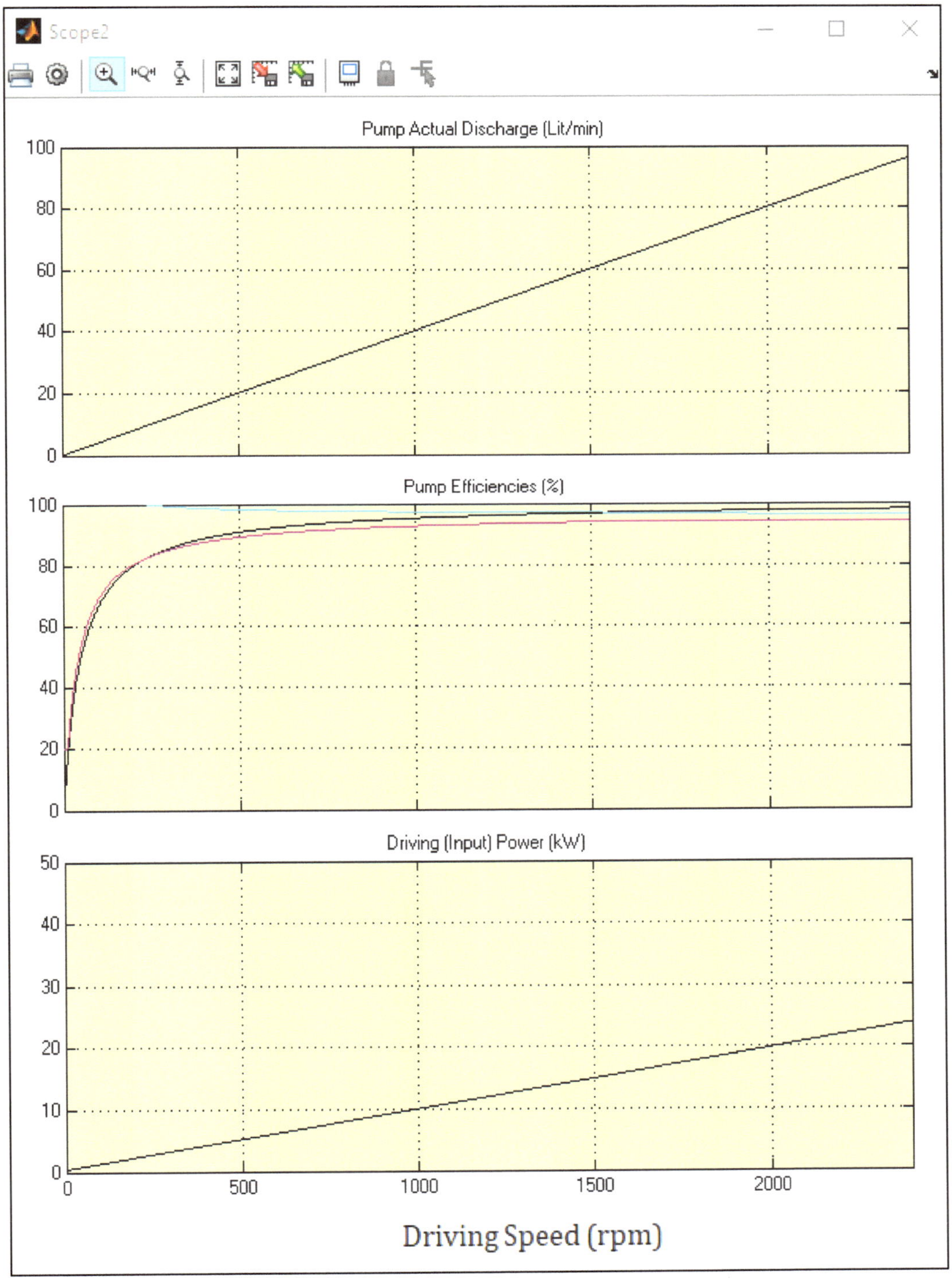

Fig. 7.24- Pump Static Characteristics versus Driving Speed (at P_p = 140 bar)

7.8- Model #03B for a Fixed Displacement Pump Running at Variable Operating Conditions Based on Given Efficiency Curves

7.8.1- Model Features and Assumptions

Same as in model 03A except that volumetric efficiency and overall efficiency curves are given instead of pump test data.

7.8.2- Simulation Model

Figure 7.25 shows the upper mask of simulation model (HSV7CH07Model03B).

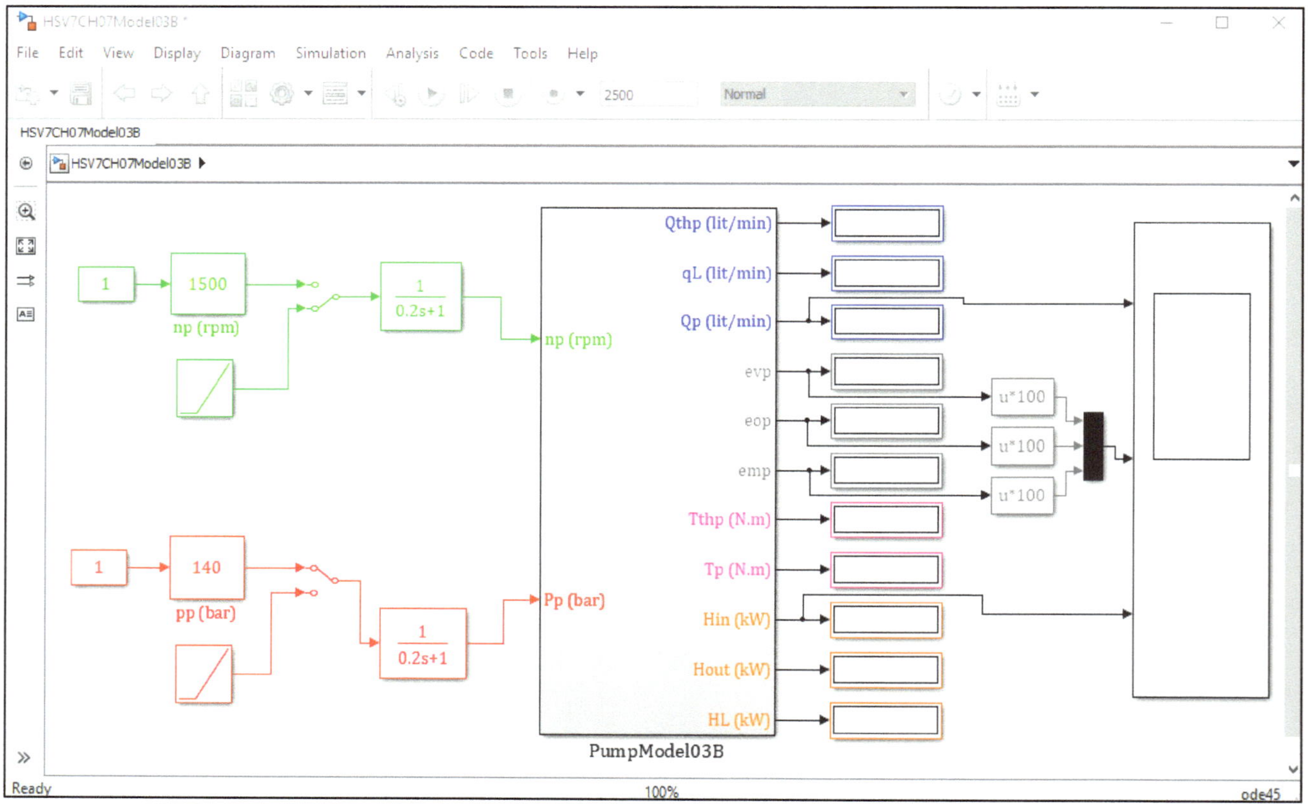

Fig. 7.25- Pump Simulation Model 03B

Figure 7.26 shows the details under the mask of simulation model 03B that is structured to override model 02B. Model 03B serves to simulate a pump that works under constant or variable speed and pressure. Two switches were added so that the model can select pump efficiencies as constants or variable from lookup table. However, if constant efficiencies are selected, this means these values are defined at whatever constant value selected at the pump speed and working pressure sider gains.

The figure shows that efficiencies were developed based on the pump pressure and then corrected based on the driving speed.

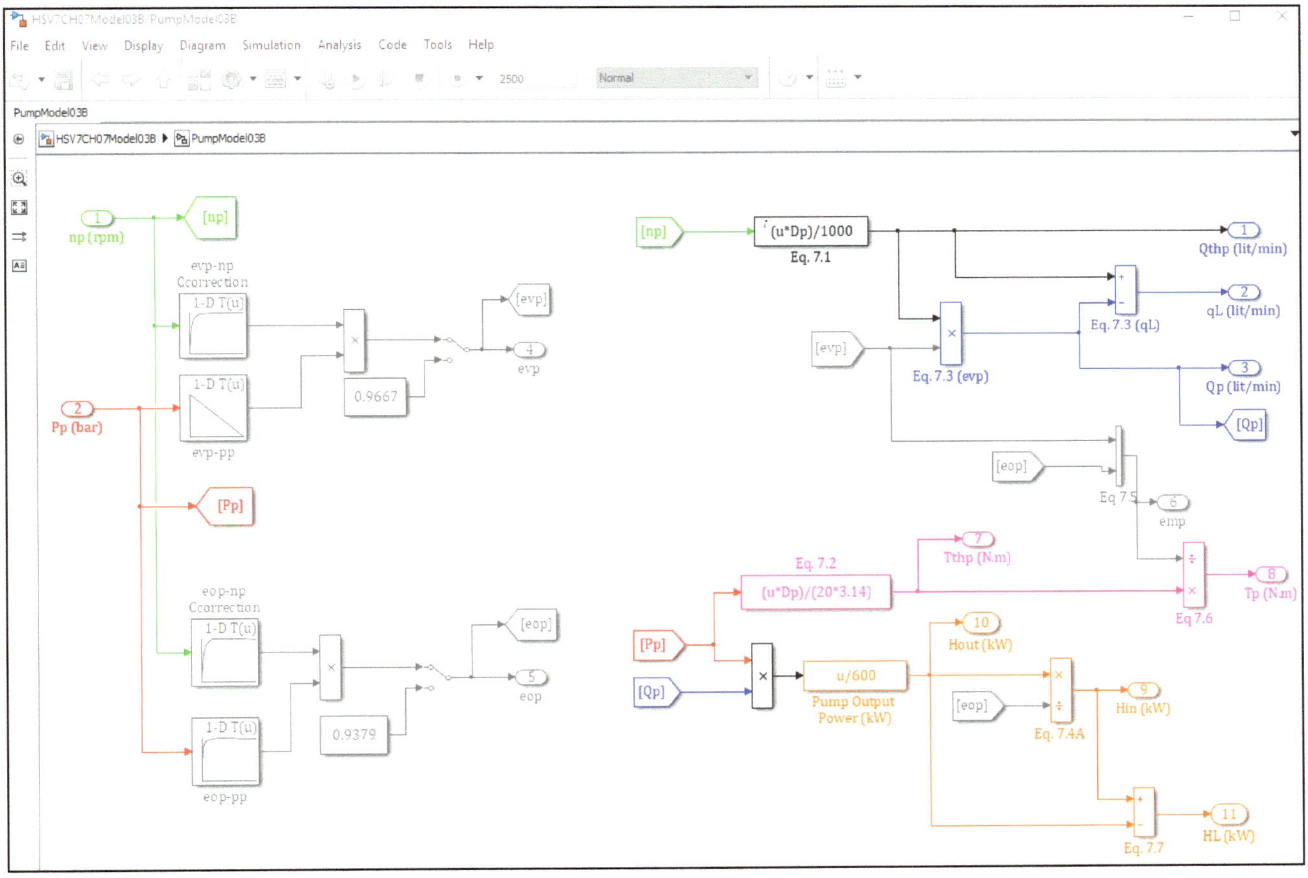

Fig. 7.26- Details under the Mask of Simulation Model 03B

7.8.3- Model Validation

Figure 7.27 shows the pump static characteristics based on give efficiency curves. The figure shows very close results as shown previously in Fig. 7.23 and 7.24. That validates the model with the new calculations. Maybe fine tuning is required for the efficiency curves lookup tables.

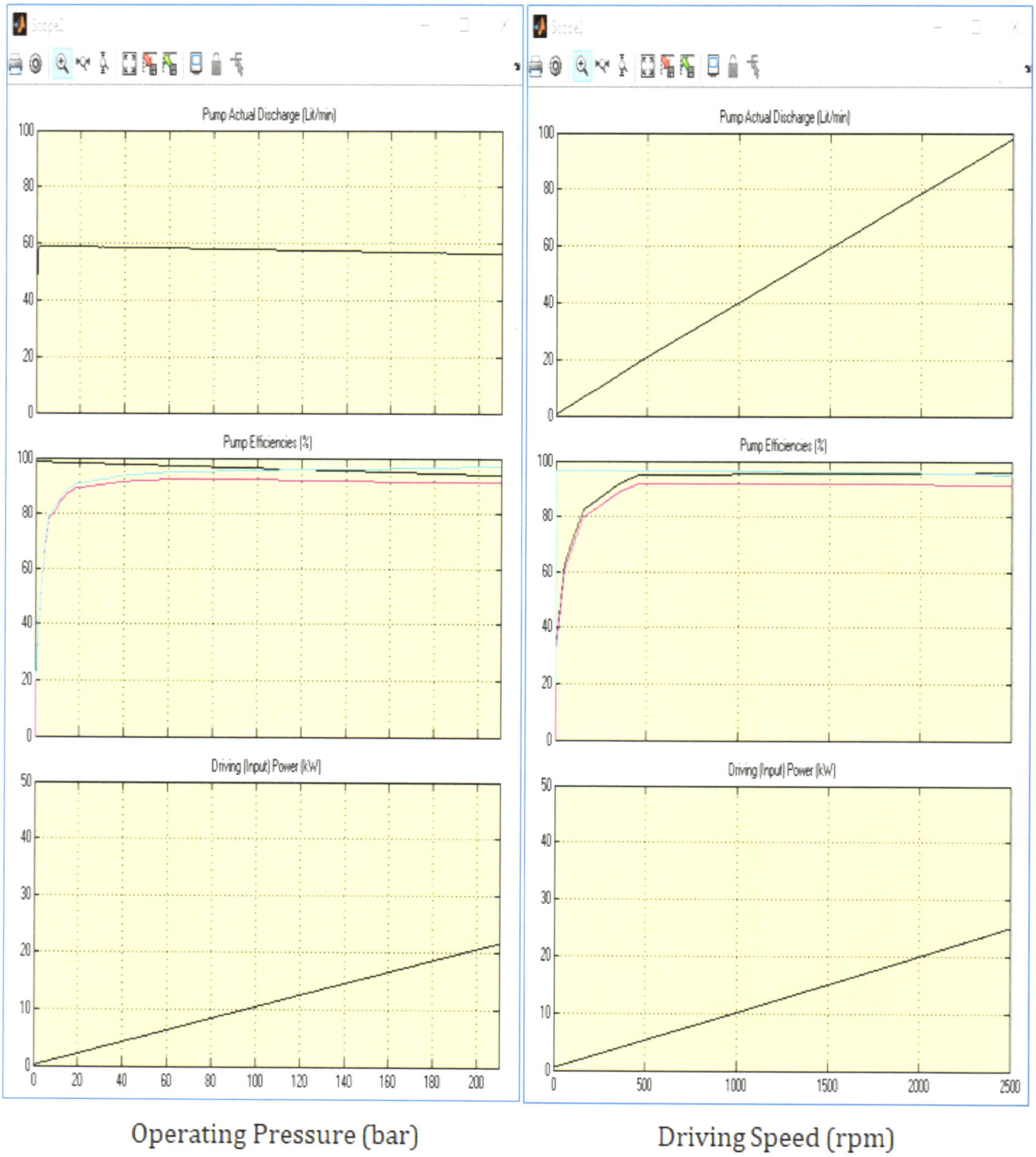

Fig. 7.27- Pump Static Characteristics Based on Given Efficiency Curves

7.9- Modeling Variable Displacement Pumps

Figure 7.28 shows common control modes of variable displacement pumps as follows:

1. Pressure-Compensated pump.
2. Displacement-Controlled pump for satisfying the variable flow demand of the application.
3. Torque-Limited (Constant-Power) pump.

There are various other control modes, such as load-sense pumps that are out of scope of this textbook. However, this textbook provides concepts that can be used to model pumps with such controllers.

If these **Q-P** performance curves are the only available information about the pump, then simple models can be developed that contain just a lookup table to provide only pump output flow versus pump output pressure. Otherwise, if full test data is available, then a special model should be developed to give full outputs. It is to be noted that these models are tied to a specific pump size and a particulaar driving speed.

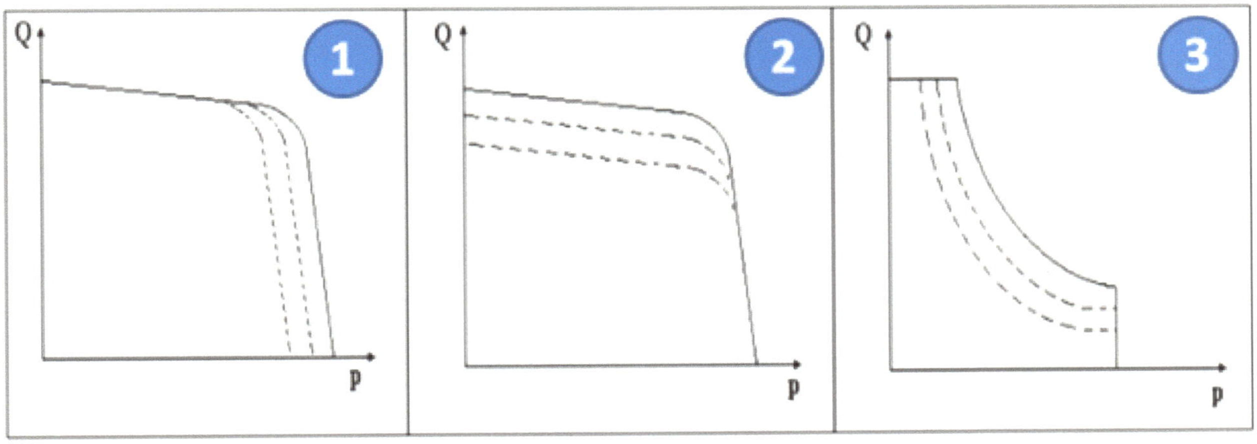

Fig. 7.28- Common Control Modes of Variable Displacement Pump (at Constant rpm)

7.10- Lumped Modeling of Pressure-Compensated Pumps

7.10.1- Static Characteristics of Pressure-Compensated Pumps

Referring to Fig. 7.28, *Pressure-Compensated Pumps* allow the user to vary the pump maximum pressure, at which the pump is de-stocked, by a controller that is actuated manually or remotely (hydraulically by a pilot pressure or electro-hydraulically by an electrical variable signal). Figure 7.29 shows two manually adjusted pressure-compensated pumps. The pump on the left side is a swash-plate axial piston type mechanism, and the pump on the right side is a vane-type mechanism. As it has been discussed previously, such pumps should be modeled based on internal structure of the pumping mechanism for purposes of pump development. For purposes of overall machine modeling, they are modeled based on the published static and dynamic characteristics, no matter what the pumping mechanism is.

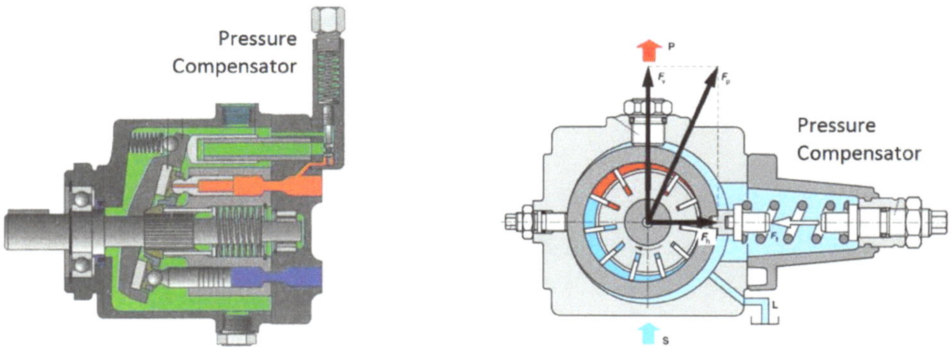

Fig. 7.29- Variable Displacement Pressure-Compensated Pumps

Figure 7.30 shows typical static characteristics of pressure-compensated pumps. Such a pump work as a fixed pump as long as the pressure is less that the *Cracking Pressure* $\mathbf{P_{CR}}$. Once pump pressure exceeds the cracking pressure, pump controller starts to de-stroke the pump. Pump continues to de-stroke within the *Override Zone*. Pump is fully de-stroked at *Cut-off* (maximum) pressure $\mathbf{P_{co}}$. Then, when $\mathbf{P_p <= P_{CR}}$, the model should treat the pump as a fixed pump. When $\mathbf{P_p > P_{CR}}$, the model should apply the control mode.

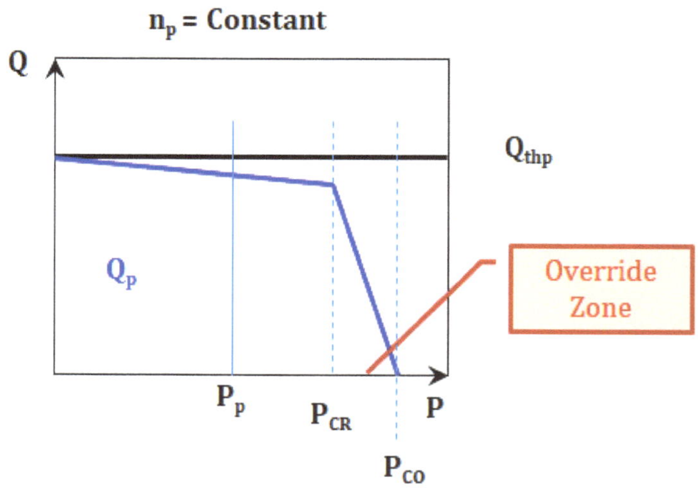

Fig. 7.30- Typical Static Characteristics of Pressure-Compensated Pumps

Figure 7.31 shows typical test data for 56 cc/rev pressure-compensated pump.

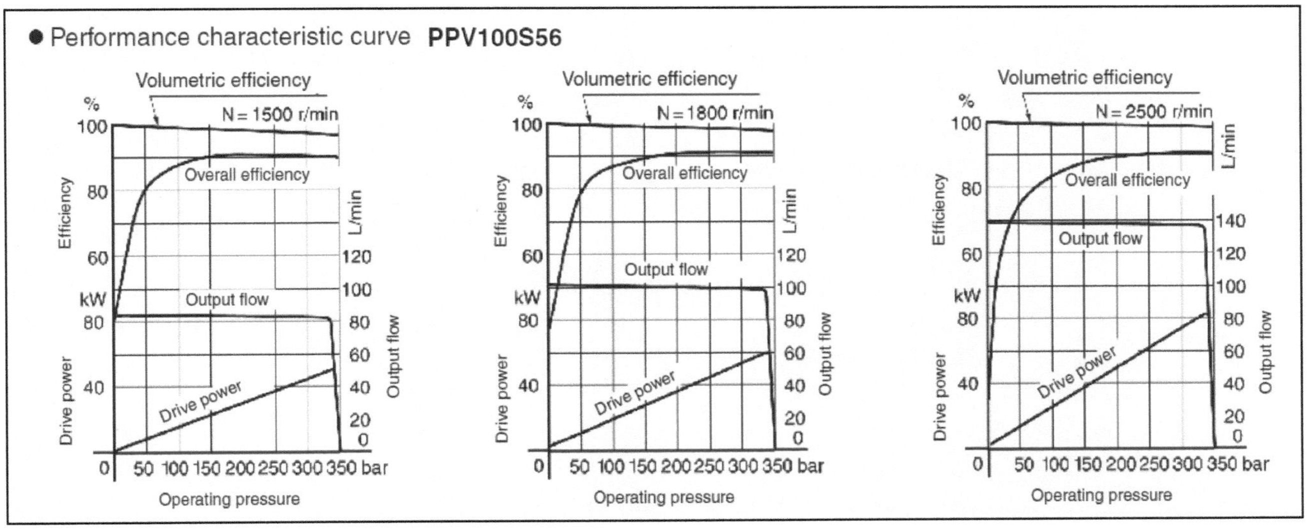

Fig. 7.31- Test Data for 56 cc/rev Pressure-Compensated Pumps (Courtesy of Hydac)

7.10.2- Model #04A for Pressure-Compensated Pumps Based on Given Test Data

7.10.2.1- Model Features and Assumptions

The model uses the generic pump data that are used in previous model.

7.10.2.2- Mathematical Model

Same set of equations 7.1 through 7.7 are still used except that a logic block was used to rout the calculation to the override zone once the working pressure exceeds the cracking pressure.

7.10.2.3- Simulation Model

Figure 7.32 shows the upper mask of the simulation model (HSV7CH07Model04A). The figure shows that the model has the following elements:

Driving Speed n_p: A slider gain is used to simulate variable driving speed. Additionally, a "Step" block is used to investigate the pump step response due to sudden change in the driving speed.

Pump Outlet Pressure P_p: A slider gain is used to simulate variable pump pressure. Additionally, a "Step" block is used to investigate the pump step response due to sudden change in the working pressure. Ramp block is still used to develop the static characteristic of the pump versus pressure.

Pump Constants.: values of cracking pressure, cut-off pressure, and pump maximum displacement are loaded to the block properties.

Dynamic Effect of Driving Speed Change (at D_p and P_p are Constants): Similar to previous model, normalized transfer function (NTF) is used to represent such an effect.

Dynamic Effect of Operating Pressure Change (at n_p = Constants): Similar to previous model, normalized transfer function (NTF) is used to represent such an effect.

Dynamic Effect of Pump Displacement Control Unit: Control unit behaves like a 2nd order system. So, a 2nd order NTF is used. For this generic pump, a reasonable natural frequency of 10 rad/s and a damping ratio of 0.7 are assigned.

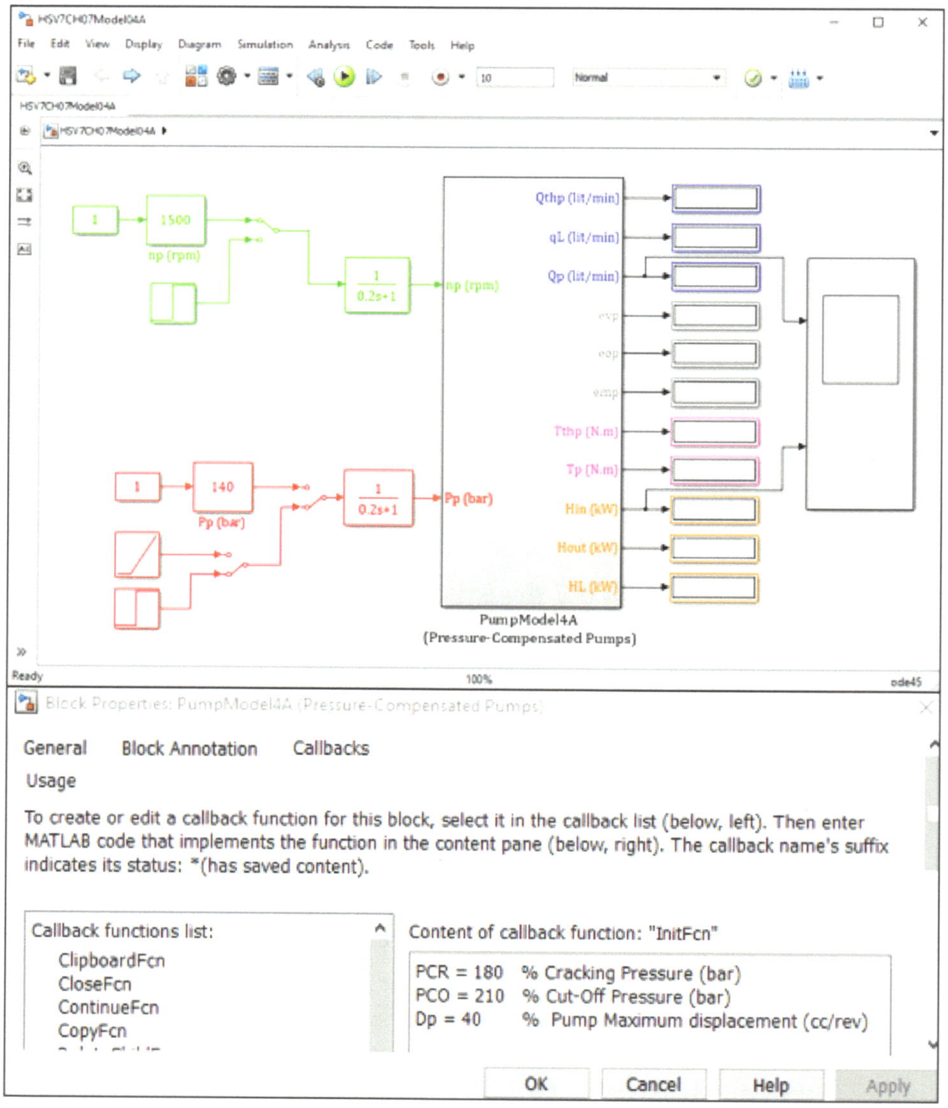

Fig. 7.32- Pump Simulation Model 04A

Figure 7.33 shows the details under the mask of the Model 04A. The model treats the pump as a fixed pump **(where $P_p \le P_{CR}$).** When the controller is activated **(where $P_{CR} < P_p \le P_{CO}$),** it applies the control action.

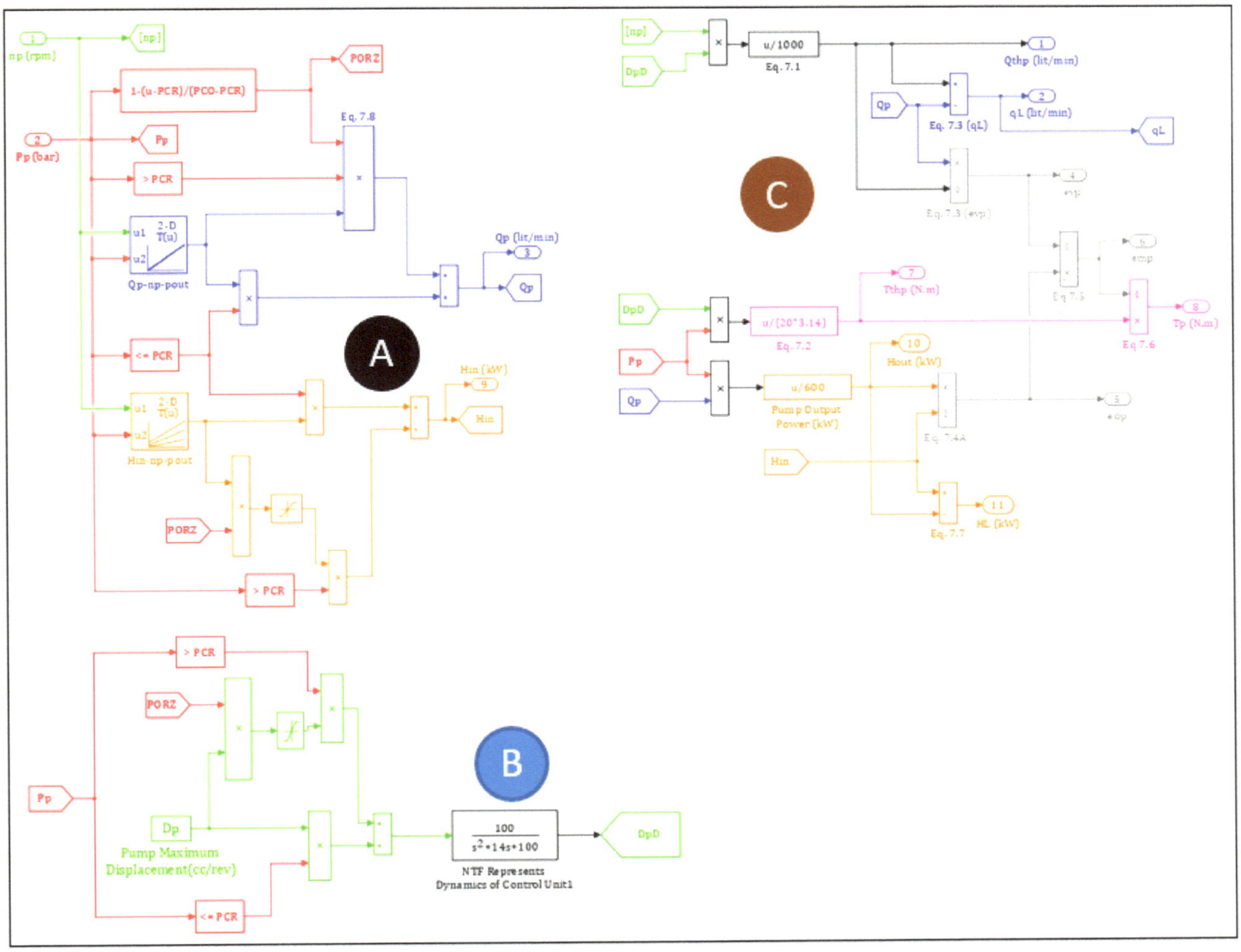

Fig. 7.33- Details under the Mask of Simulation Model 04A

Therefore, as shown in the figure, the model contains various sections as follows:

Section A (shown in Fig. 7.34): Pump Flow and Input Power

In this section, two logic blocks where used to rout the calculation based on the pump working pressure as follows:

- When $\mathbf{P_p \leq P_{CR}}$: Pump flow and input power are selected from the lookup tables as in case of a fixed pump.
- When $\mathbf{P_{CR} < P_p \leq P_{CO}}$: Pump flow and input power values are multiplied by a function that represents the pump controller. Output of the pump controller function are assigned to a global variable "PROZ", stands for Pressure in Override Zone, to be used by other sections of the model.

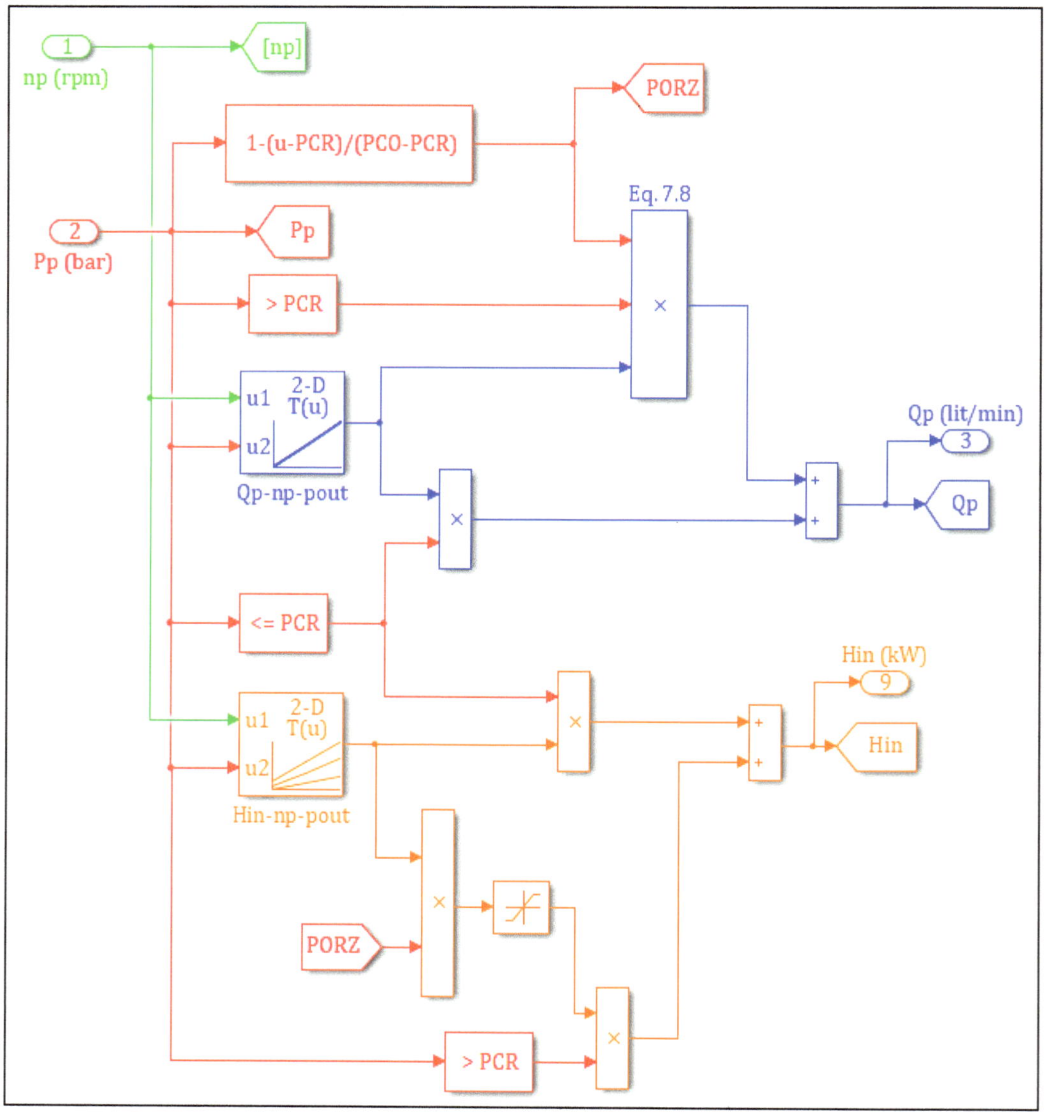

Fig. 7.34- Section A of Model 04A

Section B (shown in Fig. 7.35): Pump Displacement

In this section, same two logic blocks where used to rout the calculation based on the pump working pressure as follows:

- When $\mathbf{P_p \leq P_{CR}}$: Pump displacement is kept constant as a fixed displacement pump. The value of $\mathbf{D_p}$ is already loaded in the model properties.
- When $\mathbf{P_{CR} < P_p \leq P_{CO}}$: Pump displacement based on the controller function by multiplying pump displacement by the variable "PROZ". A saturation block is used to assign certain minimum for the pump displacement. That is because, when the pump is fully de-stroked at $\mathbf{P_p = P_{CO}}$, pump still generates flow equal internal leakage despite no flow is discharged out of the pump.

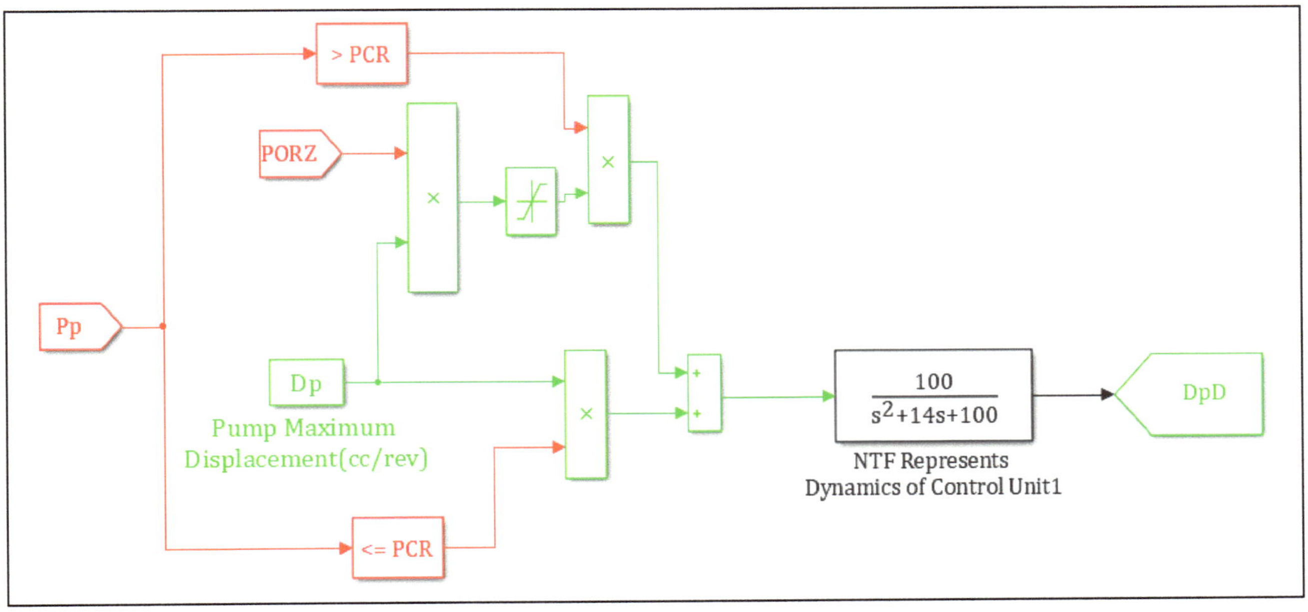

Fig. 7.35- Section B of Model 04A

Section C (shown in Fig. 7.36): Other Calculations

In this section, all other calculations are performed as in model 03A. The only difference is that pump displacement and output flow is affected by the controller action.

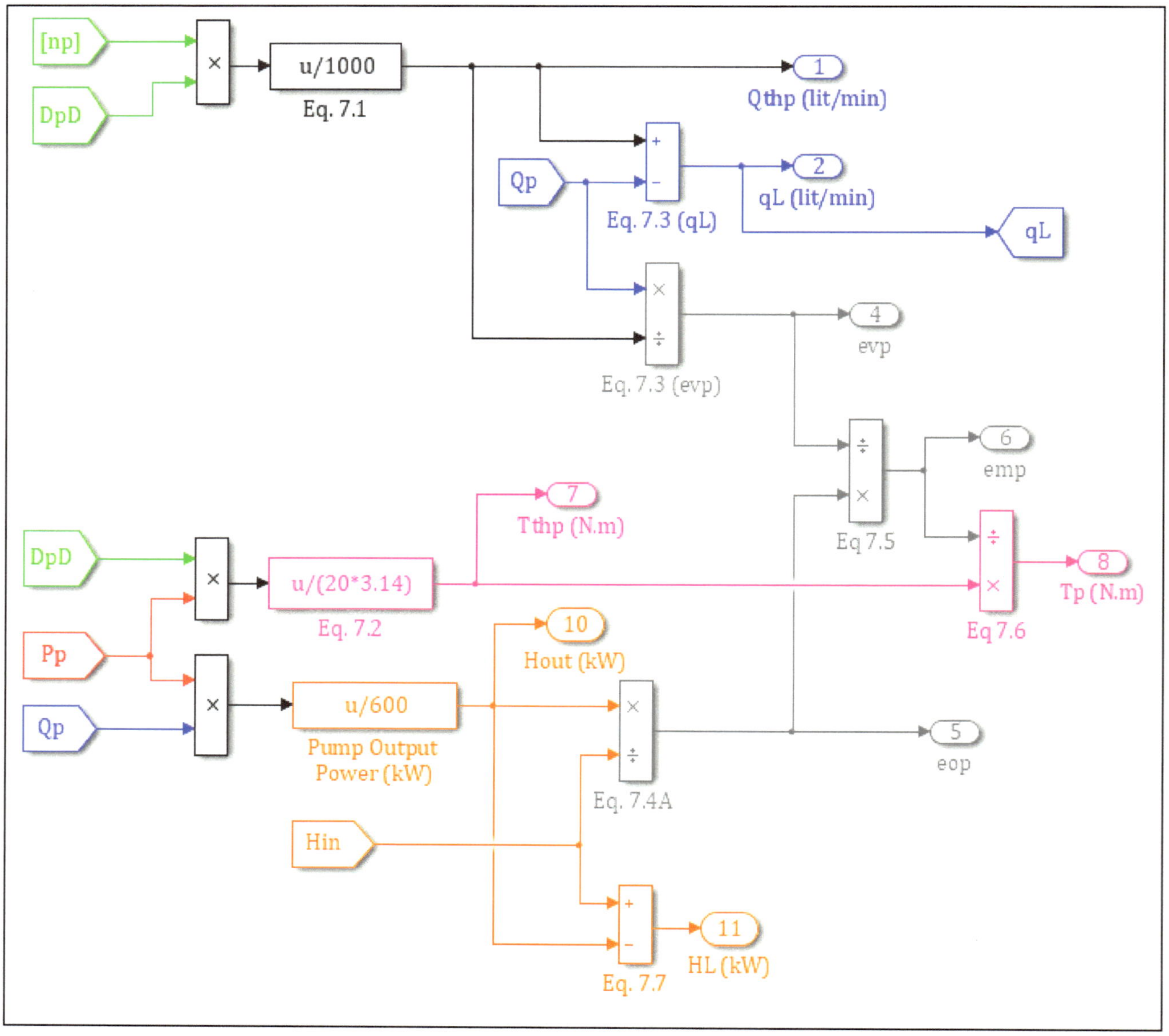

Fig. 7.36- Section C of Model 04A

7.10.2.4- Model Validation

Static Characteristics Validation: Figure 7.37 shows simulated results for a pressure-compensated pump static characteristic when the working pressure is ramped from zero to 210 bar. The figure shows that the controller starts to de-stroke the pump when the pump working pressure exceeds the cracking pressure (180 bar). Pump is fully de-stroked when the pump working pressure reaches the cut-off pressure (210 bar).

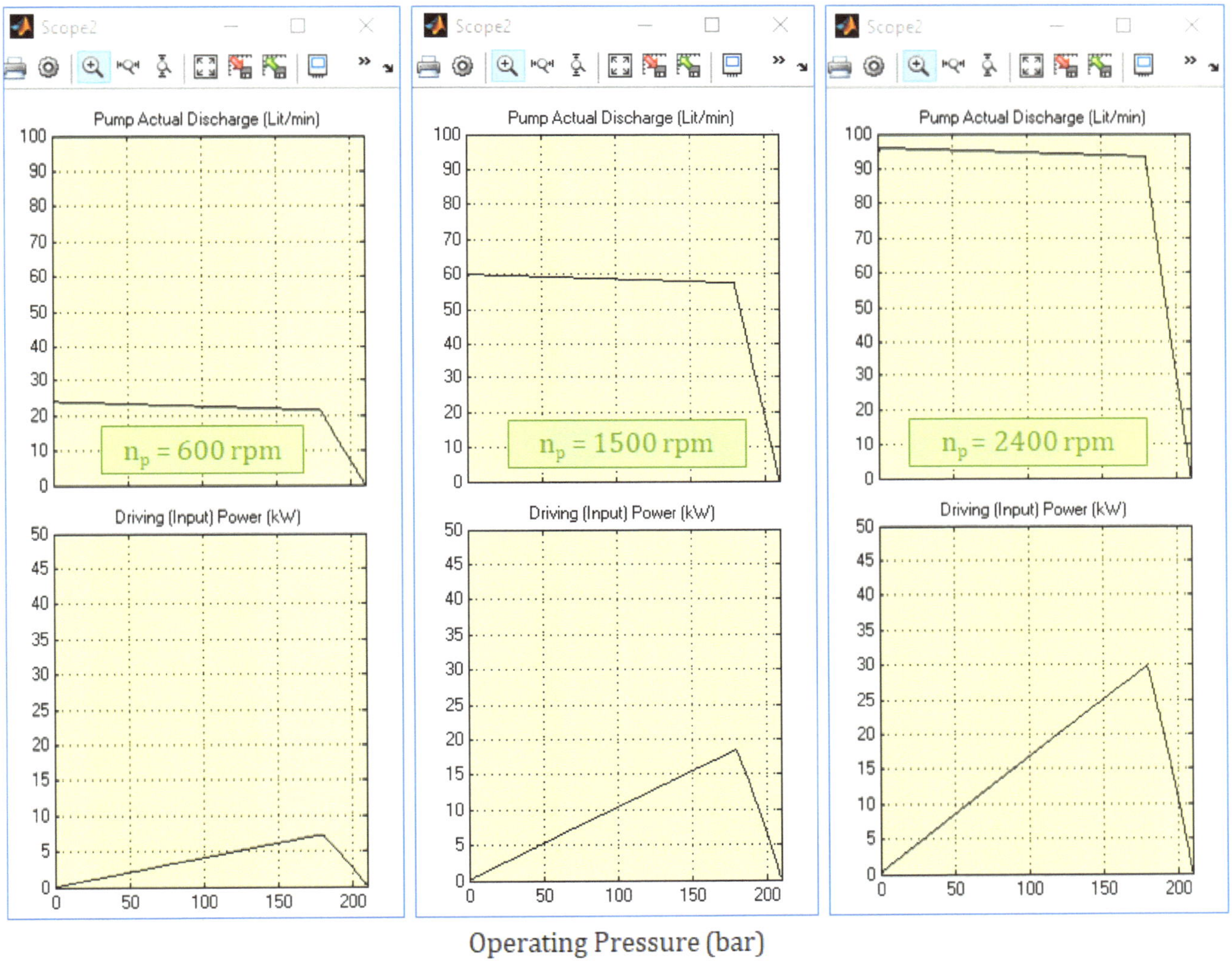

Fig. 7.37- Simulating Pressure-Compensated Pump Static Characteristics

Dynamic Characteristics Validation: Figure 7.38 shows simulated results for a pressure-compensated pump step response. The left side of the figure shows the step increase of the pump flow and input power (at constant pressure of 140 bar) when the driving speed is increased from 0 to 600 rpm and then from 600 rpm to 1500 rpm. The step response follows the dynamic characteristics assigned in the corresponding normalized transfer function.

The right side of the figure shows the step decrease of pump flow and input power (at constant driving speed of 1500 rpm) when the working pressure increased from 50 bar to 190 bar. The step response follows the dynamic characteristics assigned in the corresponding normalized transfer function.

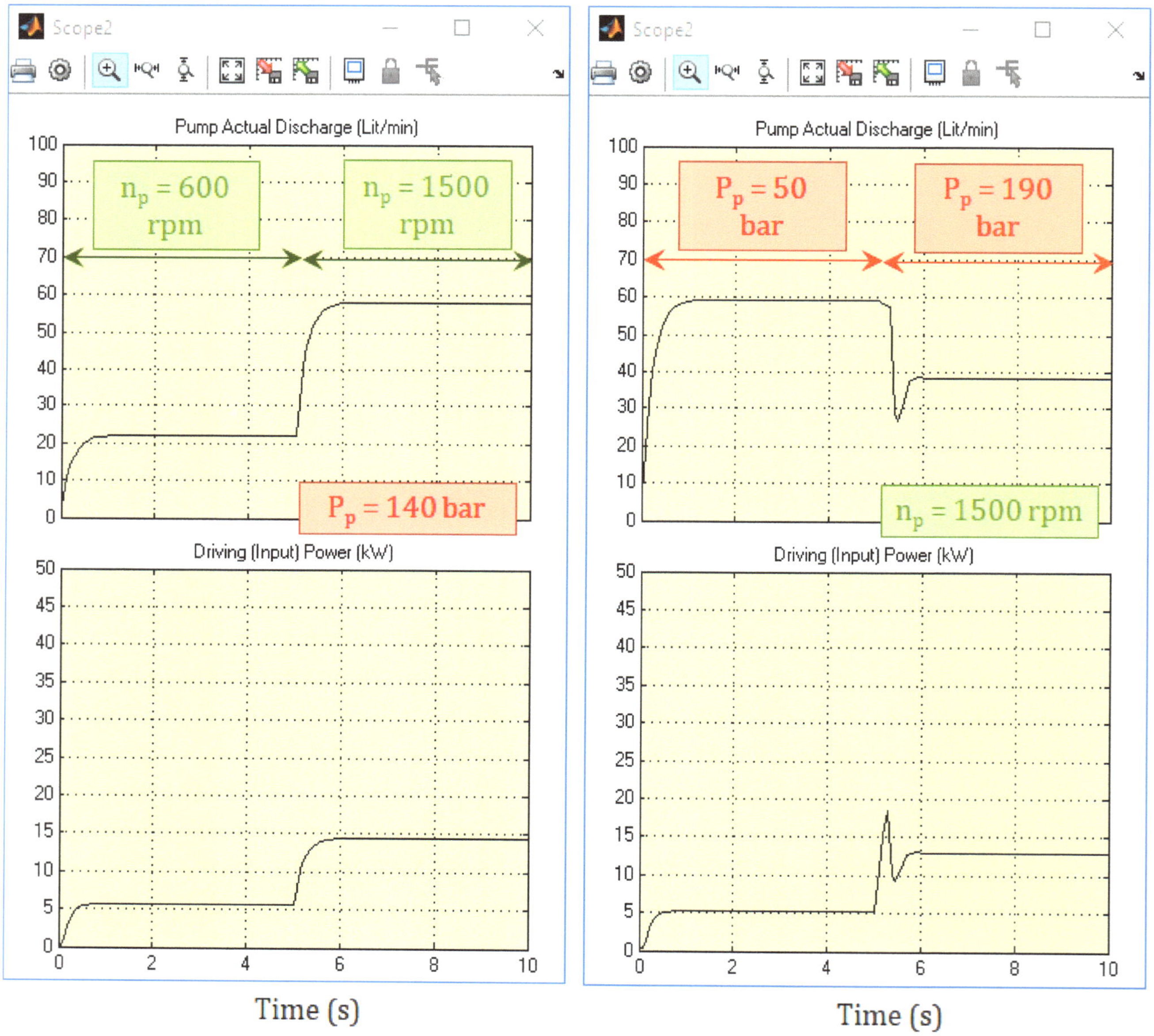

Fig. 7.38- Simulating Pressure-Compensated Pump Step Response

7.10.3- Model #04B for Pressure-Compensated Pumps Based on Given Efficiency Curves

7.10.3.1- Model Features and Assumptions

Same as in model 04A except that volumetric efficiency and overall efficiency curves are given instead of pump test data.

7.10.2.2- Simulation Model

Figure 7.39 shows the upper mask of simulation model (HSV7CH07Model04B).

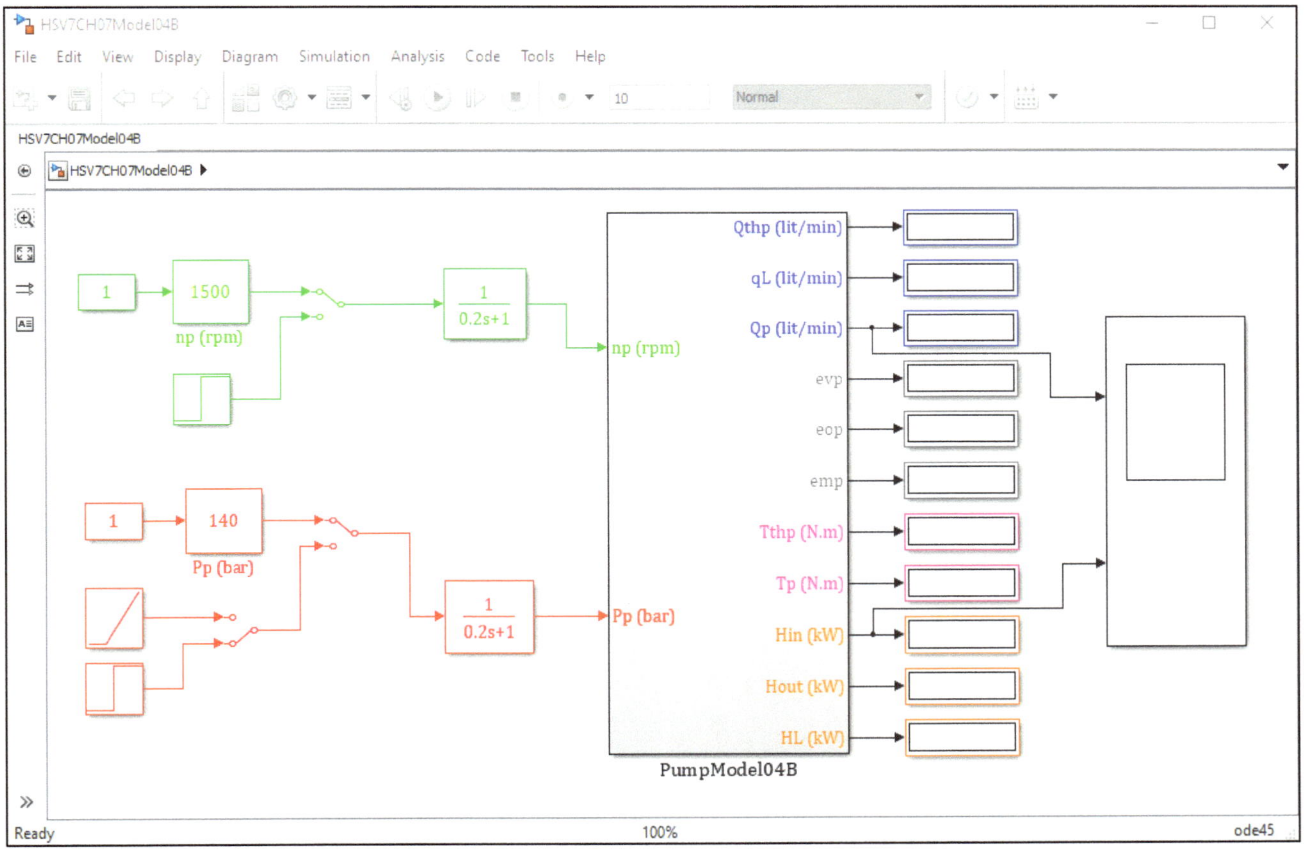

Fig. 7.39- Pump Simulation Model 04B

Figure 7.40 shows the details under the mask of the Model 04B. Like model 04A (section B), as shown previously in Fig. 7.35, the model treats the pump as a fixed pump until its controller is actuated. Unlike model 04A (section A), the given data are efficiency curves loaded into lookup tables. Some modifications have been done to section C to adapt the model for the new sequence of solving the set of Equations 7.1 through 7.7.

7.10.2.3- Model Validation

By running the model to simulate the static and step response of the pump, it results in identical results as shown in Figs. 7.37 and 7.38.

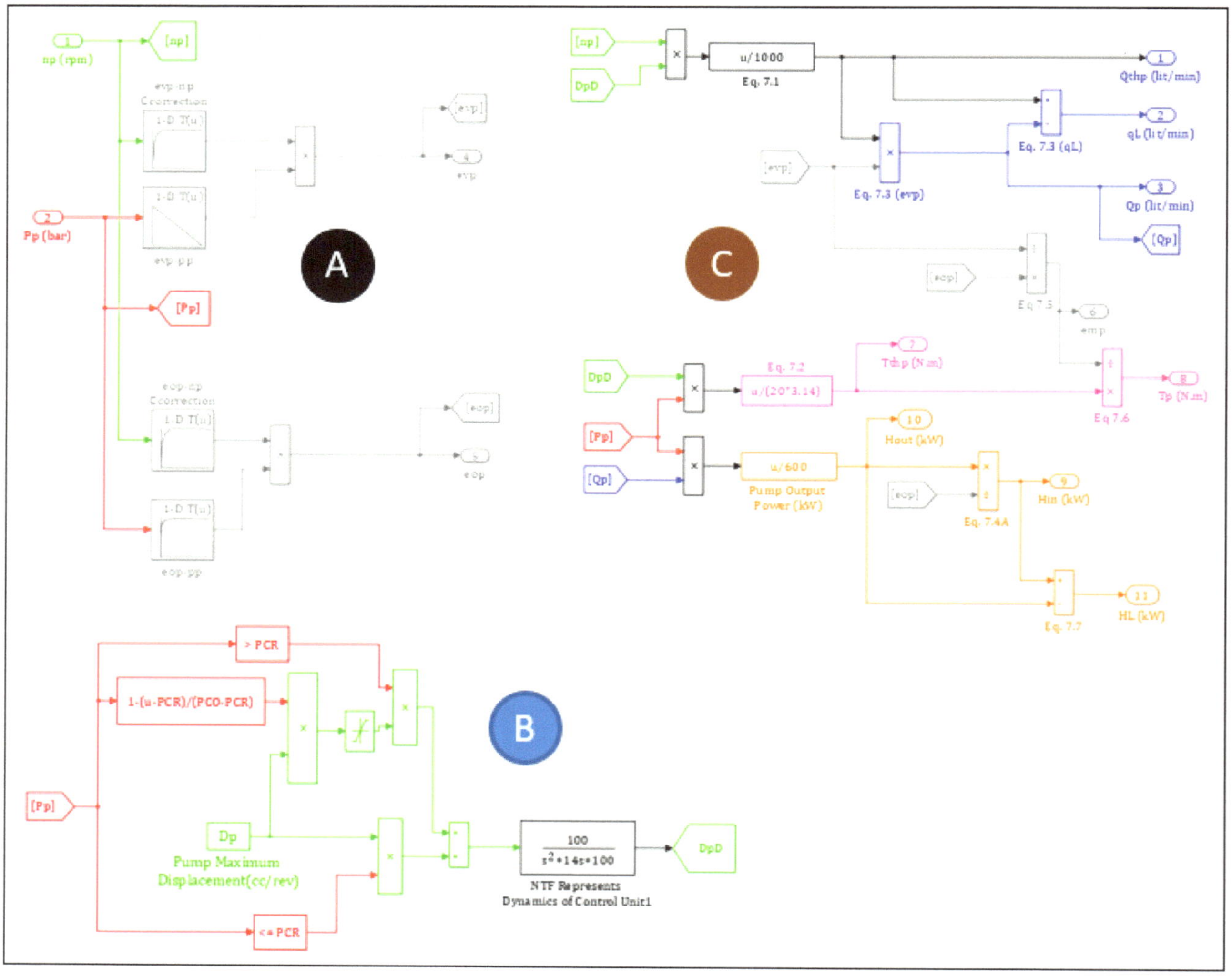

Fig. 7.40- Details under the Mask of Simulation Model 04B

7.10.4- Simplified Model #04C for Pressure-Compensated Pumps

In case if no data is available but the $\mathbf{Q_P}$-$\mathbf{P_P}$ static characteristics are available, a simplified model still can be built.

7.10.4.1- Case Study

A case study for model 04C considers a pressure-compensated pump in the Universal Fluid Power Trainer (UFPT), shown in Fig. 7.41.

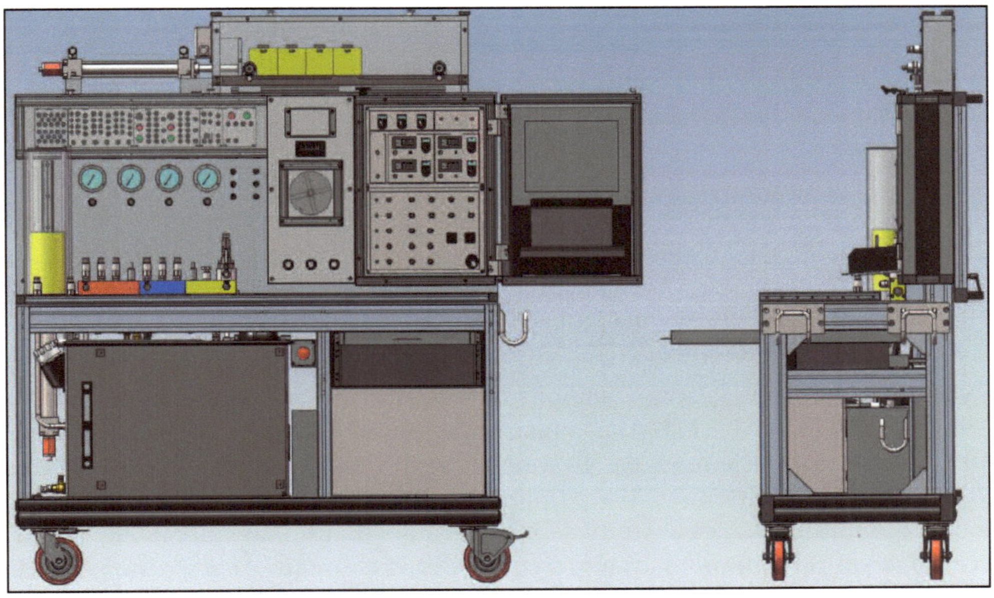

Fig. 7.41- Universal Fluid Power Trainer

7.10.4.2- Model Features and Assumptions

As shown in Fig. 7.42, the model is structured in a simplified form just to give the pump outlet flow based on the pump pressure. The model contains static and dynamic characteristics of the pump.

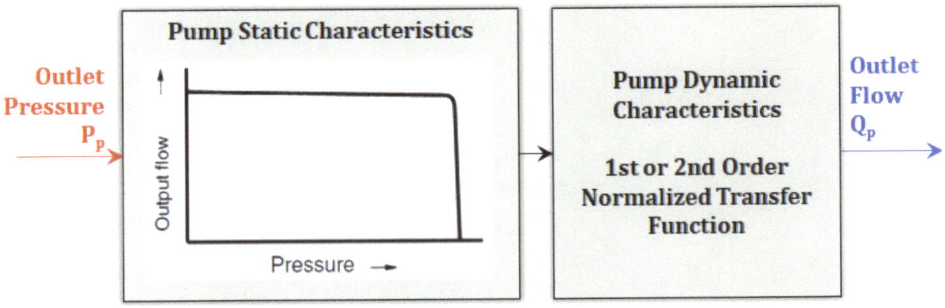

Fig. 7.42- Structure of Pump Model 04C

Pump Static Characteristics: Static characteristics should be loaded into the model in form of a lookup table. It should be published by the manufacturer. If not, it should be developed in the lab. Using simplest test machine, vary the pump outlet pressure and plot the corresponding pump flow. Figure 7.43 shows the static characteristics of the pump in this case study as developed automatically using the UFPT.

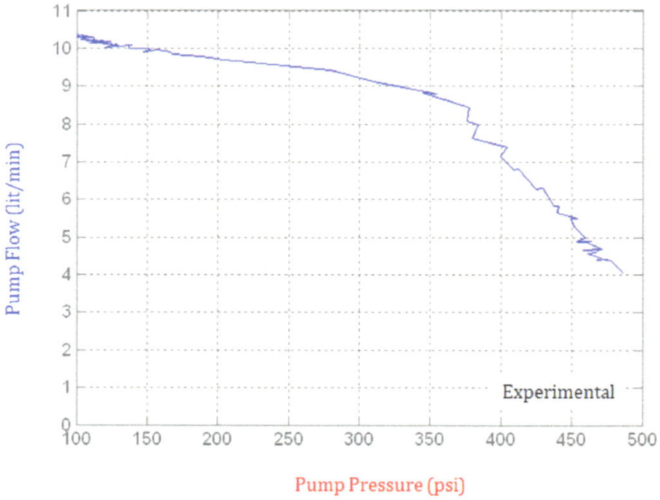

Fig. 7.43- Developed Static Characteristic of the Pump in This Case Study

Pump Dynamic Characteristics: Dynamic characteristics are modeled in form of a normalized transfer function (NTF). At least step response of the pump should be published by the manufacturer. If not, it can be reasonably assumed, e.g. first order with time constant of 20 ms. Model will have more fidelity if there is ability to develop the pump dynamic response in the lab. Figure 7.44 shows the step response of the pump in this case study as it is developed automatically using the UFPT. By analyzing the step response of the pump, the pump's controller was found to raise the pump displacement from minimum, 40%, to maximum displacement in 1.5 seconds. There is a little overshoot that can be fairly ignored. For simplifying the model, it can be assumed that the pump displacement will be dropped with the same rate and it follows a first order system. So, based on what has been discussed previously in chapter 2, a 1st order NTF represents the pump dynamics. Time constant of the NTF equals 300 ms. that is one fifth of the settling time 1.5 seconds.

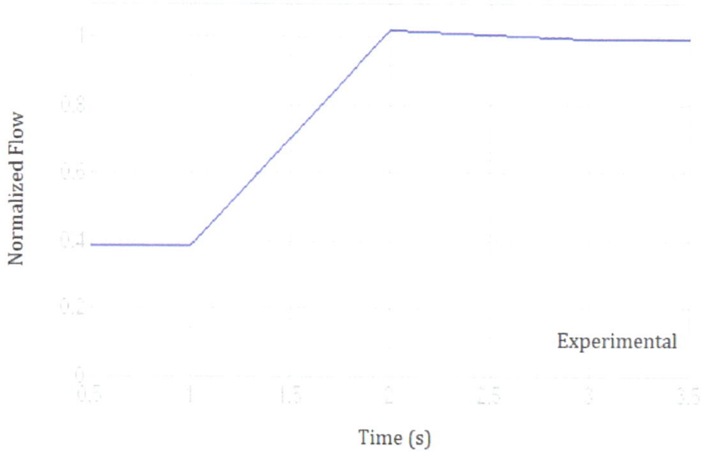

Fig. 7.44- Developed Step Response of the Pump in This Case Study

7.10.4.3- Simulation Model

Figure 7.45 shows the simulation model (HSV7CH07Model04C) that is built based on the generic structure presented in Fig. 7.42.

Static Characteristics: The model considers $\mathbf{Q_P}$-$\mathbf{P_P}$ static characteristics loaded into a lookup table.

Driving Speed: It is to be noted that this model is valid for only a specific driving speed. Otherwise, if the speed is variable such as likely in mobile machines, then the lookup table should be 3-D and the speed should be added as a varying input.

Dynamic Characteristics: The output of the static characteristic is multiplied by a first-order NTF that represents the dynamics of the control unit.

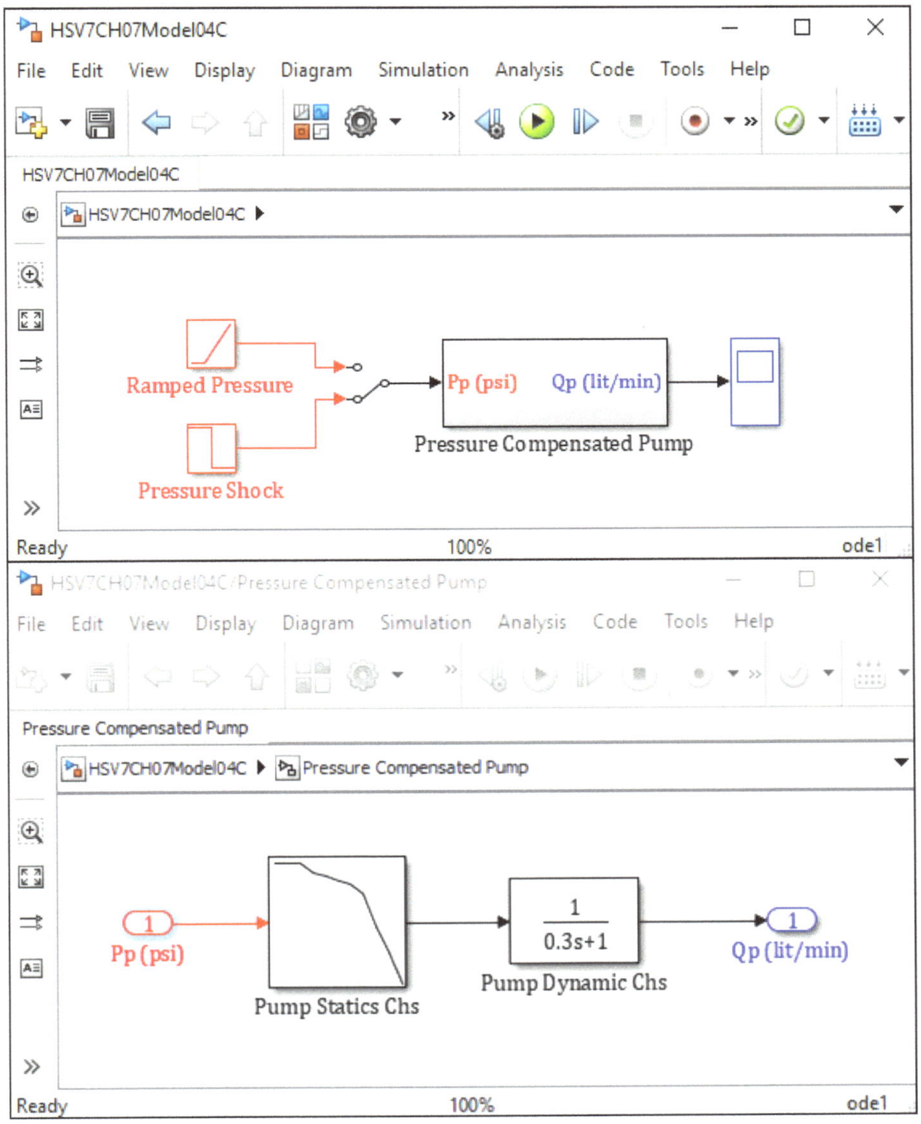

Fig. 7.45- Pump Simulation Model 04C

7.10.4.4- Model Validation

The degree of agreement between the measured and the simulated static characteristics of the pump, shown in Fig. 7.46, validates the model. The model can still be tuned if there is deviation from the tested static characteristics.

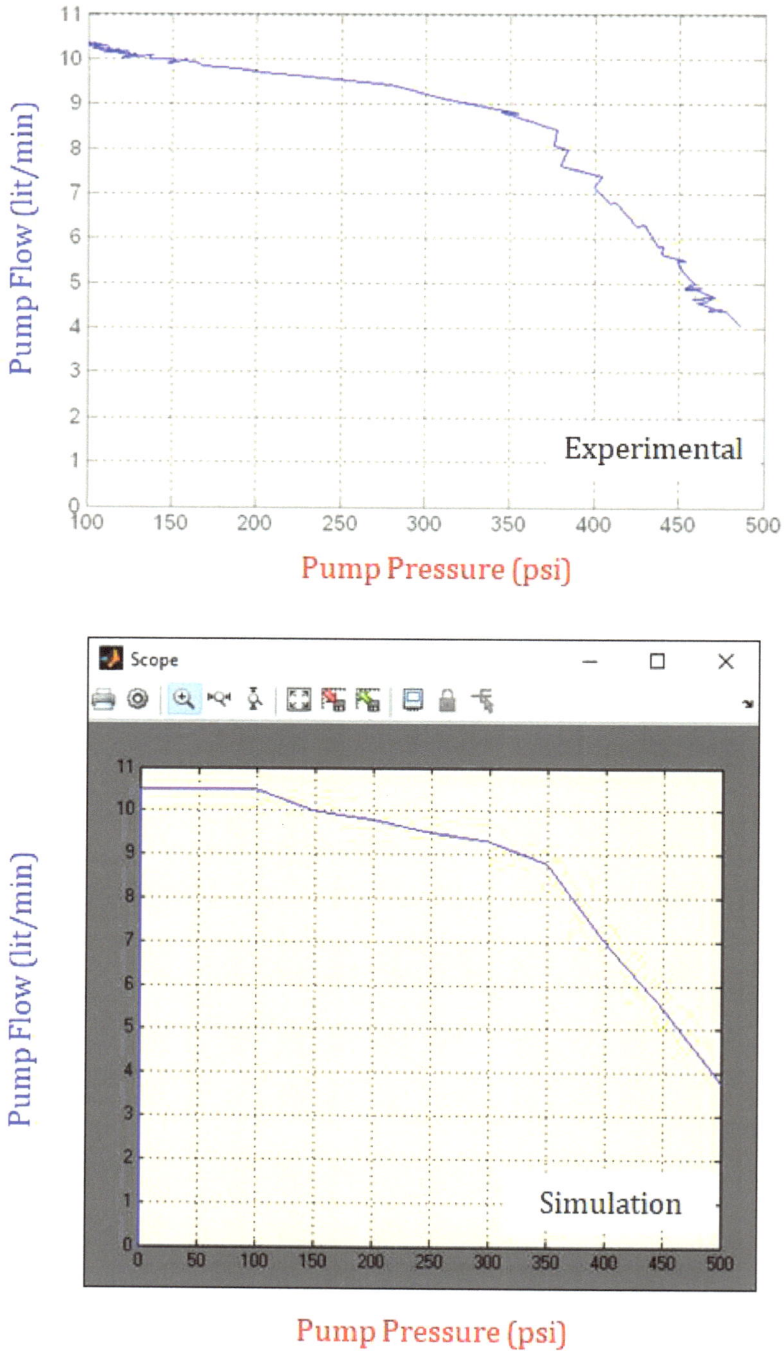

Fig. 7.46- Validating the Pump Static Characteristics

Similarly, the degree of agreement between the measured and the simulated step response of the pump, shown in Fig. 7.47, validates the model. The model can still be tuned if there is deviation from the tested step response.

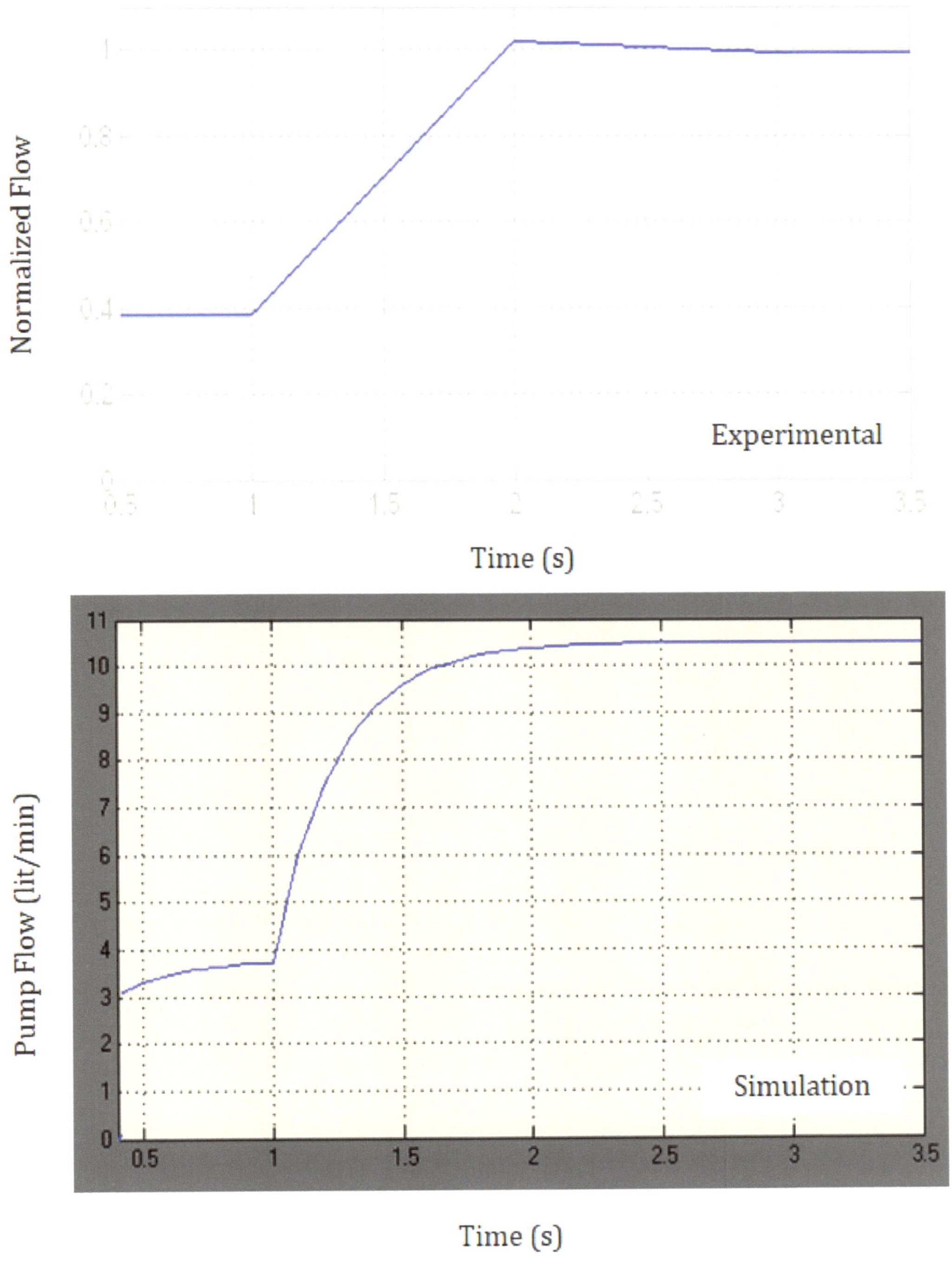

Fig. 7.47- Validating the Pump Step Response

7.11- Lumped Modeling of Displacement-Controlled Pumps

7.11.1- Characteristics of Displacement-Controlled Pumps

Displacement-Controlled Pumps allow the user to vary the pump displacement and consequently the flow by a controller that is actuated manually or remotely (hydraulically by a pilot pressure or electro-hydraulically by an electrical variable signal). Figure 7.48 shows two manually adjusted displacement-controlled pumps. The pump on the left side is of a bent-axis mechanism, in which the pump displacement varies via varying the cylinder block inclination angle. The pump on the right side is of a radial piston mechanism, in which the pump displacement varies via varying the eccentricity between the cam ring and the cylinder block. As it has been discussed previously, such pumps should be modeled based on internal structure of the pumping mechanism for purposes of pump development. For purposes of overall machine modeling, they are modeled based on the published static and dynamic characteristics, no matter what the pumping mechanism is.

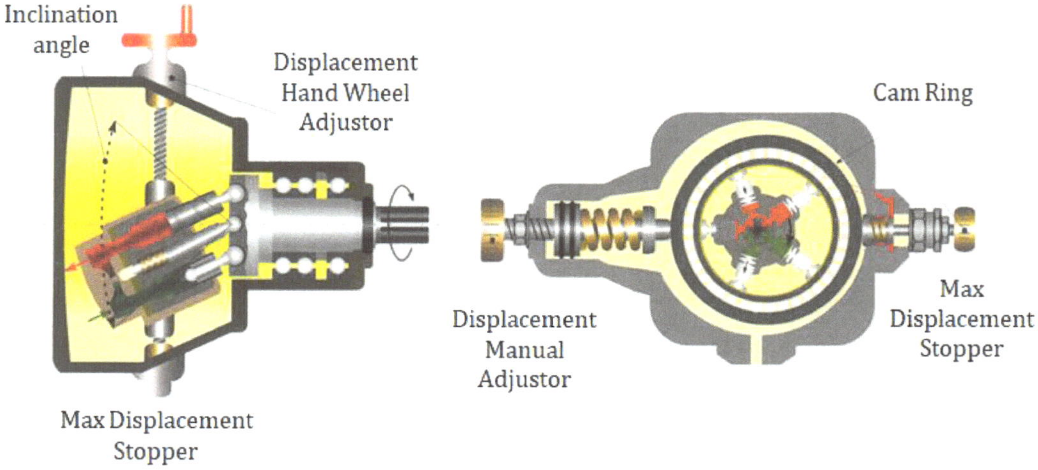

Fig. 7.48- Variable Displacement-Controlled Pumps (Courtesy of Assofluid)

Figure 7.49 shows typical static characteristics of displacement-controlled pumps. As shown in the figure, after adjusting the pump size, it works like a fixed-displacement pump. The figure shows also that both flow control and pressure control can be combined in one pump.

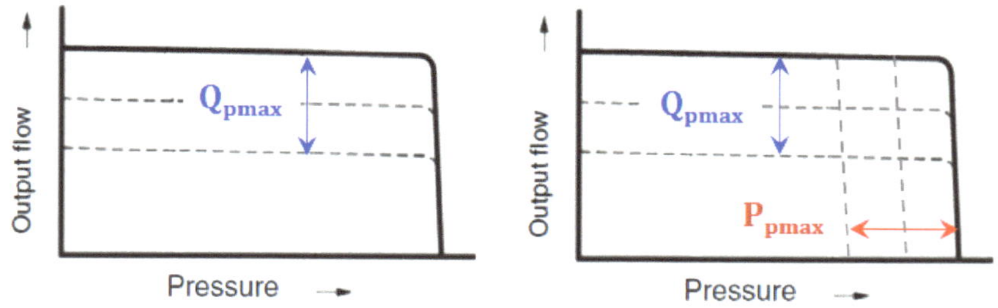

Fig. 7.49- Typical Static Characteristics of Displacement-Controlled Pumps

7.11.2- Simplified Model #05A for Displacement-Controlled Pumps

7.11.2.1- Model Features and Assumptions

In case if no data is available but the Q_p-P_p static characteristics are available, a simplified model still can be built. As shown in Fig. 7.50, the model is structured in a simplified form just to give the pump outlet flow based on the pump pressure and commanded displacement. The model contains static and dynamic characteristics of the pump.

Pump Static Characteristics: Static characteristics should be loaded in the model in form of a lookup table. It should be published by the manufacturer. If not, it should be developed in the lab.

Pump Dynamic Characteristics: Dynamic characteristics should be loaded in the model in form of a normalized transfer function (NTF). At least step response of the pump should be published by the manufacturer. If not, it can be developed in the lab or assumed.

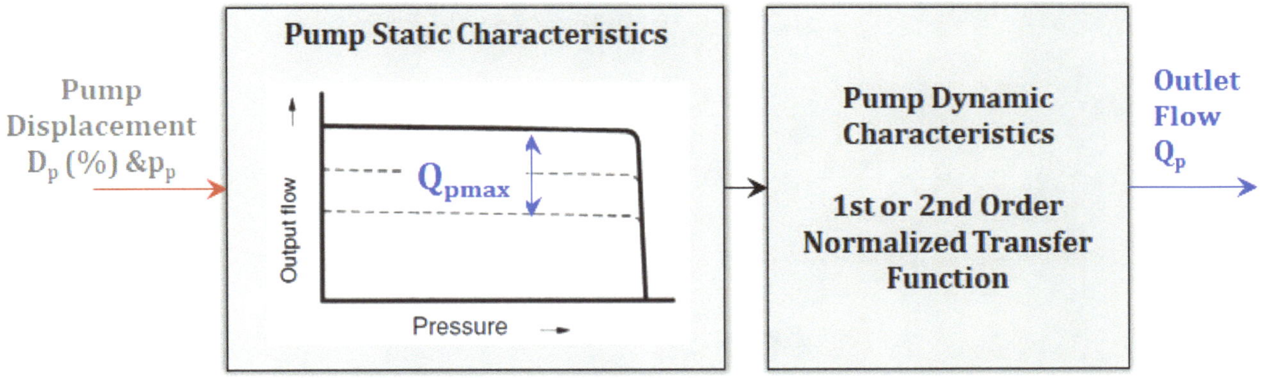

Fig. 7.50- Structure of Pump Model 05A

7.11.2.2- Simulation Model

Figure 7.51 shows the simulation model (HSV7CH07Model05A) built based on the generic structure shown in the previous figure.

Static Characteristics: The model considers Q_p-P_p static characteristics loaded into a 2-D lookup table. The output of the lookup table is multiplied by the percentage change in the pump displacement.

Driving Speed: It is to be noted that this model is valid for only a specific driving speed. Otherwise, if the speed is variable such is likely in mobile machines, then the lookup table should be 3-D and the speed should be added as a varying input.

Dynamic Characteristics: The output of the static characteristic is multiplied by a second-order NTF that represents the dynamics of the control unit. For this control unit, damping ratio is assumed equal to 0.7 and natural frequency equals to 10 rad/sec.

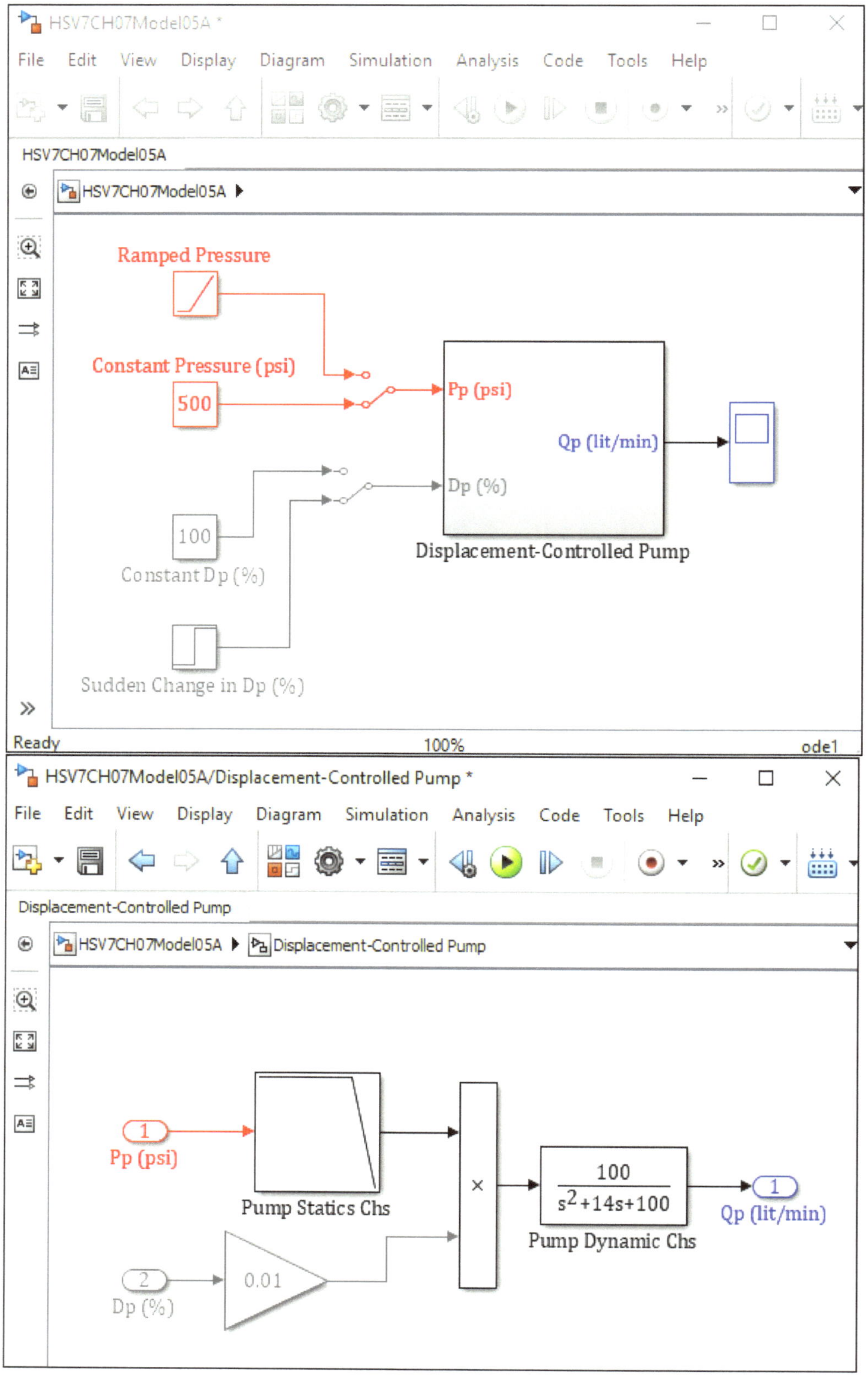

Fig. 7.51- Pump Simulation Model 05A

7.11.2.3- Model Validation

Figure 7.52 shows the simulation results after running the model. The upper part simulates the pump static characteristics at constant driving speed and pump displacement. Pump flow can always be varied by varying the percentage of pump displacement. The lower part of the model shows the pump step response at constant driving speed and working pressure. Pump is commanded first to increase the displacement from 0 to 25% and then to 75%.

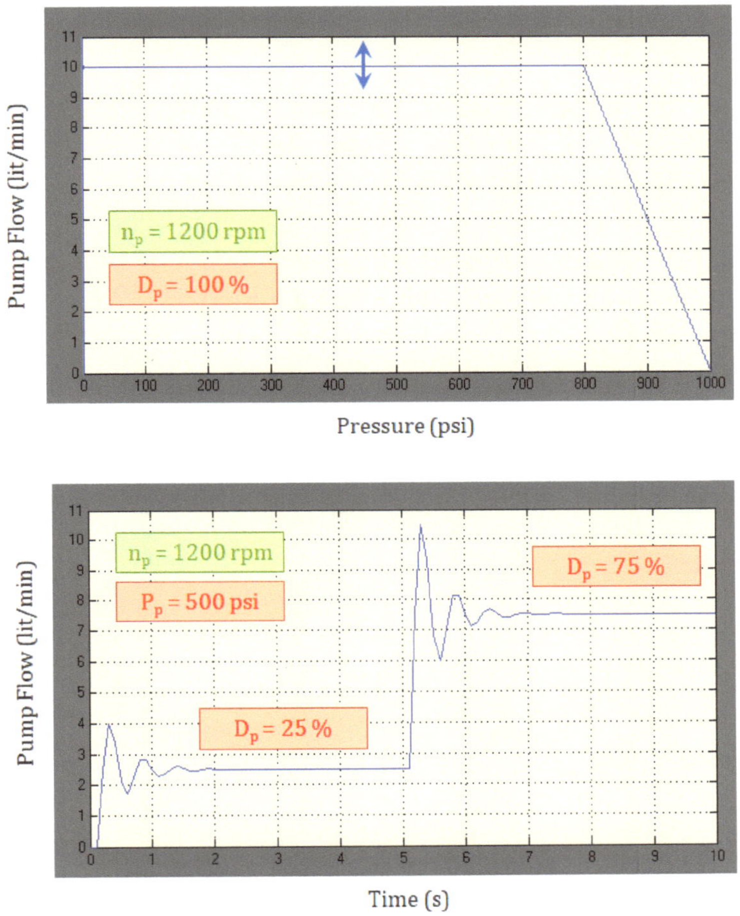

Fig. 7.52- Simulation Results from Model 05A

7.11.3- Model #05B for Displacement-Controlled Pumps Based on Given Efficiency Curves

7.11.3.1- Model Features and Assumptions

Basically, modeling such pumps requires the pump displacement as a varying input to the model. The root model 03A is not possible to be used because that model structure is built based on given test data loaded into lookup tables and this test data is tied to a specific pump displacement.

Therefore, model 03B that is built based on given efficiency curves will be used as a root to build model 05B.

Assumption: Ideally efficiency curves should be available for various pump displacements; typically, 25%, 50%, 75% and 100%. So, with varying pump displacement, the proper efficiency curve is used. However, for simplified modeling process, efficiencies can be assumed as invariant with the pump displacement.

7.11.3.2- Simulation Model

Figure 7.53 shows the upper mask of simulation model (HSV7CH07Model05B). The figure shows the only difference between model 03B and 05B. The difference is the pump displacement **D_p** is placed as an interactively varying input.

Because varying the pump displacement could be remotely as a response to an input command, dynamics of the controller should be added as a 2nd order NTF. Dynamic effects of varying the driving speed and pump pressure are kept as in model 03B.

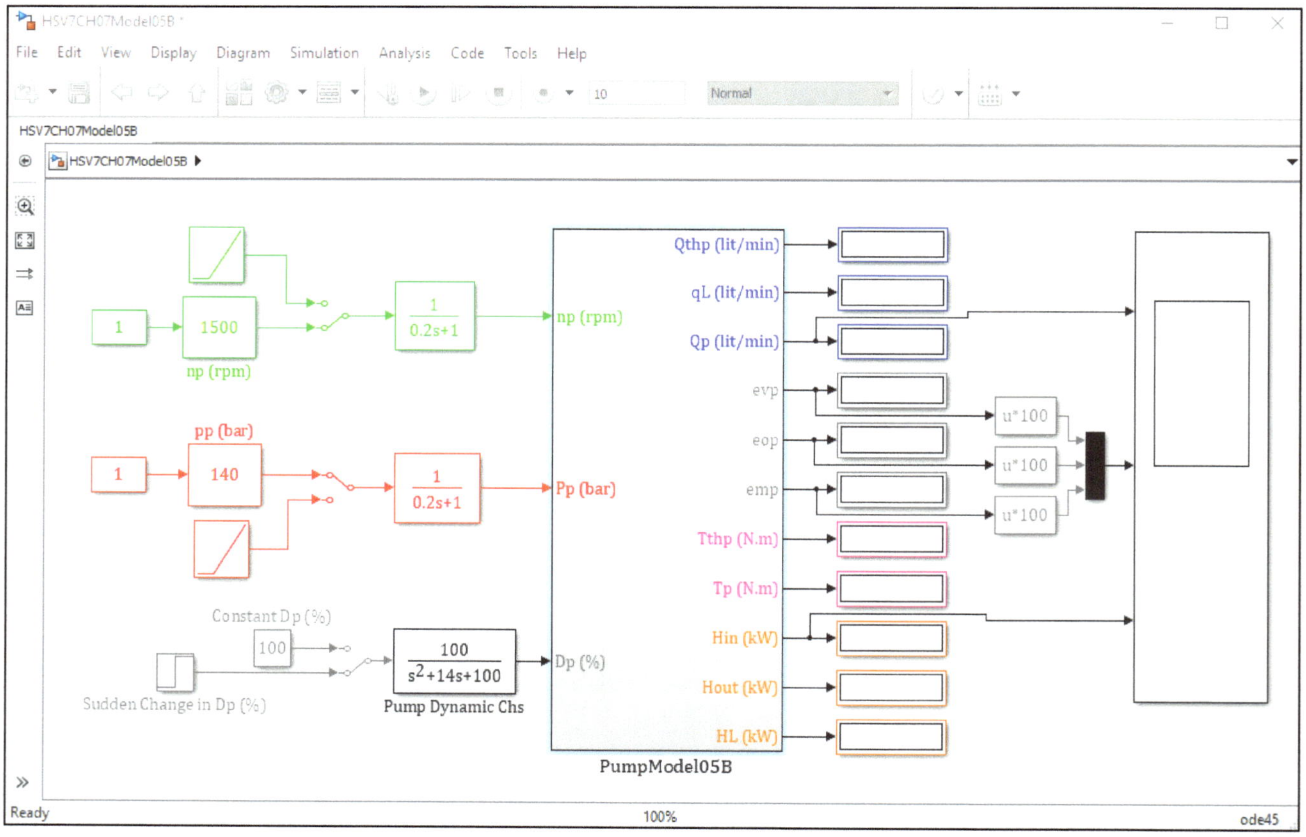

Fig. 7.53- Pump Simulation Model 05B

Figure 7.54 shows the internal structure of the model. The internal structure of the model is the same as model 03B with minor change. Instead of adding the pump displacement as a constant value to the model properties, it is used in the model input as a variable.

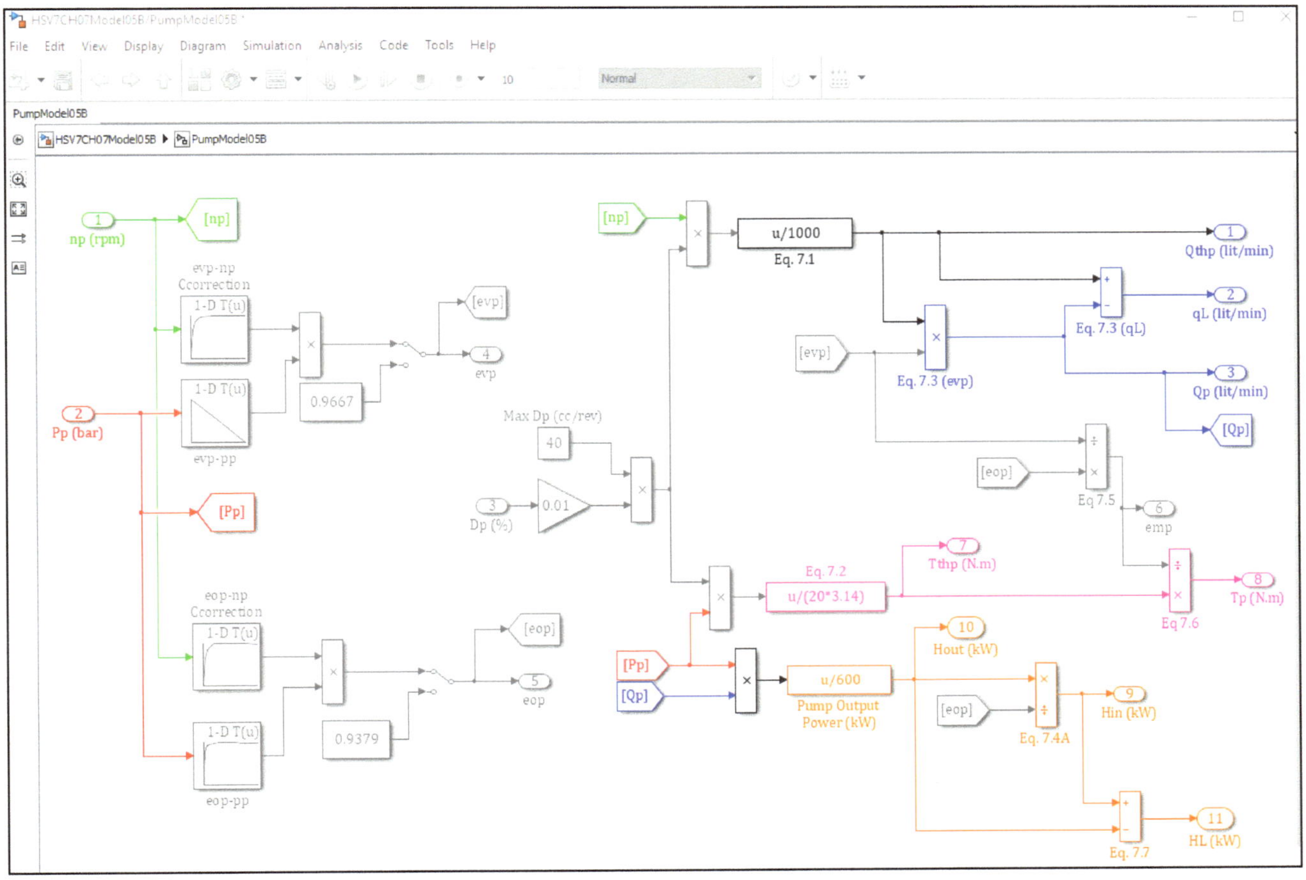

Fig. 7.54- Details under the Mask of Simulation Model 05B

7.11.3.3- Model Validation

Static Characteristics at Maximum Pump Displacement: As shown in Fig. 7.55, by running the model under the same conditions as model 03B runs, same static characteristics are developed as shown previously in Fig. 7.27.

Static Characteristics at Reduced Pump Displacement: As shown in Fig. 7.56, by running the model at various % of pump displacement, corresponding static characteristics are developed.

Dynamic Characteristics: Figure 7.57 shows the pump step response at constant driving speed and working pressure. Pump is commanded first to increase the displacement from 0 to 25% and then to 75%.

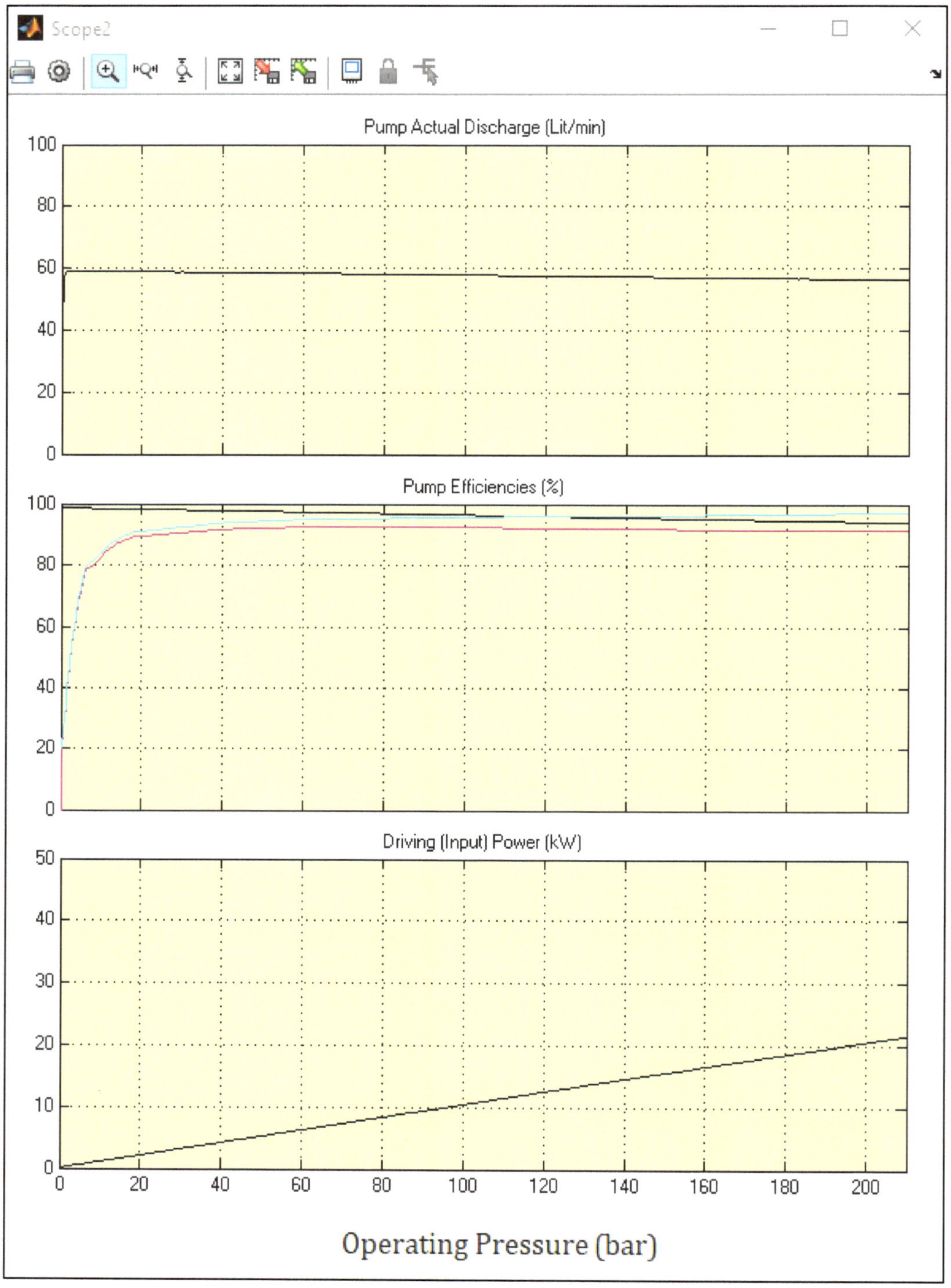

Fig. 7.55- Pump Static Characteristics at Maximum Displacement (at n_p = 1500 rpm)

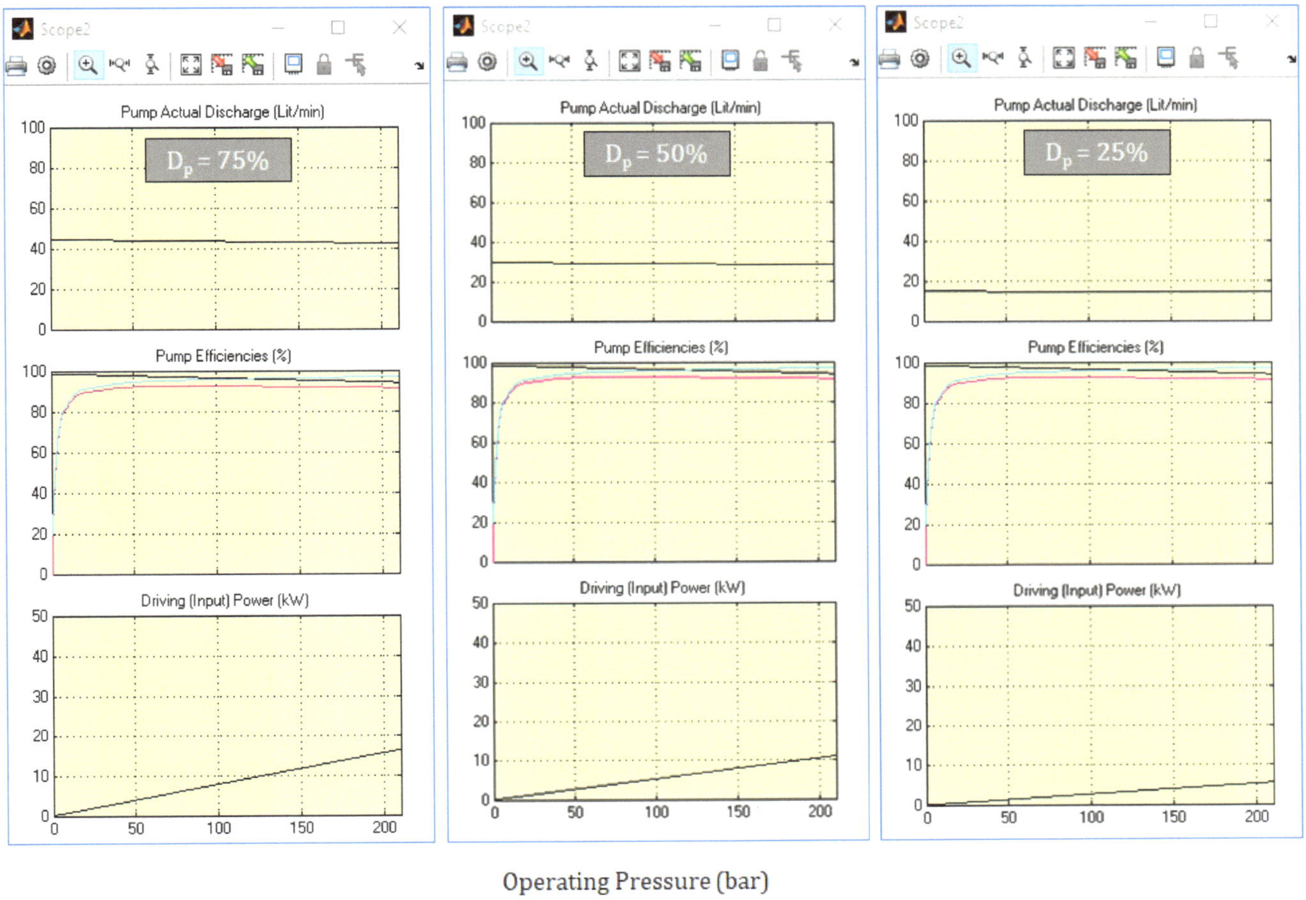

Fig. 7.56- Pump Static Characteristics at Reduced Displacement (at n_p = 1500 rpm)

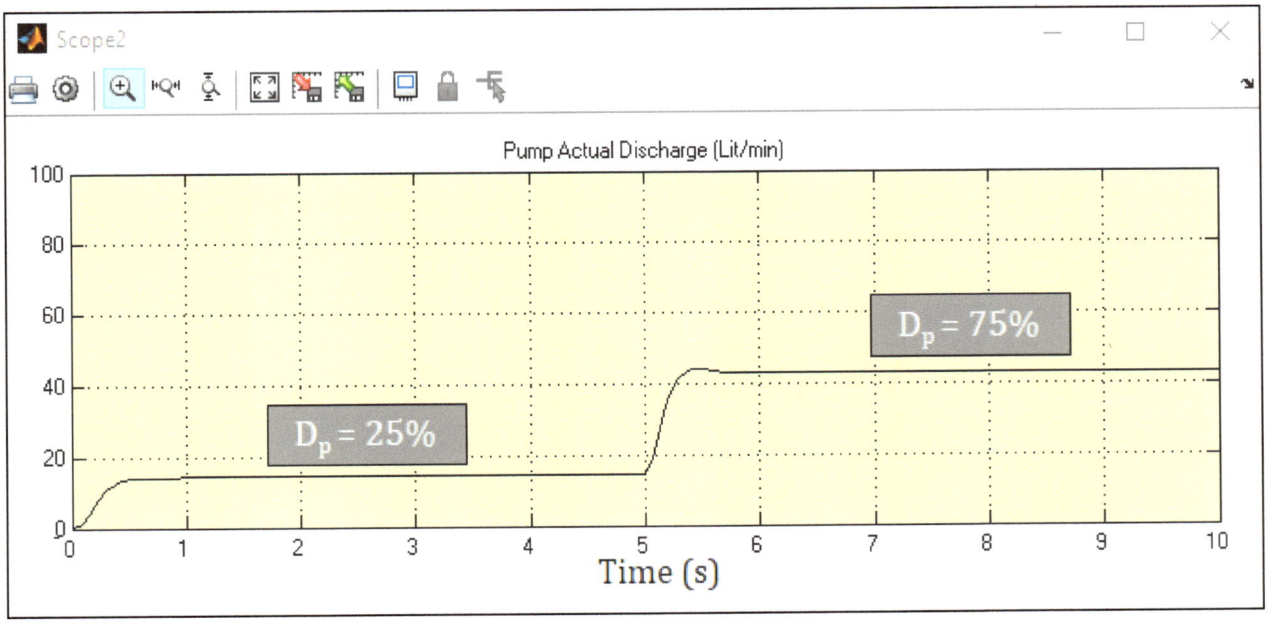

Fig. 7.57- Pump Step Response (at n_p = 1500 rpm and P_p = 140 bar)

7.12- Lumped Modeling of Torque-Limited Pumps

7.12.1- Characteristics of Torque-Limited Pumps

Torque-Limited pumps are also referred to as *Constant-Power* pumps. A controller of a torque-limited pump works to limit the withdrawal of power from the prime mover to an adjustable maximum value. To explain this control concept, as shown in Fig. 7.58 while an excavator is digging down, working pressure and consequently the power from the prime mover vary according to the variable resistance from the ground. When the pump is at maximum displacement, if the pressure is left to increase uncontrollably, the power withdrawn from the engine increases also uncontrollably and the engine may stall. Because the engine is not used to supply power for the hydraulic system only, then the power for hydraulic system must be limited to a certain maximum value to avoid engine stalling.

Fig. 7.58- Torque-Limited Pumps in Excavator

So, the controller of such pumps works to reduce the pump displacement with the increase of working pressure to comply with the constant power operation. The controller of such a pump serves to limit the maximum power that the pump can withdraw from a prime mover. The magnitude of the hydraulic power **P** out of a pump is the product of the pump outlet pressure **p** and the flow **Q**. Since the working pressure **p** is a load-dependent variable, then the hydraulic power **P** is controlled by controlling the pump flow. The controller senses the pressure based on which it adjusts the pump flow **Q**. Figure 7.59 shows a set of three **Q-P** hyperbolic curves that represents the ideal constant power operation of the pump. Curves A, B, and C go from a minimum power to intermediate to maximum power. At any point on any of these curves, e.g. 1, 2, and 3, the product of the working pressure **P** and the corresponding pump flow **Q** has the same value of the power. On point 4 and 5 the controller sets maximum flow and maximum pressure, respectively regardless the available power.

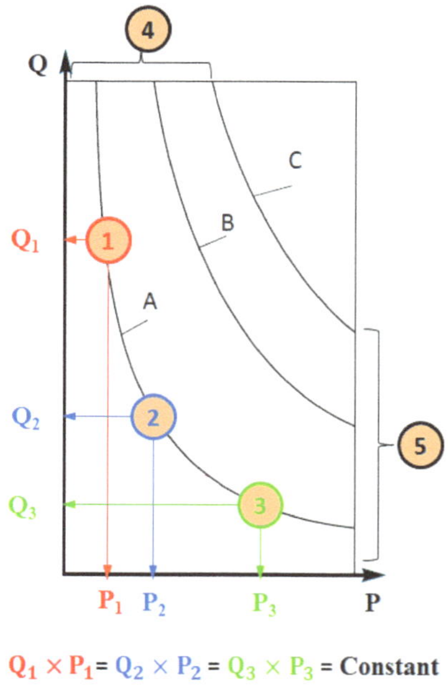

Fig. 7.59- Ideal Characteristics of Torque-Limited Pumps

7.12.1.1- Electro-Hydraulic Constant-Power Pumps

As shown in Fig. 7.60, pump displacement is controlled automatically in a closed loop based on the pump pressure variation to keep the power constant. As shown in the figure, an *Electric Control Unit* (ECU) is fed by three characterizing operational variables; $\mathbf{H_{pmax}}$, $\mathbf{Q_{pmax}}$, **and** $\mathbf{P_{pmax}}$. The ECU calculates first $\mathbf{Q_R}$ based on $\mathbf{H_{pmax}}$ and $\mathbf{P_p}$. The ECU then checks $\mathbf{Q_R}$ based on $\mathbf{Q_{pmax}}$ and $\mathbf{P_{pmax}}$. $\mathbf{Q_R}$ is used as a reference value for the closed-loop. The controller calculates the difference between $\mathbf{Q_R}$ and $\mathbf{Q_p}$. The resulting flow error $\mathbf{Q_E}$ is manipulated by the controller that generates the control signal $\mathbf{Q_C}$. The latter is used to drive the servo valve that adjusts the pump displacement to make $\mathbf{Q_p}$ equal to $\mathbf{Q_R}$. This controller provides the shown ideal hyperbolic constant-power operating curve where:

- $\mathbf{H_{pmax}}$ = Maximum Pump Power.
- $\mathbf{Q_{pmax}}$ = Maximum Pump Flow.
- $\mathbf{P_{pmax}}$ = Maximum Pump Pressure.
- $\mathbf{P_p}$ = Actual Pump Pressure.
- $\mathbf{Q_p}$ = Actual Pump Flow.
- $\mathbf{Q_R}$ = Reference Flow.
- $\mathbf{Q_E}$ = Error Flow.
- $\mathbf{Q_C}$ = Control Signal.

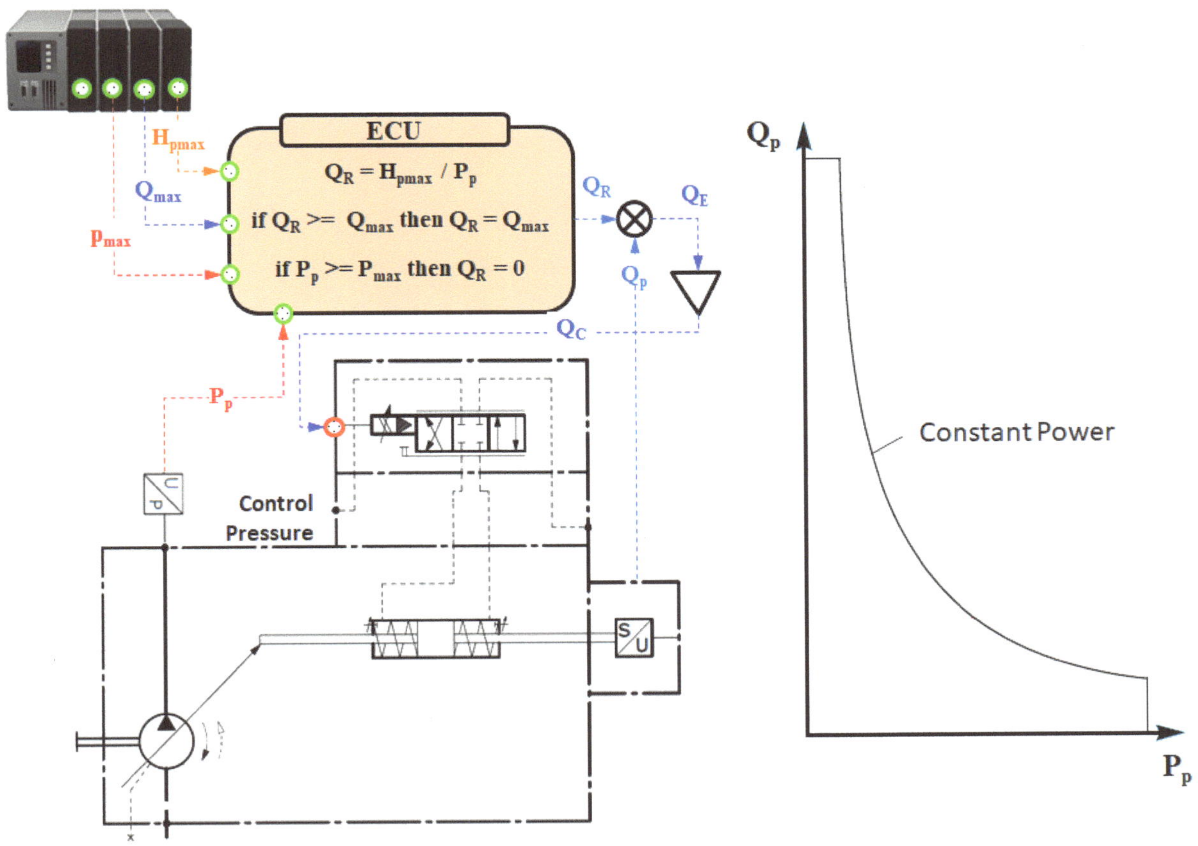

Fig. 7.60 - Electro-Hydraulic Constant-Power Controller

7.12.1.2- Hydro-Mechanical Constant-Power Pumps

As shown in Fig. 7.61, the hydro-mechanical power controller varies the pump displacement by swiveling the angle of the swash-plate against two coaxially coincided coil springs. The springs have different lengths and each, as known, follows Hooke's law. At low pressure, when the longer spring 1 is compressed, Q-P relation follows part 1 of the curve. When both springs 1 and 2 are compressed simultaneously, they act like a new tougher spring. In such a case, Q-P relation follows part 2 of the curve. The resulting overall curve, a combination of the two broken lines, approximates the ideal hyperbolic curve. The red shaded areas under the ideal curve are available power that can't be achieved because of the spring characteristics.

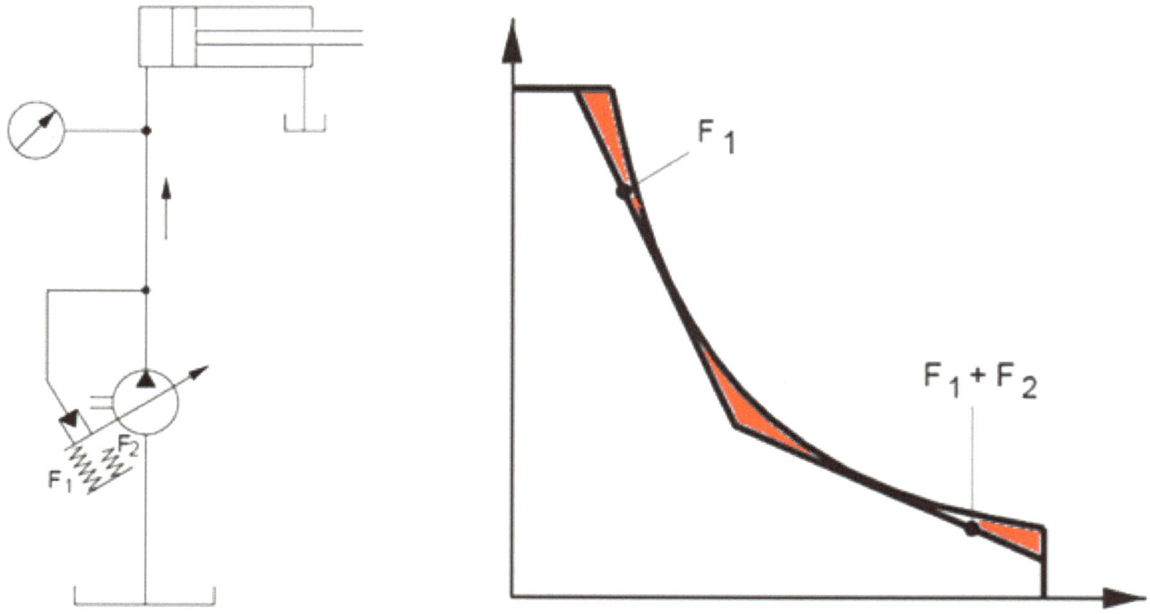

Fig. 7.61 - Hydro-Mechanical Constant-Power Controller

7.12.1.3- Modeling Approaches for Torque-Limited Pumps

As it has been shown above, operation of limiting the power to an adjustable maximum value could occur manually or remotely (hydraulically by a pilot pressure or electro-hydraulically by an electrical variable signal). In both cases, in order to achieve a constant drive torque with a varying operating pressure, the swivel angle and with it the output flow from the axial piston unit is varied so that the product of flow and pressure remains constant.

As it has been discussed previously, such pumps should be modeled based on internal structure of the pumping mechanism for purposes of pump development. For purposes of overall machine modeling, they are modeled based on the published static and dynamic characteristics, no matter what the pumping mechanism is.

7.12.2- Simplified Model #06A for Torque-Limited Pumps

7.12.2.1- Model Features and Assumptions

In case if no data is available but the Q_P-P_P static characteristics are available, a simplified model still can be built. As shown in Fig. 7.62, the model is structured in a simplified form just to give the pump outlet flow based on the pump pressure. The model contains static and dynamic characteristics of the pump.

Pump Static Characteristics: Static characteristics should be loaded into the model in form of a lookup table. It should be published by the manufacturer. If not, it should be developed in the lab.

Pump Dynamic Characteristics: Dynamic characteristics should be loaded into the model in form of a normalized transfer function (NTF). At least step response of the pump should be published by the manufacturer. If not, it can be developed in the lab or assumed.

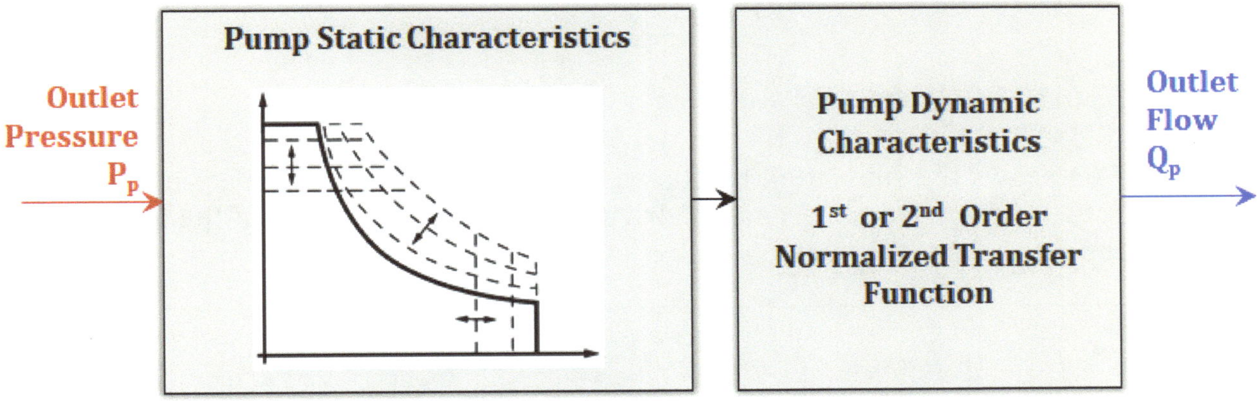

Fig. 7.62- Structure of Pump Model 06A

7.12.2.2- Simulation Model

Figure 7.63 shows the simulation model (HSV7CH07Model06A) built based on the generic structure shown in the previous figure.

Static Characteristics: The model considers Q_P-p_P static characteristics loaded into a 2-D lookup table. The output of the lookup table is multiplied by the percentage change in the pump displacement.

Driving Speed: It is to be noted that this model is valid for only a specific driving speed. Otherwise, if the speed is variable such is likely in mobile machines, then the lookup table should be 3-D and the speed should be added as a varying input.

Dynamic Characteristics: The output of the static characteristic is multiplied by a second-order NTF that represents the dynamics of the control unit. For this control unit, damping ratio is assumed equal to 0.7 and natural frequency equals to 10 rad/sec.

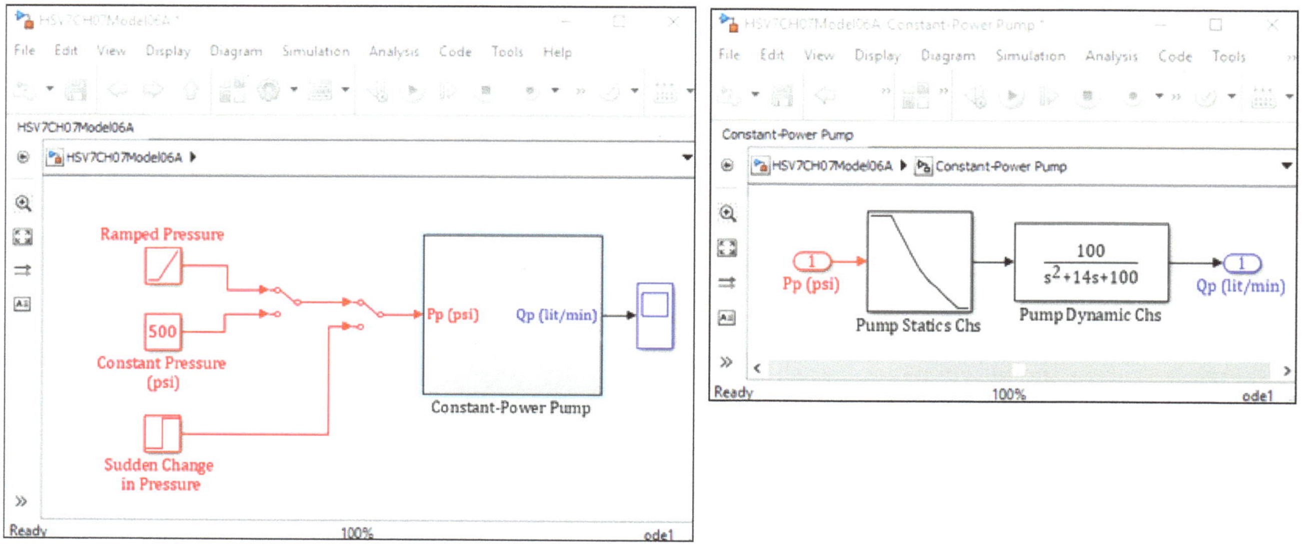

Fig. 7.63- Pump Simulation Model 06A

7.12.2.3- Model Validation

Figure 7.64 shows the simulation results after running the model. The left part simulates the pump static characteristics at constant driving speed. The right part of the model shows the pump step response of the pump

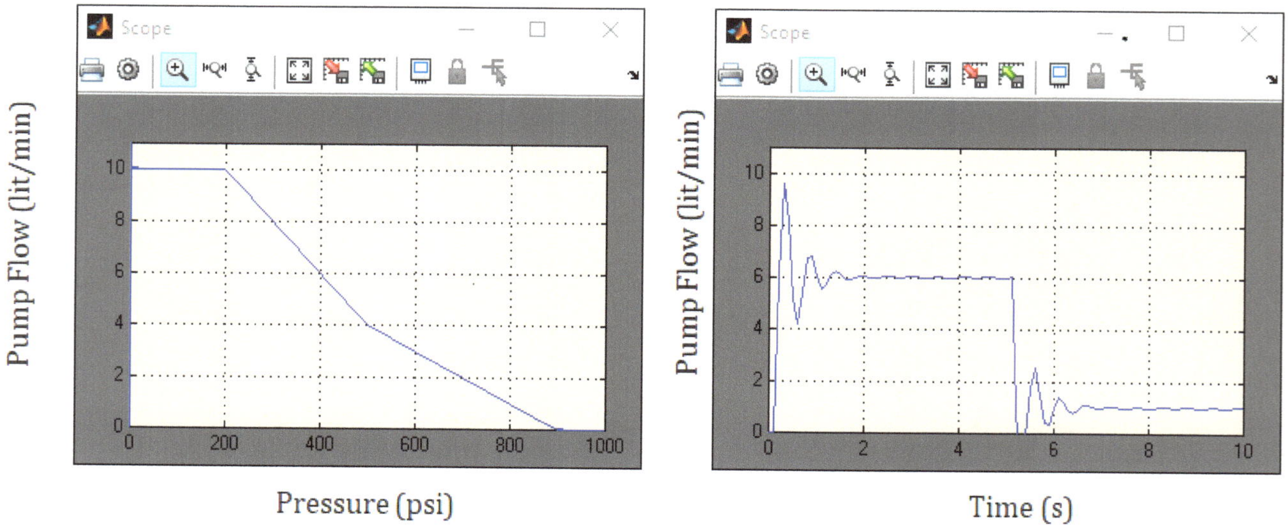

Fig. 7.64- Simulation Results from Model 06A

7.12.3- Model #06B for Torque-Limited Pumps Based on Given Efficiency Curves

7.12.3.1- Model Features and Assumptions

Model 03B is used as a root to build model 06B.

Assumption: Ideally efficiency curves should be available for various pump displacements; typically, 25%, 505, 75% and 100%. So, with varying pump displacement, the proper efficiency curve is used. However, for simplified modeling process, efficiencies can be assumed as invariant with the pump displacement.

7.12.3.2- Simulation Model

Figure 7.65 shows the upper mask of simulation model (HSV7CH07Model06B). The figure shows that pump maximum power is added as interactively varying inputs. Because pump controller is actuated by working pressure, a 2nd order NTF has been hooked to the pressure to represents the dynamics of the control unit. The other boundaries of pump operation (maximum flow = 50 lit/min and maximum pressure = 180 bar) are added as constraints to the model properties.

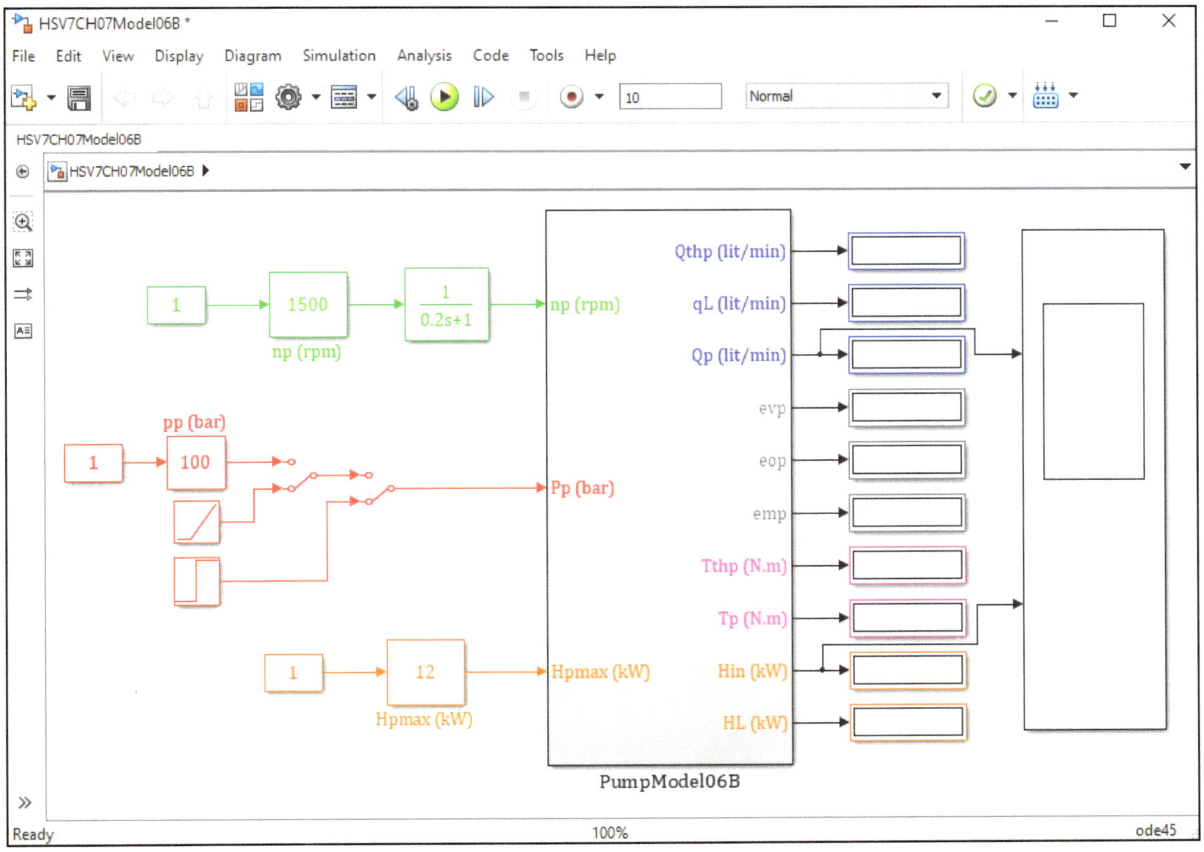

Fig. 7.65- Pump Simulation Model 06B

Figure 7.66 shows the internal structure of the model that has the following sections:

- Sections A: for efficiencies calculations
- Sections B: for power calculations
- Sections C: for flow calculations
- Sections D: for pump displacement calculations
- Sections E: for torque calculations

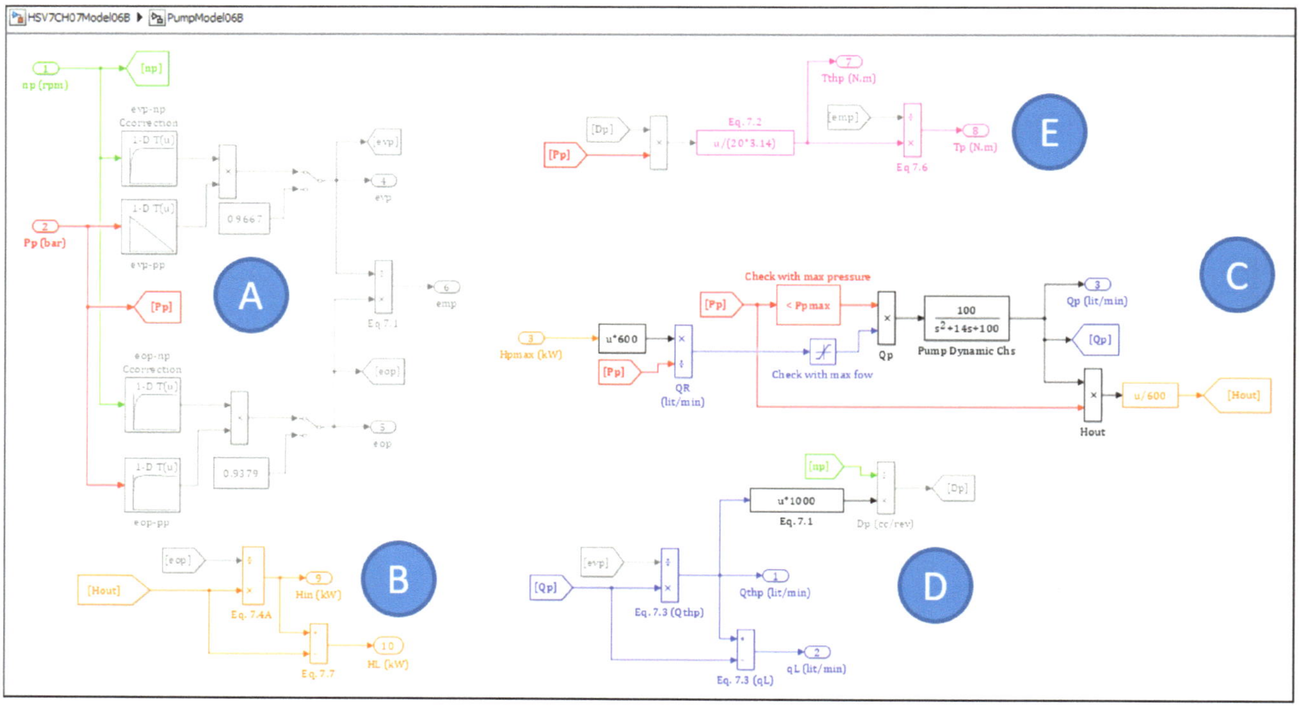

Fig. 7.66- Details under the Mask of Simulation Model 05B

7.12.3.3- Model Validation

Static Characteristics at Reduced Power: As shown in Fig. 7.67, by running the model with the assigned boundaries (maximum flow = 50 lit/min and maximum pressure = 180 bar) static characteristics are simulated for various reduced power levels.

Dynamic Characteristics: Figure 7.68 shows the pump step response where pump pressure jumps from 50 bar to 150 bar. The response shows that the controller kept the power constant at the assigned maximum power of 10 kW.

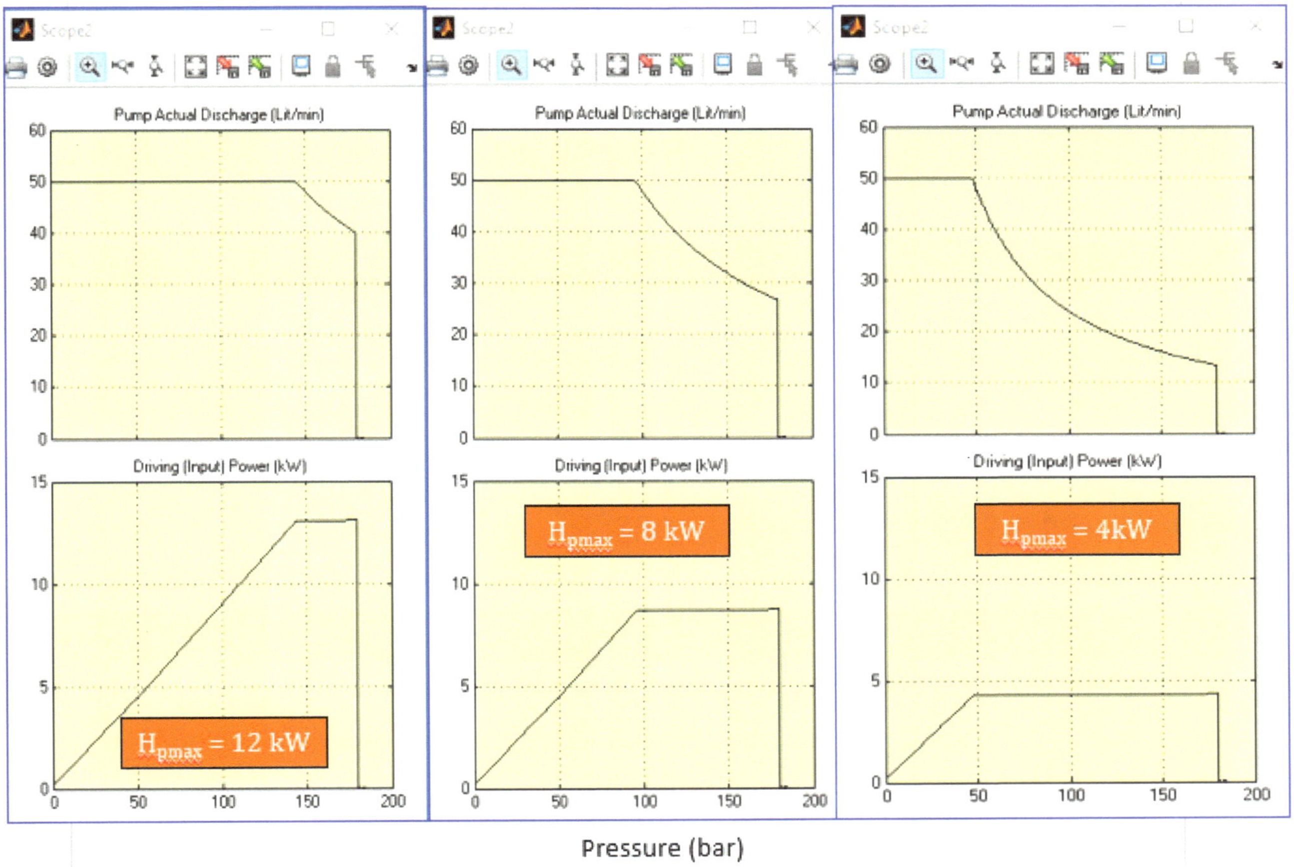

Fig. 7.67- Pump Static Characteristics at Reduced Power (at n_p = 1500 rpm)

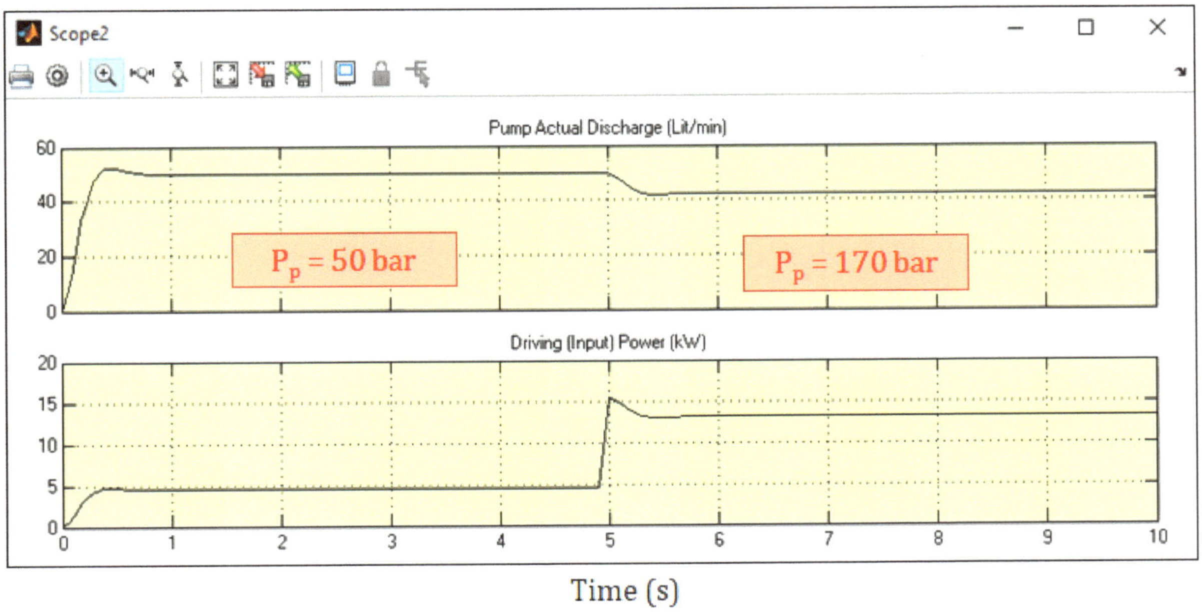

Fig. 7.68- Pump Step Response (at n_p = 1500 rpm)

Chapter 8
Modeling of Hydraulic Motors

Objectives

This chapter presents the lumped modeling concept as applied for fixed and variable displacement motors. This chapter considers situations where a motor works under a constant or variable torque and inlet flow. Models for two-position control, proportional control, and torque-limited motors are developed.

Brief Contents

Chapter 8: Modeling of Hydraulic Motors

8.1- Lumped Model Structure of a Unidirectional Hydraulic Motor

Figure 8.1 shows a lumped model structure of a unidirectional hydraulic motor. The model contains the following features:

- Assuming incompressible flow, where effect of oil bulk modulus is ignored, motor outlet flow assumed equal inlet flow. Inlet flow is also used, with the static and dynamic characteristics of the motor, to calculate the motor speed.

- External *Load Torque* **T_L** is used, with the static and dynamic characteristics of the motor, to calculate the differential pressure across the motor. With the knowledge of the motor outlet pressure, inlet pressure is calculated.

- Motor static and dynamic characteristics are loaded into the model properties.

- Motor lumped model takes different ascending levels based on available information, assumptions, and the typical applications in which the motor runs. Inputs and Outputs of the model depend on the model level.

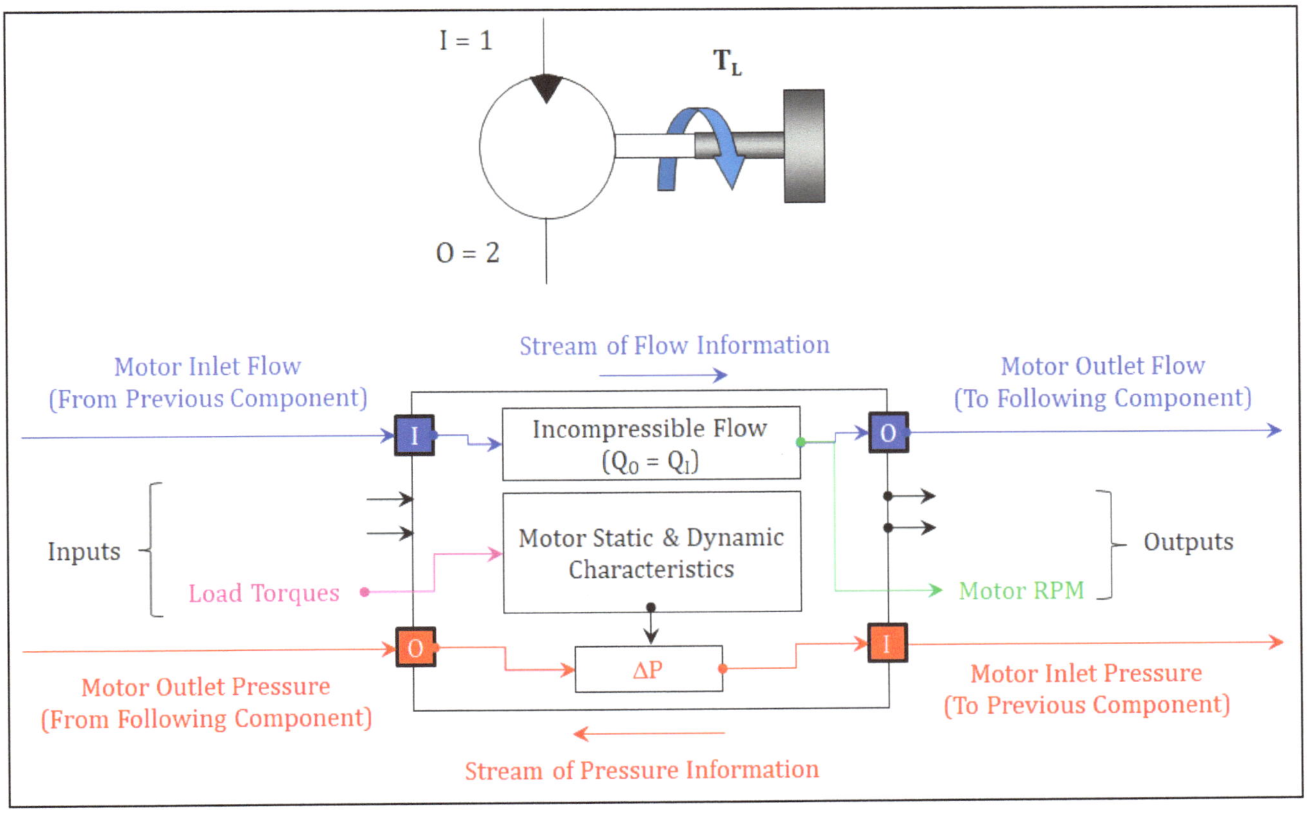

Fig. 8.1- Lumped Model Structure of a Unidirectional Hydraulic Motor

8.2- Lumped Model Structure of a Bidirectional Hydraulic Motor

Unlike a unidirectional motor model, Fig. 8.2 shows that a model for a bidirectional motor has an identifier **U** that is used to identify the motor direction of rotation, based on which flow and pressure on each physical port is assigned. If **U** is positive, then port 1 is the inlet port and port 2 is the outlet port (**$P_1 = P_I$, $Q_1 = Q_I$, $P_2 = P_O$ and $Q_2 = Q_O$**) and vise-versa.

The logic for the identifier **U** can be performed in different way, such as using "If" statements. However, as shown previously in Chapter 7 (Fig. 7.3), the logic can be structured using Simulink logic blocks. For bidirectional motor, outlet flow splits into return flow and case drain.

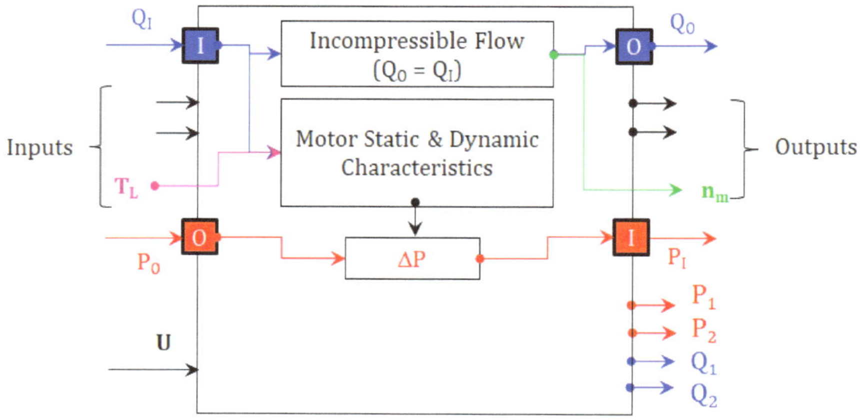

Fig. 8.2- Lumped Model Structure of a Bidirectional Hydraulic Motor

8.3- Modeling Fixed Displacement Motors

Part of the challenge in modeling hydraulic components in general is that these components work under various scenarios depending on the applications they are used for. For example, hydraulic motors can work under fixed or variable working pressure and output speed. Ideally for modeling a motor, full range of test data should be available. Unfortunately, this is not the case all the time. Therefore, this chapter introduces various scenarios depending on working conditions and available information.

The following sections cover modeling static and dynamic characteristics for fixed displacement motors first and then variable displacement motors. So, model level and complexity will be ascendingly increased to cover various scenarios. As shown in Fig. 8.3, static characteristics for a 71 cc/rev fixed displacement motor (A4FM from Rexroth) are reported in form of test data. As shown in the figure, motors are typically tested where motor actual flow and differential pressure requirements are reported versus input motor output speed and motor torque.

It is to be noted that, this test data is used as follows:

- For the purpose of motor sizing for an application, for a predefined torque and speed, determine the differential pressure and flow requirements, respectively.
- For the purpose of motor modeling, for a predefined torque and inlet flow, determine the differential pressure and output speed, respectively.

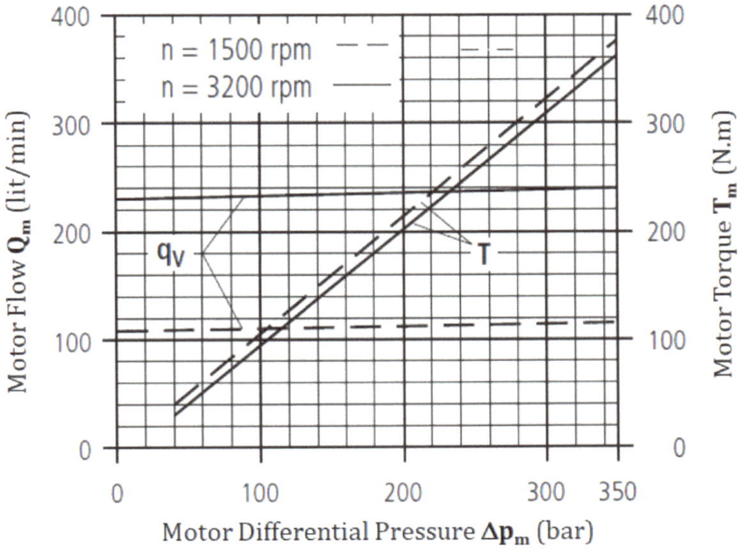

Fig. 8.3- Static Characteristics of a Fixed Displacement Motor in form of Test Data (Motor A4FM Series 71 – Courtesy of Bosch Rexroth)

Alternatively, as shown in Fig. 8.4., motor characteristics are reported in form of motor efficiencies versus rpm and pressure.

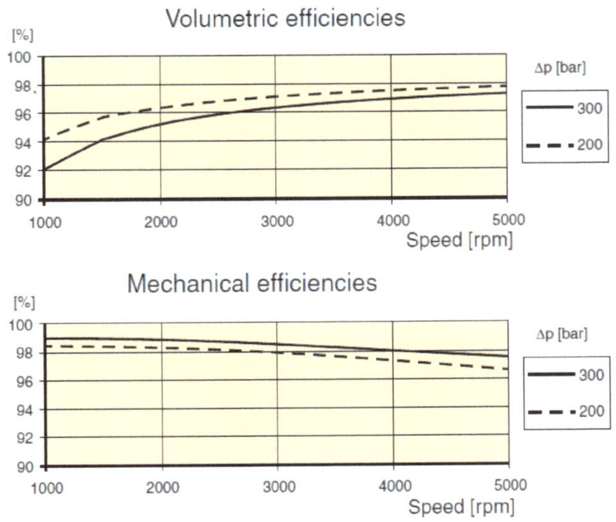

Fig. 8.4- Static Characteristics of a Fixed Displacement (Series 11) Motor in form of Efficiency Curves (Courtesy of Parker)

8.4- Model #01 for an Ideal Fixed Displacement Motor

8.4.1- Model Features and Assumptions

Ideal Motor: In some applications, where motors run continuously at constant rpm under low torque, there is no sense to perform the efficiency calculations. So, in this case, the model represents just a an ideal (100% efficient) motor. Then, no need for any test data.

8.4.2- Motor Theoretical RPM

Based on the 100% efficient motor assumption, Fig. 8.5 shows that theoretical motor speed $\mathbf{n_{thm}}$ is function only of motor size $\mathbf{D_m}$ and motor flow $\mathbf{Q_m}$ and is not affected by the differential pressure across the motor $\mathbf{\Delta P_m}$ that is function of the motor torque $\mathbf{T_m}$.

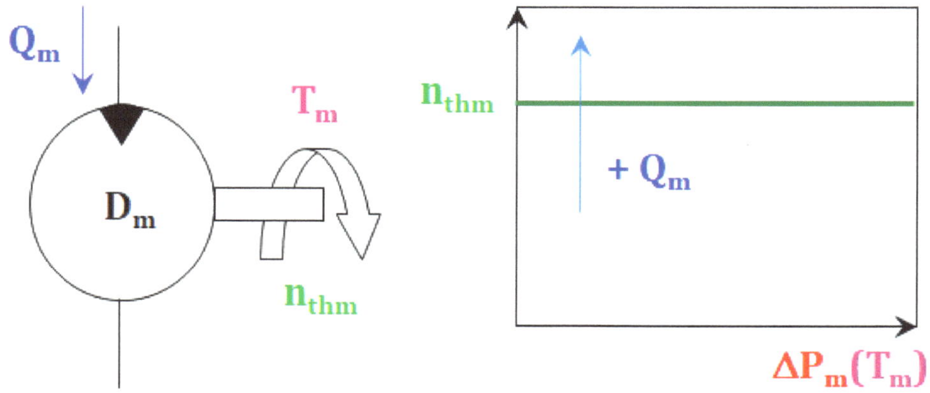

Fig. 8.5- Motor Theoretical RPM Characteristics

Therefore, Eq. 8.1 is used to calculate the motor theoretical flow as follows

$$n_{thm} = Q_m / D_m \quad \mathbf{8.1}$$

Equations 8.1A and 8.1B represent Eq. 8.1 with the units being balanced in English and Metric systems of units; respectively.

$$n_{thm}(\text{rpm}) = \frac{231 \times Q_m(\text{gpm})}{D_m\left(\frac{\text{in}^3}{\text{rev}}\right)} \quad \mathbf{8.1A}$$

$$n_{thm}(\text{rpm}) = \frac{1000 \times Q_m(\text{lit/min})}{D_m\left(\frac{\text{cc}}{\text{rev}}\right)} \quad \mathbf{8.1B}$$

8.4.3- Motor Torque Calculation

Torques act on the motor shaft are classified as follows:

- **Based on the Source:** *Internal or External.*
- Internal means from inside the motor, e.g. frictional torques.
- External means from outside the motor, e.g. external loading torque.
- **Based on the Sign:** *Positive (Resistive) or Negative (Assistive).*
- Positive (Resistive) means it opposes the motor rotation.
- Negative (Assistive) means it drives the motor shaft make it work like a pump.
- **Based on the Motor Motion:** *Passive or Active.*
- Passive means it equals zero when the motor stops, e.g. frictional torques.
- Active means it has a magnitude (positive or negative) even when the motor stops.

Figure 8.6 shows examples of hydraulic motor loading conditions as follows:

1. A motor drives an external load against frictional surface. Load is external, passive, and positive no matter the motor direction of rotation.
2. A motor drives a drum of a winch to left and lower a mass.
3. A motor drives a boom in a swinging motion. Load is external, active, and positive in one direction and negative in the other direction of motor rotation.
4. Hydraulic motors are used to drive loads in mobile machines. Load is external, active, and switches from positive to negative during the same motor direction of rotation.

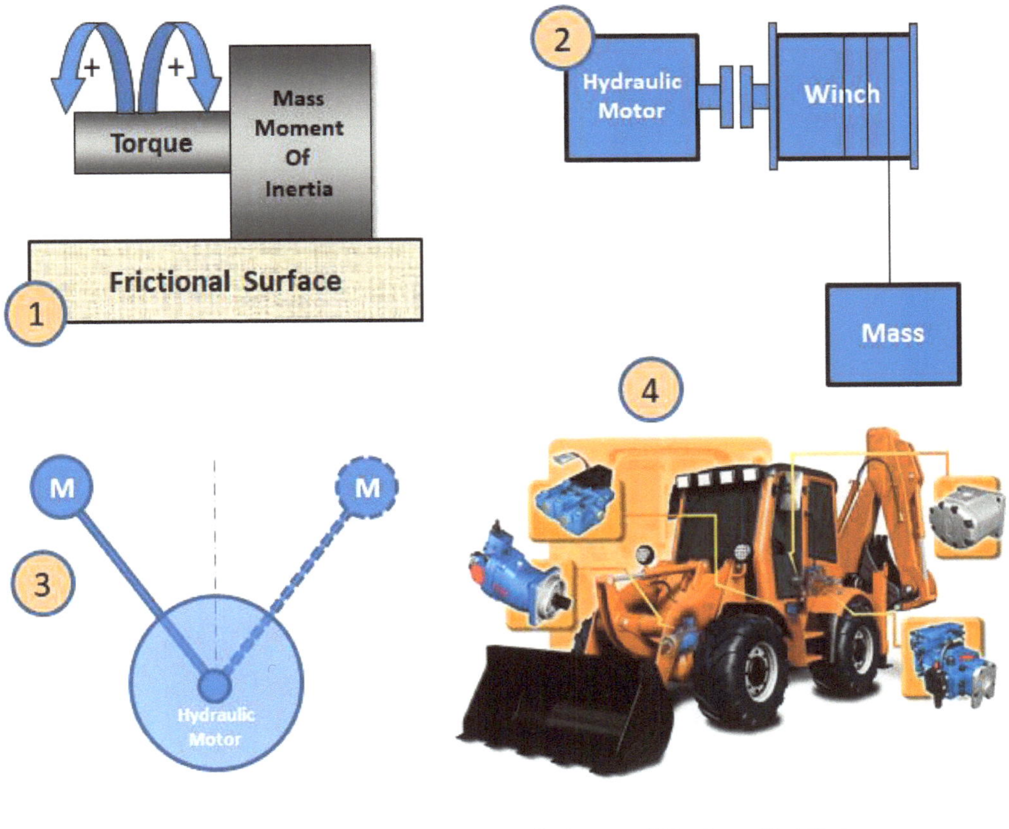

Fig. 8.6- Various Loading Conditions of Hydraulic Motors

Referring to Fig. 8.7, Eq. 8.1C through 8.1F show calculations of instantaneous motor torque.

$$\mathbf{T_V(N.m) = \omega\,(1/s) \times k_V\left(\frac{N.m}{1/s}\right) = \frac{2\pi n(rpm)}{60} \times k_V\left(\frac{N.m}{1/s}\right)} \quad \mathbf{8.1C}$$

$$\mathbf{T_I(N.m) = J\,(kg.m^2) \times \dot{\omega}\ (1/s^2)} \quad \mathbf{8.1D}$$

$$\mathbf{T_F(N.m) = F_F\,(N) \times R\,(m) = F_N\,(N) \times k_F(-) \times R\,(m)} \quad \mathbf{8.1E}$$

$$\mathbf{T_m = T_L + T_V + T_I + T_F} \quad \mathbf{8.1F}$$

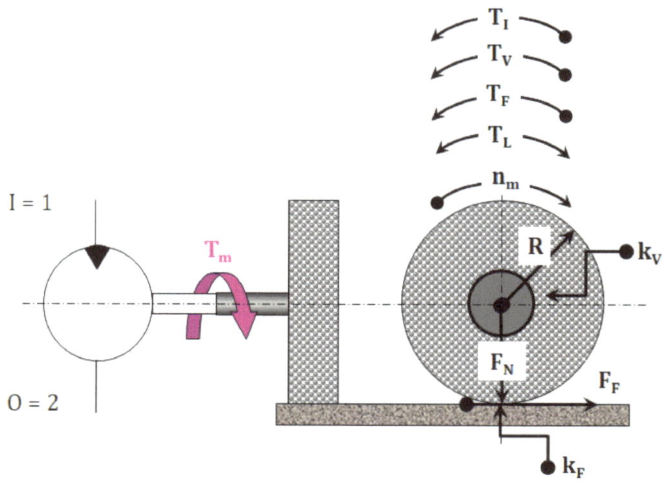

Fig. 8.7- Hydraulic Motor Torque Calculation

Where:

- **T_m (Motor Torque):** Torque generated by the motor to drive the external load torque and overcome all other torques acting on the motor. This motor torque results in differential pressure across the motor.
- **T_L (External Load Torque):** External - positive or negative - passive or active. Provided by load model. If loading conditions are not the same in both motor directions, then identifier **U** is used to select the loading conditions accordingly.
- **T_I (Inertial Torque):** External - positive - passive. Inertial torque is due to equivalent mass moment of inertia including the rotating mechanism of the motor.
- **T_V (Viscous Friction Torque):** Internal - positive – passive (already multiplied by N).
- **T_F (Dry Friction Torque):** External - positive – passive (Applicable only when |n|>0).
- **J:** Equivalent mass moment of inertia of the rotating mass.
- **ω:** Angular speed of the motor.
- **k_V:** Viscous friction Coefficient.
- **k_F:** Dry friction Coefficient.
- **F_F:** Friction Force.
- **F_N:** Normal Force.
- **R:** Radius of rotation.

8.4.4- Theoretical Differential Pressure Across the Motor

As shown in Fig. 8.8, a motor is a device that converts the input hydraulic power into output mechanical power. Therefore, starting with the power balance of a 100% efficient motor, Eq. 8.2 shows how to calculate the theoretical differential pressure across the motor $\mathbf{\Delta P_{thm}}$ as function of the motor displacement $\mathbf{D_m}$ and motor torque $\mathbf{T_m}$.

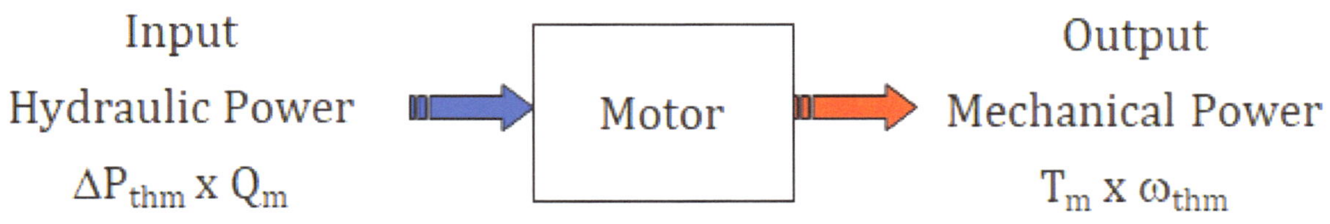

Fig. 8.8- Power Balance in an Ideal Motor

$$\Delta P_{thm} \times Q_m = T_m \times \omega_{thm} \rightarrow \Delta P_{thm} \times D_m \times n_{thm} = T_m \times 2\pi n_{thm} \rightarrow \Delta P_{thm} = \frac{2\pi T_m}{D_m} \qquad 8.2$$

Equations 8.2A and 8.2B represent Eq. 8.2 with the units being balanced in English and Metric systems of units; respectively.

$$\Delta P_{thm}\,(psi) = \frac{2\pi \times T_m(lb.in)}{D_m(in^3/rev)} \qquad 8.2A$$

$$\Delta P_{thm}\,(bar) = \frac{20\pi \times T_m(N.m)}{D_m(cc/rev)} \qquad 8.2B$$

Figure 8.9 shows that the theoretical differential pressure $\mathbf{\Delta P_{thm}}$ across the motor is function only of the motor size $\mathbf{D_m}$ and motor torque $\mathbf{T_m}$ and is not affected by the motor speed $\mathbf{n_m}$ that is function of the motor flow $\mathbf{Q_m}$.

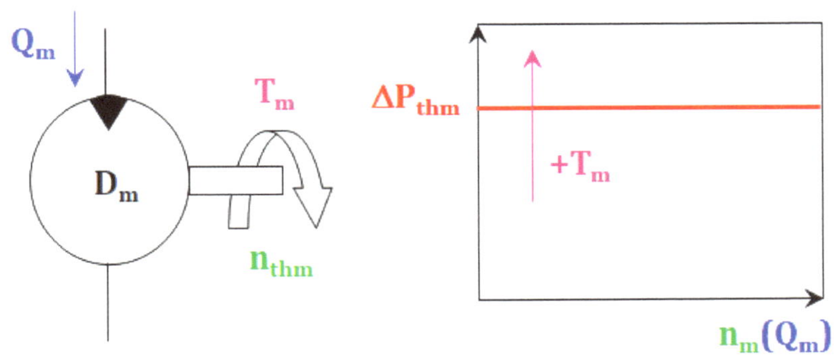

Fig. 8.9- Motor Theoretical Differential Pressure Characteristics

8.4.5- Simulation Model

Figure 8.10 shows simulation model (HSV7CH08Model01) for ideal fixed displacement unidirectional motor built using Matlab-Simulink. As shown in the figure, metric system of units was used. The figure shows the upper mask and the interior details of the model.

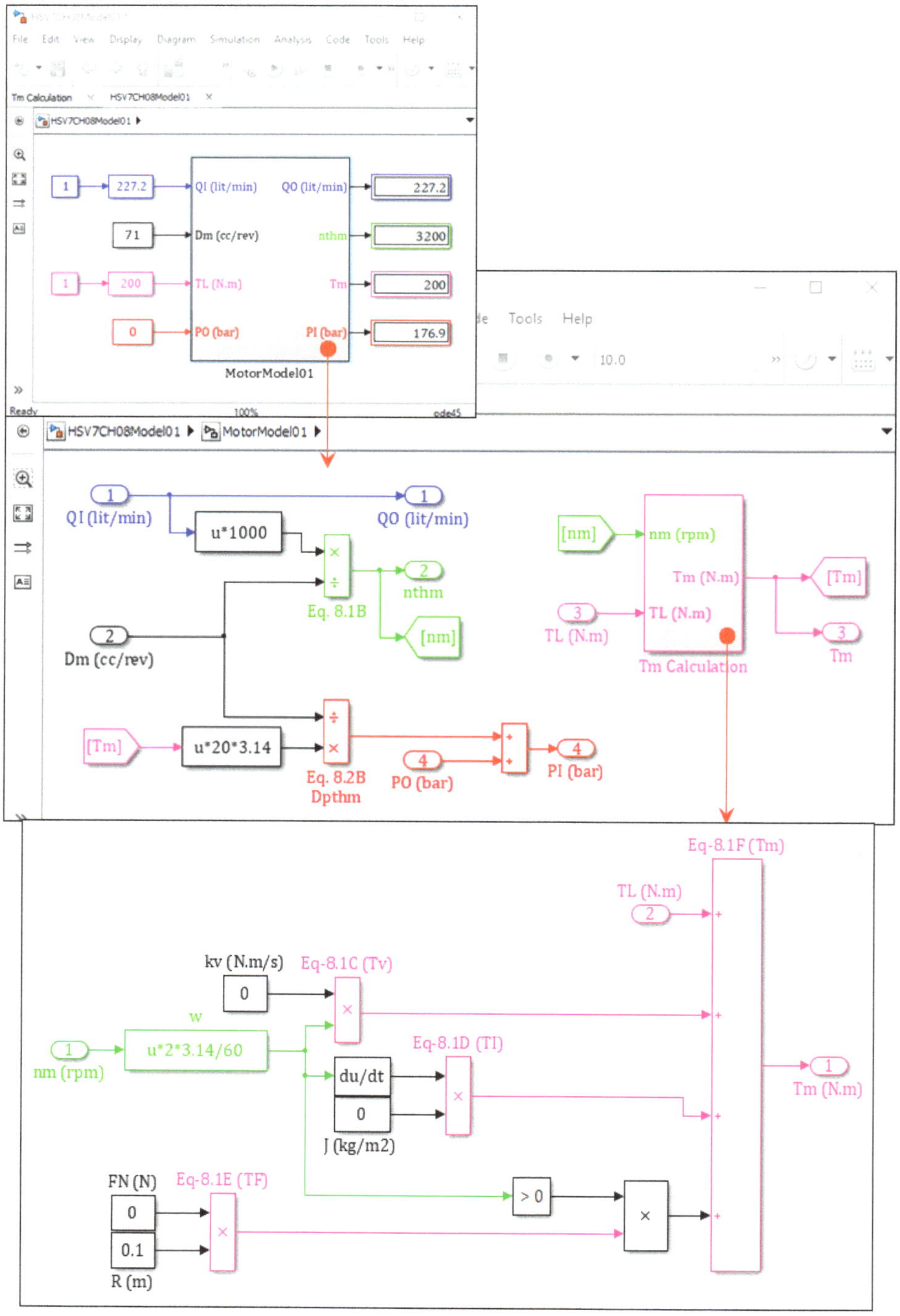

Fig. 8.10- Motor Simulation Model 01

8.4.6- Model Validation

Figure 8.11 shows case study when based on the test data given previously in Fig. 8.3. The model shows that, in order to achieve motor torque of 200 N.m and 3200 rpm motor speed:

- Theoretical differential pressure ΔP_{thm} (176.9 bar) < actual motor differential pressure ΔP_m (200 bar) reported by the test data. The difference is due to internal friction.
- Theoretical motor flow Q_{thm} (227.2 lit/min) < actual motor flow Q_m (240 lit/min) reported by the manufacturer. The difference is due to motor internal leakage.
- Only external load torque acts on the motor shaft and no considerable mass or friction.

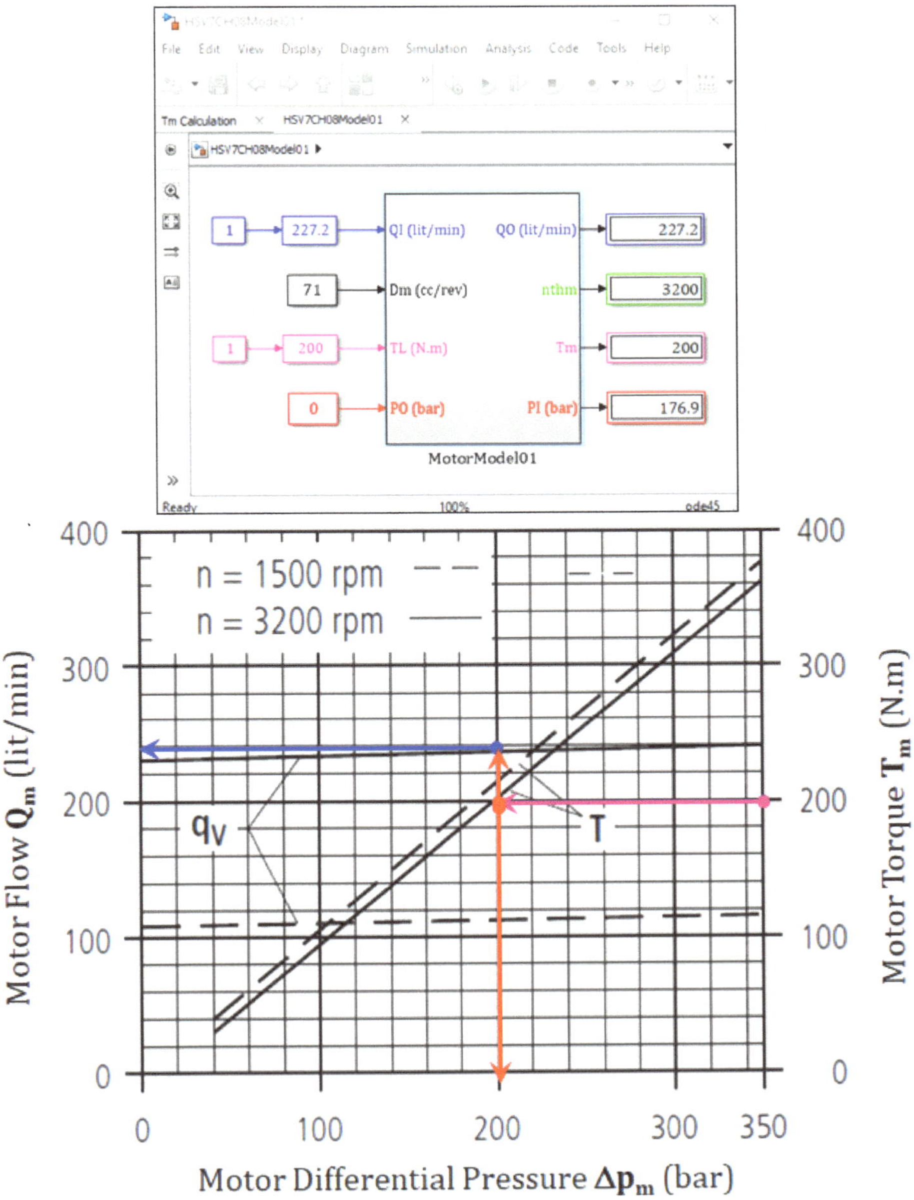

Fig. 8.11- Validation of Motor Model 01

8.5- Model #02A for a Fixed Displacement Motor Running at Constant Operating Conditions Based on Given Test Data

8.5.1- Model Features and Assumptions

Practical Motor Working at Constant Operating Conditions: In some applications, motors run at constant torque and receive constant flow. Under these conditions, motor works under constant efficiencies. In this case, only actual motor rpm and differential pressure (at this flow and torque) are needed to model the motor. Alternatively, two of the three motor efficiencies predefined at rpm and pressure are needed to model the motor.

8.5.2- Motor Actual Speed and Volumetric Efficiency

As shown in Fig. 8.12, at a specific differential pressure across the motor that is function of the motor torque, and a specific displacement:

- **On Curve A:** supplying theoretical flow $\mathbf{Q_{thm}}$ WILL NOT result in theoretical speed $\mathbf{n_{thm}}$. So, $\mathbf{Q_{thm}}$ results in actual motor speed $\mathbf{n_m} < \mathbf{n_{thm}}$. That is because a portion of the theoretical flow is leaking forward from high pressure at the motor inlet port to low pressure at the motor outlet port.

- **On Curve B:** As shown previously in Eq. 8.1, in order to achieve $\mathbf{n_{thm}}$, motor internal leakage $\mathbf{q_{Lm}}$ must be compensated (added to $\mathbf{Q_{thm}}$). So, actual flow $\mathbf{Q_m} > \mathbf{Q_{thm}}$.

- So, Eq. 8.3 is used to solve for the motor actual speed $\mathbf{n_m}$ or motor *Volumetric Efficiency* $\mathbf{e_{vm}}$ or leakage flow of the motor $\mathbf{q_{Lm}}$ depending on the available information.

$$e_{vm} = Q_{thm}/Q_m = Q_{thm}/(Q_{thm} + q_{Lm}) = n_m/n_{thm} \qquad 8.3$$

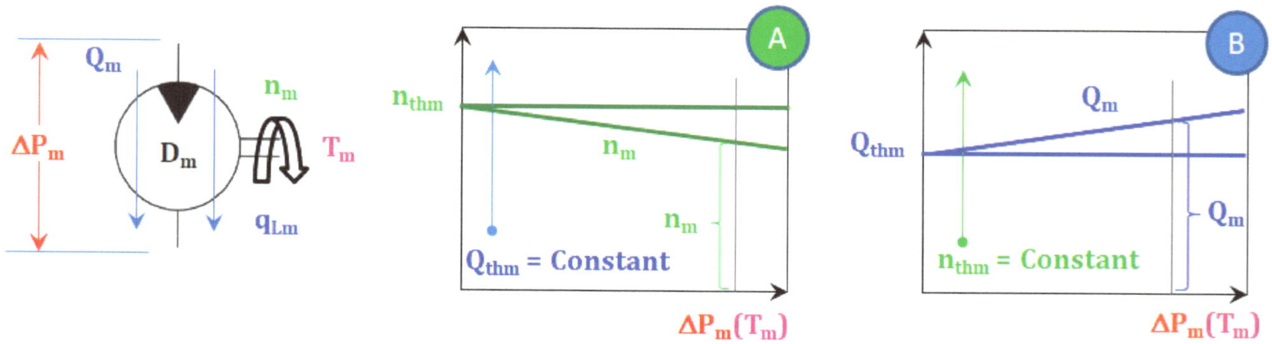

Fig. 8.12- Leakage Characteristics of a Fixed Motor

8.5.3- Motor Actual Torque and Mechanical Efficiency

As shown in Fig. 8.13, at a specific motor speed that is function of the inlet flow, for a hydraulic motor of a specific displacement:

- **On Curve A:** Applying theoretical torque $\mathbf{T_{thm}}$ WILL NOT result in theoretical differential pressure $\mathbf{\Delta P_{thm}}$. So, $\mathbf{T_{thm}}$ results in actual motor differential pressure $\mathbf{\Delta P_m} > \mathbf{\Delta P_{thm}}$.That is because additional incremental differential pressure $\Delta \mathbf{p_f}$ is needed to overcome the motor internal friction.

- **On Curve B:** As shown previously in Eq. 8.2, in order to achieve $\mathbf{\Delta P_{thm}}$, torque losses due to motor internal friction $\mathbf{T_{fm}}$ must be compensated (subtracted from $\mathbf{T_{thm}}$). So, that the actual net torque on the shaft $\mathbf{T_m} < \mathbf{T_{thm}}$.

- So, Eq. 8.4 is used to solve for the motor actual differential pressure $\mathbf{\Delta P_m}$ or motor *Mechanical Efficiency* $\mathbf{e_{mm}}$ or torque losses of the motor $\mathbf{T_{fm}}$ depending on the available information.

$$\mathbf{e_{mm} = T_m / T_{thm} = (T_{thm} + T_{fm})/T_{thm} = \Delta P_{thm}/\Delta P_m} \qquad \mathbf{8.4}$$

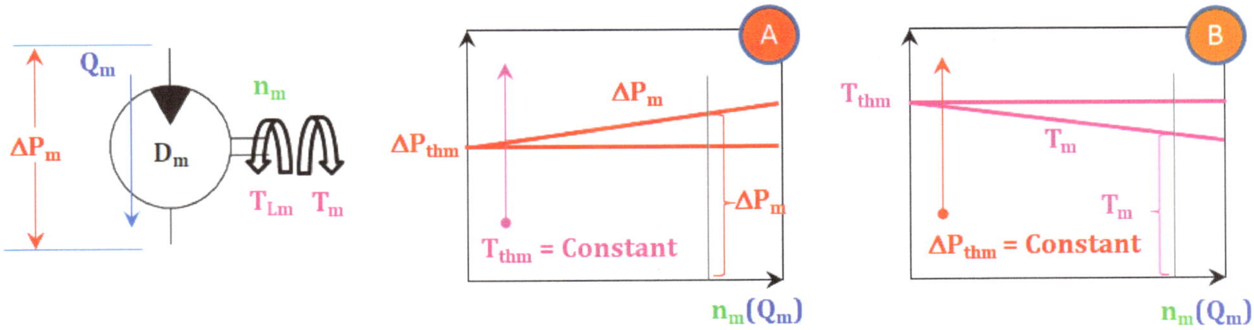

Fig. 8.13- Friction Characteristics of a Fixed Motor

8.5.4- Motor Overall Efficiency and Power

Equations 8.5A and 8.5B show calculations of motor *Overall Efficiency* and Power in English and Metric systems of units; respectively.

$$\mathbf{e_{om} = \frac{Motor\ Output\ (Shaft)Power\ [H_{outm}]}{Motor\ Input\ (Fluid)Power[H_{inm}]} = \frac{1714 \times H_{outm}(Hp)}{\Delta P_m(psi) \times Q_m(gpm)} = e_{vm} \times e_{mm}} \qquad \mathbf{8.5A}$$

$$\mathbf{e_{om} = \frac{Motor\ Output\ (Shaft)Power[H_{outm}]}{Motor\ Input\ (Fluid)Power[H_{inm}]} = \frac{600 \times H_{outm}(kW)}{\Delta P_m(bar) \times Q_m(lit/min)} = e_{vm} \times e_{mm}} \qquad \mathbf{8.5B}$$

Equation 8.6 is used to calculate the power losses $\mathbf{H_{Lm}}$ that is converted into heat in the motor.

$$\mathbf{H_{Lm} = H_{inm} - H_{outm} = H_{inm} - H_{inm} \times e_{Om} = H_{inm}(1 - e_{Om})} \qquad \mathbf{8.6}$$

8.5.5- Simulation Model

Figure 8.14 shows simulation model (HSV7CH08Model02A) for a practical fixed displacement unidirectional motor built using Matlab-Simulink. The model is structured based on having motor test data (motor outlet speed and differential pressure as defined at the given inlet flow and torque acting on the motor shaft). Because the given test data is tied to this specific motor displacement, it is loaded as constants in the model properties. They can be shown on the model page as constant for easier investigations and model debugging.

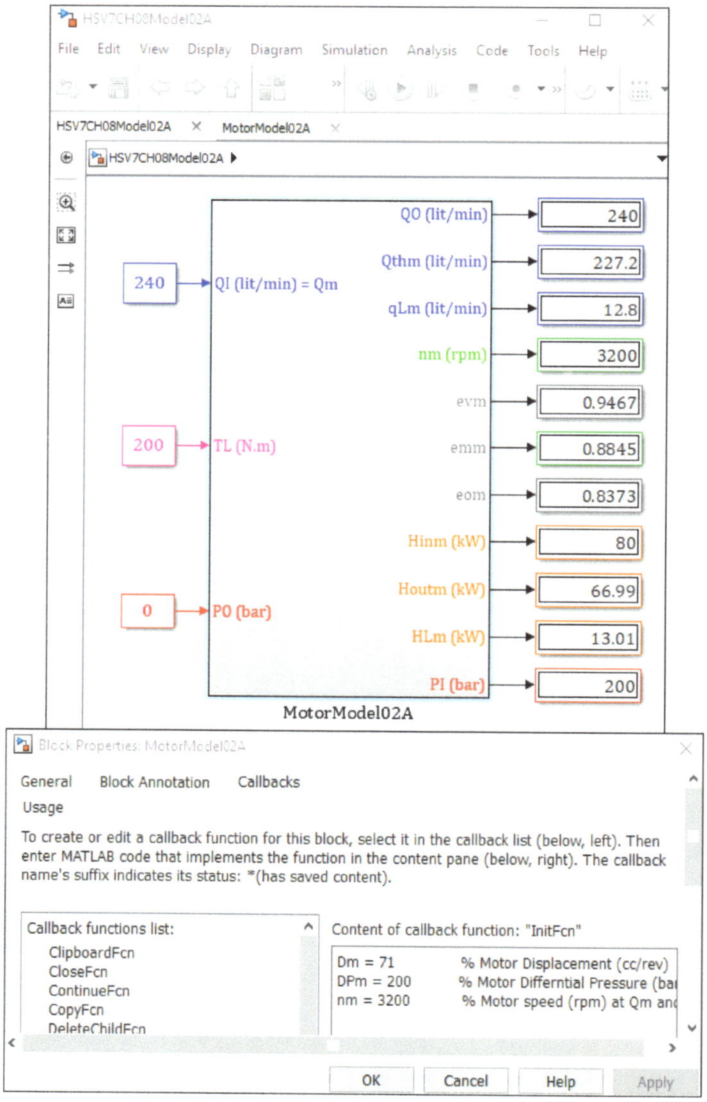

Fig. 8.14- Motor Simulation Model 02A

Figure 8.15 shows the details under the mask of the simulation model. The model solves Eq. 8.1 through 8.6 in the following sequence:

- Given Output Test data (n_m and Δp_m), at Input Operating Conditions (Q_m and T_m), and motor size D_m:
- Eq. 8.1 $\rightarrow$ n_{thm} and T_m
- Eq. 8.2 $\rightarrow$ Δp_{thm}
- Eq. 8.3 $\rightarrow$ e_{vm} and q_{Lm}
- Eq. 8.4 $\rightarrow$ e_{mm} and T_{Lm}
- Eq. 8.5 $\rightarrow$ H_{out} and e_{om}
- Eq. 8.6 $\rightarrow$ H_{Lm}

It is to be noted that, for unidirectional motors, output flow equals inlet flow because the internal leakage is routed to the outlet port. In Bidirectional motors, outlet flow equals to the theoretical flow (less than actual flow) because motor leakage is routed to the case drain.

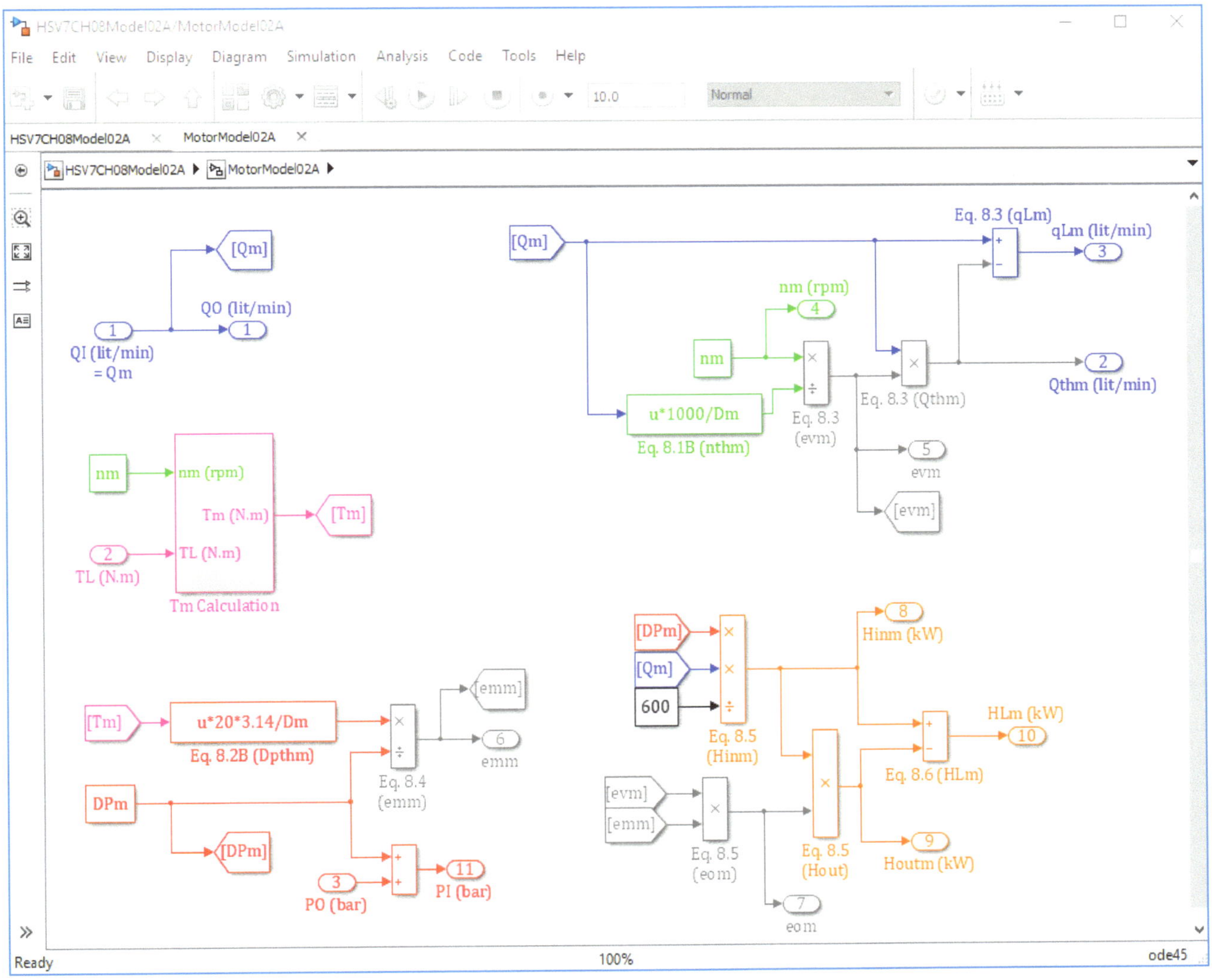

Fig. 8.15- Details under the Mask of Simulation Model 02A

8.5.6- Model Validation

Figure 8.16 shows that the model is validated based on other software (Hydraulic Components Sizing Calculator).

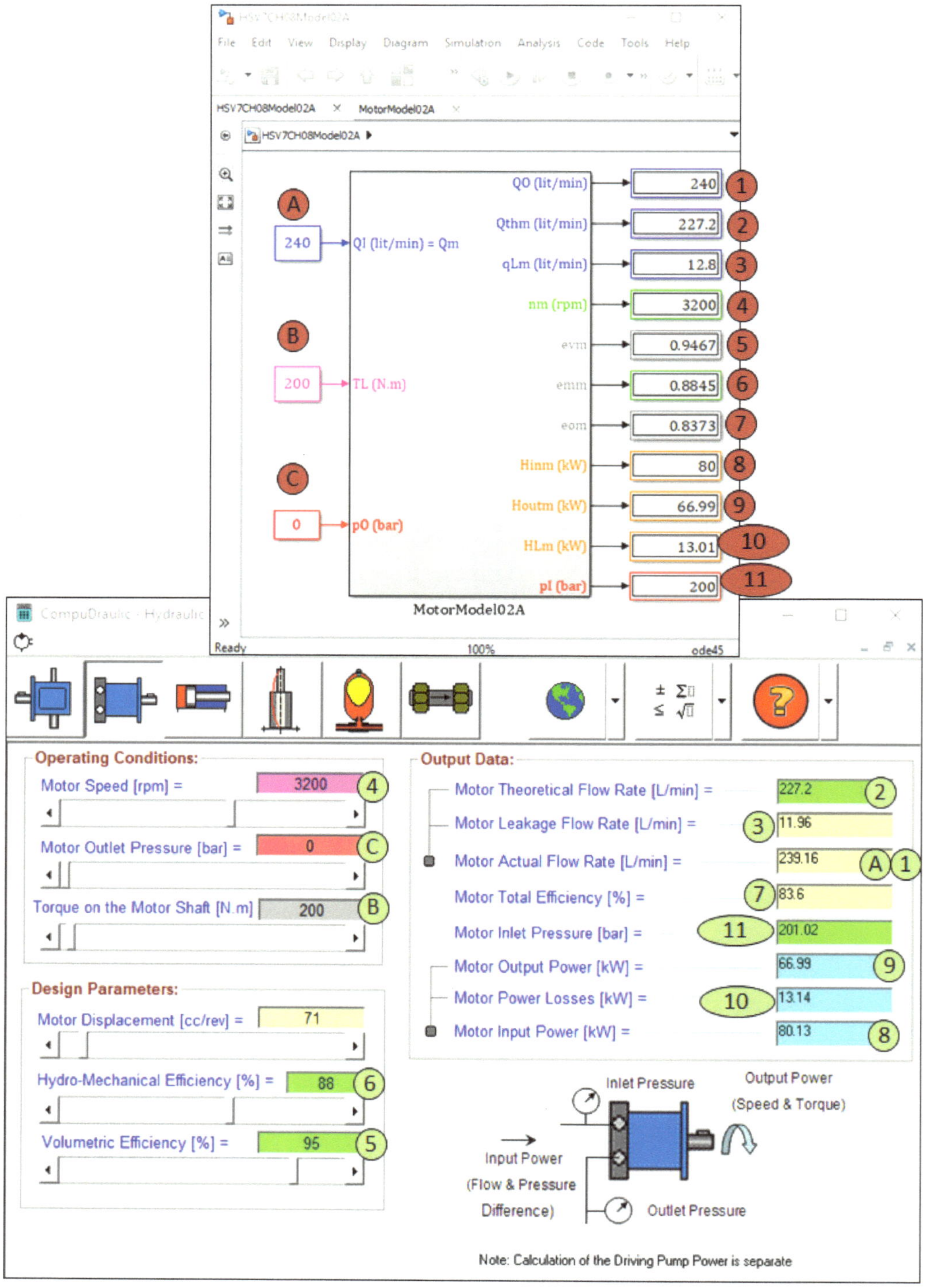

Fig. 8.16- Validation of Model 02A

8.6- Model #02B for a Fixed Displacement Motor Running at Constant Operating Conditions Based on Given Efficiency Values

8.6.1- Model Features and Assumptions

Same as in model 02A except that volumetric efficiency and mechanical efficiency are given instead of motor test data.

8.6.2- Simulation Model

Figure 8.17 shows the upper mask of simulation model (HSV7CH08Model02B). When the model is fed by the same efficiencies shown in model 02A, it produces the same other results. This validates the model.

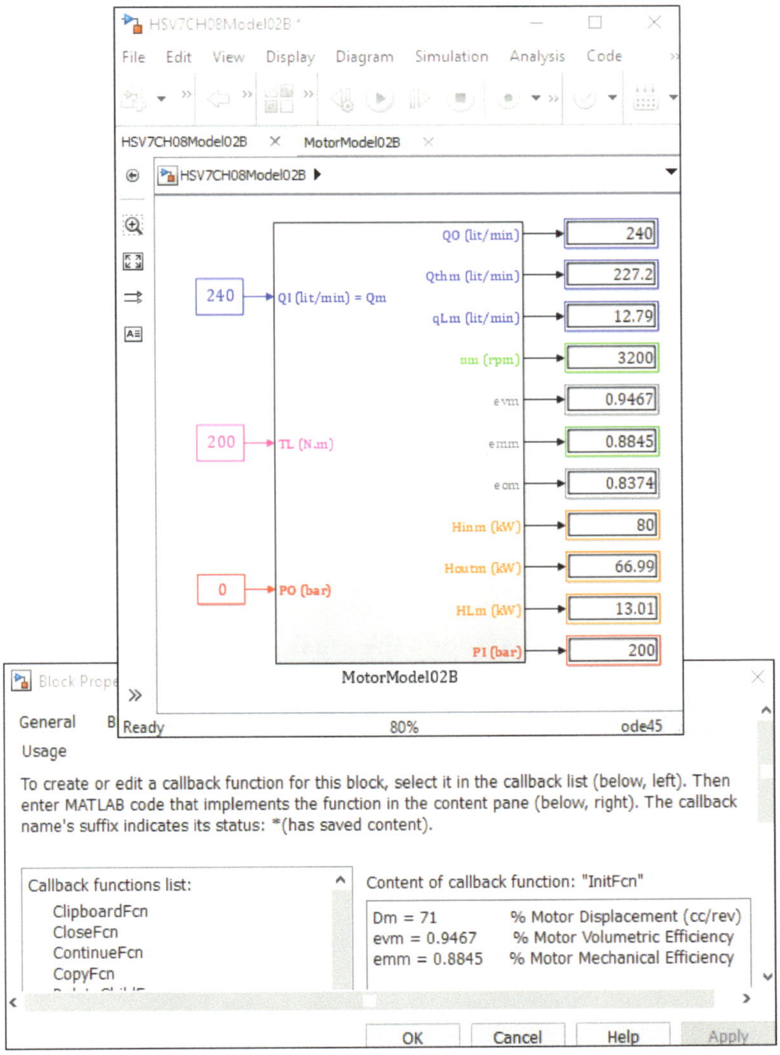

Fig. 8.17- Pump Simulation Model 02B

Figure 8.18 shows the details under the mask of simulation model 02B. The model solves the same set of Eq. 8.1 through 8.6.

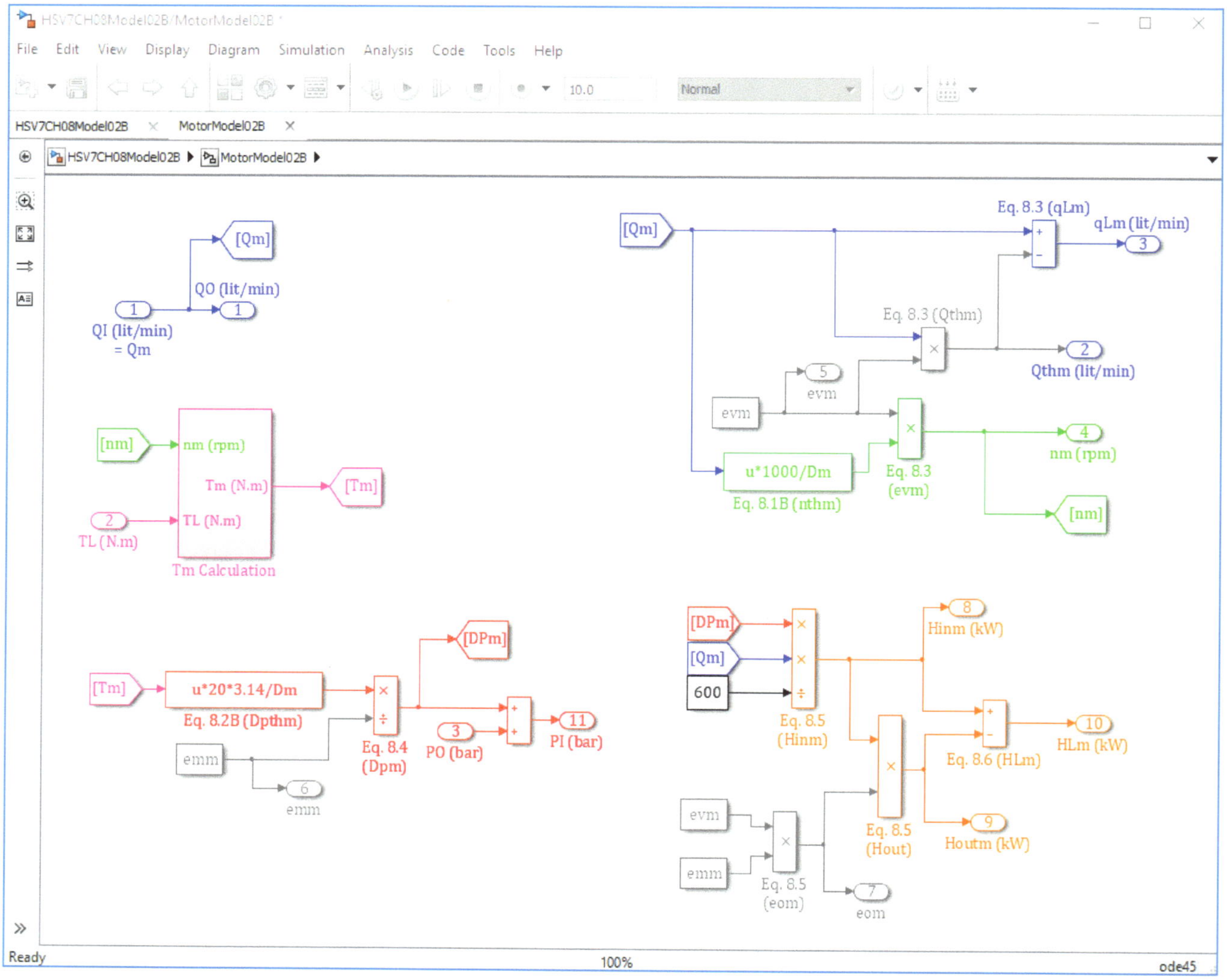

Fig. 8.18- Details under the Mask of Simulation Model 02B

8.7- Model #03A for a Fixed Displacement Motor Running at Variable Operating Conditions Based on Given Test Data

8.7.1- Model Features and Assumptions

Practical Motor Working at Variable Operating Conditions: In some applications, a motor may work under variable driving torque and inlet flow. This means that motor efficiencies are variable and model outputs must be updated interactively with changes in driving torque and inlet flow.

8.7.2- Mathematical Model

Same set of equations 8.1 through 8.6 are still used except that motor test data are loaded to the model in the form of two 2-D lookup tables instead of being given as constant values.

8.7.3- Simulation Model

Figure 8.19 shows the upper mask of the simulation model (HSV7CH08Model03A). The figure shows that the model has the following elements:

Motor Inlet Flow Q_m: For a motor that receives variable inlet flow, it receives flow from the upstream model of a control valve or a variable displacement pump. However, for investigating the motor performance under variable inlet flow, a "Slider Gain" is temporarily used to simulate variable flow. Additionally, a "Ramp" block is used to develop the motor steady state characteristics versus the inlet flow.

Motor Load Torque T_L: Similarly, for a motor that works under variable torque, it receives torque from the downstream model of the load. However, for investigating the motor performance under variable torque, a "Slider Gain" is temporarily used to simulate the load. Additionally, a "Ramp" block is used to develop the motor steady state characteristics versus the torque.

Motor Displacement D_m: It is very important to remember that the lookup tables are assigned for that specific motor size So, if the motor size is changed, then the lookup table must be updated accordingly. That is why, rather than having the motor size as a variable input to the model, it has been kept inside the model properties as a motor constant.

Dynamic Effect of Motor Torque Change (at D_m and Q_m are Constants): Motor torque is used to calculate the differential pressure and the internal leakage. So, due to modulus of elasticity of the driving shaft and how the internal leakage is developed within the motor clearances, it can't be assumed that these changes occur instantaneously with variation of the torque. Modeling this effect is very exhaustive and depends on specific motor design. However, in view of machine modeling, it can be feasibly represented by a fist order normalized transfer function with a reasonable time constant (e.g. 20 ms).

Dynamic Effect of Inlet Flow Change (at D_m and T_p are Constants): Inlet flow is used to calculate the motor speed. Due to filling coefficient, fluid hydrodynamic effect, and inertial effect of motor rotating mass, it can't be assumed that the resulting motor speed varies instantaneously with variation in motor inlet flow. Therefore, similarly, dynamic effect of inlet flow can be feasibly represented by a fist order normalized transfer function with a reasonable time constant.

The figure shows that, by having same Q_m and T_m as in model 02A, model 03A generates the same resulting data with a bit of tuning required in the motor test data lookup tables.

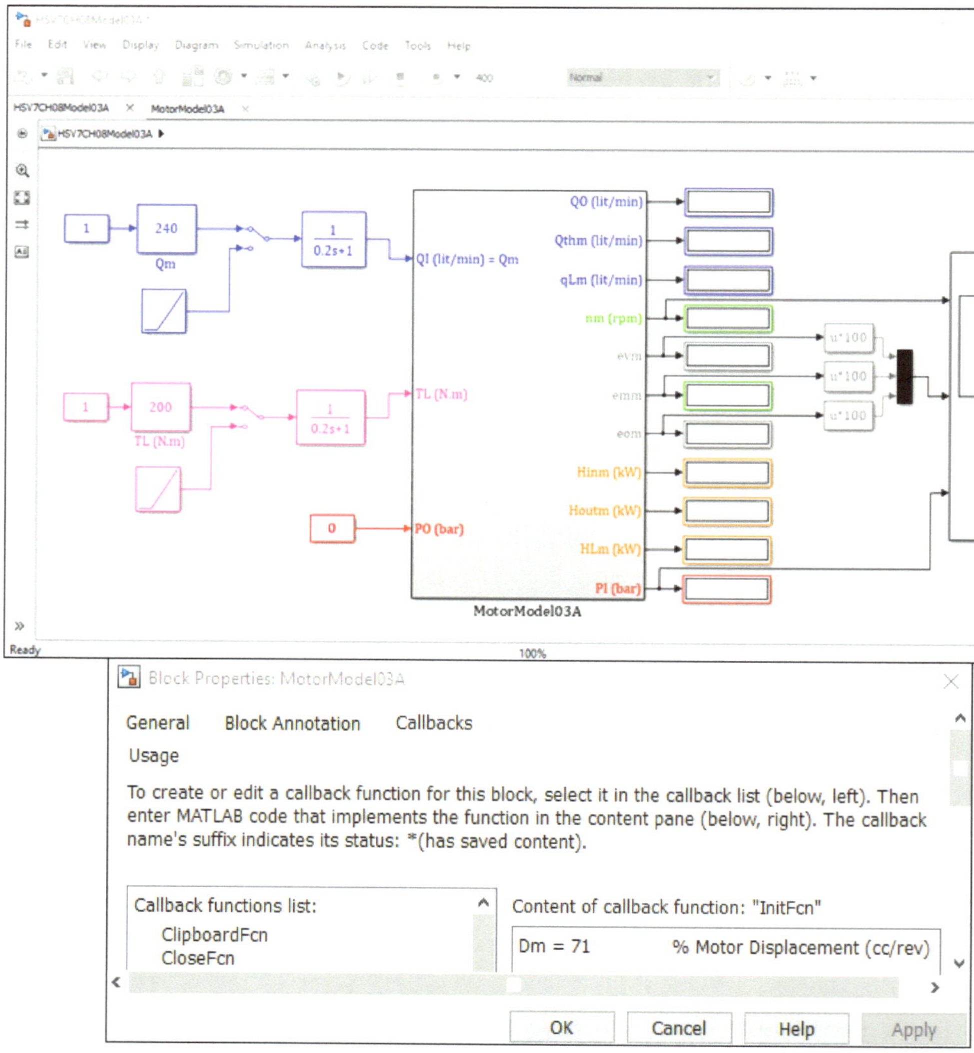

Fig. 8.19- Motor Simulation Model 03A

Figure 8.20 shows the details under the mask of simulation model 03A that is structured to override model 02A. Model 03A serves to simulate a motor that works under constant or variable flow and torque.

Two switches were added so that the model can select motor speed and differential pressure as constants or variable from a lookup table. However, if constant speed and differential pressure are selected, this means these values are defined at whatever constant values selected at the motor inlet flow and torque slider gains.

The figure shows how the model solves Eq 8.1 through 8.6. Equation numbers are listed accordingly on the model blocks.

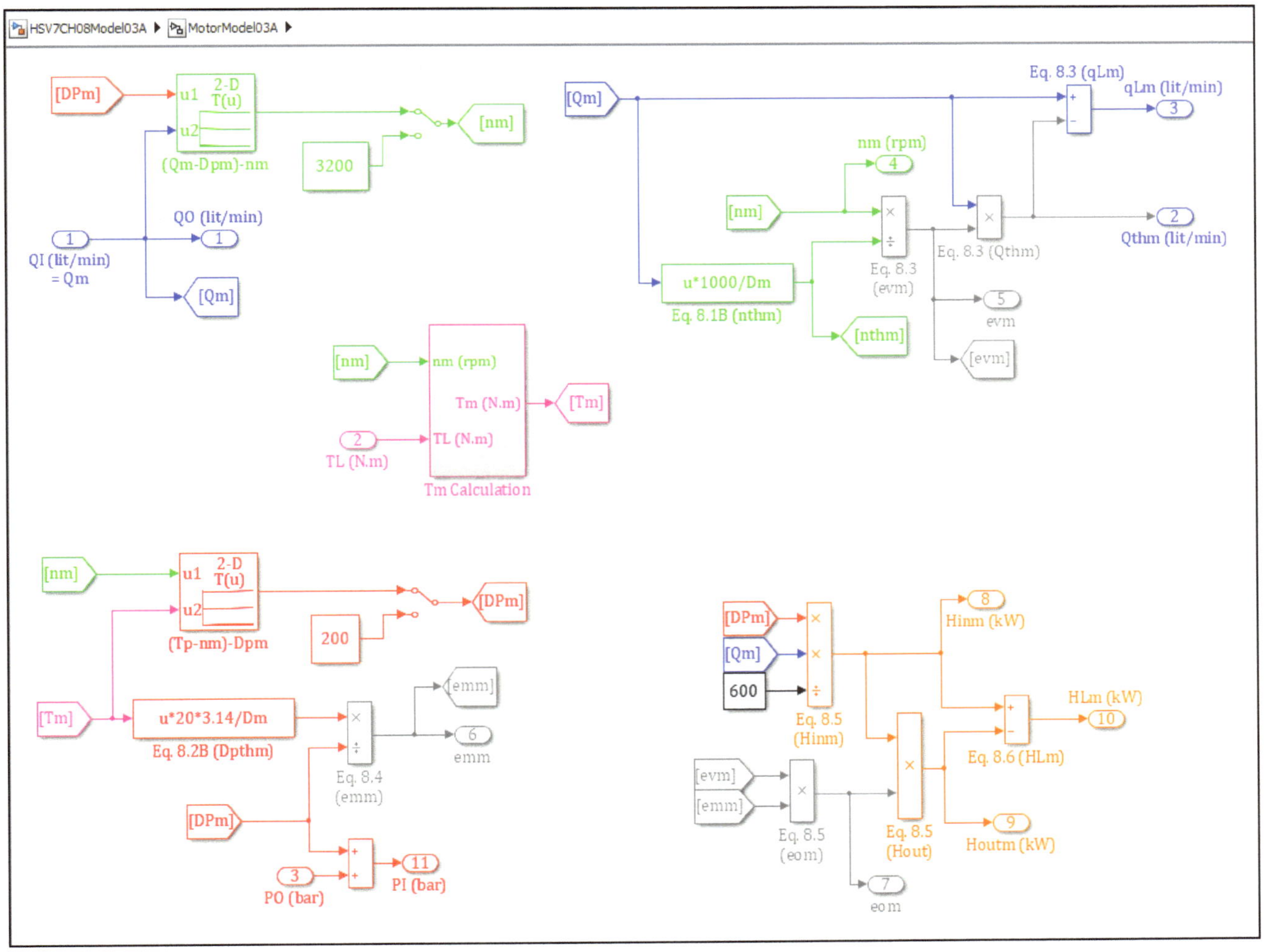

Fig. 8.20- Details under the Mask of Simulation Model 03A

Figures 8.21 and 8.22 show that the lookup tables were built based on motor characteristics previously given in Fig. 8.3. Since the motor characteristics are linear, then three points are just fine to build the lookup table. Algorithm of interpolation and extrapolation in the lookup tables are set to be "linear".

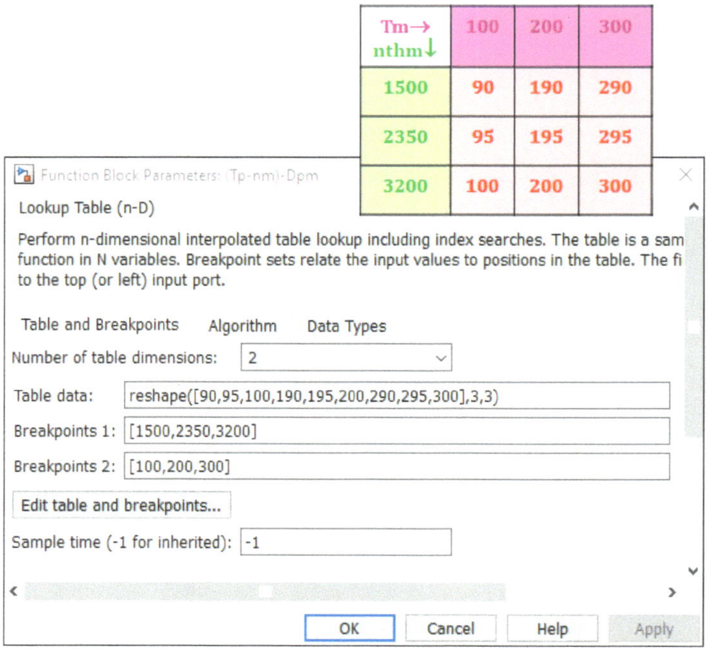

Tm→ nthm↓	100	200	300
1500	90	190	290
2350	95	195	295
3200	100	200	300

Fig. 8.21- Lookup Table for Motor Pressure-Torque Characteristics

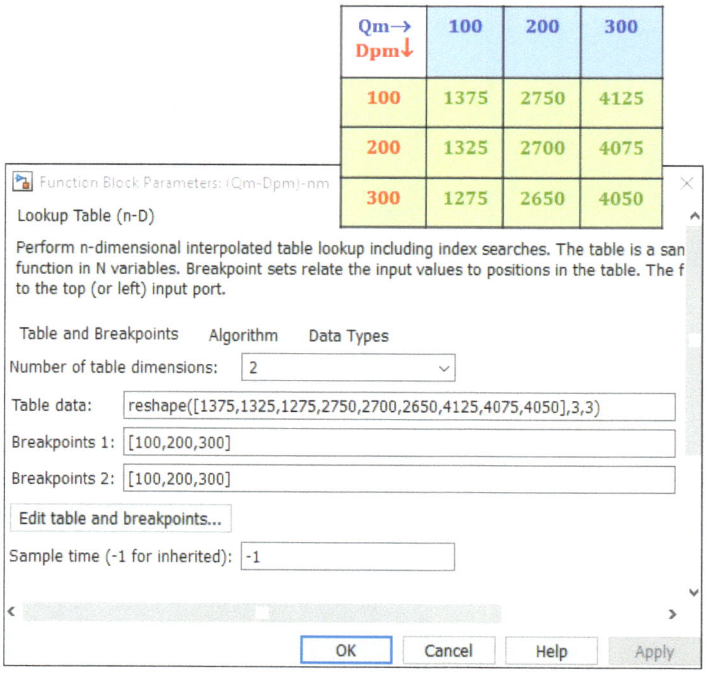

Qm→ Dpm↓	100	200	300
100	1375	2750	4125
200	1325	2700	4075
300	1275	2650	4050

Fig. 8.22- Lookup Table for Motor Speed-Flow Characteristics

8.7.4- Model Validation

Figure 8.23 shows the motor static characteristics versus inlet flow (motor speed) at constant torque. As shown in the figure, for typical fixed displacement motors, volumetric efficiency increases, and mechanical efficiency decreases with the increase of inlet flow (motor speed).

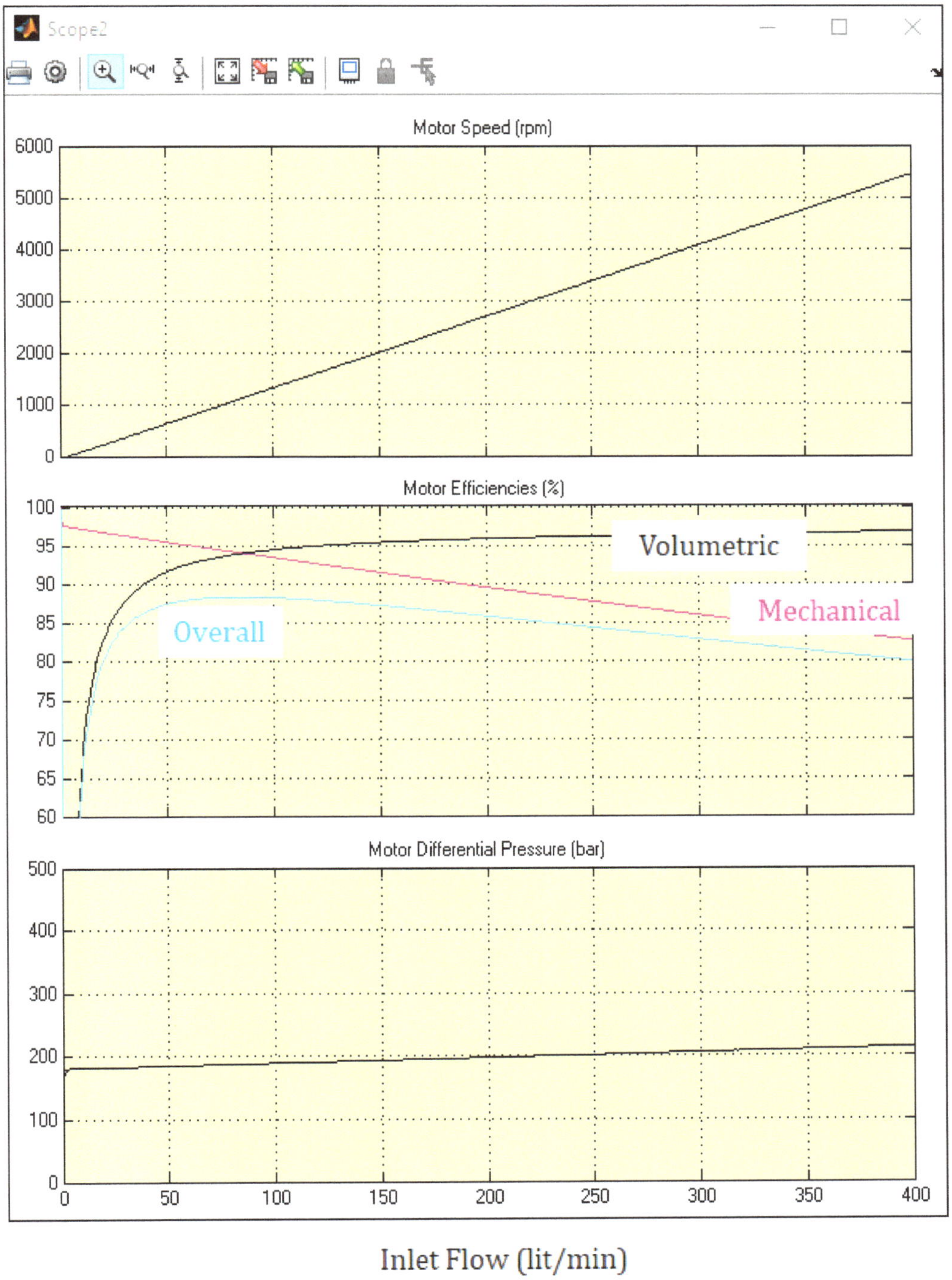

Fig. 8.23- Motor Static Characteristics versus Inlet Flow (at T_m = 200 N.m.)

Figure 8.24 shows the motor static characteristics versus motor torque (differential pressure) at constant flow. As shown in the figure, for typical fixed displacement motors, volumetric efficiency decreases, and mechanical efficiency increases with the increase of motor torque (differential pressure).

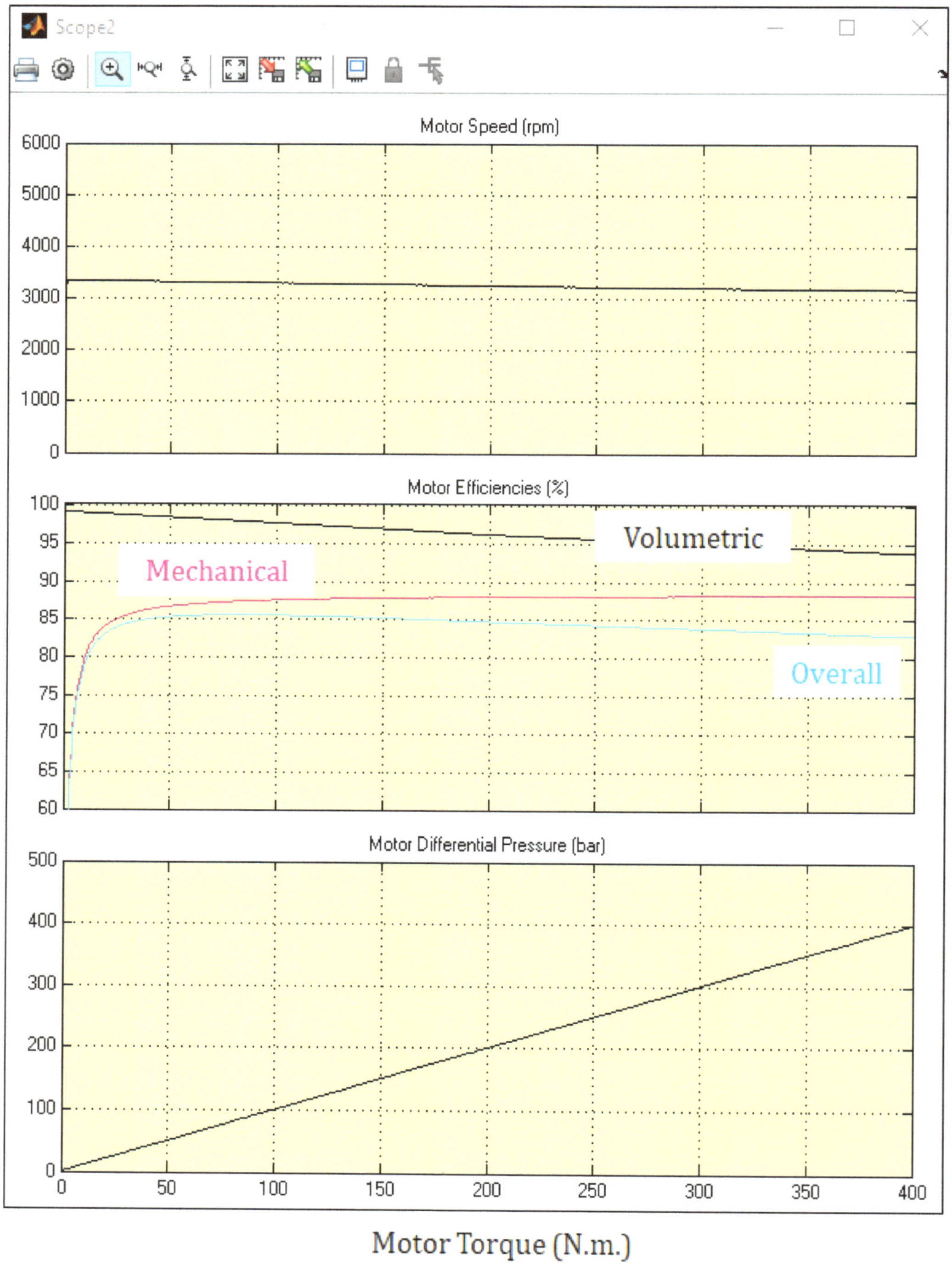

Fig. 8.24- Motor Static Characteristics versus Motor Torque (at Q_m = 240 l/min)

8.8- Model #03B for a Fixed Displacement Motor Running at Variable Operating Conditions Based on Given Efficiency Curves

8.8.1- Model Features and Assumptions

Same as in model 03A except that volumetric efficiency and mechanical efficiency curves are given instead of motor test data.

8.8.2- Simulation Model

Figure 8.25 shows the upper mask of simulation model (HSV7CH08Model03B). The figure shows that, by having the same $\mathbf{Q_m}$ and T_m as in model 02A, model 03A generates the same resulting data with a bit of tuning required in the motor efficiency curves lookup tables.

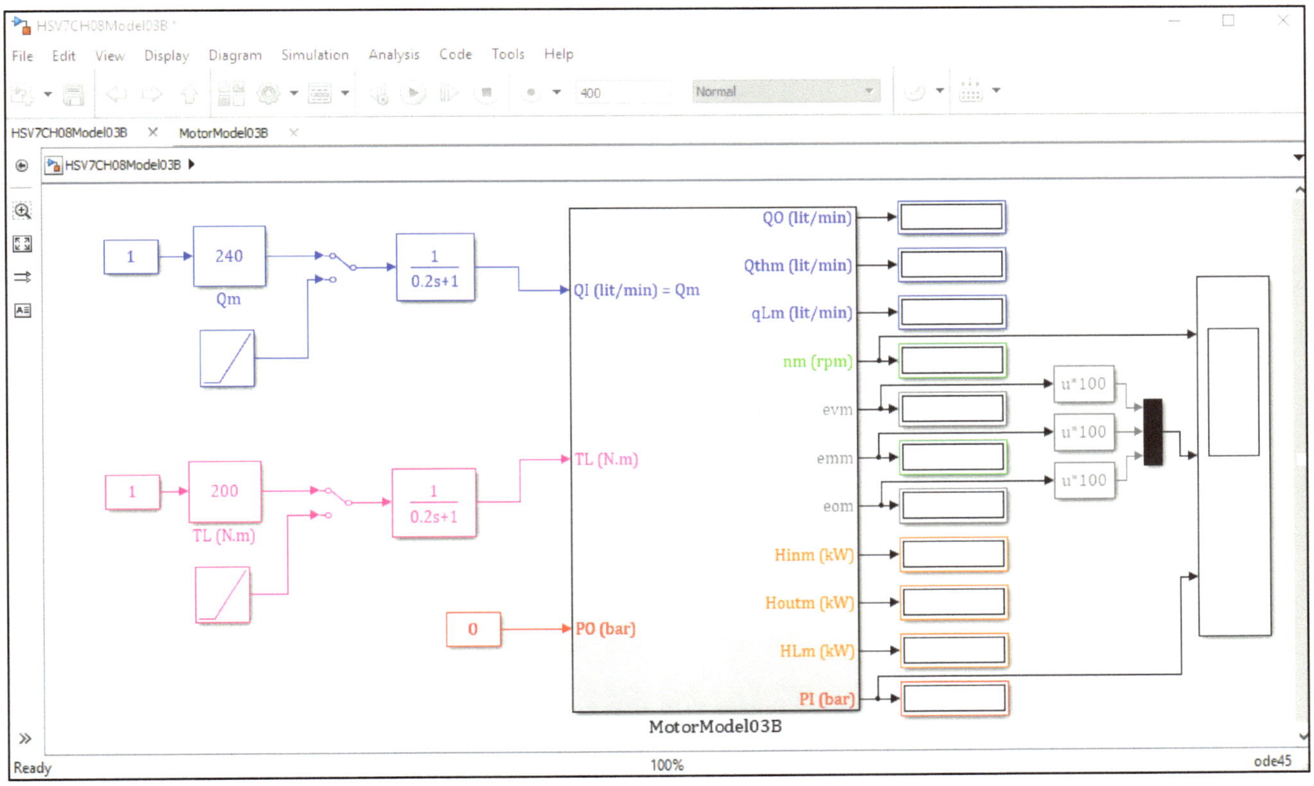

Fig. 8.25- Motor Simulation Model 03B

Figure 8.26 shows the details under the mask of simulation model 03B that is structured to override model 02B. Model 03B serves to simulate a motor that works under constant or variable flow and torque. Two switches were added so that the model can select motor efficiencies as constants or variable from lookup table. However, if constant efficiencies are selected, this means these values are defined at whatever constant values selected at the motor inlet flow and torque slider gains.

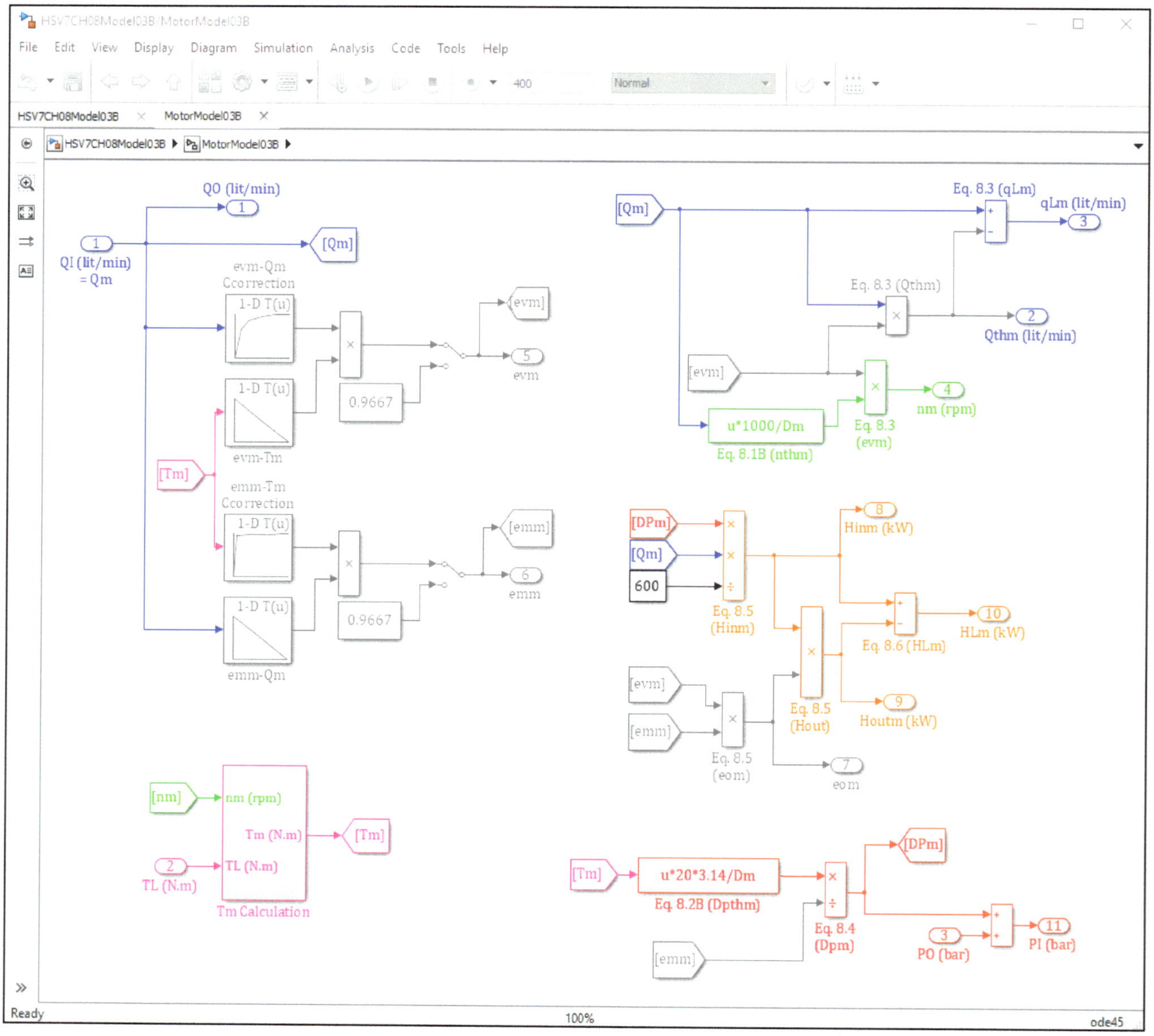

Fig. 8.26- Details under the Mask of Simulation Model 03B

8.8.3- Model Validation

Figure 8.27 shows the motor static characteristics based on give efficiency curves. The figure shows very close results as shown previously in Fig. 8.23 and 8.24. These results validate the model with the new calculations. Maybe fine tuning is required to the efficiency curves lookup tables. It is to be noted that, for more accurate results when efficiency curves are given, lookup tables require more points because some of efficiencies are highly nonlinear.

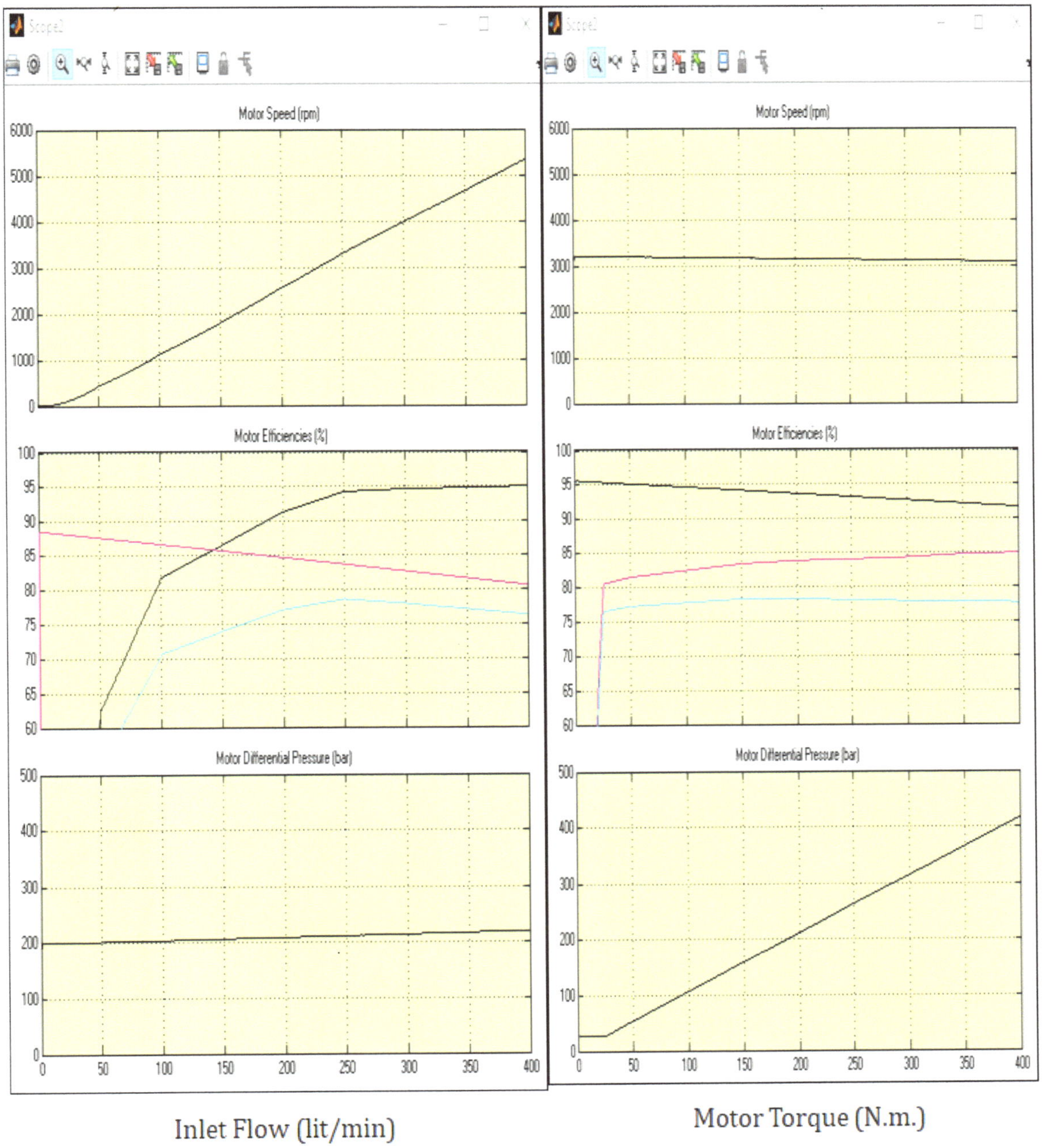

Fig. 8.27- Motor Static Characteristics Based of Given Efficiency Curves

8.9- Modeling Variable Displacement Motors

Modeling Approach: Figure 8.28 shows two mechanisms of axial-piston motors. The motor on the left side is a swash-plate type mechanism, and the one on the right side is a bent-axis mechanism. As it has been previously discussed, such motors should be modeled based on their internal structure for purposes of motor development. For purposes of overall machine modeling, they are modeled based on the published static and dynamic characteristics, no matter what the motoring mechanism is.

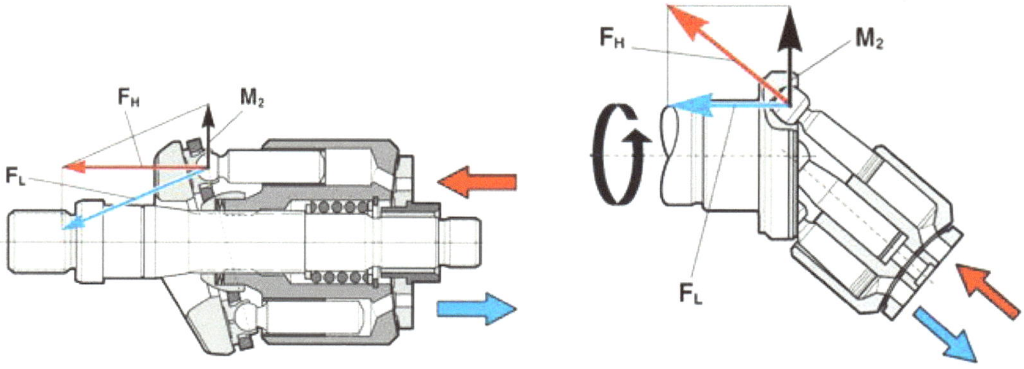

Fig. 8.28- Variable Displacement Motor Mechanisms (Courtesy of Bosch Rexroth)

Control Modes: Figure 8.29 shows common control modes of variable displacement motors:

1. **Two Position Control:** A two-position control, where max and min displacements are governed by a controller command. This is equivalent to pressure-compensated pump controllers.
2. **Proportional Control:** This controller satisfies the variable speed demand of an application such as in hydrostatic transmission. This is equivalent to displacement-controlled pumps.
3. **Torque-Limiting (Constant-Power) Control:** This controller takes care of assuring motor power (torque) won't exceed certain maximum set value. This is equivalent to torque-limiting (constant-Power) controlled pumps.
4.

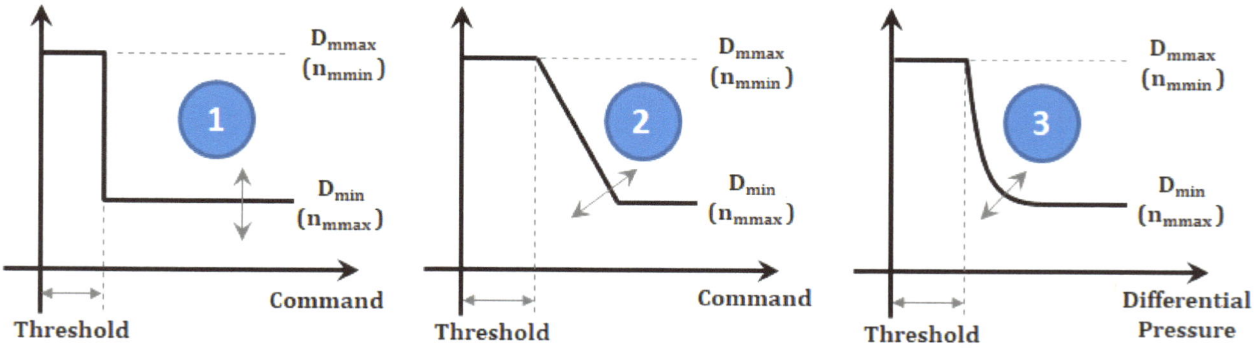

Fig. 8.29- Common Control Modes of Variable Displacement Motors

Controlling Command: To vary the motor displacement, a command signal is required to actuate the controller. Controller is actuated manually or remotely (hydraulically by a pilot pressure or electro-hydraulically by an electrical variable signal).

Model Complexity: If these performance curves are the only available information about the motor, then simple models can be developed that contain just a lookup table to provide only motor speed versus the command. Otherwise, if full test data is available, then a special model should be developed to give full outputs.

Min-Max Speed: Varying hydraulic motor displacement results in varying motor speed. It is to be noted that minimum displacement of a hydraulic motor should be reviewed with the motor's manufacturer because it corresponds to maximum allowable speed above which the motor runs inefficiently. Similarly, maximum displacement of a hydraulic motor should be reviewed with the motor manufacturer because it corresponds to minimum allowable speed below which the motor runs erratically. That is why, as shown in Fig. 8.30, some manufacturers use mechanical stroke (displacement) adjustors in order to physically limit motor displacement within the permissible range.

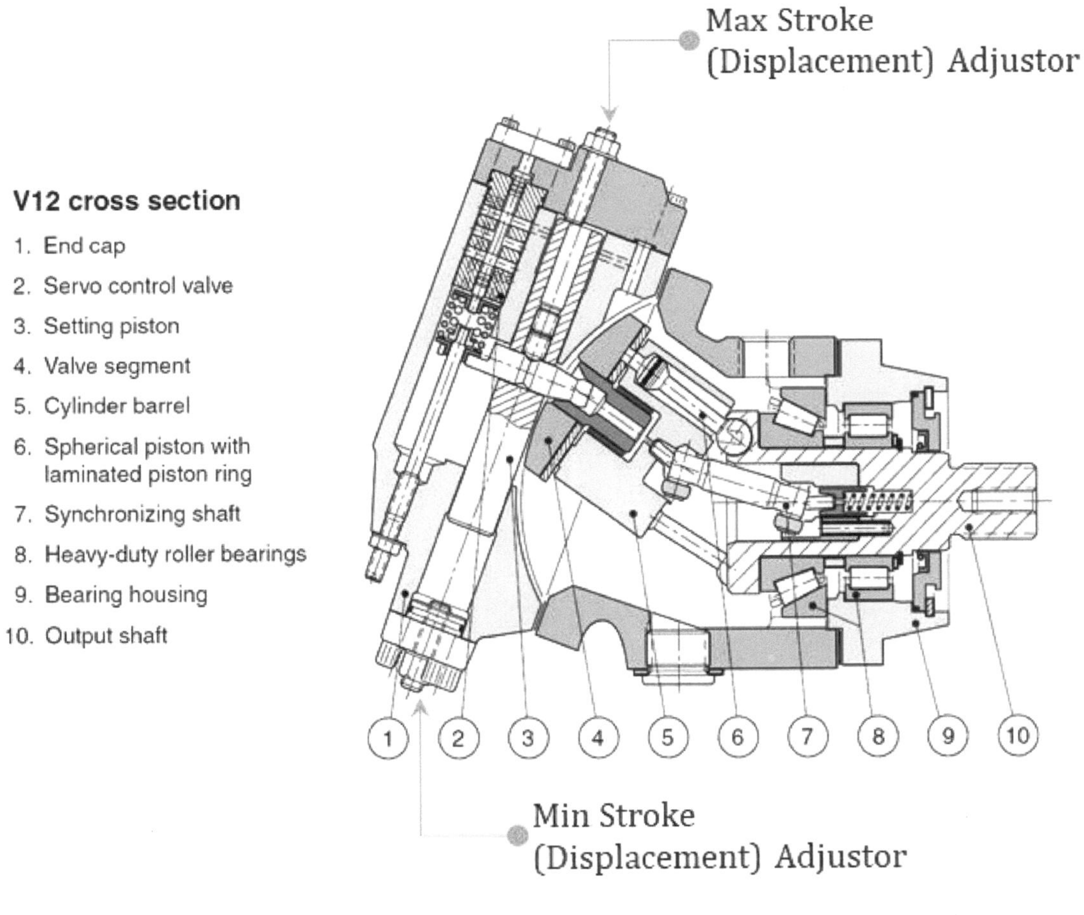

Fig. 8.30- Limiting Displacement of a Variable Motor (Courtesy of Parker)

8.10- Model #04 for Variable Displacement Motors

8.10.1- Model Features and Assumptions

Easiest rout to model a variable displacement motor is based on given efficiency curves, assuming these efficiencies are valid for all range of motor displacements. Otherwise, multi-dimension lookup table should be used to determine efficiencies at various displacements.

8.10.2- Simulation Model

Figure 8.31 shows the upper mask of simulation model (HSV7CH08Model04). In this model, motor displacement is located as an interactively changing variable. Command motor displacement could be by a direct command or could be pressure-related. So, the model should be retrofitted accordingly to reflect that situation. It is to be reminded that a saturation block is used to limit the max-min displacement as discussed previously.

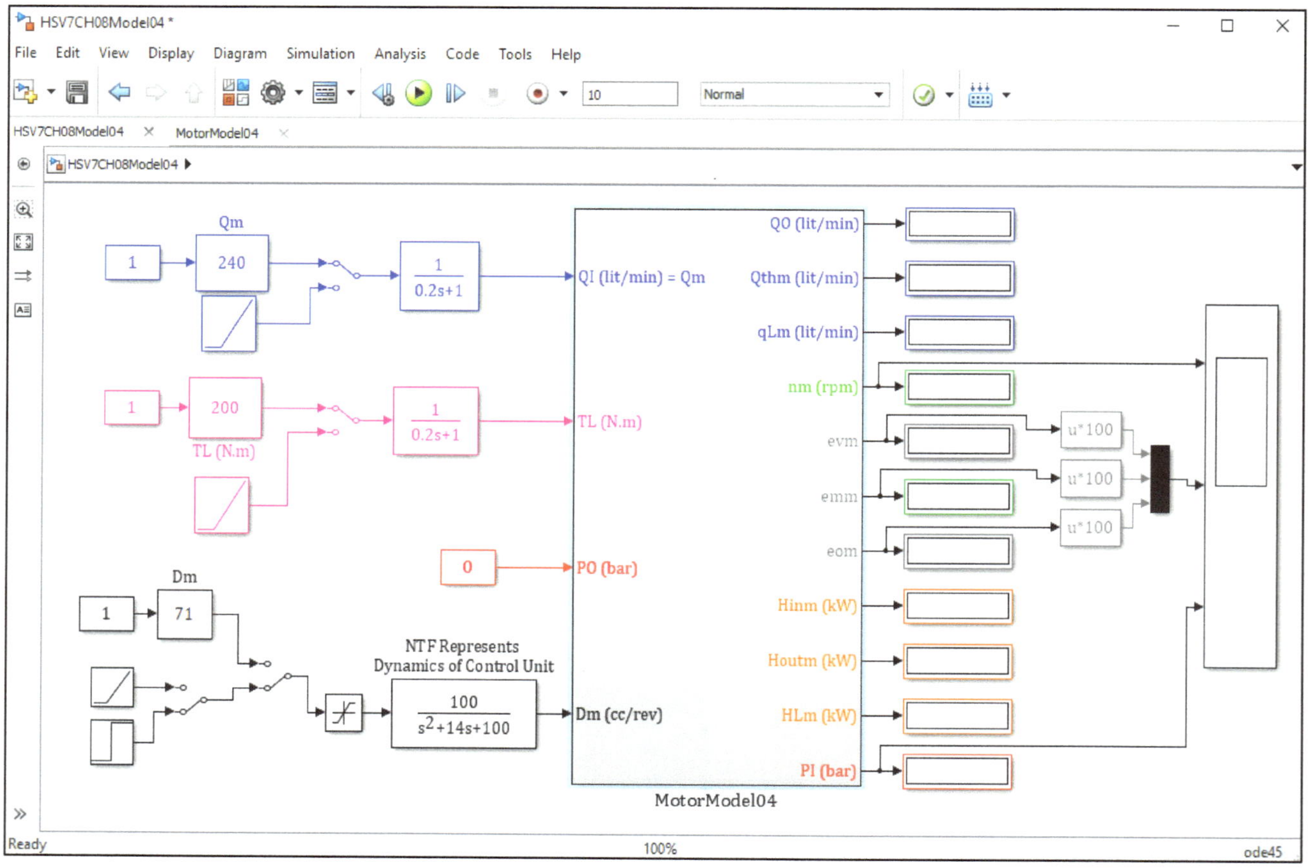

Fig. 8.31- Motor Simulation Model 04

Figure 8.32 shows the details under the mask of the Model 04. This model has the same structure as model 03B except that the motor displacement is now an input rather than a constant in the model properties. This model is applicable for either 2-position controlled motors or proportional controlled motor, not for torque-limited motors.

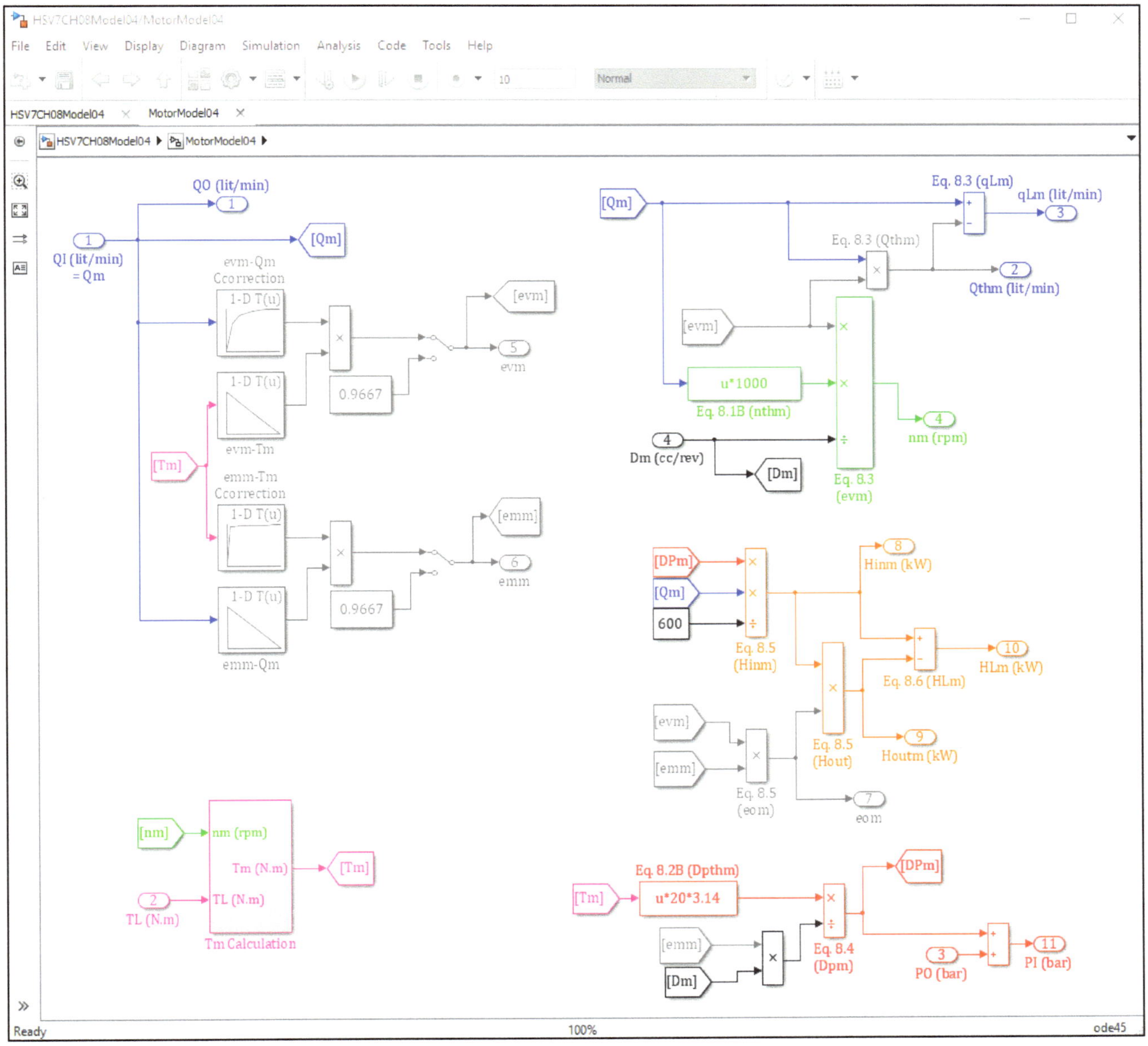

Fig. 8.32- Details under the Mask of Simulation Model 04

8.10.3- Model Validation

Static Characteristics: Figure 8.33 shows static characteristics of a variable displacement motor at constant flow (240 lit/min) and constant torque (200 N.m) where the displacement is limited between 40-70 cc/rev. Obviously, motor speed and differential pressure decrease with the increase of motor displacement.

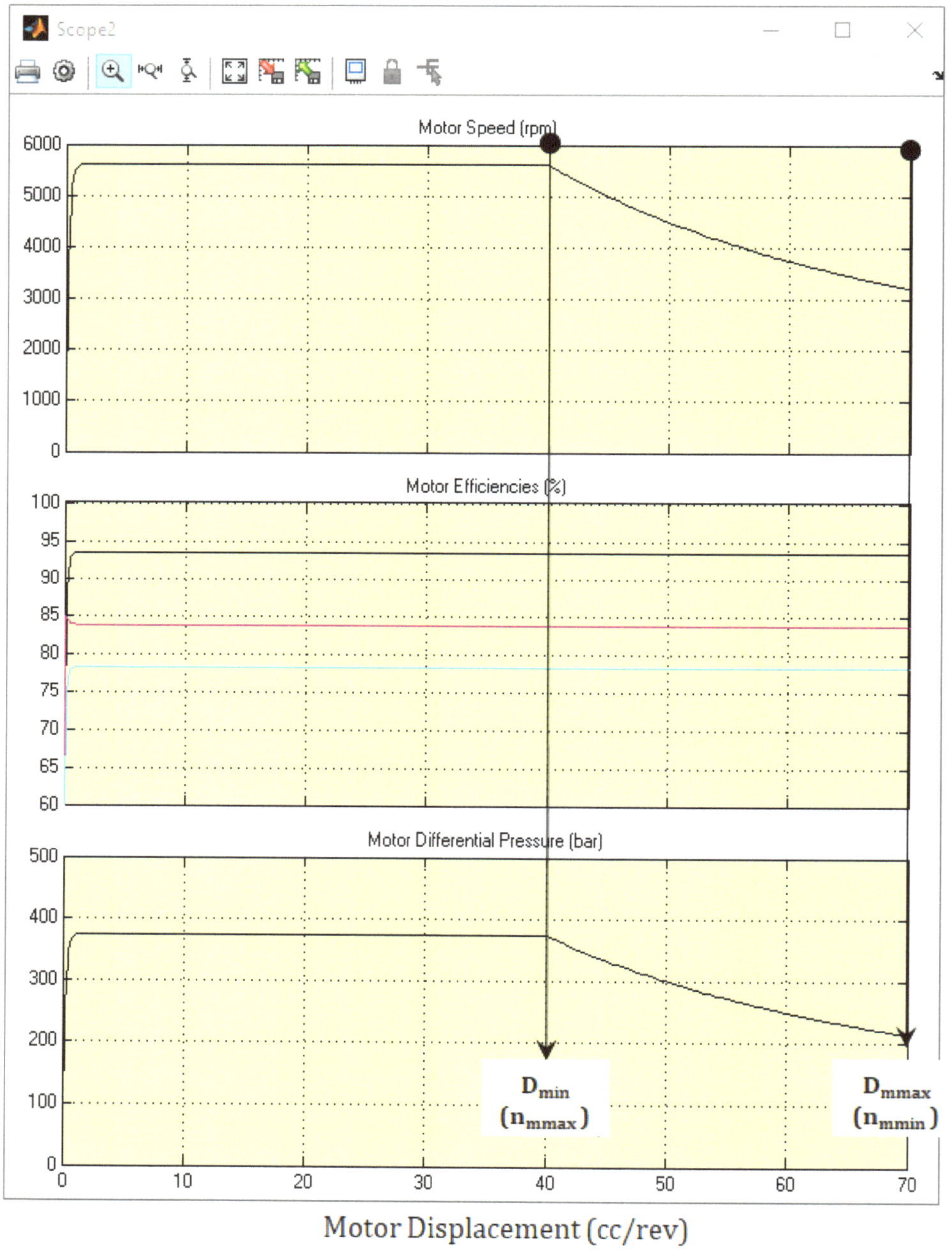

Fig. 8.33- Static Characteristics of Variable Motor at Constant Flow and Torque

Dynamic Characteristics: Figure 8.34 shows the step response of a variable displacement motor at constant flow (240 lit/min) and constant torque (200 N.m) where the displacement is increased from 40 to 60 cc/rev. The figure shows that displacement increased dynamically based on the parameters loaded to the second-order normalized transfer function that represents the dynamics of the control unit.

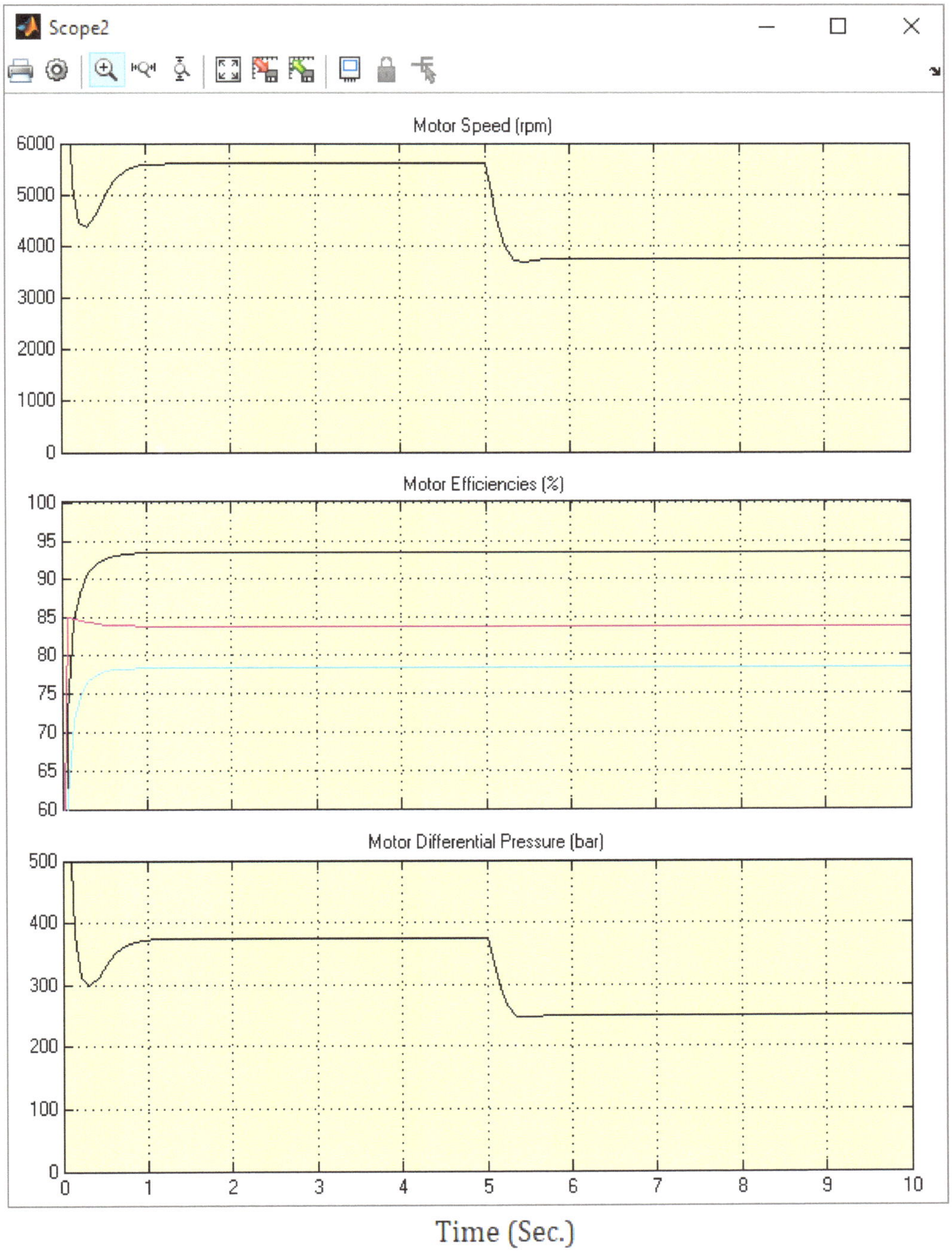

Fig. 8.34- Step Response of Variable Motor at Constant Flow and Torque

8.11- Simplified Model #05 for Valve-Controlled Fixed-Displacement Bidirectional Hydraulic Motor

8.11.1- Case Study

In case if no data is available about the motor, a simplified model still can be built. A case study for model 05 considers a servo valve-controlled hydraulic fixed-displacement motor in the Universal Fluid Power Trainer (UFPT), shown previously in Chapter 7 (Fig. 7.41).

8.11.2- Model Features and Assumptions

As shown in Fig. 8.35, the model is structured in a simplified form just to give the motor speed based on input signal to a servo control valve. The model contains motor static and dynamic characteristics.

Motor Static Characteristics: Static characteristics are loaded into the model in form of a lookup table. It should be published by the manufacturer. If not, it should be developed in the lab.

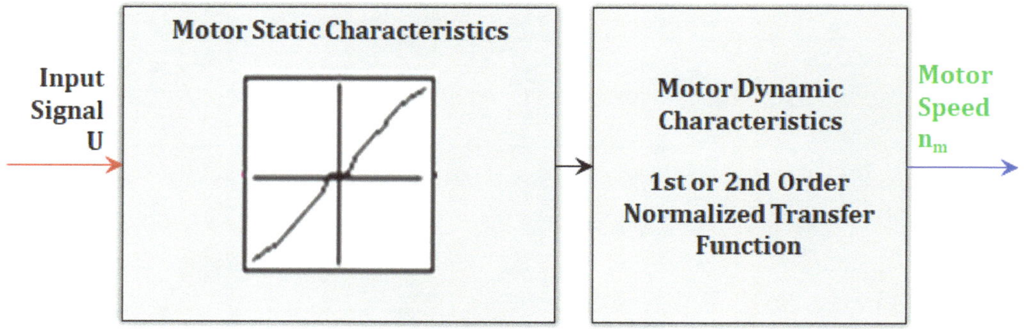

Fig. 8.35- Structure of Motor Model 05

Motor Dynamic Characteristics: Dynamic characteristics are modeled in form of a normalized transfer function (NTF). At least step response of the motor should be published by the manufacturer. If not, it can be assumed. Model will have more fidelity if there is ability to develop the motor dynamic response in the lab. Figure 8.36 shows the step response of the motor in this case study as it is developed automatically using the UFPT. Motor step response shows that it behaves as a 1st order NTF with 50 ms. Time Constant. Assuming loading conditions are identical in both motor directions of rotation, so the presented step response can be used for both motor directions of rotation.

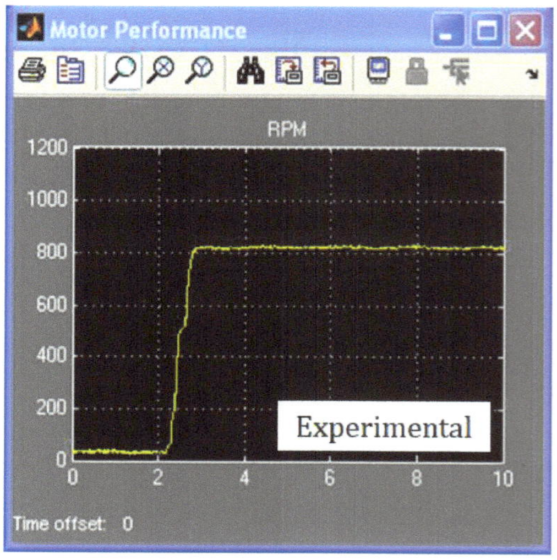

Fig. 8.36- Developed Step Response of the Motor

8.11.3- Simulation Model

Figure 8.38 shows the simulation model (HSV7CH08Model05) that is built based on the generic structure presented previously in Fig. 8.35.

Static and Dynamic Characteristics: The model considers **U-n$_{\mathbf{m}}$** static characteristics loaded into a lookup table. The output of the static characteristic is multiplied by a first-order NTF that represents the dynamics of the motor with the integrated load.

Bidirectional Operation: The model uses the sign of the command U as an identifier to determine the flow and pressure at the physical ports of the motor.

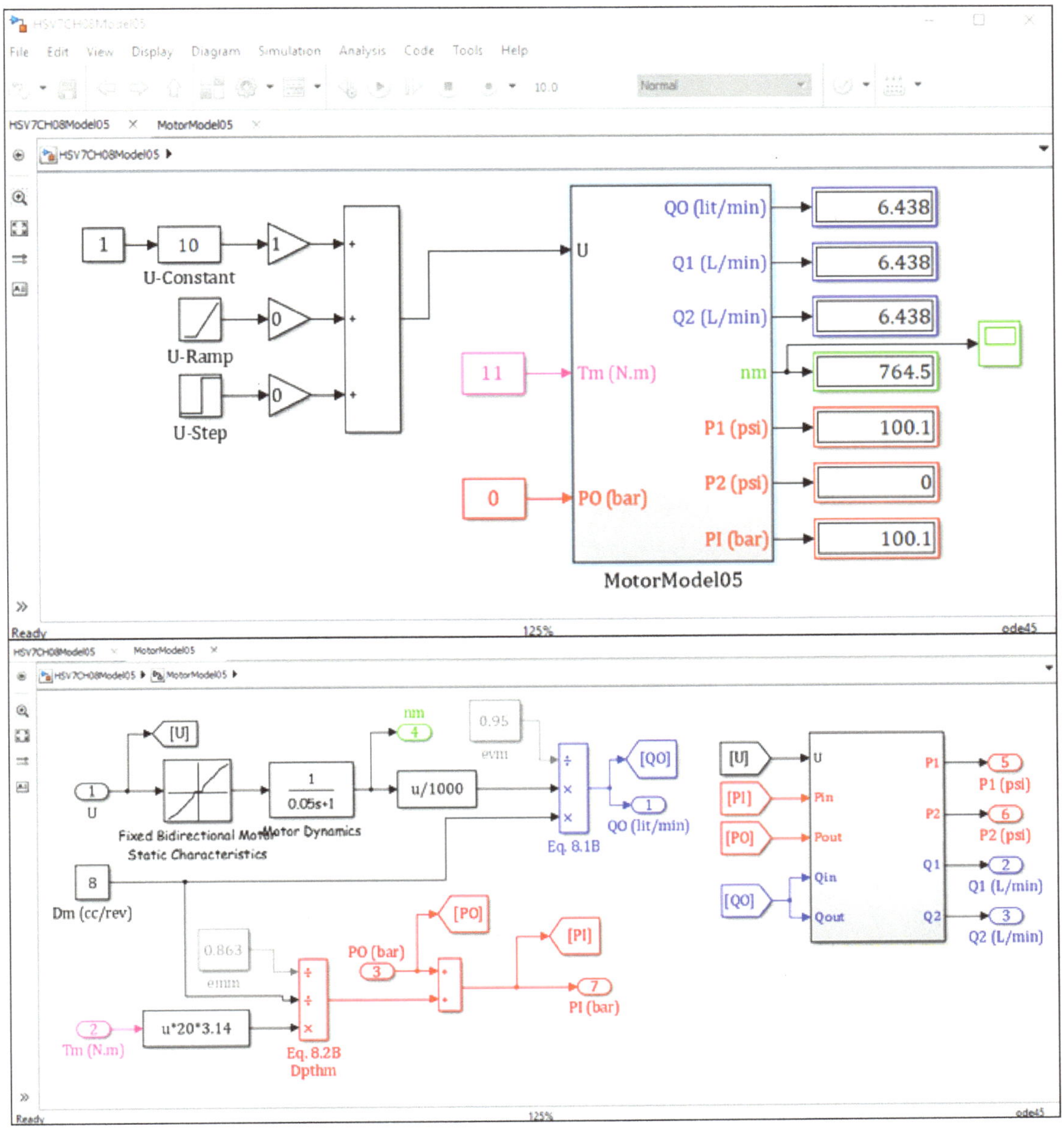

Fig. 8.37- Motor Simulation Model 05

8.11.4- Model Validation

As shown in Fig. 8.38, deviation between simulation and experimental static characteristics can always be compensated by tuning the lookup table.

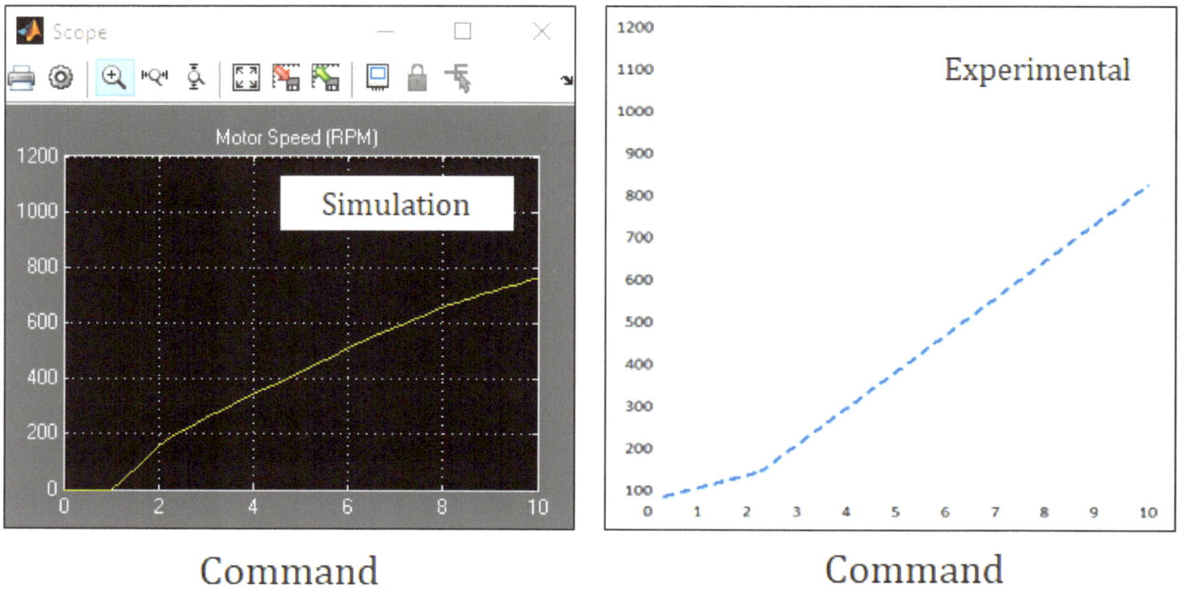

Fig. 8.38- Validating the Motor Static Characteristics

Similarly, as shown in Fig. 8.39, deviation between simulation and experimental dynamic results can always be compensated by tuning the transfer function time constant.

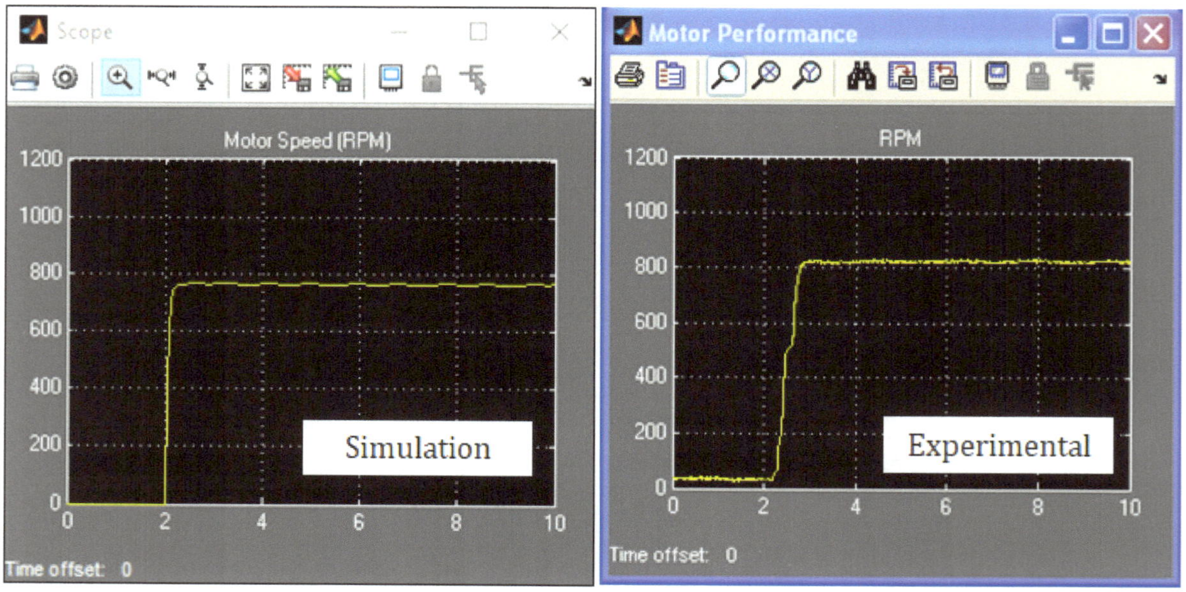

Fig. 8.39- Validating the Motor Step Response

Figure 8.40 shows the motor performance at maximum command. The figure shows that by having a torque of 11 (N.m) acting on the motor shaft, at motor speed of 764 rpm, results in 100 bar differential pressure. This data matches the test data reported by the motor's manufacturer shown in Fig. 8.41.

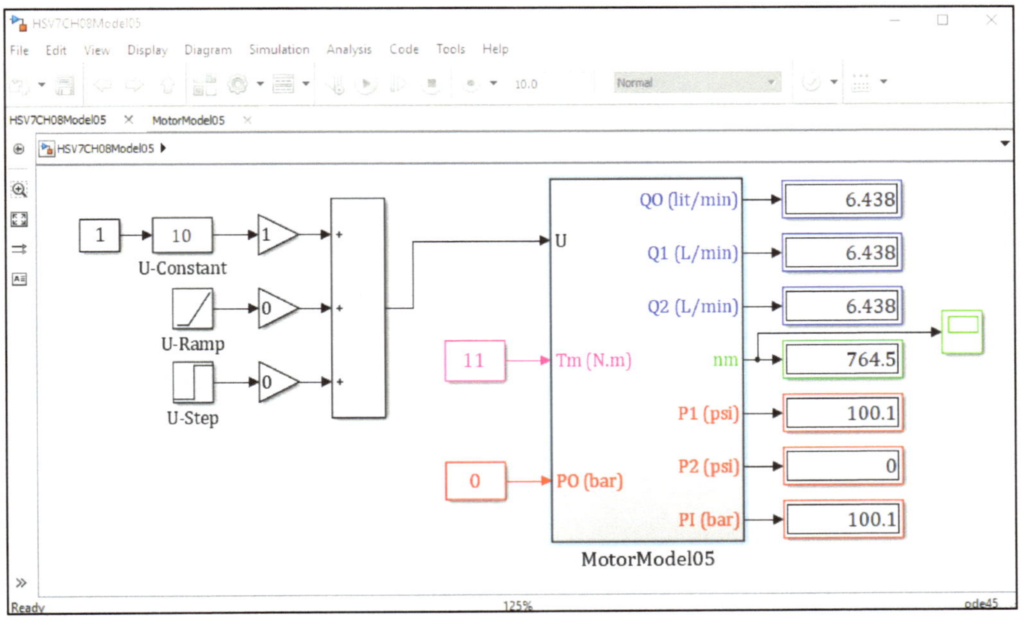

Fig. 8.40- Motor Performance at Maximum Command

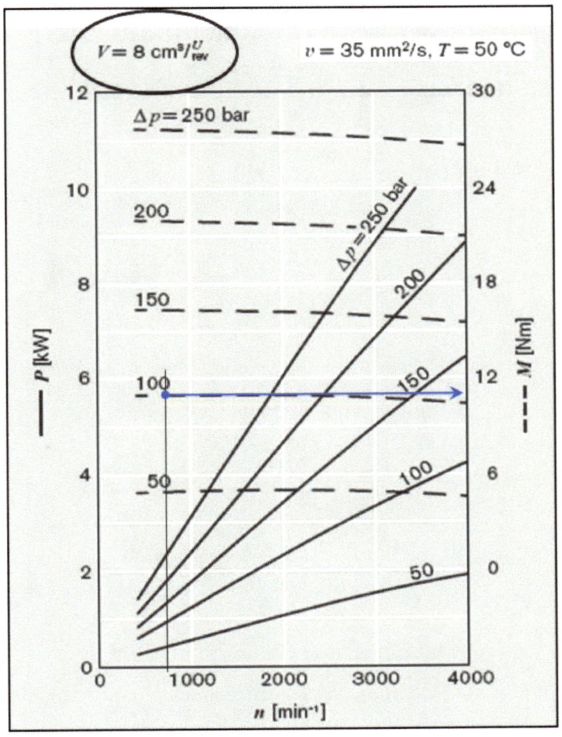

Fig. 8.41- Motor Published Test Data

Chapter 9
Modeling of Hydraulic Cylinders

Objectives

This chapter presents the lumped modeling concept as applied for double-acting hydraulic cylinders. This chapter considers situations where a cylinder works under a constant or variable external load and inlet flow. Calculations for cylinder slowing due to leakage, cylinder drift due to oil compressibility, and pressure increase due to thermal expansion are presented.

Brief Contents

Chapter 9: Modeling of Hydraulic Cylinders

9.1- Lumped Model Structure of a Double-Acting Hydraulic Cylinder

Modeling Approach: Figure 9.1 shows two different configurations of hydraulic cylinders. The cylinder on the left side is a mill-type cylinder, and the one on the right side is a tie-rod. As it has been previously discussed, such cylinders should be modeled based on their internal structure for purposes of cylinder development. For purposes of overall machine modeling, they are modeled based on lumped modeling concepts, no matter what the cylinder mechanism is,

Fig. 9.1- Hydraulic Cylinder Structural Configurations

Figure 9.2 shows a lumped model structure of a double-Acting hydraulic cylinder. The model contains the following features:

- Assuming incompressible, where effect of oil bulk modulus is ignored, cylinder area ratio and direction of cylinder motion are used to calculate the outlet flow.
- External *Load Force* $\mathbf{F_L}$ is used, to calculate the differential pressure across the cylinder. With the knowledge of the cylinder outlet pressure, inlet pressure is calculated.
- Cylinder dimensions and dynamic characteristics are used to calculate cylinder acceleration, velocity, and travel displacement.
- An identifier **U** is used to identify the cylinder direction of motion and, accordingly, working conditions at piston side (1) and rod side (2).

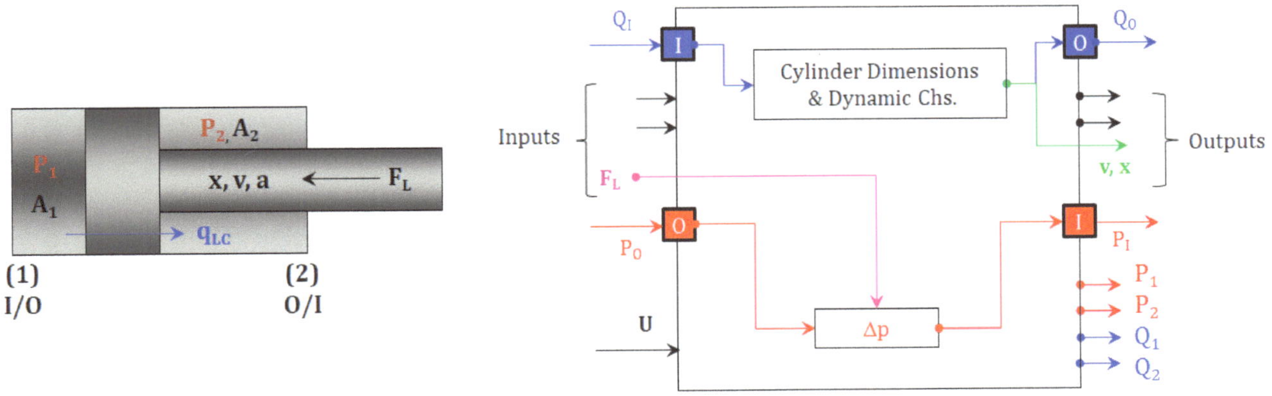

Fig. 9.2- Lumped Model Structure of a Double-Acting Hydraulic Cylinder

9.2- Lumped Model #01 for Hydraulic Cylinder

9.2.1- Mathematical Model

9.2.1.1- Cylinder Effective Areas

This section presents the calculation of piston side and rod side effective areas. Each of these effective areas could be inlet or outlet depending on the sign of identifier **U**. If **U** is positive, then piston side is the inlet side and cylinder is extending. If **U** is negative, then rod side is the inlet side and cylinder is retracting.

Equations 9.1A and 9.1B shows the calculation of piston side area with the units being balanced in English and Metric systems of units; respectively.

$$\mathbf{A_C(in^2) = \frac{\pi}{4} D_C^2\,(in)} \qquad \mathbf{9.1A}$$

$$\mathbf{A_C(cm^2) = \frac{\pi}{400} D_C^2\,(mm)} \qquad \mathbf{9.1B}$$

Equations 9.1C and 9.1D shows the calculation of rod side area with the units being balanced in English and Metric systems of units; respectively.

$$\mathbf{A_R(in^2) = \frac{\pi}{4}\left[D_C^2\,(in) - D_R^2\,(in)\right]} \qquad \mathbf{9.1C}$$

$$\mathbf{A_R(cm^2) = \frac{\pi}{400}\left[D_C^2\,(mm) - D_R^2\,(mm)\right]} \qquad \mathbf{9.1D}$$

9.2.1.2- Cylinder Inlet Pressure

By applying force balance on the cylinder piston,

$$\mathbf{P_I \times A_I - P_O \times A_O = F_C \rightarrow P_I = [F_C + P_O \times A_O]/A_I} \qquad \mathbf{9.2}$$

Where **P** is pressure and **A** is effective area. Equations 9.2A and 9.2B represent Eq. 9.2 with the units being balanced in English and Metric systems of units; respectively.

$$\mathbf{P_I\,(psi) = \frac{F_C(lb) + P_O(psi) \times A_O(in^2)}{A_I(in^2)}} \qquad \mathbf{9.2A}$$

$$\mathbf{P_I\,(bar) = \frac{0.1 \times F_C(N) + P_O(bar) \times A_O(cm^2)}{A_I(cm^2)}} \qquad \mathbf{9.2B}$$

9.2.1.3- Cylinder Internal Leakage

Equations 9.3A and 9.3B shows the calculation of cylinder internal leakage with the units being balanced in English and Metric systems of units; respectively. Leakage coefficient $\mathbf{k_L}$ is used to investigate the effect of leakage on the cylinder performance.

$$\mathbf{q_{LC}\,(gpm) = (P_I - P_O)(psi) \times k_L\,(gpm/psi)} \qquad \mathbf{9.3A}$$

$$\mathbf{q_{LC}\,(lit/min) = (P_I - P_O)(bar) \times k_L\,[(lit/mi)/bar]} \qquad \mathbf{9.3B}$$

As shown in the previous equations, internal leakage $\mathbf{q_{LC}}$ is assumed flow from high pressure chamber to the low-pressure chamber. In some cases, such as when extending a differential hydraulic cylinder, the cylinder works like a pressure intensifier. So, in this case, rod side (outlet) pressure is higher than piston side (inlet) pressure. However, the equation is still valid because the sign of the internal leakage changes accordingly.

9.2.1.4- Cylinder Speed

Equations 9.4A and 9.4B shows the calculation of cylinder velocity with the units being balanced in English and Metric systems of units; respectively. Cylinder displacement and acceleration are calculated by integrating/differentiating the speed.

$$\mathbf{v\,(ft/s) = \frac{0.321\,[Q_I(gpm) - q_{LC}(gpm)]}{A_I(in^2)}} \qquad \mathbf{9.4A}$$

$$\mathbf{v\,(cm/s) = \frac{1000\,[Q_I(lit/min) - q_{LC}(lit/min)]}{60 \times A_I(cm^2)}} \qquad \mathbf{9.4B}$$

9.2.1.5- Cylinder Outlet Flow

Unlike hydraulic motors, because of unequal effective areas across the piston and internal leakage, outlet flow is different from the inlet flow. Equations 9.5A and 9.5B shows the calculation of piston side area with the units being balanced in English and Metric systems of units; respectively. For synchronous cylinders, if internal leakage equal zero, $\mathbf{Q_O = Q_I}$.

$$\mathbf{Q_O(gpm) = \frac{v\,(ft/s) \times A_O(in^2)}{0.321} + q_{LC}} \qquad \mathbf{9.5A}$$

$$\mathbf{Q_O(lit/min) = \frac{60 \times v(cm/s) \times A_O(cm^2)}{1000} + q_{LC}} \qquad \mathbf{9.5B}$$

9.2.1.6- Cylinder Force

Forces acting on the cylinder rod are classified as follows:

- **Based on the Source:** *Internal or External.*
- Internal means from inside the cylinder, e.g. frictional forces.
- External means from outside the cylinder, e.g. external loading force.
- **Based on the Sign:** *Positive (Resistive) or Negative (Assistive).* Negative force are commonly known as overrunning load.
- Positive (Resistive) means it opposes the cylinder movement.
- Negative (Assistive) means it drives the cylinder rod.
- **Based on the Cylinder Motion:** *Passive or Active.*
- Passive means it equals zero when the cylinder stops, e.g. frictional forces.
- Active means it has a magnitude (positive or negative) even when the cylinder stops.

Figure 9.3 shows examples of hydraulic cylinders loading conditions as follows:

1. A cylinder pushes an external mass against frictional surface. Load is external, passive, and positive no matter the cylinder direction of motion.
2. A cylinder drives a vertical mass. Load is external and active in both directions. Load is positive when the cylinder extends and negative when the cylinder retracts.
3. A cylinder drives a boom in a swinging motion. Load is external, active, and switches from positive to negative during the same cylinder stroke.
4. Cylinder drives a boom arm in mobile machines.

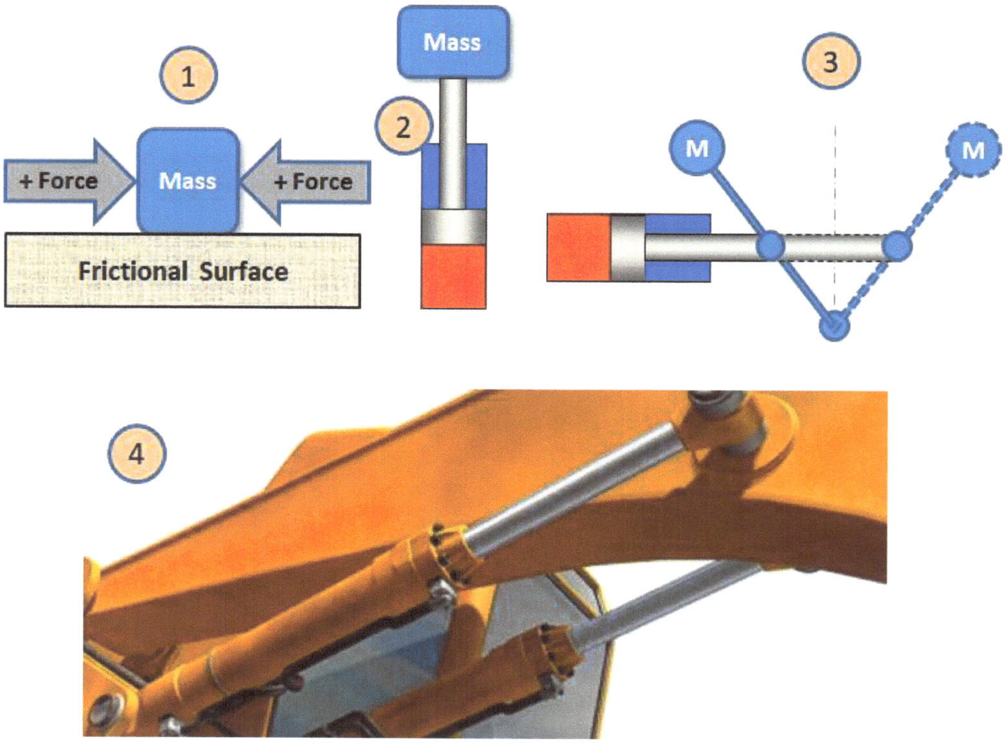

Fig. 9.3- Various Loading Conditions of Hydraulic Motors

Referring to Fig. 9.4, Eq. 9.6 through 9.9 show calculations of instantaneous cylinder force in Metric system of units.

$$\mathbf{F_V(N) = v\,(cm/s) \times k_V\left(\frac{N}{cm/s}\right)} \quad 9.6$$

$$\mathbf{F_I(N) = M\,(kg) \times \dot{v}\,(cm/s^2) \times 0.01} \quad 9.7$$

$$\mathbf{F_F(N.m) = F_N\,(N) \times k_F(-)} \quad 9.8$$

$$\mathbf{F_C = F_L + F_V + F_I + F_F} \quad 9.9$$

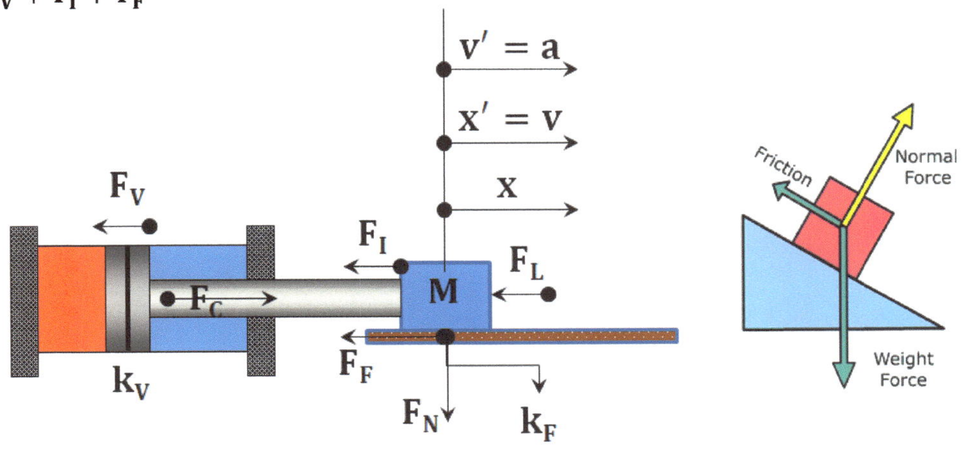

Fig. 9.4- Hydraulic Cylinder Force Calculation

Where:

- **F_C (Cylinder Force):** Force generated by the cylinder to drive the external load and overcome all other forces acting on the cylinder rod. This cylinder force results in differential pressure across the cylinder piston.
- **F_L (External Force):** External - positive or negative - passive or active. Provided by load model. If loading conditions are not the same in both cylinder directions, then identifier **U** is used to select the loading conditions accordingly.
- **F_I (Inertia):** External - positive - passive. Inertia is due to equivalent moving mass including the cylinder piston and rod.
- **F_V (Viscous Friction Force):** Internal - positive – passive (already multiplied by v).
- **F_F (Dry Friction Force):** External - positive - passive. Applicable only when |v|>0.
- **M:** Equivalent moving mass.
- **v:** Cylinder velocity.
- **k_V:** Viscous friction Coefficient.
- **k_F:** Dry friction Coefficient.
- **F_F:** Friction Force.
- **F_N:** Normal Force. As shown in the figure, if the mass moves horizontally, then **F_N** equals the weight of the mass. Otherwise, **F_N** is calculated based on the cylinder inclination.

9.2.2- Simulation Model

Figure 9.5 shows the upper mask of simulation model (HSV7CH09Model01). The figure shows the interactive inputs/outputs of the model. The figure shows also the dynamic effect of the inlet flow and the external load. The model includes in its properties the actual dimensional and design parameters of the cylinder in the Universal Fluid Power Trainer (UFPT) shown in the previous figure.

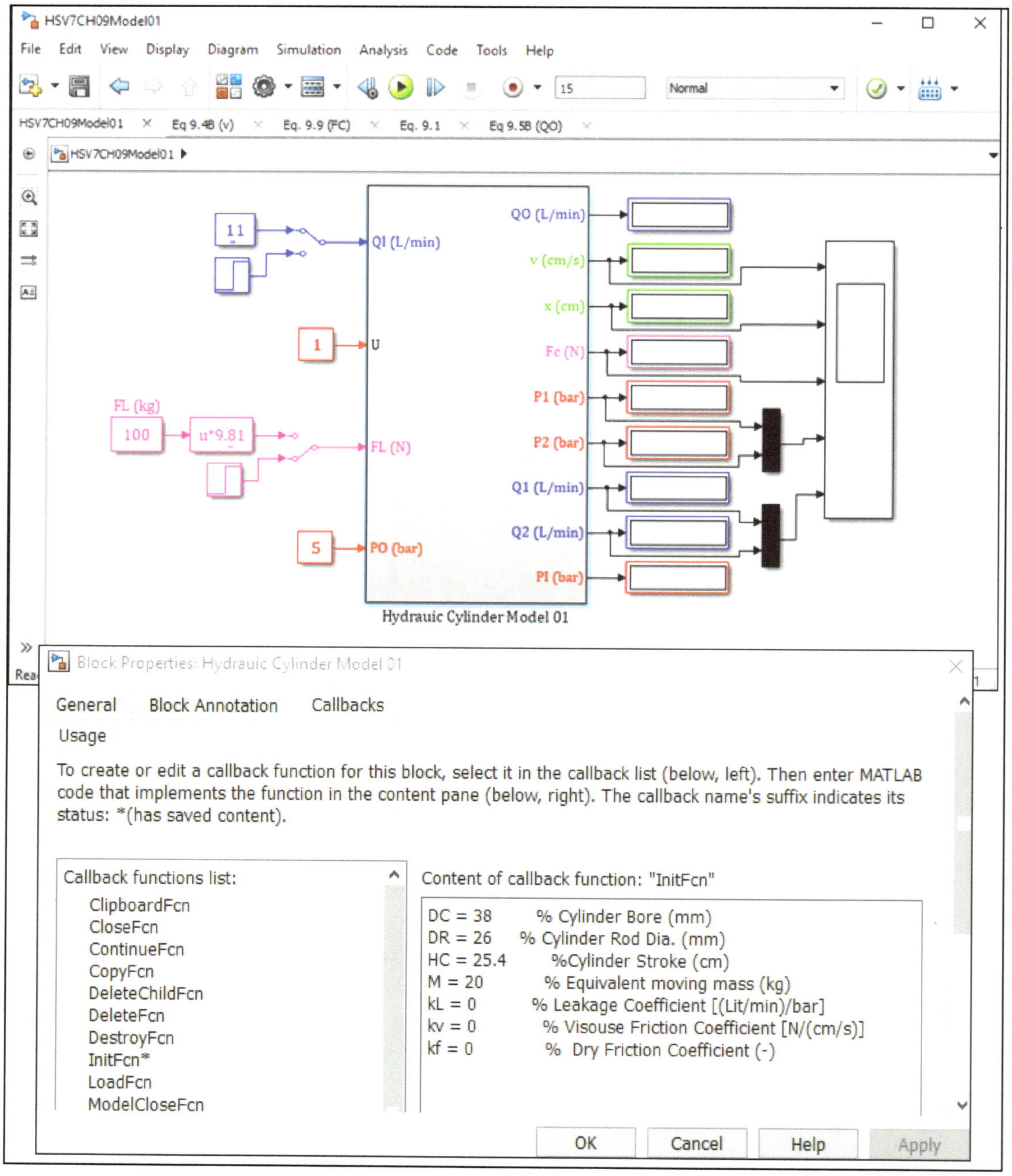

Fig. 9.5- Cylinder Simulation Model 01

Figure 9.6 shows the details under the mask of Model 01. The model is built to solve the set of equations in the same sequence they are presented in the mathematical model using metric system of units. Figure 9.7 shows the details of solving Eq. 9.1 and assigning inlet/outlet areas.

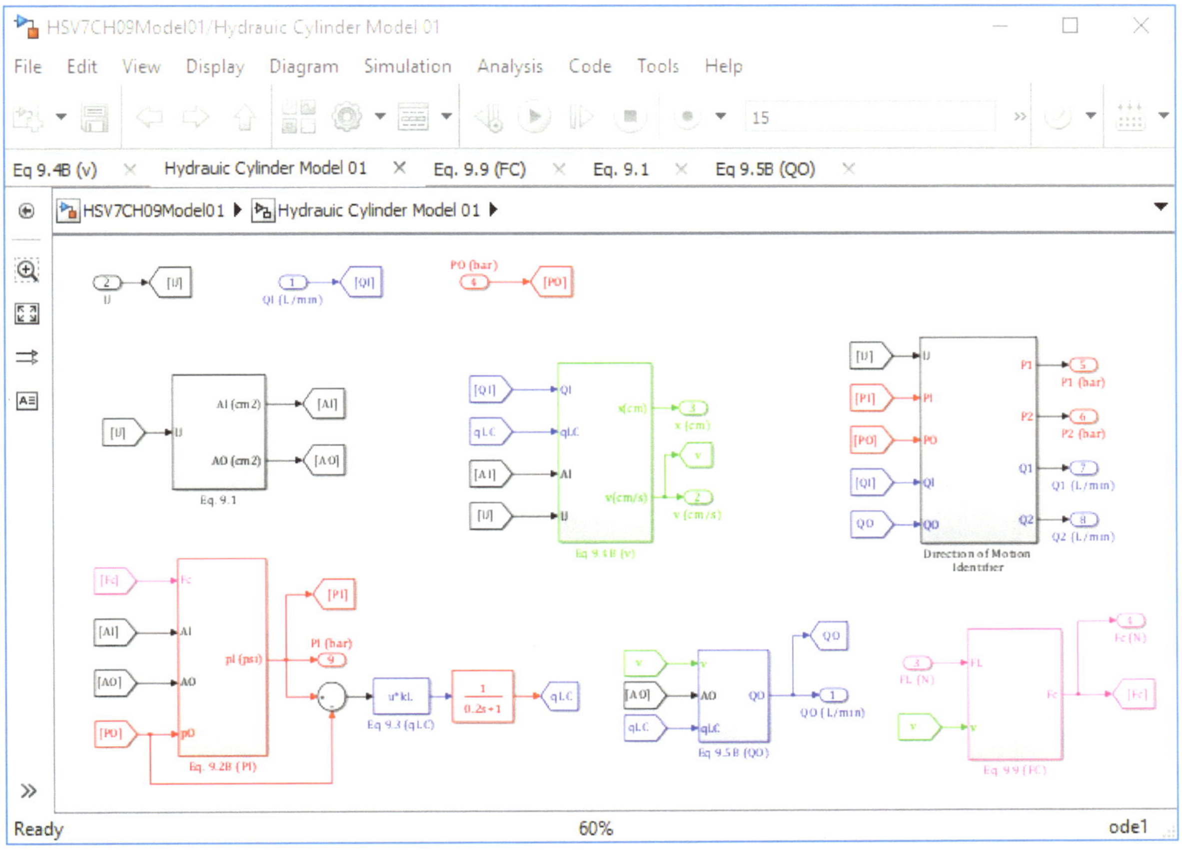

Fig. 9.6- Details under the Mask of Model 01

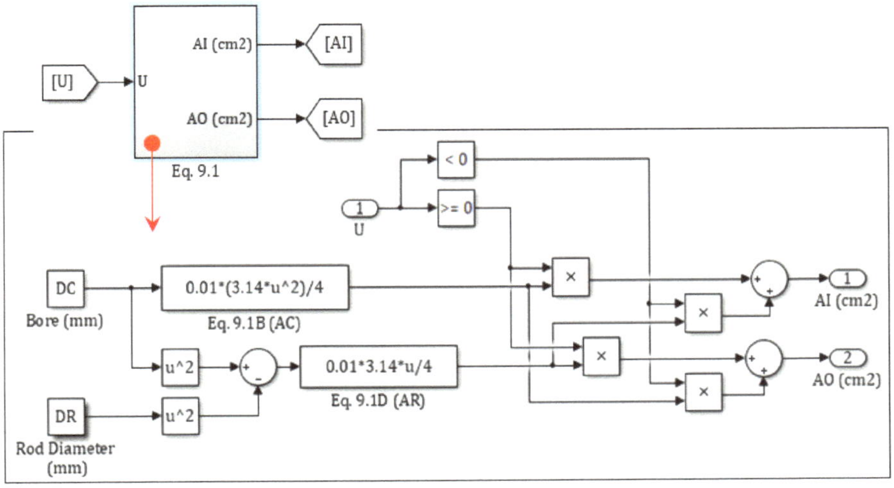

Fig. 9.7- Details of Solving Equation 9.1

Modeling the Dynamic Effect of Cylinder Load: Cylinder load is used to calculate the differential pressure and the internal leakage. So, due to various sealing packages, it can't be assumed that leakage occur instantaneously with the change of the differential pressure. Modeling this effect is very exhaustive and depends on specific cylinder seal package design. However, in view of machine modeling, it can be feasibly represented by a 1st order NTF with a reasonable time constant (200 ms.). Figure 9.8 shows the details of Eqs. 9.2 and 9.3.

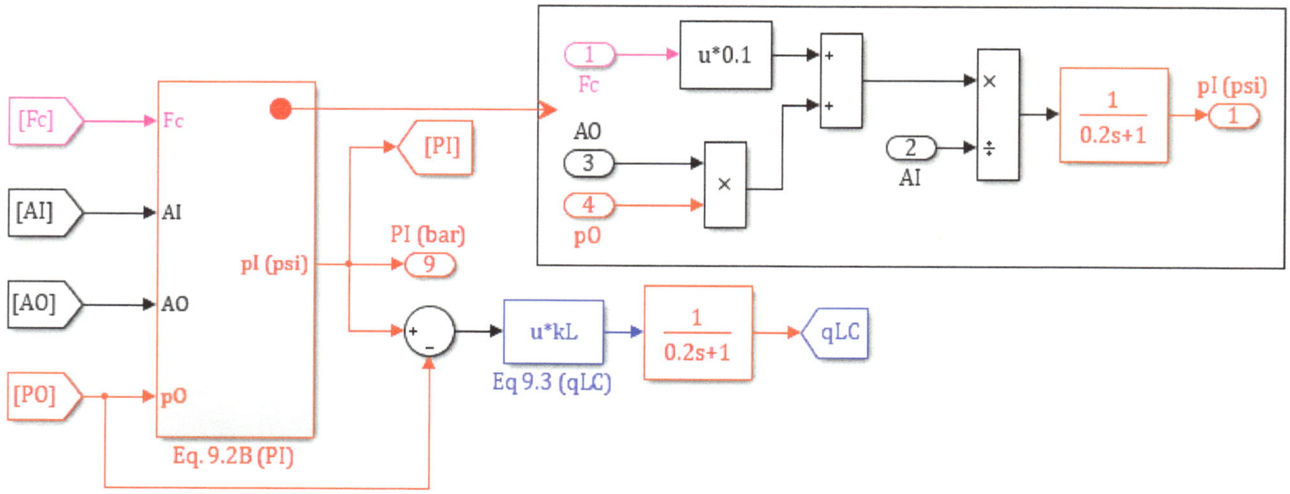

Fig. 9.8- Details of Solving Equations 9.2 and 9.3

Modeling the Dynamic Effect of Inlet Flow: Inlet flow is used to calculate the cylinder velocity. Due to fluid compressibility and inertial effect of the equivalent moving mass, it can't be assumed that the resulting cylinder velocity varies instantaneously with the variation of the inlet flow. Therefore, dynamic effect of inlet flow can be feasibly represented by a 1st order NTF with a reasonable time constant (e.g. 200ms).

Given: The hydraulic cylinder, shown below in Fig. 9.9, is extended suddenly. Dynamics of the cylinder have been captured as shown in the figure. The cylinder was found performing like a first-order system and reached a constant velocity of 6.3 inches/s (16 cm/s) in 1 second. So, cylinder dynamics is represented in the model by a first-order normalized transfer function of time constant equal 1/5 = 0.2 s.

Figure 9.10 shows the details of solving Eq. 9.4. The figure shows that cylinder displacement **x** is calculated by differentiating cylinder velocity **v**. Saturation block is used to limit the displacement to be as maximum as the cylinder stroke **Hc**. Based on the calculated cylinder velocity.

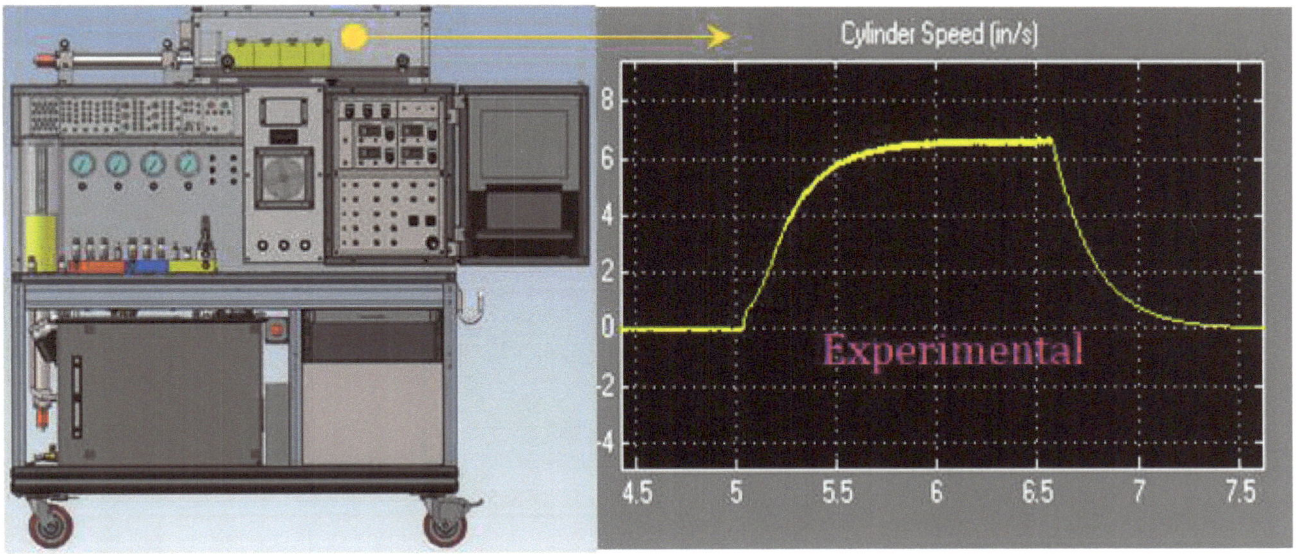

Fig. 9.9 – Cylinder Dynamics

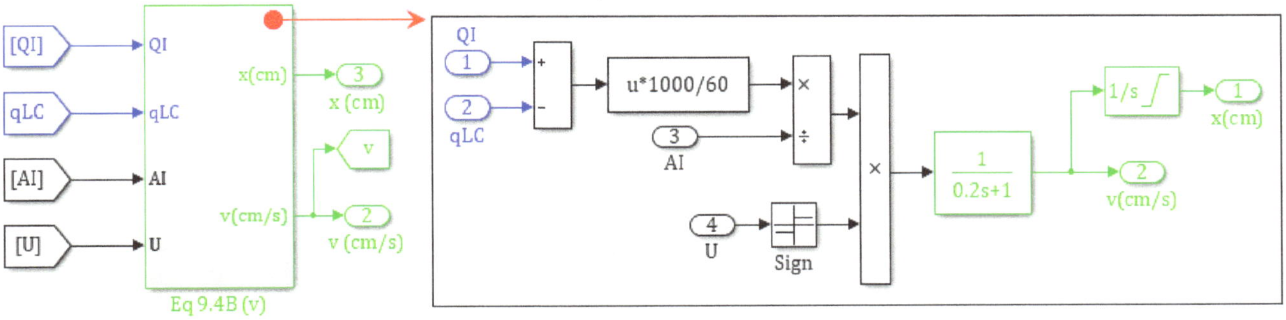

Fig. 9.10- Details of Solving Equation 9.4

Modeling the Outlet Flow: Fig. 11. shows the details of solving Eq. 9.5 to provide the outlet flow. Figure 9.12 shows the details of solving Eqs. 9.6 through 9.9 to provide the cylinder force F_C. Figure 9.13 shows the details of determining the operating conditions at the cylinder physical ports using the identifier **U**.

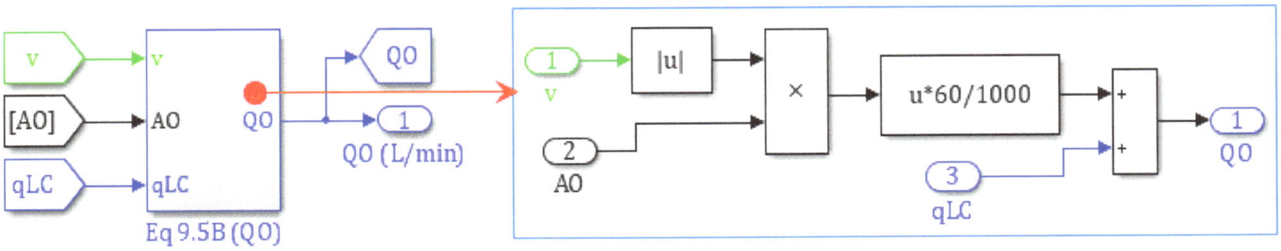

Fig. 9.11- Details of Solving Equation 9.5

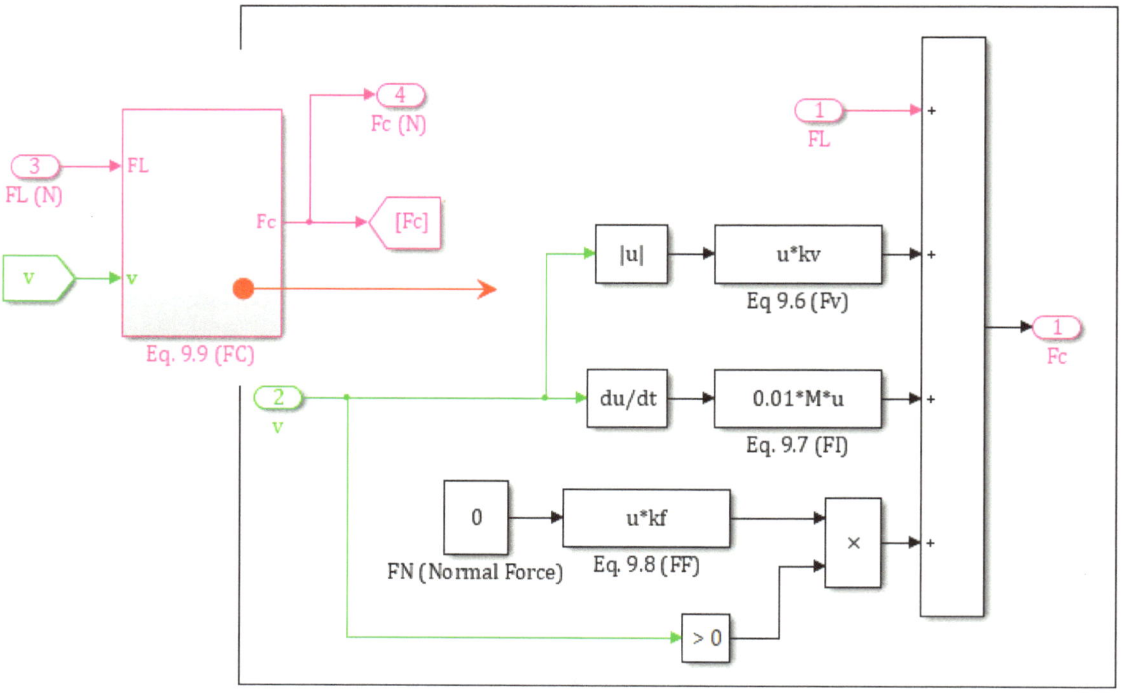

Fig. 9.12- Details of Solving Equations 9.6 through 9.9

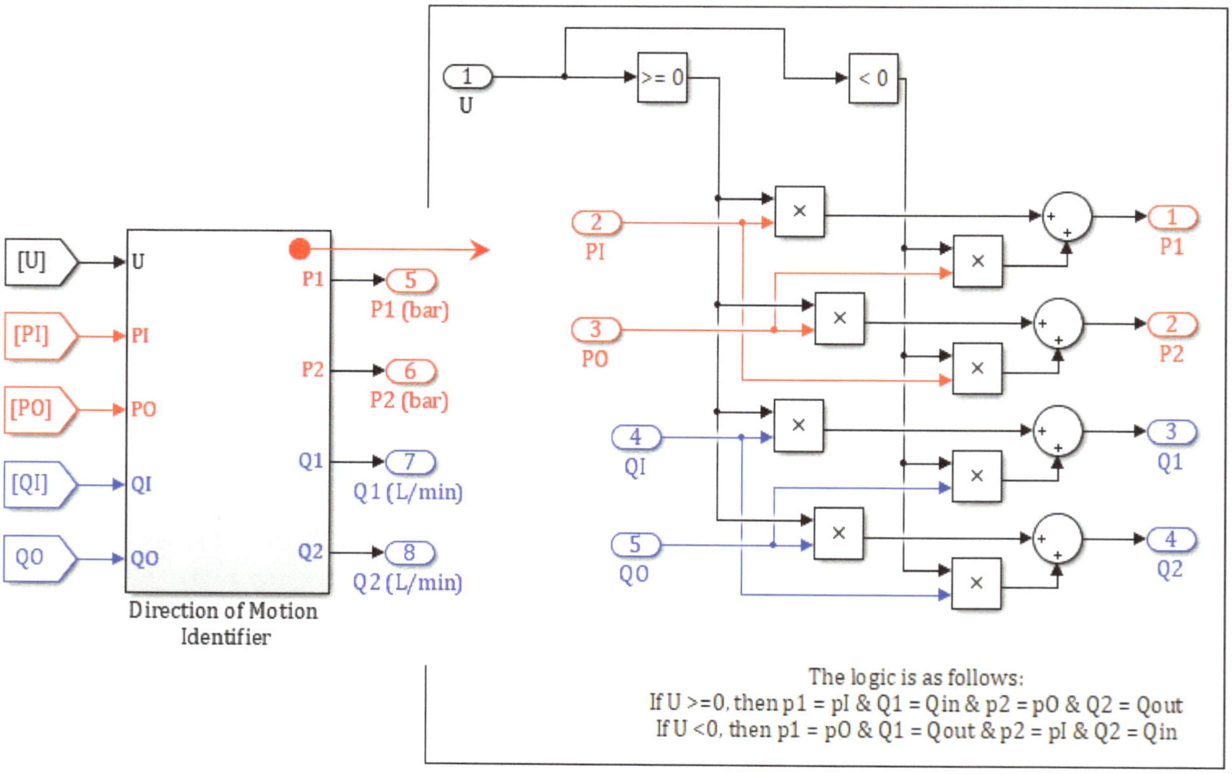

Fig. 9.13- Details of Determining the Operating Conditions at the Cylinder Physical Ports

9.2.3- Model Validation

Steady State Validation: Assuming the cylinder moves horizontally, this section provides several steps to validate the model. Figure 9.14 shows a case when the cylinder is extending. Agreement between the simulated results and the results from Hydraulic Components Sizing Calculator (HCSC) software validates the model.

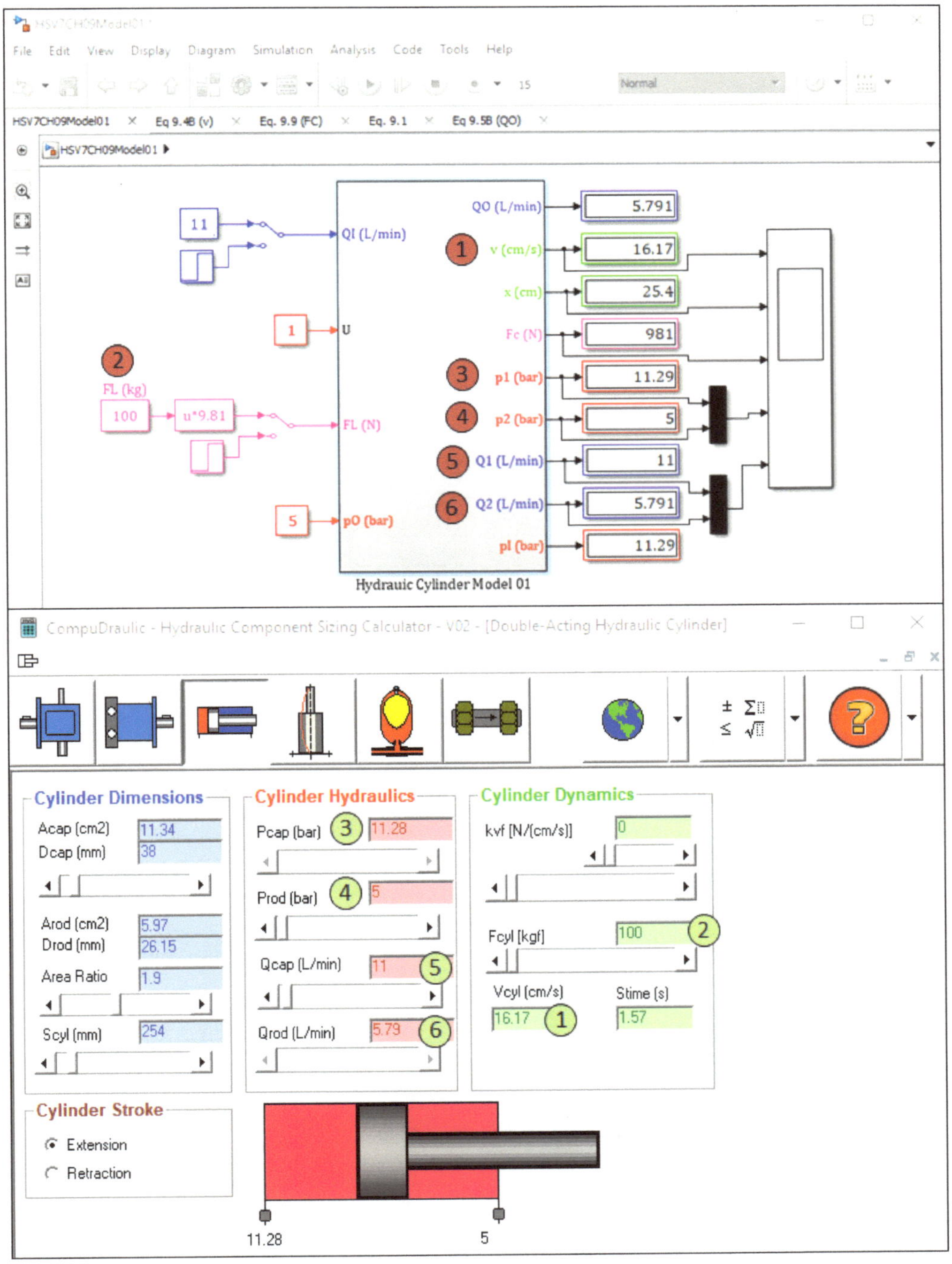

Fig. 9.14- Model Validation when the Cylinder Extending

Figure 9.15 shows a case when the cylinder is retracting. Agreement between the simulated results and the results from HCSC software validates the model.

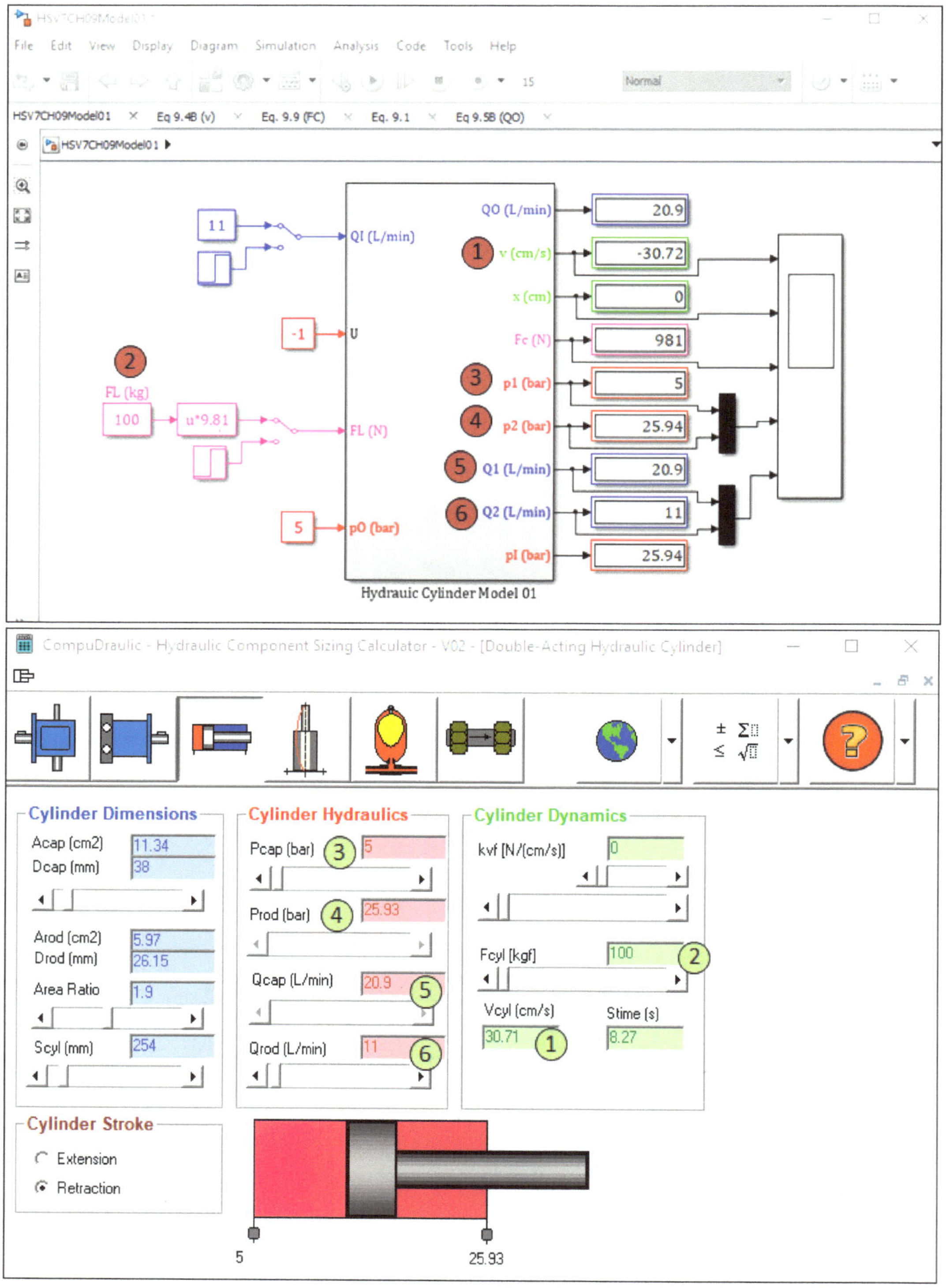

Fig. 9.15- Model Validation when the Cylinder Retracting

Validating the Dynamic Effect of Inlet Flow: Figure 9.16 (left side) shows a case when inlet flow is suddenly supplied to the piston side of the cylinder at constant external load. Cylinder speed and displacement are dynamically affected according the NTF assigned to inlet flow. The right side of the figure shows that, by adjusting the inlet flow in the model, simulation results agree with the experimental results. While simulation of the cylinder displacement is subjected to a saturation block based on the cylinder stroke, the rest of simulations assumed that the cylinder continues to move with long stroke.

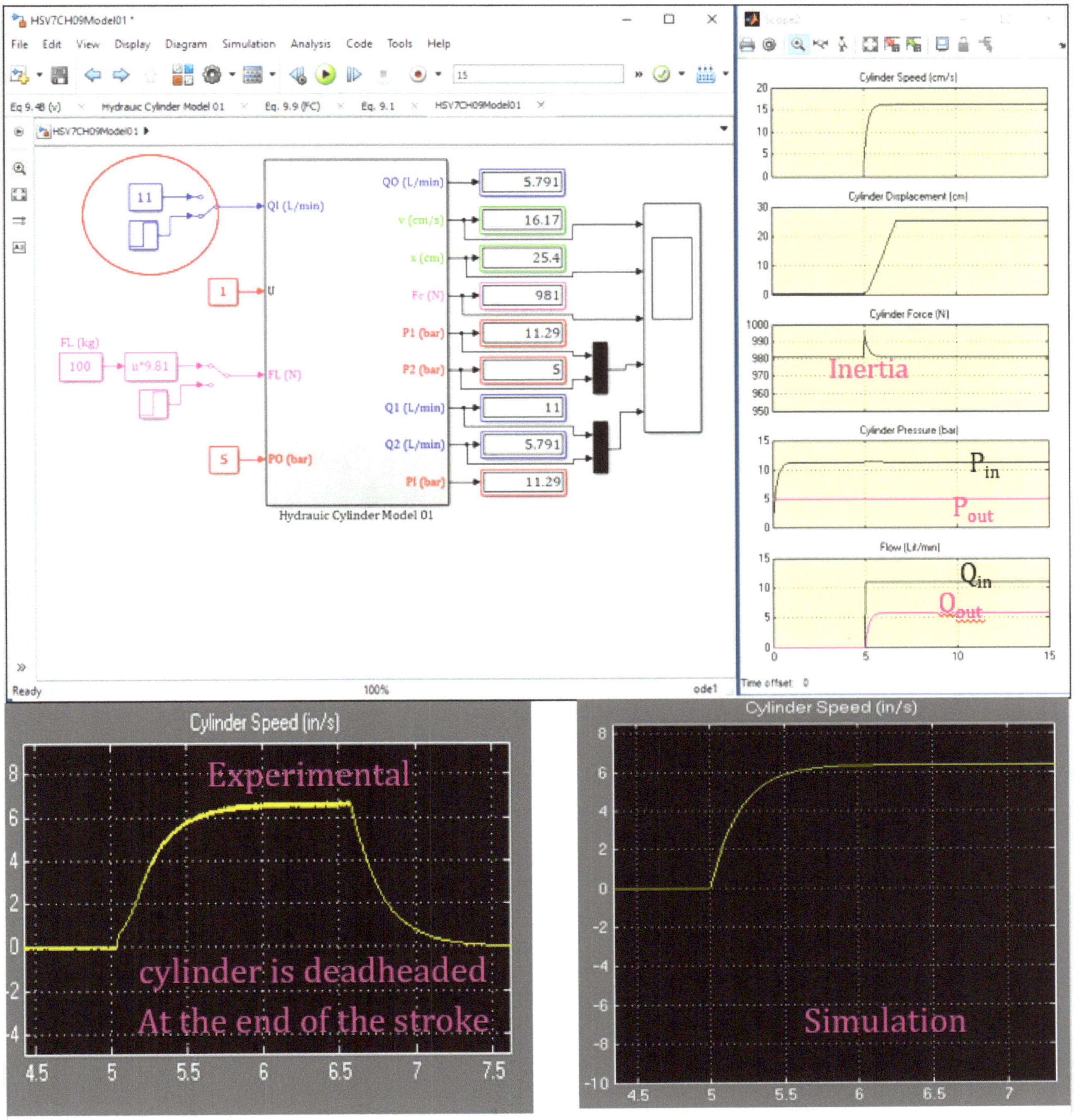

Fig. 9.16- Dynamic Effect of Inlet Flow during Cylinder Extension

Modeling the Dynamic Effect of External load: Figure 9.17 (left side) shows a case when external load is suddenly increased on the cylinder rod at constant inlet flow. Cylinder inlet and outlet pressures are dynamically affected according the NTF assigned to inlet flow. For the first five seconds, because of the area ratio and no external load, less pressure at the piston side balances pressure at the rod side. When the external load is applied, pressure at the piston side increases accordingly.

Modeling the Leakage Effect: Figure 9.17 (right side) shows same previous case assuming a leakage coefficient $\mathbf{k_L}$ = 1 [(lit/min)/bar]. The figure show that, when the external load is increased suddenly, differential pressure across the piston increases, leakage increases accordingly, and cylinder decelerates to a lower speed. Despite of that, outlet flow increased because leakage from piston side is added to flow displaced by the piston at the rod side.

In both figures, While simulation of the cylinder displacement is subjected to a saturation block based on the cylinder stroke, the rest of simulations assumed that the cylinder continues to move with long stroke.

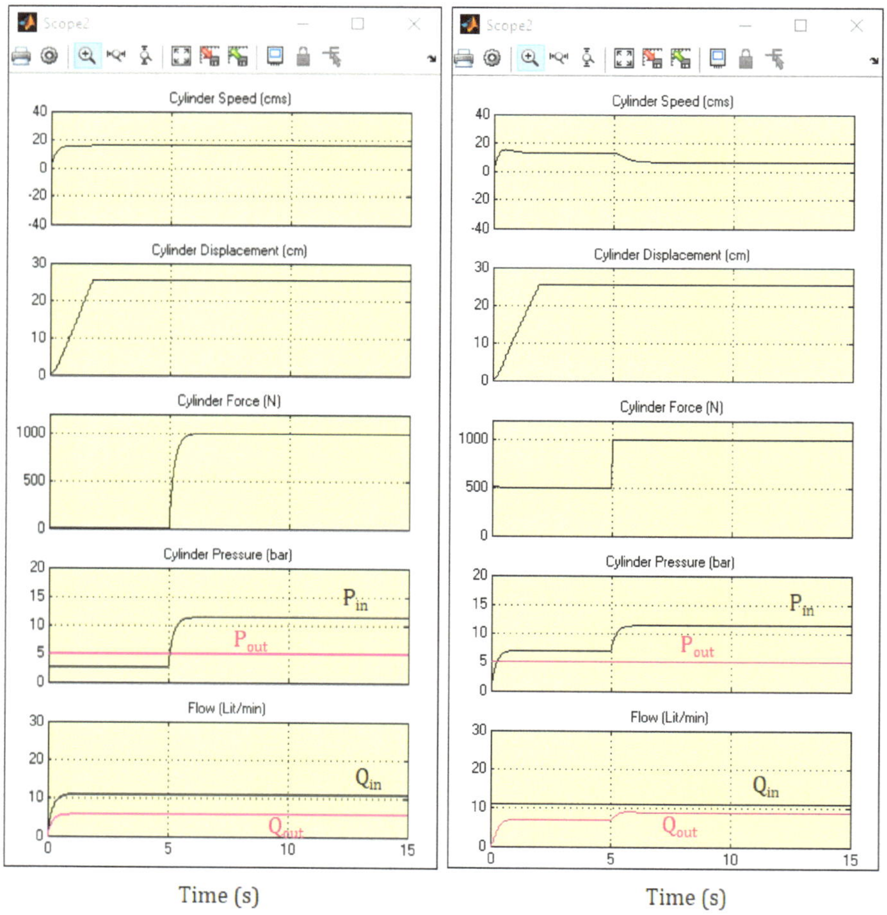

Fig. 9.17- Dynamic Effect of External Load and Cylinder Leakage

9.3- Modeling of Cylinder Drifting due to Oil Bulk Modulus

Figure 9.18 shows that, due to oil compressibility, cylinder drifts with the increase of the load. Load may be an external force or weight of a vertical mass. Referring to Eq. 5.1 shown previously in Chapter 5, *Cylinder Drift* can be calculated in absolute value using Eq. 9.10. Equations 9.10A and 9.10B shows the calculation of cylinder drift with the units being balanced in English and Metric systems of units; respectively.

$$\Delta v = \frac{V_0 \Delta P}{\beta} = \textbf{Cylinder Area (A)} \times \textbf{Cylinder Drifting (L)}$$

$$\rightarrow \textbf{Cylinder Drifting (L)} = \frac{V_0 \Delta P}{A\beta} = \frac{H\Delta p}{\beta} = \frac{HF}{A\beta} \qquad 9.10$$

$$\rightarrow L\,(mm) = \frac{10 \times H\,(cm)\,\Delta P(bar)}{\beta\,(bar)} \qquad 9.10A$$

$$\rightarrow L\,(in) = \frac{H\,(in) \times \Delta P(psi)}{\beta\,(psi)} \qquad 9.10B$$

Example: If incompressible fluid is used, then β = infinity and Cylinder Drift = 0. A case when Bulk Modulus = 20,000 bar, an increase in pressure of 200 bar causes cylinder drift of 1% of the cylinder stroke.

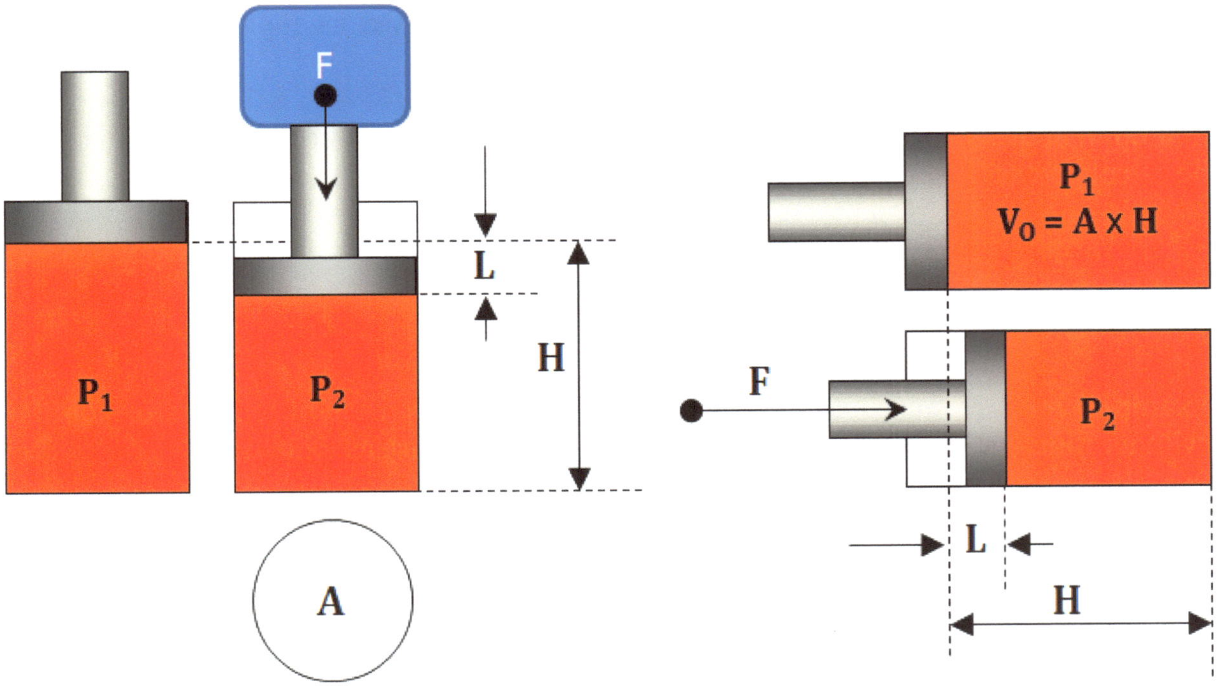

Fig. 9.18- Cylinder Drift Modeling

9.4- Modeling of Pressure Increase due to Thermal Expansion

Definition of Fluid Thermal Expansion: If a confined volume of liquid is exposed to an increase in temperature, the liquid is thermally expanded intensifying the pressure.

Mathematical Expression: Equation 9.11 shows that the *Coefficient of Thermal Expansion* $\mathbf{K_{THE}}$ is the ratio between the volumetric strain $\mathbf{\Delta V/V_O}$ of the fluid and the change in the temperature $\mathbf{\Delta T}$.

$$\mathbf{K_{THE}}\left[\frac{1}{{}^{0}C}\right] = \frac{\Delta V/V_O}{\Delta T} \rightarrow \Delta V = K_{THE}\, V_O\, \Delta T \qquad 9.11$$

Table 9.1 shows values of KTHE for three basic types of hydraulic fluids.

Hydraulic Fluid	Mineral Oil	Phosphate Ester	Water Glycol
Coefficient of Thermal Expansion K_{THE} $1/{}^{0}C$	0.0005	0.00041	0.00034

Table 9.1- Coefficient of Thermal Expansion of Hydraulic Fluids

As shown by the previous equation, a change in the temperature, causes an incremental oil volume increase $\mathbf{\Delta V}$ which added to the original volume $\mathbf{V_O}$. If this volume increase did not find a space to expand in, it increases the pressure. A practical example is when a cylinder is deadheaded and the surrounding temperature increases.

Combining Eq. 9.11 and Eq. 5.1, cylinder pressure increases due to thermal expansion can then be calculated as follows:

$$\Delta v = \frac{V_O \Delta p}{\beta} = K_{THE}\, V_O\, \Delta T \rightarrow \Delta P = \beta \times K_{THE} \times \Delta T \qquad 9.12$$

Example: A case when Bulk Modulus = 20,000 bar, an increase of temperature of 50 ${}^{0}C$ in a cylinder filled with mineral oil causes an increase of pressure as follows:

$$\Delta P = \beta \times K_{THE} \times \Delta T = 20,000 \times 0.0005 \times 50 = 500 \text{ bar}$$

This example shows that pressure can increase to a limit that can damage the cylinder because of oil thermal expansion. That is why built in thermal pressure relief valves are used to limit the pressure inside the cylinder.

Chapter 10
Modeling of Hydraulic Valves

Objectives

This chapter presents the lumped modeling concept as applied for hydraulic valves. This chapter considers modeling at least one pressure control valve, one flow control valve, and one directional control valves. Models for electro-hydraulic proportional and servo valves are also developed.

Brief Contents

Chapter 10: Modeling of Hydraulic Valves

10.1- Introduction to Hydraulic Valve Modeling

As shown in Fig. 10.1, hydraulic valves are classified based on their control mission into three main valve families as follows:

***Pressure Control Valves* (PCV):** This family of valves controls the magnitude of the transmitted hydraulic power by controlling the pressure in a part of the hydraulic system. By controlling the pressure, ultimately the load exerted by a hydraulic actuator is controlled. The load of a hydraulic actuator could be the force of a linear actuator or torque of a motor.

***Flow Control Valves* (FCV):** This family of valves controls the magnitude of the transmitted hydraulic power by controlling the flow in a part of the hydraulic system. By controlling the flow, ultimately the speed of a hydraulic actuator is controlled. The speed of a hydraulic actuator could be the linear velocity of a cylinder or the rpm of a hydraulic motor.

***Directional Control Valves* (DCV):** In most of the hydraulic systems the pump feeds more than one actuator, so DCV are required to manage the power distribution to the actuators or to change the direction of motion of an actuator.

Under each valve family, there are various family members that are performing related control function. Any of these valves could be structured differently and could work hydro-mechanically or electro-hydraulically. As it has been previously discussed, such valves should be modeled based on their internal structure for purposes of cylinder development. For purposes of overall machine modeling, they are modeled based on lumped modeling concepts, no matter what the valve mechanism is. This chapter is not intended and can't provide a model for every hydraulic valve. The chapter rather presents lumped concepts of valve modeling as applied to the main valve of each valve family.

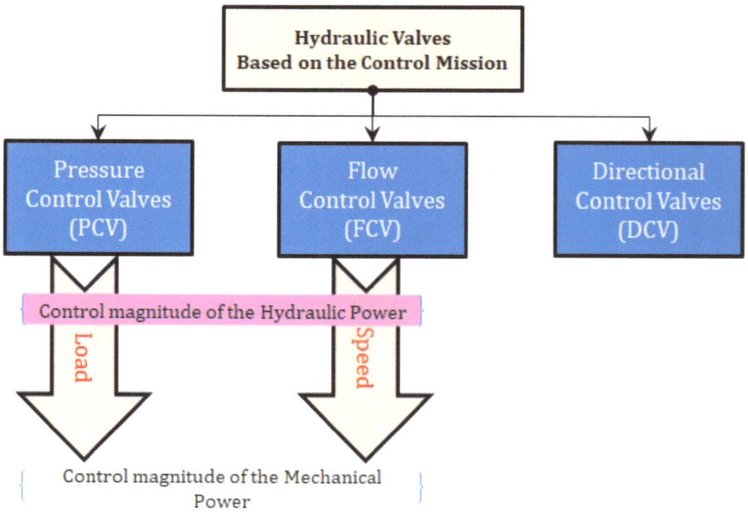

Fig. 10.1- Hydraulic Valves

10.2- Lumped Model #01 for Pressure Relief Valve Based on Linear Characteristics

10.2.1- Mathematical Model

As shown in Fig. 10.2, two pressure values characterize a *Pressure Relief Valve* (PRV). The first value is called the *Cracking Pressure* or *Critical Pressure*, $\mathbf{P_{cr}}$. It is the value that balances the spring force and at which the valve is about to open. Below the cracking pressure, all the pump flow $\mathbf{Q_p}$ is directed towards the system $\mathbf{Q_S}$. Any further increase of the pressure will partially open the valve so that the flow is distributed between the system and the PRV. When the system pressure exceeds the cracking pressure, the flow through the relief valve $\mathbf{Q_{RV}}$ increases on account of system flow $\mathbf{Q_S}$.

The second characterizing pressure is the maximum pressure that is called *Cut-Off Pressure*, $\mathbf{P_{co}}$. When the maximum setting pressure reached, the valve will swallow the full flow of the pump comfortably and no flow will be directed to the system.

At the time of adjusting a PRV, the pump outlet line is to be blocked and the full flow of the pump goes through the PRV. So, the pressure setting value would be the cut-off value. The valve starts to open early at the cracking pressure because the valve is not receiving the full flow of the pump until the pump outlet is blocked and the pump flow has nowhere to go but through the PRV.

The pressure zone between the cracking pressure and the cut-off pressure is defined as the *Override Zone* of the valve. Typical % value of override zone $\mathbf{P_{OR}}$ equals (15-20) % of $\mathbf{P_{cr}}$. for direct operated PRVs and less than 10% for $\mathbf{P_{co}}$.

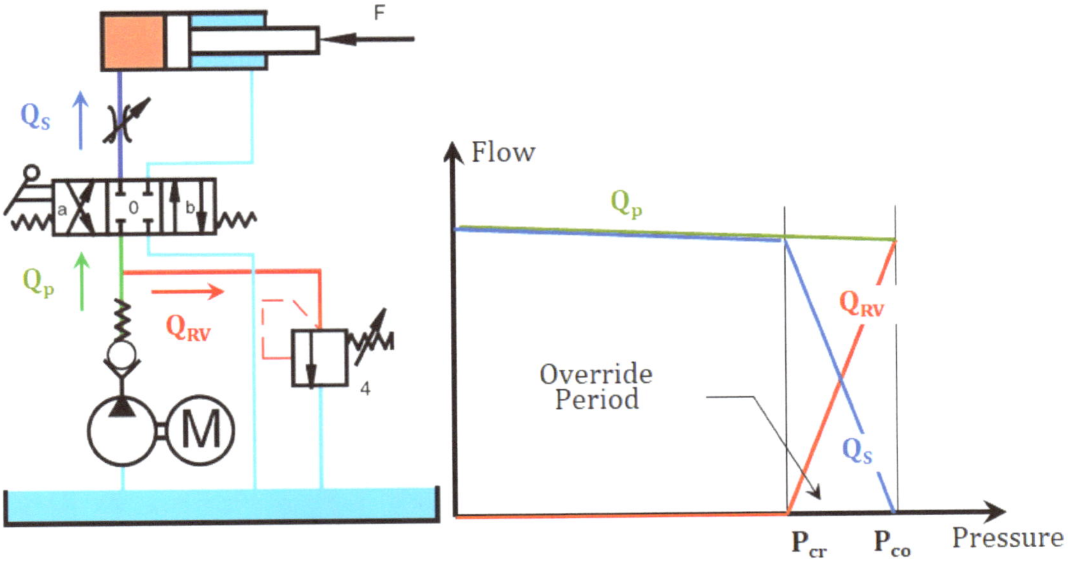

Fig. 10.2- Pressure Relief Valve Characteristics

The following set of equations solves for the static characteristics of PRVs.

$$P_{CO} = P_{CR} + P_{CR} \times \frac{P_{OR}}{100} = P_{CR}(1 + \frac{P_{OR}}{100}) \qquad 10.1$$

$$\text{If } P_p \leq P_{CR} \rightarrow Q_{RV} = 0 \qquad 10.2$$

$$\text{If } P_{CR} < P_p < P_{CO} \rightarrow Q_{RV} = Q_p \times \frac{P_p - P_{CR}}{P_{CO} - P_{CR}} \qquad 10.3$$

$$\text{If } P_p \geq P_{CO} \rightarrow Q_{RV} = Q_p \qquad 10.4$$

$$Q_S = Q_p - Q_{RV} \qquad 10.5$$

10.2.2- Simulation Model

Figure 9.6 shows the upper mask of simulation model (HSV7CH10Model01). The figure shows the interactive inputs/outputs of the model. The figure shows also three different scenarios to simulate the system pressure as constant, ramping and step change.

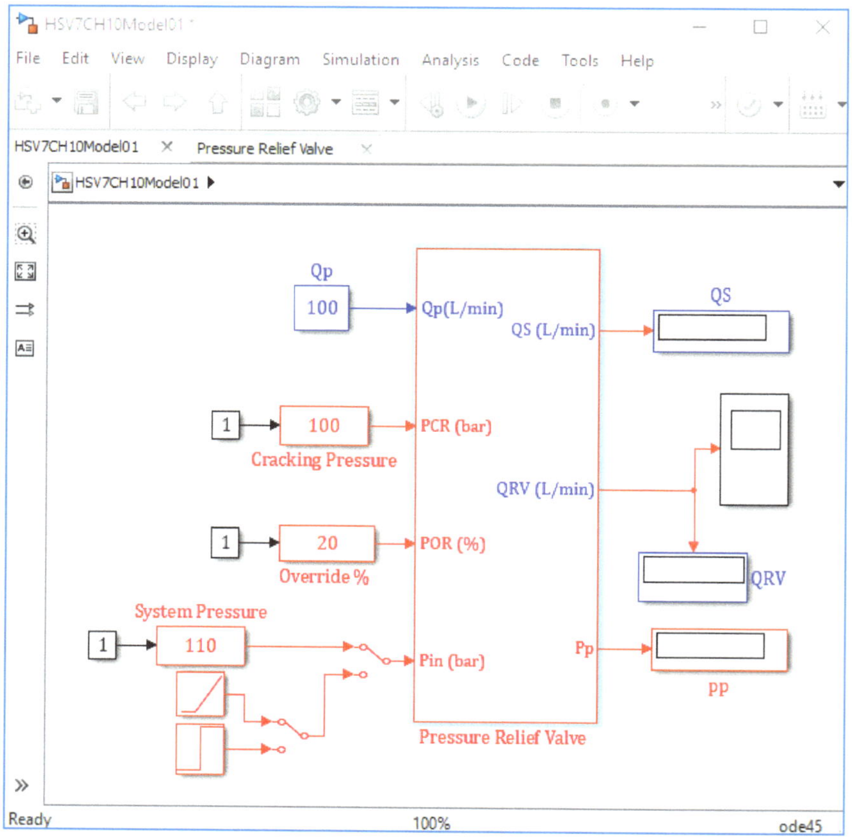

Fig. 10.3- Model 01 for Pressure Relief Valve Simulation

Figure 10.4 shows the details under the mask of Model 01. The model is built to solve the set of equations in the same sequence they are presented in the mathematical model using metric system of units. An NTF is added to represent the PRV valve dynamics as second-order system.

The figure shows that, whatever the value of the inlet pressure to the PRV model, pump pressure has a maximum value equal to the cut-off pressure. Saturation block is used to perform this logic.

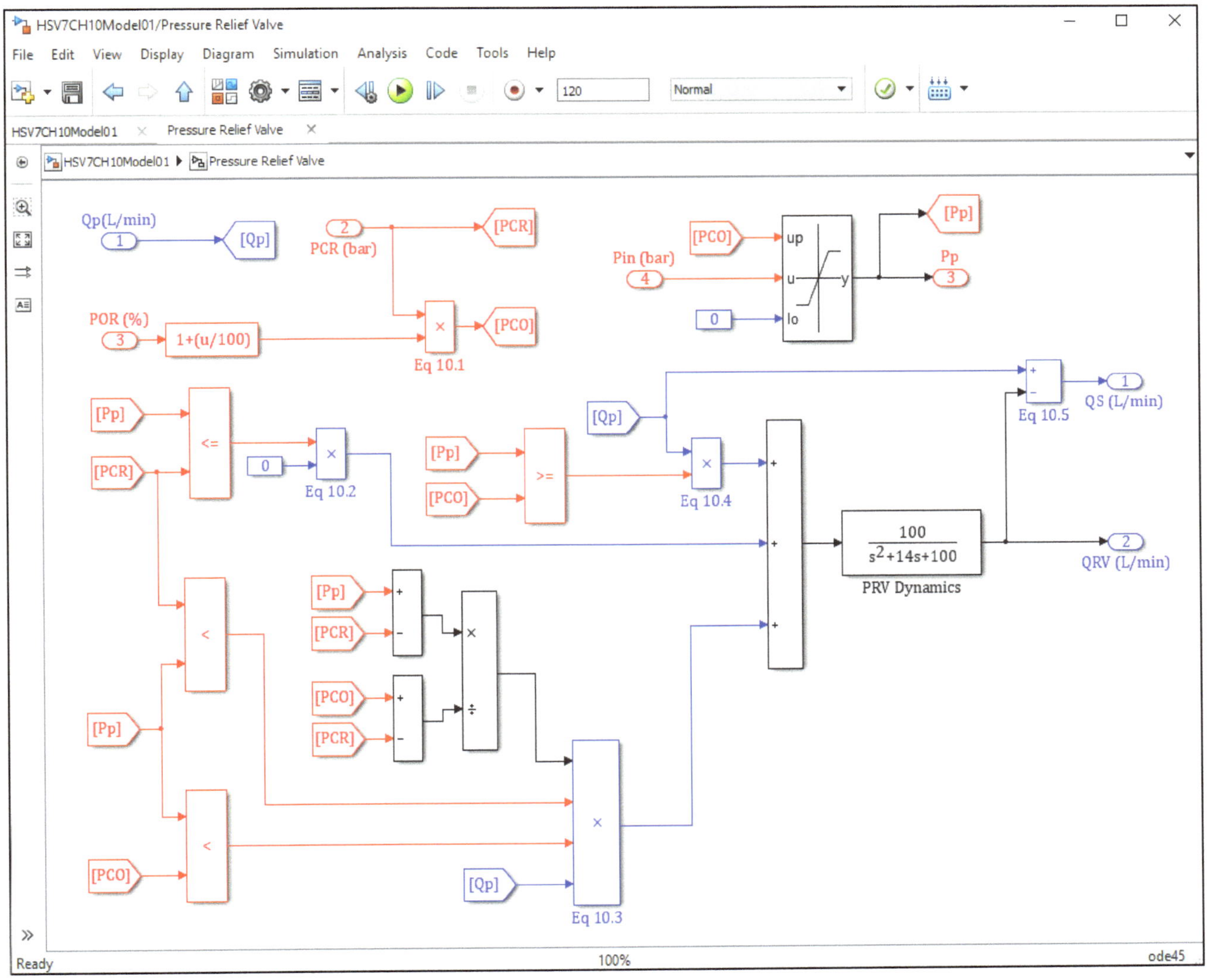

Fig. 10.4- Details under the Mask of Model 01

10.2.3- Model Validation

Modeling the Valve Steady State: Figure 10.5 (left side) shows a case when the system pressure increases gradually from zero up to the cut-off value (120 bar). The figure shows that no flow is passing through the PRV until the pressure increases above the cracking pressure (100 bar). Once the pressure reaches the cut-off value, full flow of the pump (100 lit/min) passes through the PRV.

Modeling the Valve Dynamics: Figure 10.6 (right side) shows a case when pump pressure increases suddenly from a value lower than the cracking pressure to a pressure at the middle of the override zone (110 bar). This results in half of the pump flow passing through the PRV.

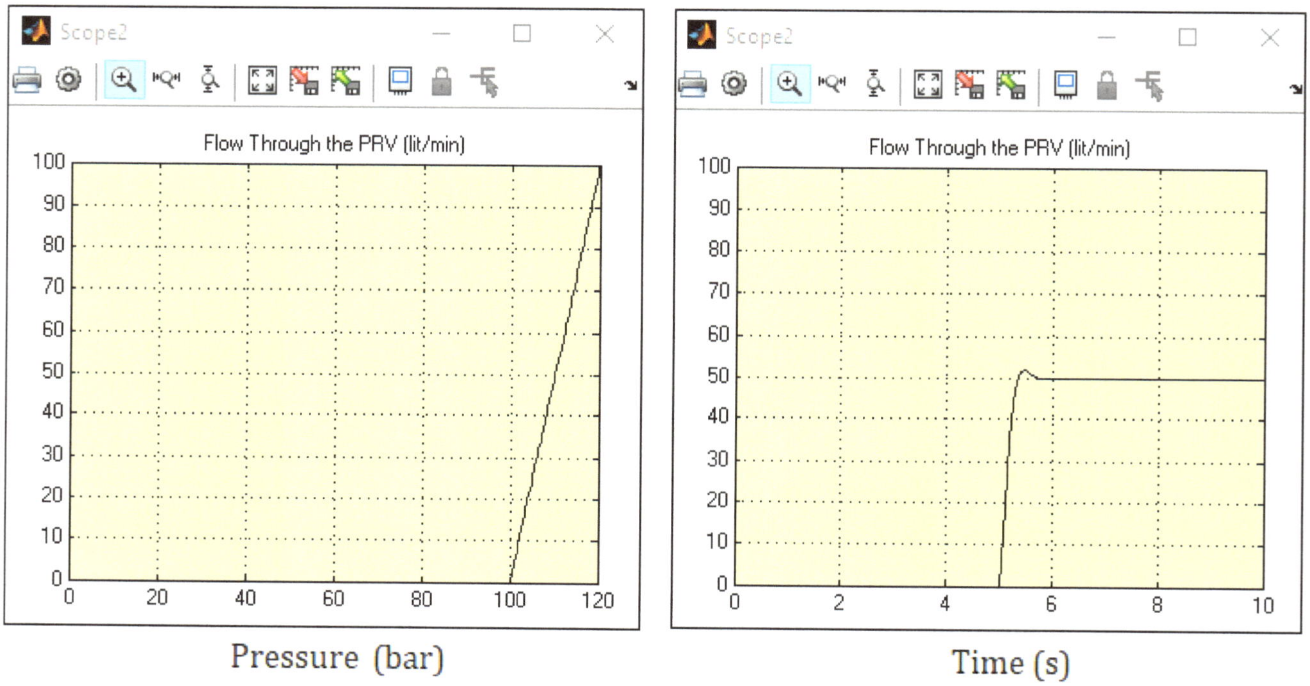

Fig. 10.5- Validation of Model 01

10.3- Lumped Model #02 for Pressure Relief Vale Based on Nonlinear Characteristics

As shown in Fig. 10.6, PRVs may have nonlinear characteristics. So, easiest way to model such valve is to load the characteristics into a lookup table. As shown in the figure, simulation model (HSV7CH10Model02) is built to model such valves. Figure 10.7 shows the details of model 02. In this model no need for the cracking pressure or override zone since data is loaded to a lookup table. But this model is valid only for this valve size. If the valve size is changed, lookup table must be updated accordingly. Simulation results can always be tuned for better results.

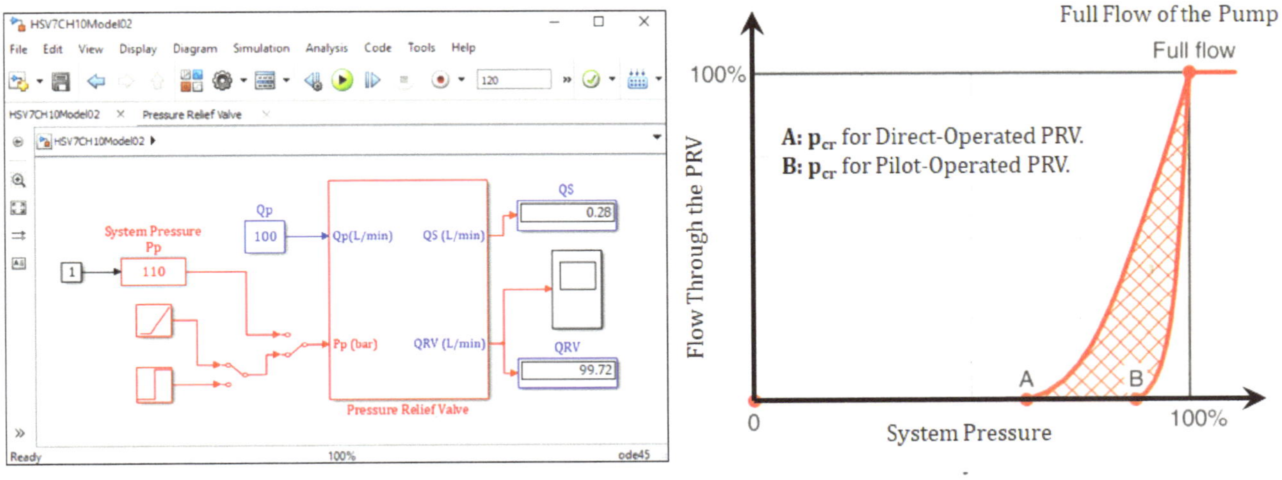

Fig. 10.6- Modeling of Pressure Relief Valves Based on Nonlinear Characteristics

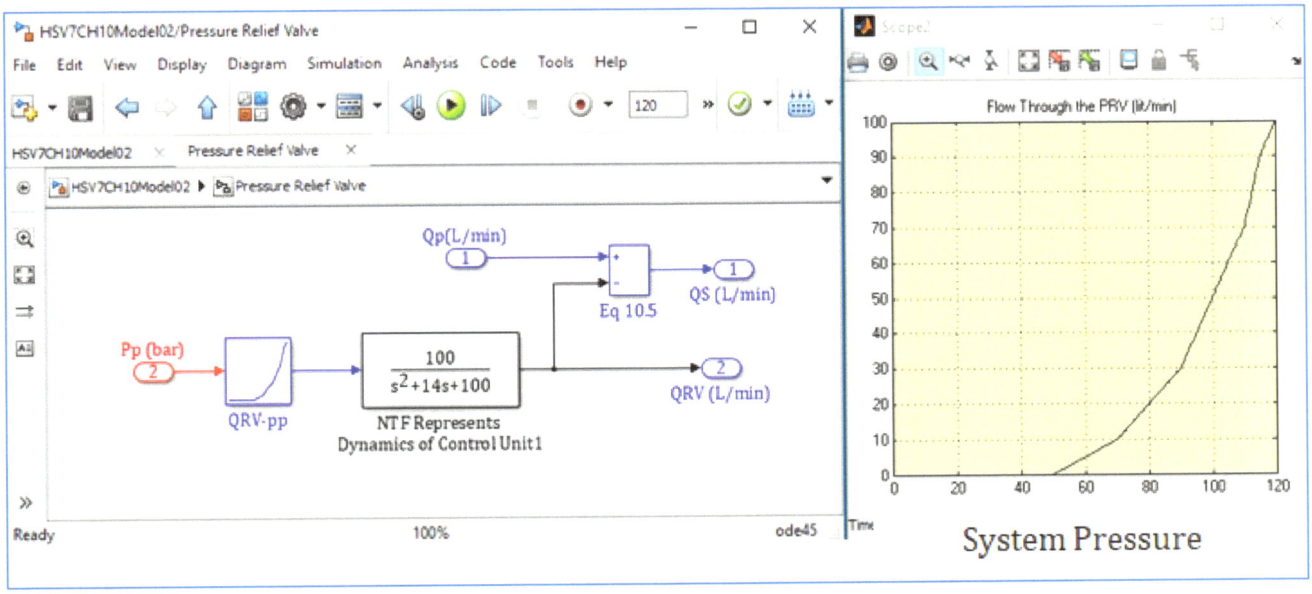

Fig. 10.7- Details and Validation of Model 02

10.4- Lumped Model #03 for Flow Control Valves

10.4.1- Model Structure and Assumptions

As shown in Fig. 10.8, a *Flow Control* Valve in its simplest form is a variable orifice. Variation of the throttling coefficient is based on an input command that could be done manually (by an adjustor), or remotely (by mechanical, hydraulic or electric signal).

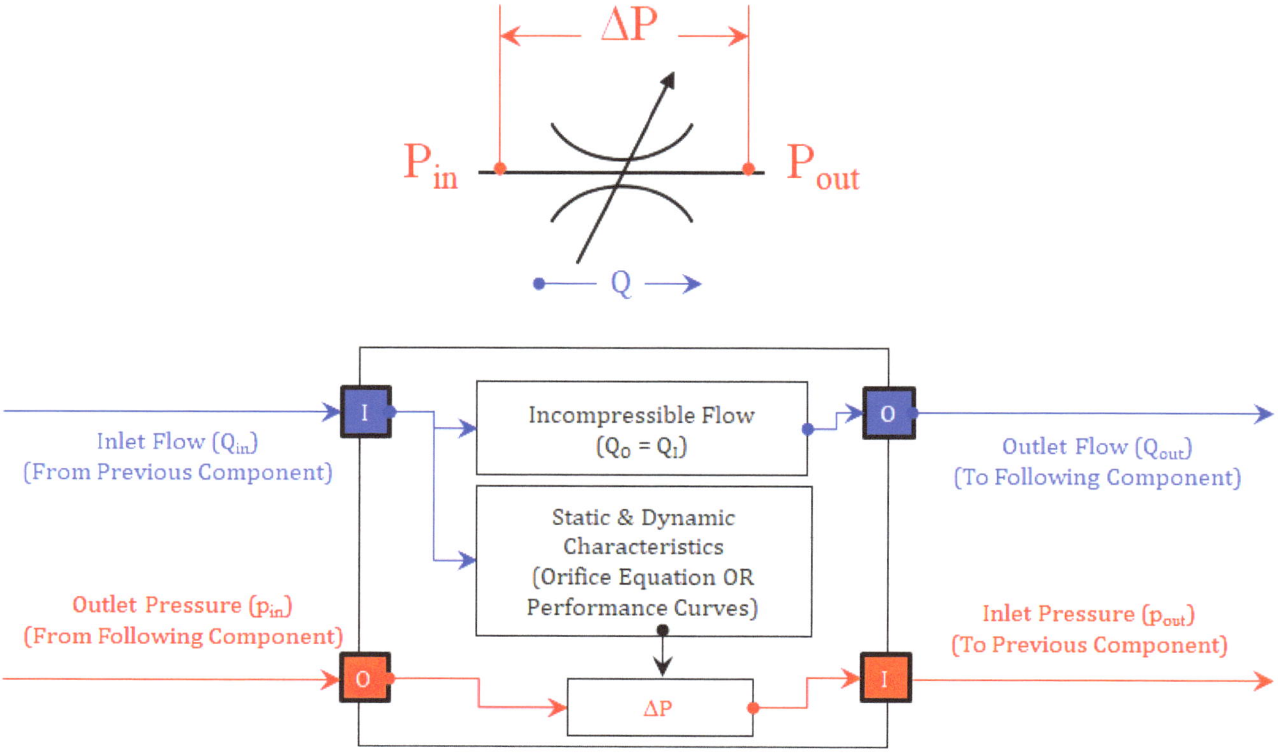

Fig. 10.8- Lumped Model Structure of Flow Control Valve

10.4.2- Mathematical and Simulation Model

As shown in Fig. 10.9, simulation model (HSV7CH10Model03) is built based on retrofitting model (HSV7CH06Model03), that previously presented in Chapter 03 simulating a fixed orifice, to make it function as variable orifice. Model for hydraulic fluid is added because the specific gravity is part of the orifice equation. Figure 10.10 shows the details of model 03. A saturation block is used to limit the minimum orifice diameter to avoid singularity in the model. An NTF is added (if needed) to the model to represent the valve dynamics in case if a remote input signal is used. It is to be noted that, if the orifice equation is used, the model is good for any valve size. If modeling is based on given static characteristics, then this model is valid only for a specific valve size. If the valve size is changed, lookup table must be updated accordingly. Alternatively, since this is a variable orifice, a set of characteristics curves are required for every % of the valve closure.

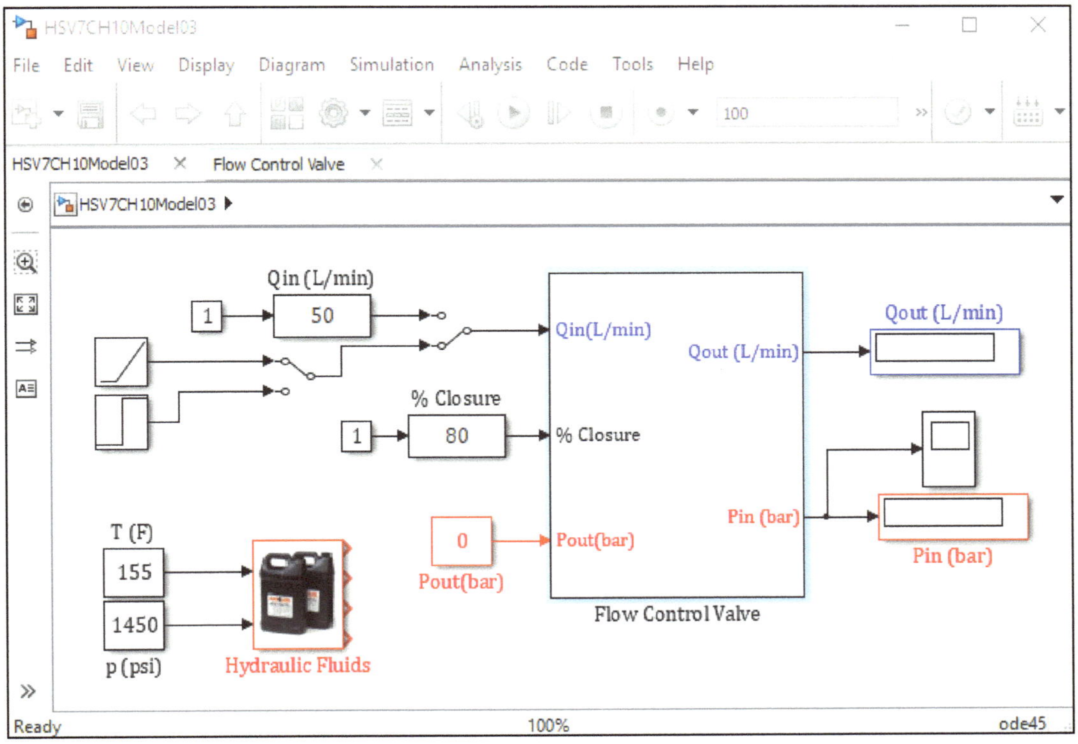

Fig. 10.9- Model 03 for Flow Control Valve

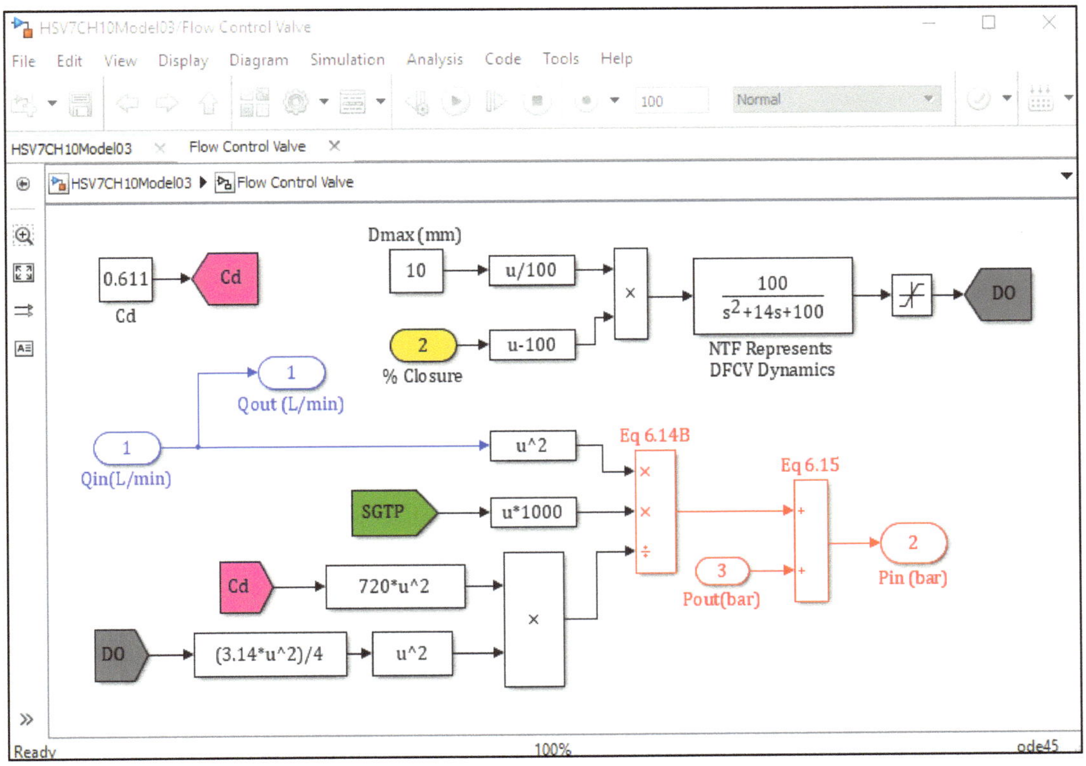

Fig. 10.10- Details under the mask of Model 03

10.4.3- Model Validation

However, Fig. 10.11 shows the static characteristics of the valve based on flow increasing gradually from 0 to 100 lit/min for various % closures of the valve that has a sharp-edged short orifice with a maximum orifice diameter (10 mm diameter).

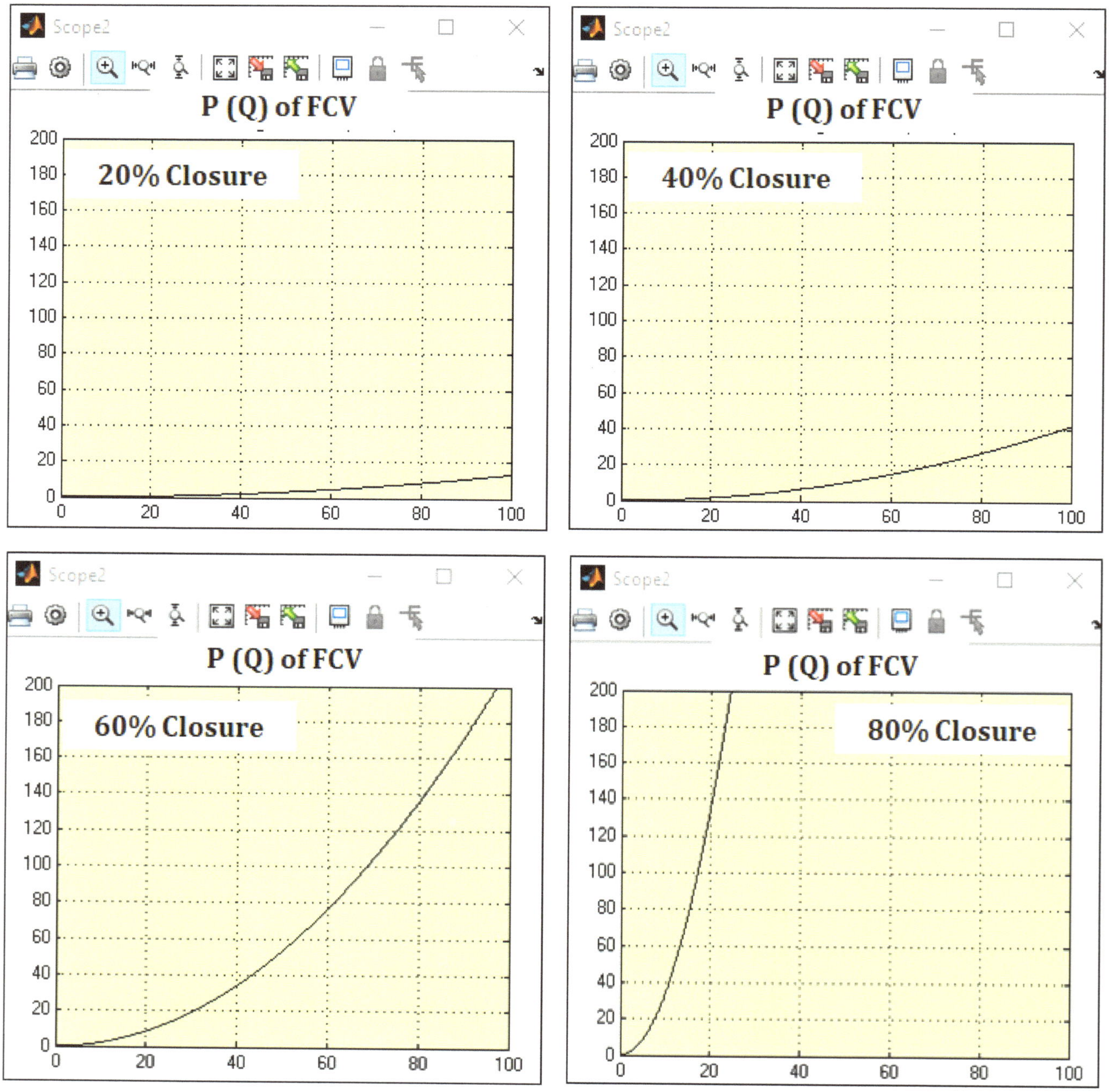

Fig. 10.11- Validation of Model 03

10.5- Lumped Model #04 for Check Valves

10.5.1- Mathematical and Simulation Model

A *Check Valve* in its simplest form is a one-way valve that allows flow in one direction only when outlet pressure is lower than inlet pressure. Simulation model (HSV7CH10Model04), shown in Fig. 10.12, just calculates the differential pressure across the valve based on the inlet flow when the valve is opened. This differential pressure is added to the outlet pressure $\mathbf{P_{out}}$ to calculate inlet pressure $\mathbf{P_{in}}$ assuming the flow continues to pass through the valve. Maximum inlet $\mathbf{P_{inmax}}$ pressure to the check valve is received from the component model before the check valve model. The status of the check valve is determined based on $\mathbf{P_{out}}$ as compared to $\mathbf{P_{inmax}}$. If $\mathbf{P_{out} < P_{inmax}}$, then the valve is open (status = 1) and vise-versa (status = 0).

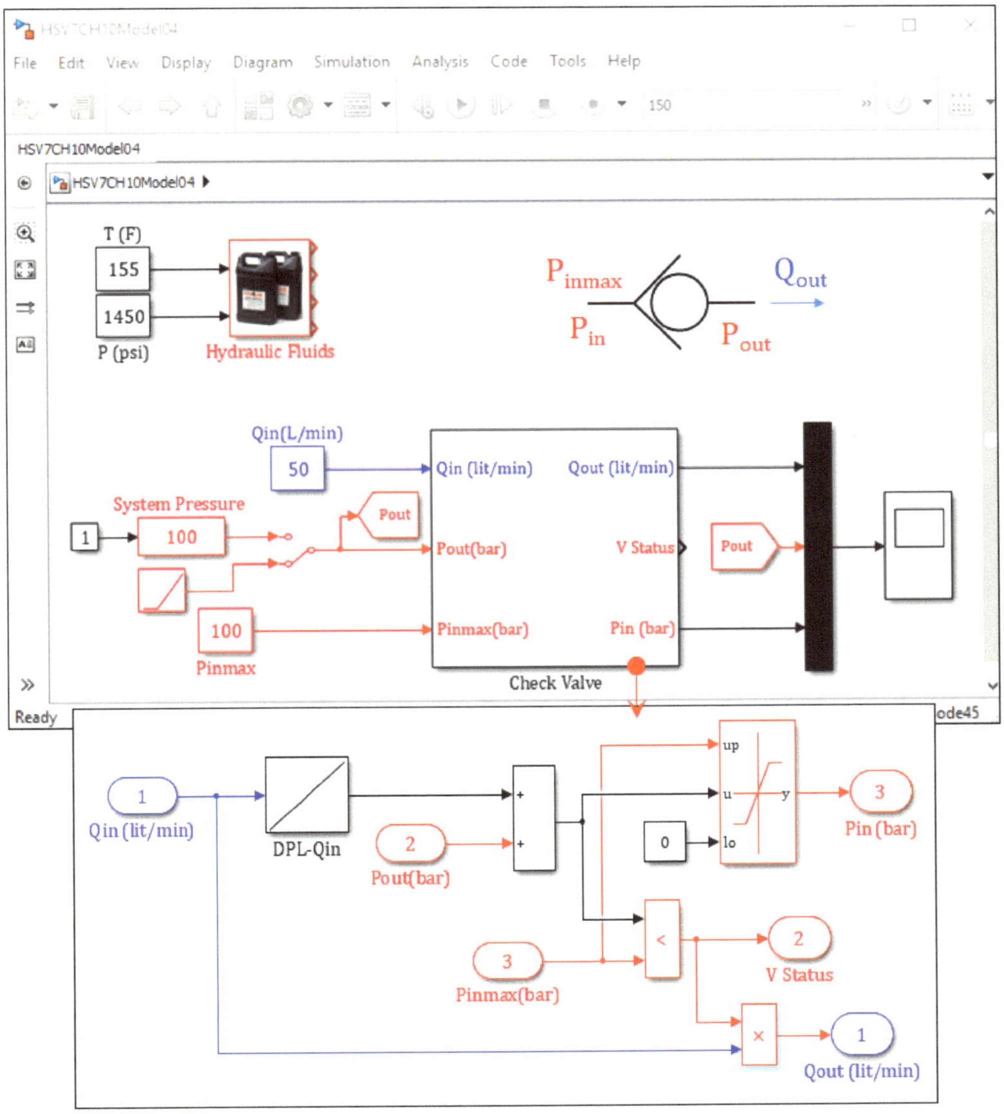

Fig. 10.12- Model 04 for Flow Check Valve

10.5.2- Model Validation

As shown in Fig. 10.13, by ramping the outlet pressure **P_{out}**, inlet pressure **P_{in}** increases by the amount of the differential pressure across the valve. Once the **P_{out}** exceeds the given maximum inlet pressure **P_{inmax}**, the valve closes and the flow across the valve drops to zero.

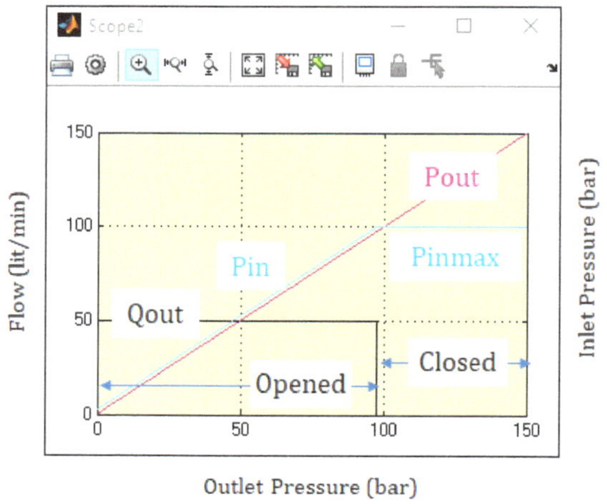

Fig. 10.13- Validation of Model 04

10.5.3- Validating the Models for Check Valve and PRV together

As shown in Fig. 10.14, simulation model (HSV7CH10Model05) is constructed by combining model 01 and model 03.

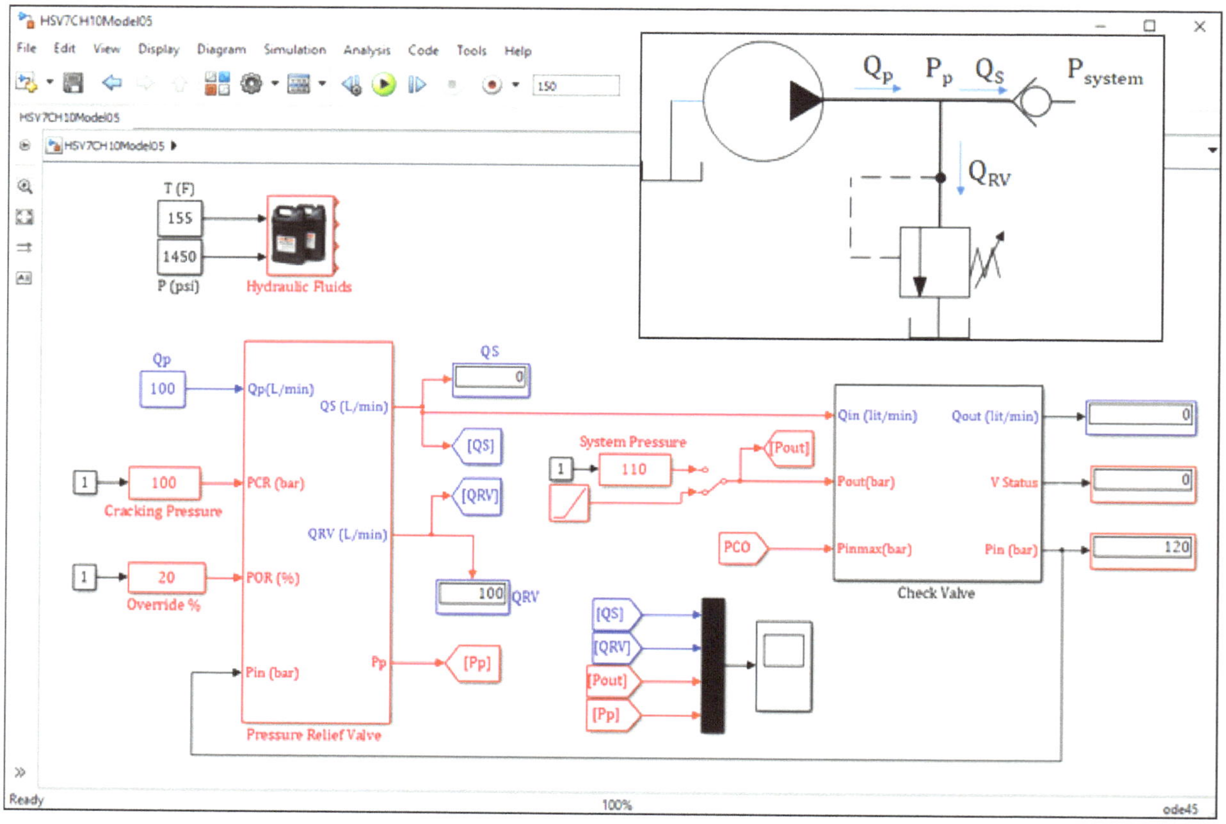

Fig. 10.14- Model 05 for Validating the Check Valve and PRV Operation

As shown in Fig. 10.15, when the system pressure (downstream the check valve) is ramped, pump pressure **P_p** is ramped accordingly including the differential pressure across the check valve. Once **P_p** falls in the override zone of the PRV, flow through the relief valve **Q_{RV}** increases on the account of the system flow that passes through the check valve **Q_S**. When the system pressure exceeds the cut-off pressure, check valve closes, full flow passes through the PRV, and **P_p** is limited to maximum of the cut-off pressure.

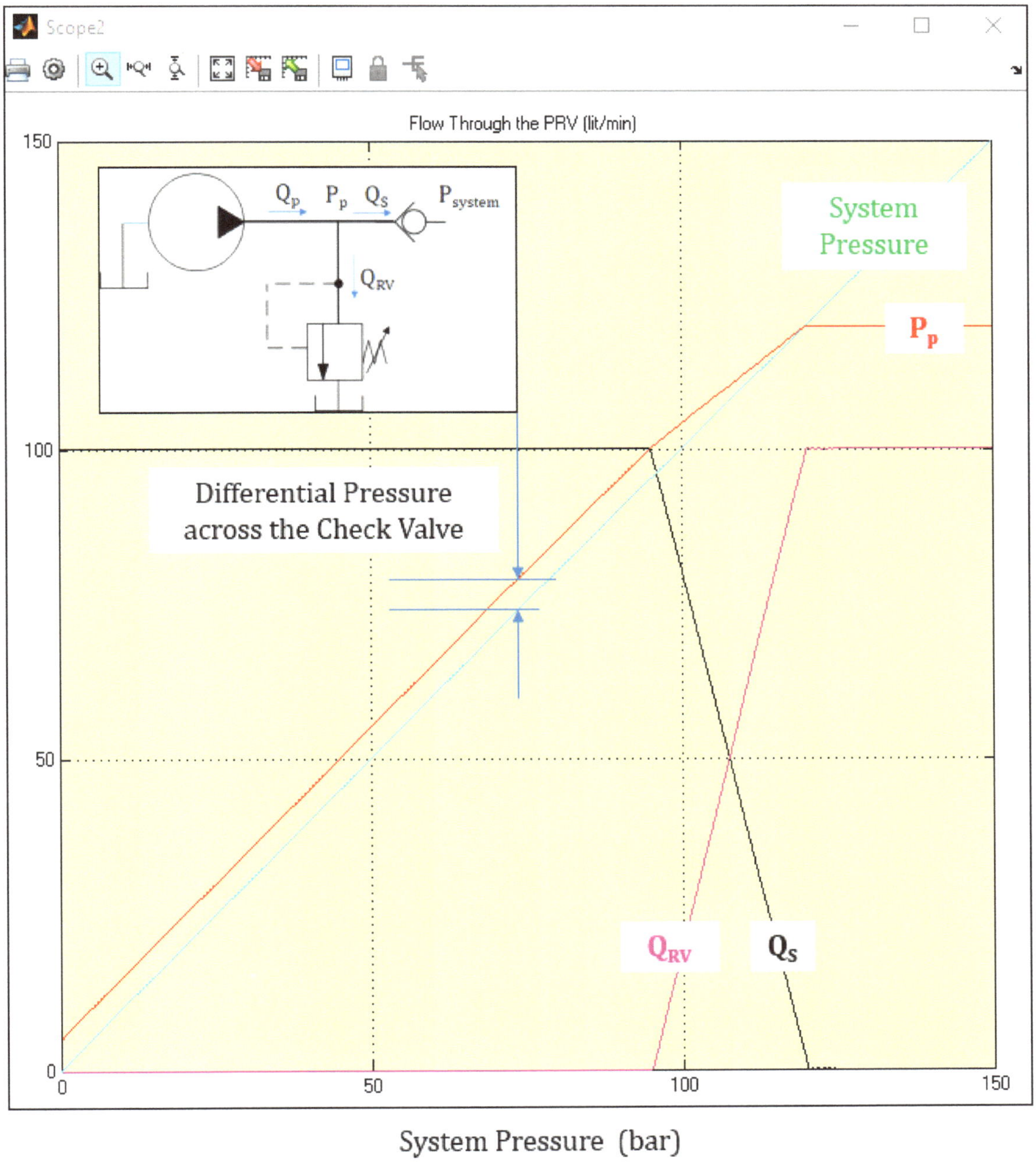

Fig. 10.15- Simulation Results for Model 04

10.6- Lumped Model #06 for Continuous Directional Control Valves

10.6.1- Model Structure and Assumptions

As is it has been mentioned at the beginning of this chapter, there are a wide variety of valves and this chapter is not intended and can't provide a model for every hydraulic valve. So, this section provides lumped modeling concept for 4/3 a continuous (Proportional or Servo) directional control valve based on the valve flow gain. Such valves are used for directional and speed control as they receive variable input signal.

As shown in Fig. 10.16, conceptually a 4/3 continuous valve connects four ports (**P→A** and **B→T** OR **P→B** and **A→T**) based on the differential pressure across the vale and the spool displacement $\mathbf{X_S}$ that is function of the sign and magnitude of variable input signal $\mathbf{I_N}$.

So, one way to develop this valve test data is to plot *Nominal* (*Rated*) *Flow* $\mathbf{Q_N}$ versus *Nominal* (*Rated*) *Differential Pressure* $\mathbf{\Delta P_N}$ at constant *Nominal (Rated) input signal* $\mathbf{I_N}$. This is known as the *Flow-Pressure Sensitivity* performance curves. Alternatively, *Flow Gain* performance curves are developed. Flow gain is a plot of $\mathbf{Q_N}$ versus $\mathbf{I_N}$ at constant $\mathbf{\Delta P_N}$. Flow gain is more commonly known as "test data".

As shown in Fig. 10.17, flow gain of a continuous valve depends on the design of the metering edges of the valve spool. As shown in the figure, flow gain may be linear or nonlinear.

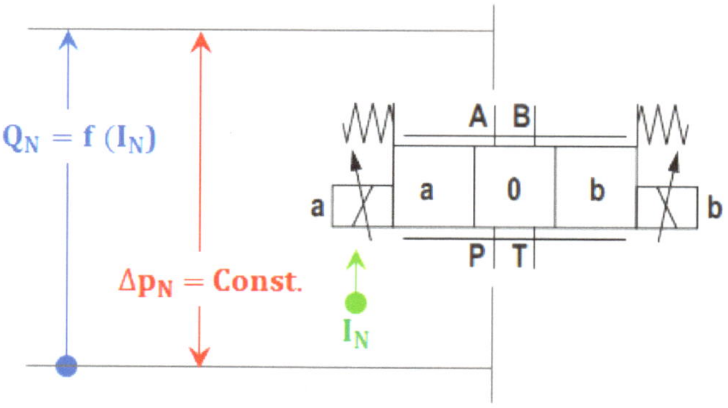

Fig. 10.16- Continuous Valve Operation

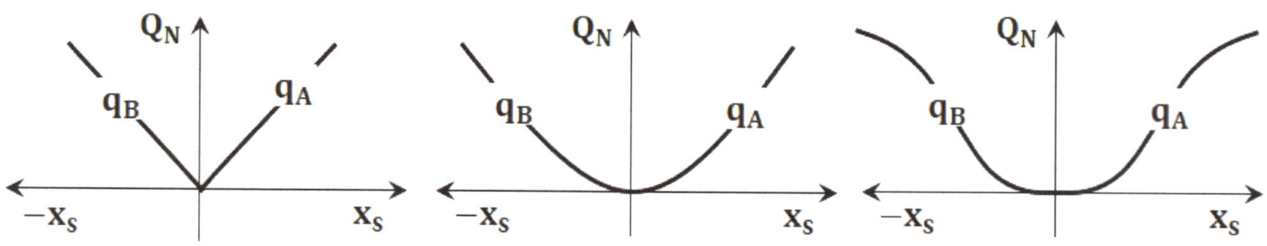

Fig. 10.17- Flow Gain Depends on Metering Edges Design

As shown in Fig. 10.18, flow gain of a continuous valve also depends on spool-sleeve the lapping conditions. The figure shows the two most common lapping conditions for directional valves. The condition on the left is known as *Overlapped* which is used for *Closed-Center* valves. In this condition, dimension of spool land is larger than the undercut. This condition results in a discontinuous flow gain with a *Dead Zone* of (5-20) % of the spool stroke. The condition on the right side is known as *Zero-Lapped*. In this condition, dimension of spool land is same as the undercut. This condition result in a continuous flow gain.

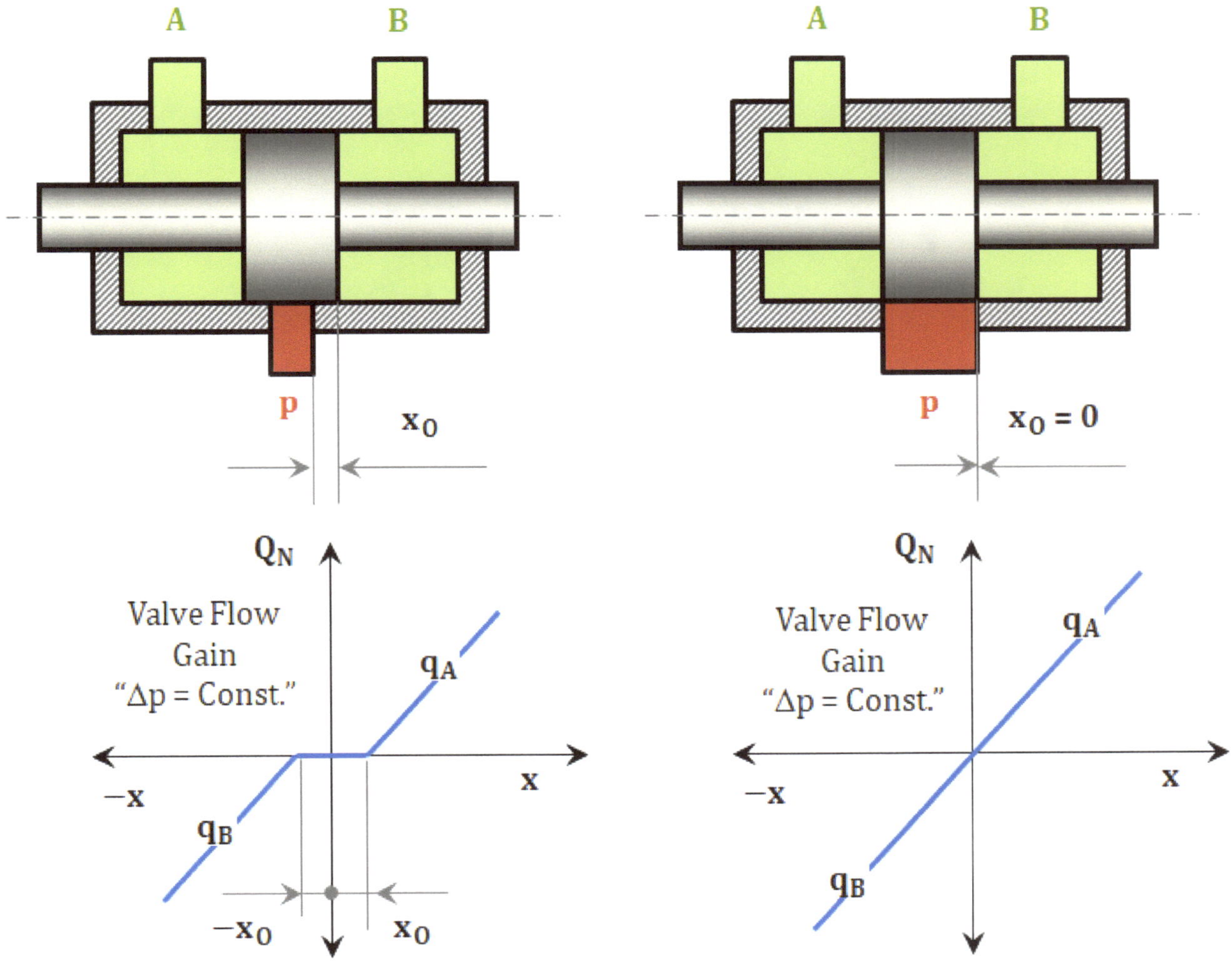

For $|x| \leq x_0, Q_N = 0$

For $|x| > x_0, Q_N = q_A$ or q_B

For $|x| > x_0, Q_N = q_A$ or q_B

Fig. 10.18- Flow Gain Depends on Lapping Conditions

As shown in Fig. 10.19, when the valve receives an input signal $\mathbf{I_N}$, the spool moves a displacement $\mathbf{x_s}$, and two pairs of the valve ports are connected accordingly simultaneously. The figure shows an example when a continuous valve is connected to a pump, a tank and an actuator. When the valve receives a positive signal, port **P** and **A** are connected, and port **B** and **T** are connected simultaneously. This connection is reversed when the valve receives a negative signal. The valve receives at port **P** pump flow $\mathbf{Q_{pump}}$ and passes it through port **A** or **B** as actuator inlet flow $\mathbf{Q_{AI}}$ that is affected by actuator inlet side pressure $\mathbf{P_{AI}}$. The actuator generates flow from the outlet side $\mathbf{Q_{AO}}$ and passes it through port **B** or **A** to port **T** as tank flow $\mathbf{Q_{tank}}$ that is affected by tank pressure $\mathbf{P_{tank}}$. So, as shown in the figure, the valve hypothetically comprises two variable flow control valves that should be modeled simultaneously using the same model structure shown in Fig. 8.8 presented previously in Chapter 8.

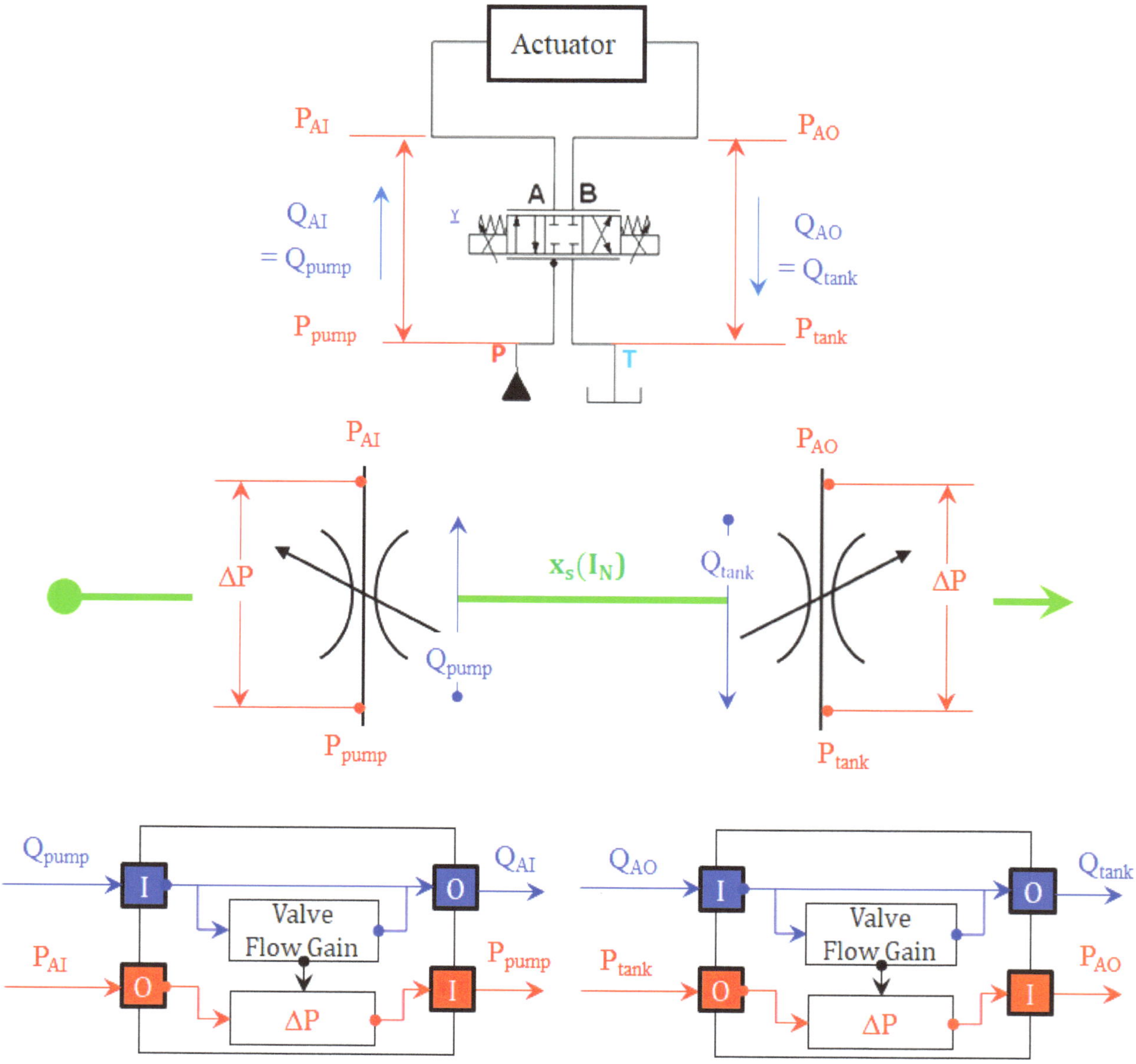

Fig. 10.19- Lumped Mode Structure of Proportional and Servo Valves

10.6.2- Developing Static Characteristics (Flow Gain) of a Continuous Valve

Valve flow gain should be available as part of the valve data sheet. If not, it can always be developed experimentally on a valve test stand. Commonly, the spool of a 4/3 continuous valve is symmetric and results in identical flow gain for both directions. Otherwise, four flow gain curves must be developed, one for each of the following flow paths **P→A**, **P→B**, **B→T**, and **A→T**.

Figure 10.20 shows flow gains developed experimentally at Δp_N = 100 psi for an overlapped symmetric-spool *Proportional* valve (left side) and a symmetric-spool zero-lapped *Servo* valve (right side). These valves are mounted on the UFPT shown previously in Fig. 7.41.

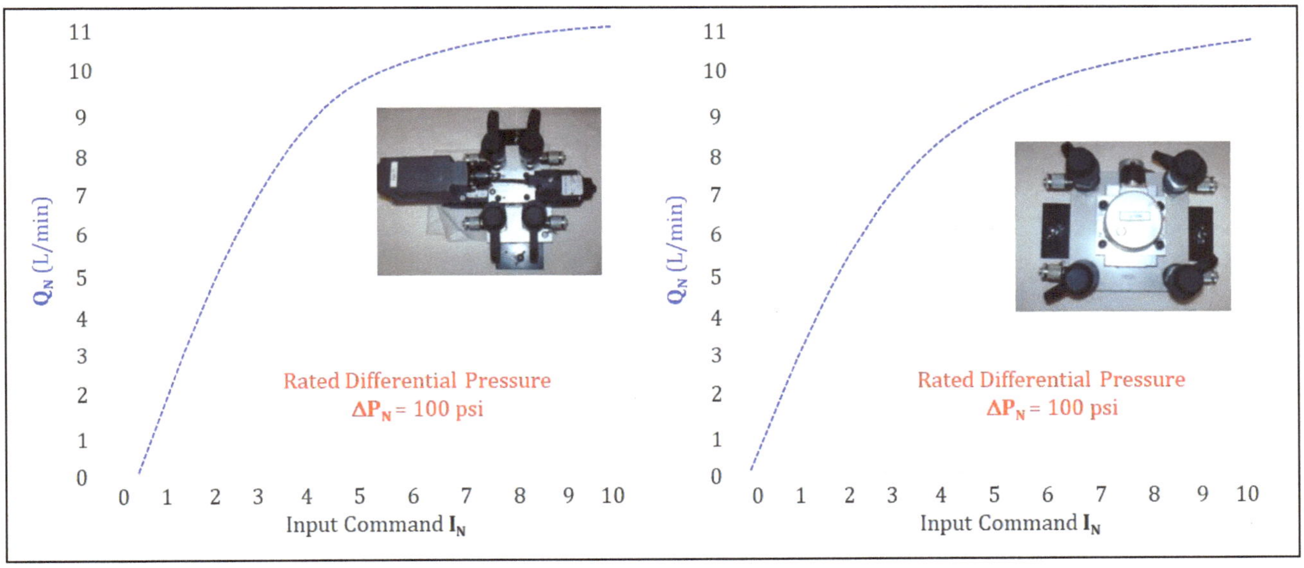

Fig. 10.20- Experimentally Developed Flow Gain for a Proportional and Servo Valve

10.6.3- Developing Dynamic Characteristics of a Continuous Valve

Commonly, the spool of a 4/3 continuous valve is symmetric and results in identical dynamic characteristics in both directions. Otherwise, separate dynamic characteristics for each side of the spool movement must be developed. Spools of the above-mentioned valves are spring-centered and surrounded by a damping oil film. So, dynamics of these valves are represented by 2nd order NTF.

By reviewing the data sheets of the valves, it was found that:

- Proportional Valve: Natural Frequency = 40 Hz (251rad/s) and damping ratio = 1.
- Servo Valve: Natural Frequency = 110 Hz (690 rad/s) and damping ratio = 0.2.

10.6.4- Mathematical Model Based on the Flow Gain

Figure 10.21 shows the upper mask of simulation model (HSV7CH10Model06) built based on the model structure and assumptions discussed in the previous section. The upper enclosure line contains the connecting ports with the pump and inlet side of the actuator. The lower enclosure line contains the connecting information with the outlet side of the actuator and the tank.

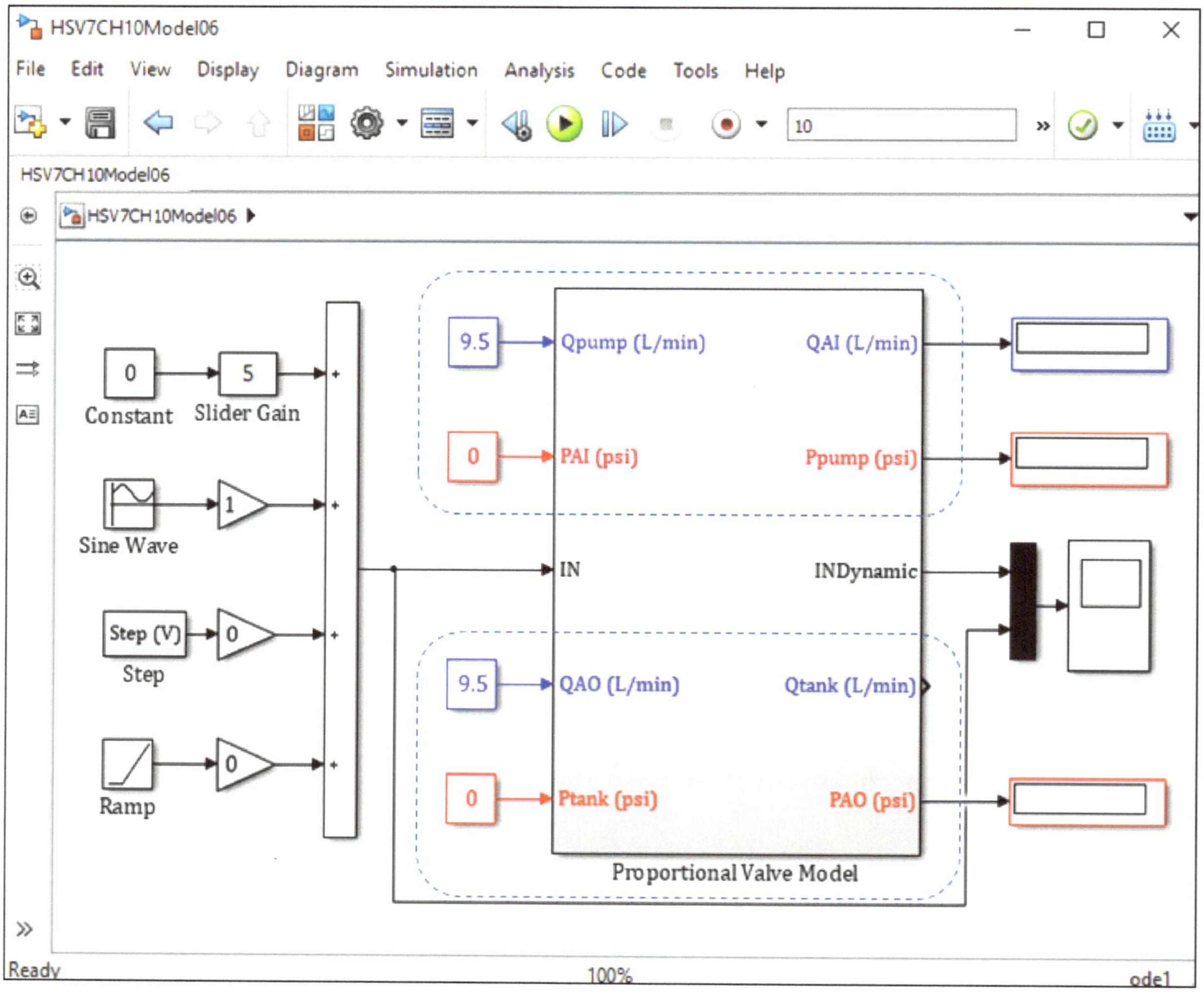

Fig. 10.21- Model 06 for 4/3 Proportional or Servo Valve

Figure 10.22 shows the details under the mask of model 06. The figure shows the sequence of calculation as follows:

1- **Defining the Nominal (Rated) Differential Pressure, ΔP_N:** in this step, the differential pressure at which the static characteristic is developed is assigned to a global variable so that it can be accessed from anywhere in the model.

2- **Modeling the Dynamic Characteristics:** When the valve receives an input signal $\mathbf{I_N}$, the spool moves dynamically a displacement $\mathbf{x_S}$, and all other working conditions are affected accordingly. Despite that the spool dynamic is assumed to be symmetric in this case, two NTFs are used just in case this model is used for dissymmetric spool dynamics. Natural frequency and damping ratio for the proportional valve are used to construct the NTF. Information for servo valve is used in a separate model. With the logics built in, one NTF is considered at a time based on the sign of $\mathbf{I_N}$. It is assumed that the slope of $\mathbf{x_S}$-$\mathbf{I_N}$ relation = 1 in this model. otherwise, actual slope is used.

3- **Determining $\mathbf{Q_N}$ assuming $\mathbf{\Delta P = \Delta P_N}$:** When the valve receives positive $\mathbf{I_N}$, port **P** is connected to Port **A**. So $\mathbf{Q_N}$ for this flow path is generated from the upper capsulated box. This box, as shown in Fig. 10.23, contains two lookup tables with logic that one of them is used at a time based on the sign of $\mathbf{I_N}$. Simultaneously, port **B** is connected to port T. So, $\mathbf{Q_N}$ for this pass is generated from one of the two lookup tables contained in the lower capsulated box. When the valve receives negative $\mathbf{I_N}$, lookup tables **P→B** and **A→T** are used automatically. It is to be noted that, despite the valve spool is assumed to be symmetric in this case, one lookup table is assigned to represent flow gain for each individual flow path in case this model is used for dissymmetric spool.

4- **Determining Actual Differential Pressure ΔP:** When the valve receives $\mathbf{Q_N}$ at $\mathbf{I_N}$, it generates $\mathbf{\Delta P_N}$. But it receives $\mathbf{Q_{pump}}$ through one flow path and $\mathbf{Q_{AO}}$ through the other flow path. So, what would be the actual differential pressure **ΔP** on each pass? To answer this question, we should refer to Eq. 6.13 (presented previously in Chapter 6), it can be briefed as $\mathbf{Q \,\alpha\, I\, (\Delta p)^{0.5}}$. Then, Eq. 10.7 can be used to determine **ΔP**.

5- **Determining $\mathbf{P_{pump}}$ and $\mathbf{P_{AO}}$:** By applying Eq. 10.6 to both legs of the valve, Eqs. 10.8A and 10.8B calculates pump pressure and actuator's outlet pressure; respectively.

6- **Saturation Blocks:** Saturation blocks were used on the flow and pressure calculation because, to avoid unrealistic overshoot of the pressure and flow. For example, during the first iteration, the model receives surge flow that may result is a very large differential pressure. This is just a virtual numerical value that won't happen in reality.

$$\text{Since } Q \propto I\sqrt{\Delta P} \;\rightarrow \frac{Q}{Q_N} = \sqrt{\frac{\Delta P}{\Delta P_N}} \rightarrow \Delta P = \Delta P_N \left[\frac{Q}{Q_N}\right]^2 \qquad 10.7$$

$$P_{pump} = P_{AI} + \Delta P_N \left[\frac{Q_{AI} = Q_{pump}}{Q_N}\right]^2 \qquad 10.8A$$

$$P_{AO} = P_{tank} + \Delta P_N \left[\frac{Q_{AO} = Q_{tank}}{Q_N}\right]^2 \qquad 10.8B$$

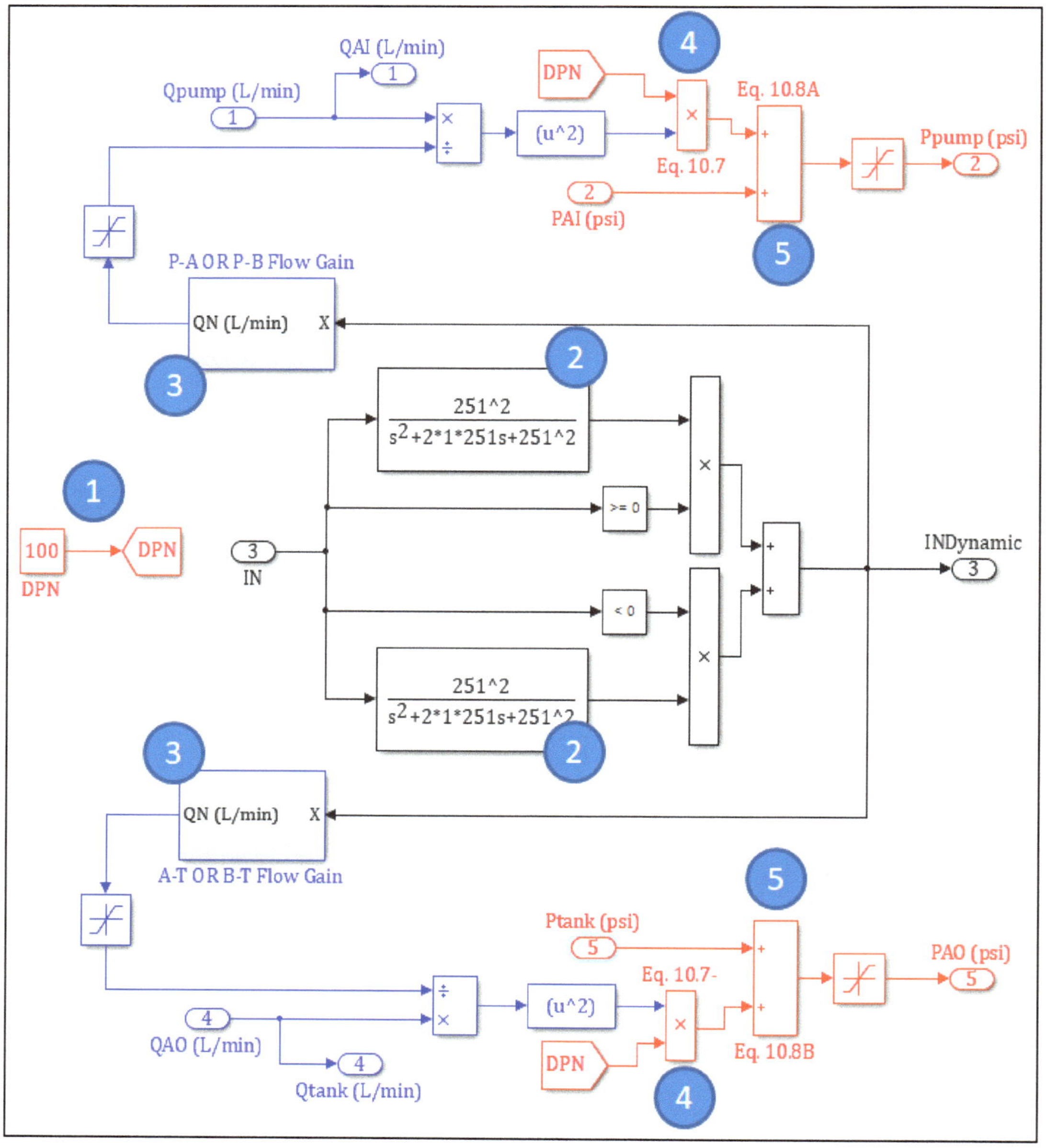

Fig. 10.22- Sequence of Calculations Under the Mask of Model 06

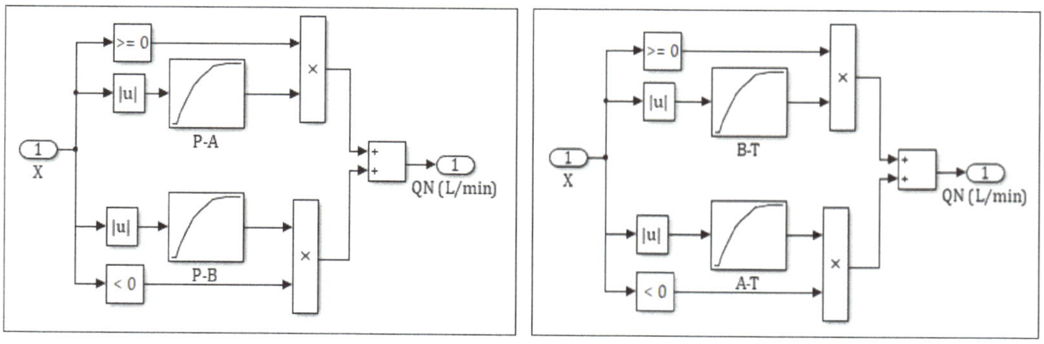

Fig. 10.23- Flow Gain for Four-Way Valve

10.6.5- Model Validation

Point-by-Point Flow Gain Validation: As shown in Fig. 10.24, it is assumed that that valve drives a synchronous cylinder so that, **$Q_{AO} = Q_{AI}$**. It also assumed that **$P_{AI} = 0$** and **$P_{tank} = 0$**. Then, if each of the valve flow paths receives **$Q = Q_N$** at **$I = I_N$**, it should result in **$\Delta P = \Delta P_N = 100$** psi. The test is made based on **$Q_N = 9.5$** at **$I_N = 5$**. Similarly, each point on the flow gain can be checked point-by-point.

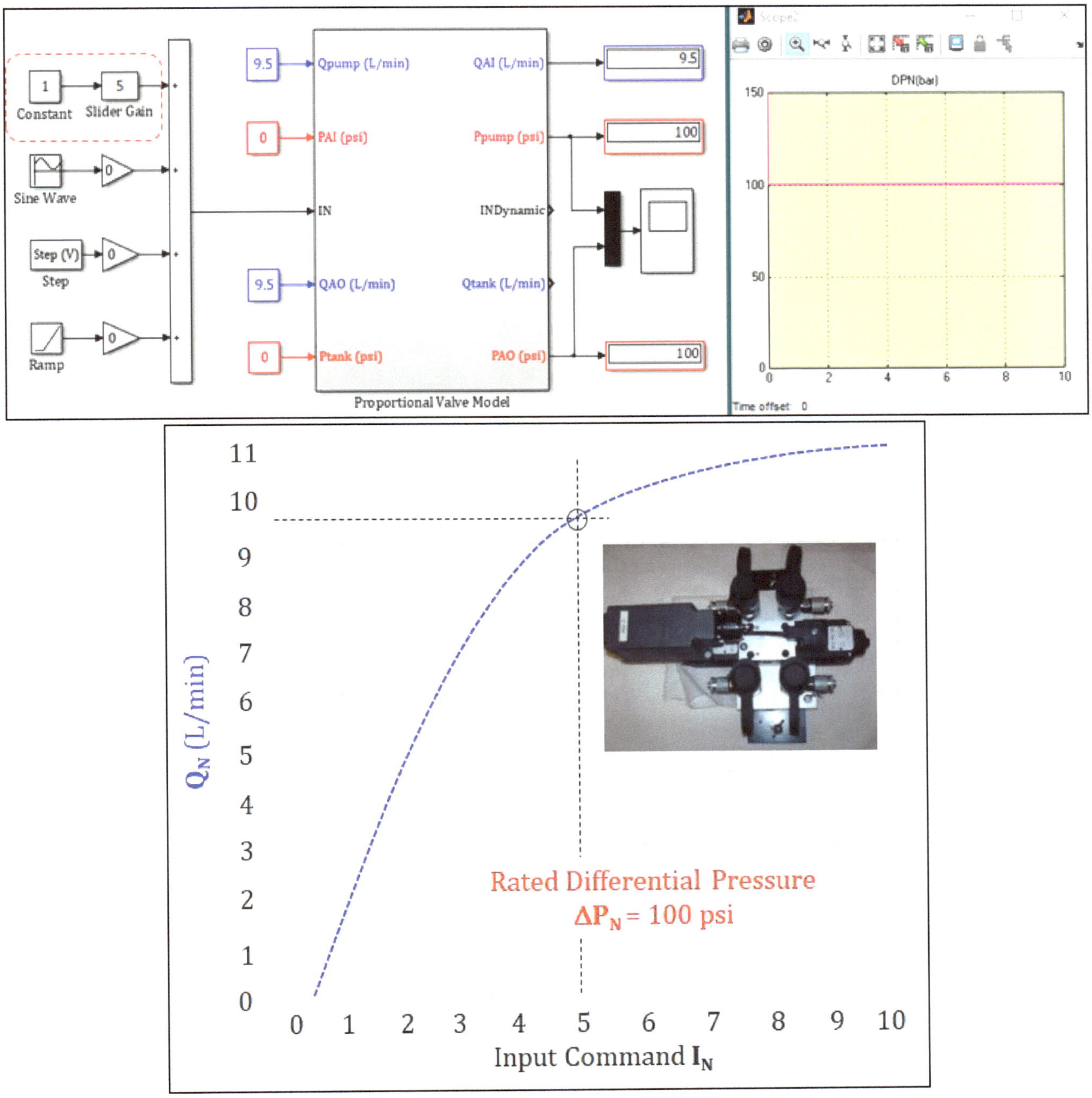

Fig. 10.24- Point-by-Point Flow Gain Validation

Flow Gain Curve Validation: An alternative method to validate all the points of the flow gain curve automatically is shown in Fig. 10.25. In this method, while $\mathbf{I_N}$ is ramped from 0-10. Simultaneously, the corresponding $\mathbf{Q_N}$ is generated from the relevant flow gain and used as input flow. This results in constant $\mathbf{\Delta P_N = 100}$ psi over the range of $\mathbf{I_N}$. In the very beginning pressure overshoots during the valve dead zone because the flow is surging but the valve is not yet opened. Saturation block is used to limit the overshoot to 150 psi.

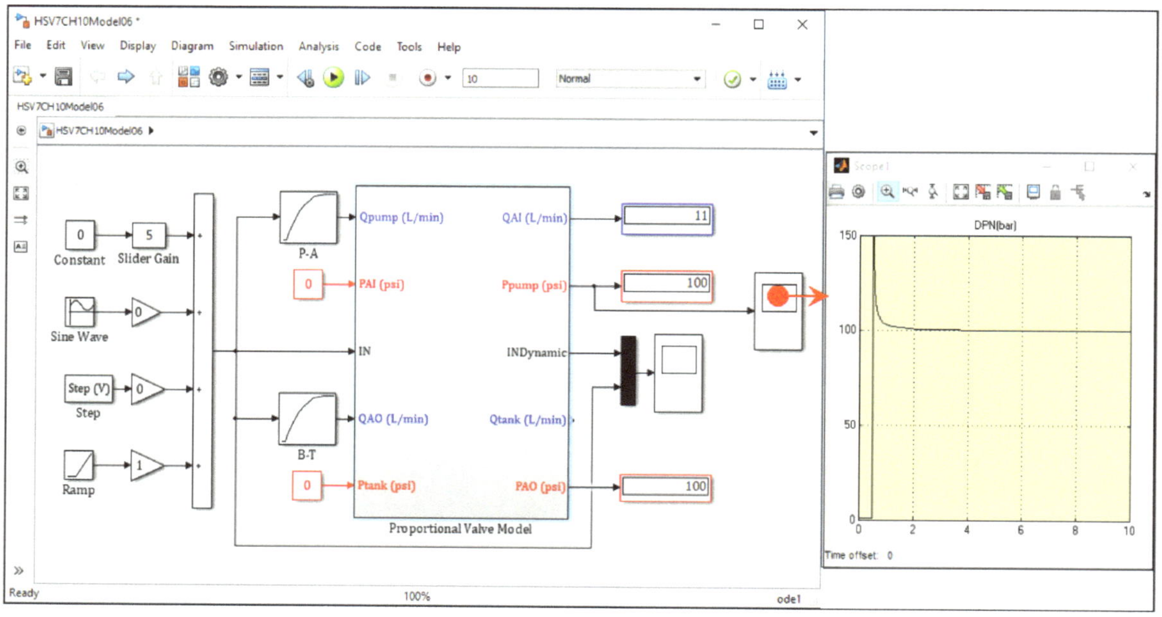

Fig. 10.25- Flow Gain Curve Validation

Step Response Validation: Figure 10.26 shows step response of the valve based on the parameters used to structure the NTF.

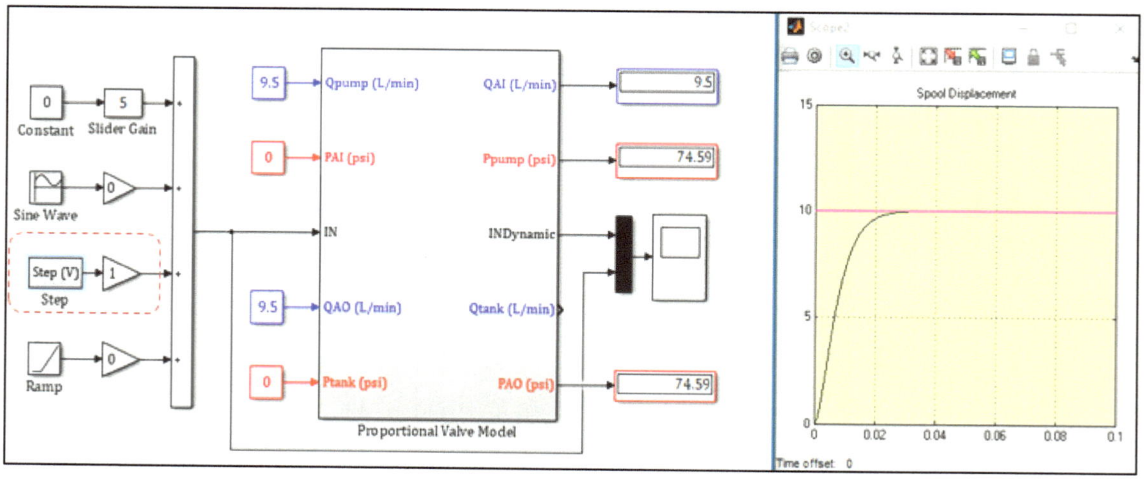

Fig. 10.26- Step Response of the Valve

Frequency Response Validation: Figure 10.27 shows step response of the valve based on the parameters used to structure the NTF. The figure shows that, when the frequency of the exciting signal = the natural frequency of the valve = 40 Hz (251 rad/s), phase shift was found equals 90 degrees and amplitude ratio = -70% = -3dB. That validates the model based on the frequency response, shown in Fig. 10.28 for valve size 3, that is reported by the valve manufacture.

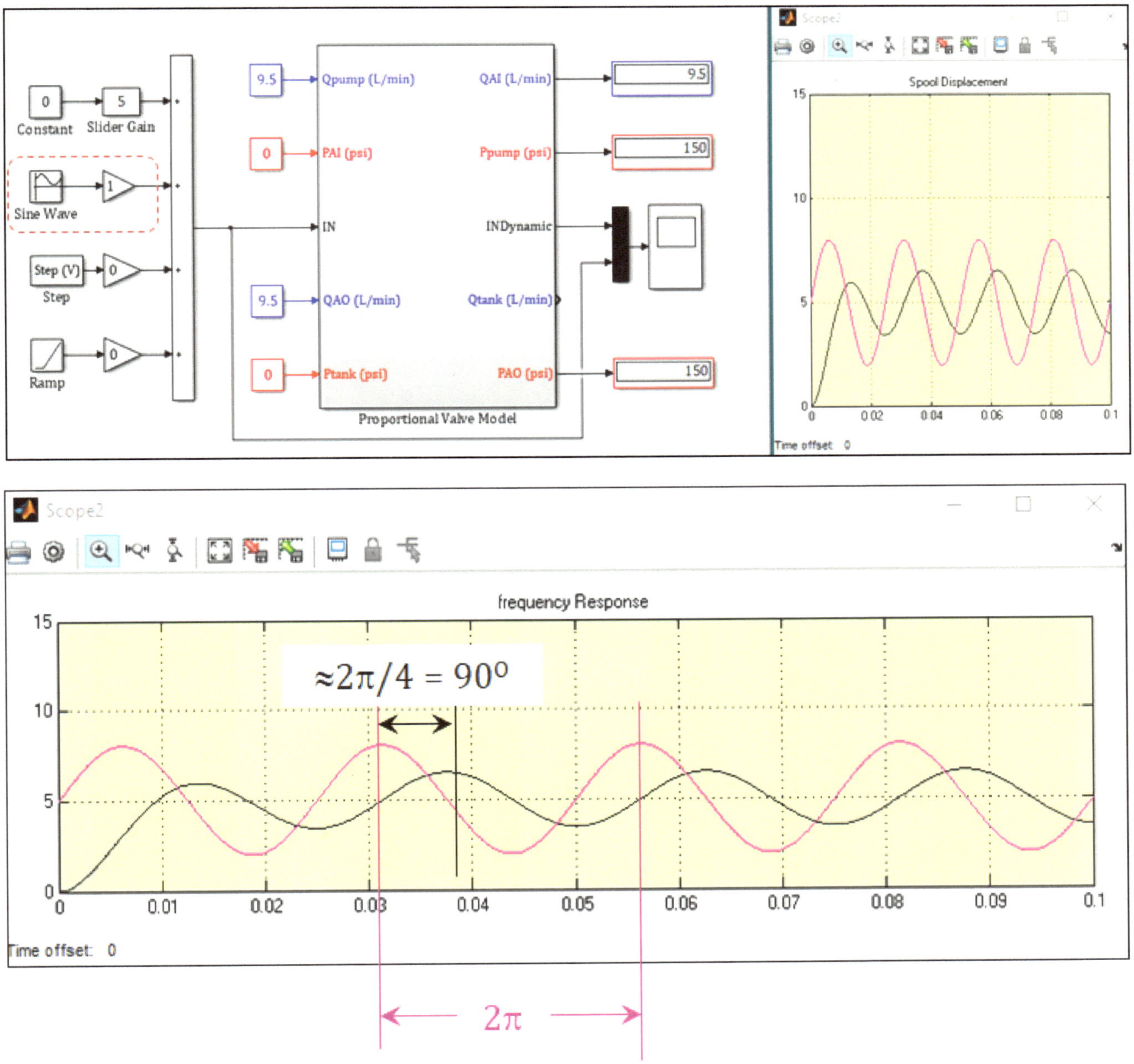

Fig. 10.27- Frequency Response Validation

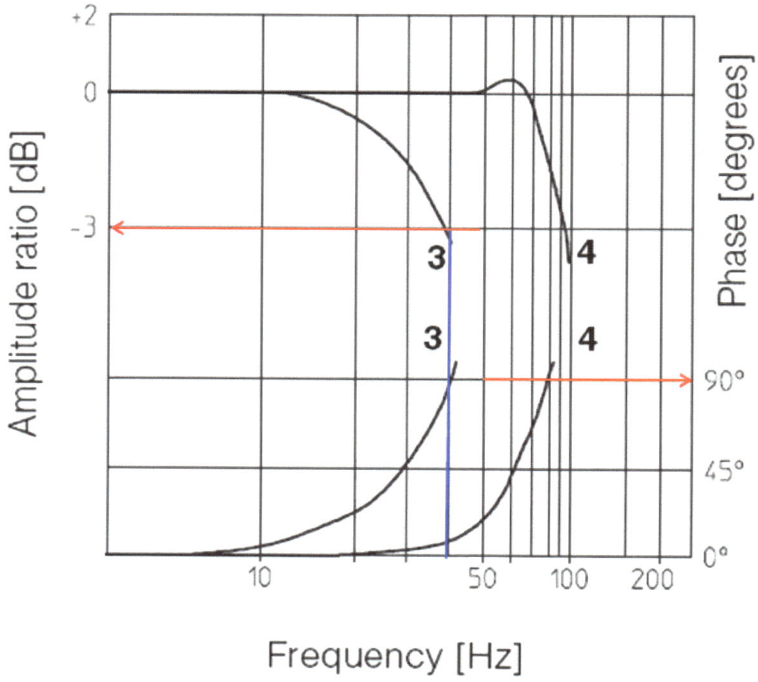

Fig. 10.28- Frequency Response Reported for Proportional Valve

Chapter 11
Modeling of Hydraulic Control System

Objectives

This chapter utilizes the lumped component models previously built to build system models. In this chapter, there is no additional math models to be developed. This chapter assembles system models from component models. After validating system models, they can be used as reference models for purposes of system design or investigating effects of operating conditions on system performance. This chapter presents models for electrohydraulic cylinder position control, electrohydraulic motor speed control, and hydraulic loading system.

Brief Contents

Chapter 11: Modeling of Hydraulic Control Systems

This is the compilation chapter for all the effort that has been accomplished in the previous chapters. The component lumped models are assembled to build a system model. In all previous chapters, Matlab-Simulink R2013b is used. In this chapter, Matlab-Simulink R2020a is used.

11.1-Modeling of Electro-Hydraulic Cylinder Position Control System

In this section, a simulation model for an EH *Cylinder Position Control* system is assembled from hydraulic component lumped models that were built previously. System step response and harmonic response are simulated and validated based on actual system response.

11.1.1- Simulation Model

Figure 11.1 shows the hydraulic circuit diagram for an electro-hydraulic cylinder position control system. Physical ports of the cylinder are assigned as 1 at piston side, and 2 at cylinder rod side. Effective areas are assigned as $\mathbf{A_1}$ is the full-bore area, and $\mathbf{A_2}$ is the annulus area at the rod side. Cylinder speed **v** and displacement **x** are positive when the cylinder extends, and negative when the cylinder retracts. When the cylinder extends, flow entering the piston side $\mathbf{Q_1}$ side and flow leaving the rod side $\mathbf{Q_2}$ are positive. When the cylinder retracts, flow entering the rod side $\mathbf{Q_2}$ side and flow leaving the piston side $\mathbf{Q_1}$ are negative. Pressure readings at physical points are assigned on the figure.

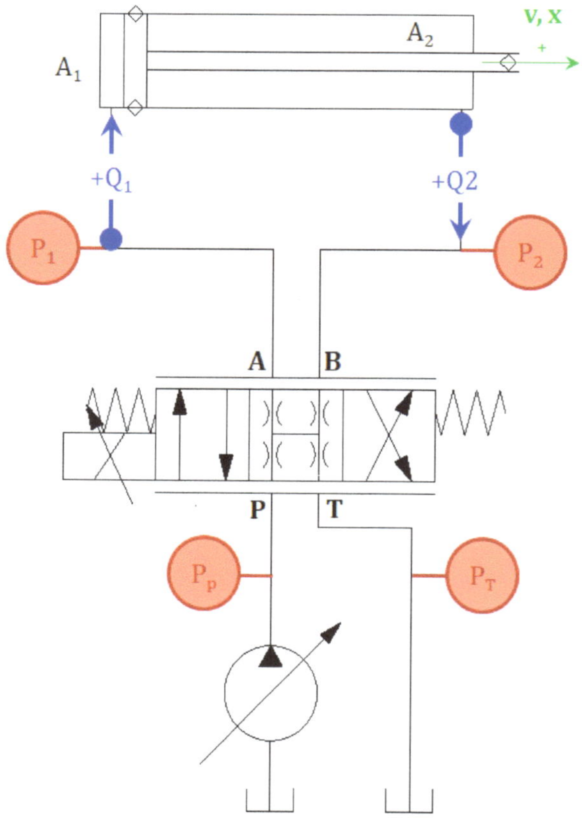

Fig. 11.1- Circuit Diagram for EH Cylinder Position Control System

Figure 11.2 shows the control diagram for the hydraulic system described above. The diagram shows a *Closed-Loop Negative-Feedback* control system. The system contains an Electric Control Unit (ECU) that is also referred to as an *Amplifier*. The system contains a signal generator that generates reference (required destination) cylinder position.

A signal calibration unit converts the reference position into an electronic *Reference Signal* **I_R**. The ECU compares **I_R** versus the *Feedback* signal **I_{FB}** that represents the actual cylinder position. The resulting *Error Signal* **I_E** is handled by the controller that generates a *Control Signal* **I_C**, based on which the control valve drives the cylinder towards the reference position. Cylinder position is measured by a sensor. The output of the sensor is calibrated and sends *Feedback Signal* to the controller. In this specific case calibration factors = 1.

Figure 11.3 shows the simulation model (HSV7CH11Model01R2020) that matches the hydraulic system and the control system described above. It is to be noted that, starting this model uses MatlabSimulink (R2020a). The system contains the model for a pressure compensated pump (1), a proportional valve (2), and a hydraulic cylinder (3).

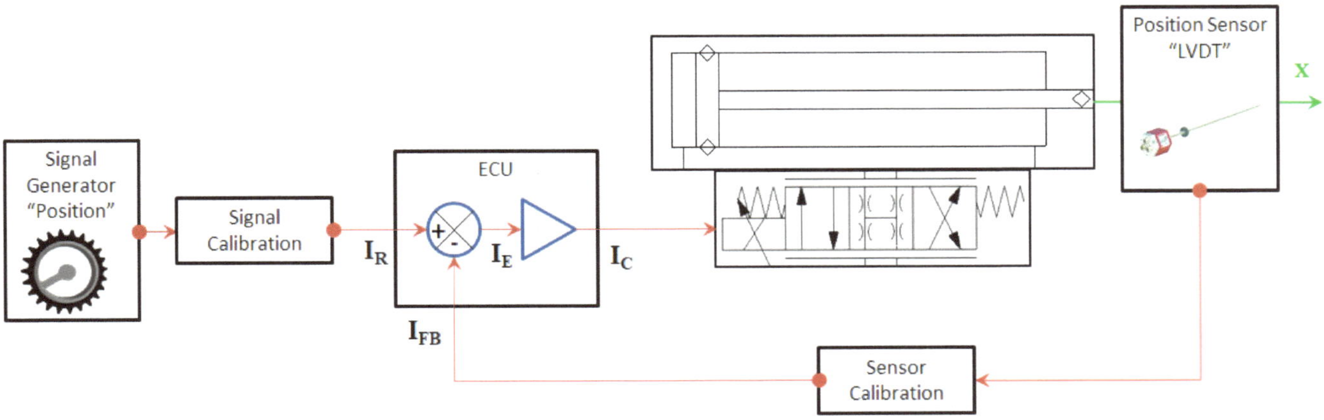

Fig. 11.2- Circuit Diagram for EH Cylinder Position Control System

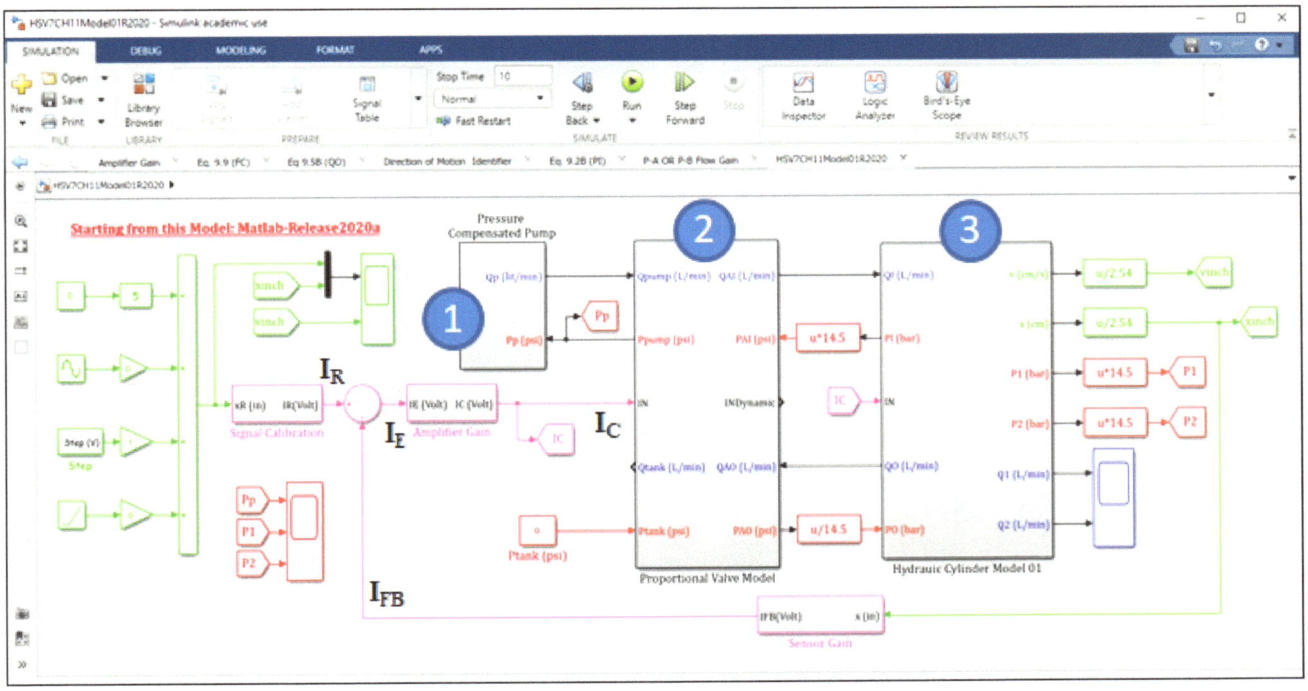

Fig. 11.3- Simulation Model for EH Cylinder Position Control System

11.1.2- System Performance Simulation

11.1.2.1- Step Response

Figure 11.4 shows step response of the system. The figure shows three sections from the left, dynamics, flow, and pressure. Here is the interpretation of the system step-response.

- **Time 0-2 s.:** Control system receives 0 input signal, no flow pass through the valve, no pressure at both cylinder sides, and pump is fully de-stroked at max pressure of 1000 psi.
- **Time 2-6 s.:** Control system receives input signal = 10 in (max stroke), the pump is stroked suddenly, the cylinder is driven and settles at the commanded destination in approximately 1 s. Because this is a type 1 system, PD controller is used. Hence, once the cylinder is settled at the commanded destination, control valve returns to zero position, the pump is then de-stroked again at maximum pressure.
- **Tim 6 s.:** Cylinder is commanded to return to half stroke. The valve reverses the flow, and accordingly the cylinder motion. When the cylinder returns to half stroke and stops there, the pump is de-stroked again at maximum pressure.
- **Notes:** Every time the pump is stroked, surge flow passes through the valve resulting in pressure spikes. Flow change pattern is similar to dynamic pattern because the flow is used to calculate the speed. Flow magnitude is affected by the area ratio.

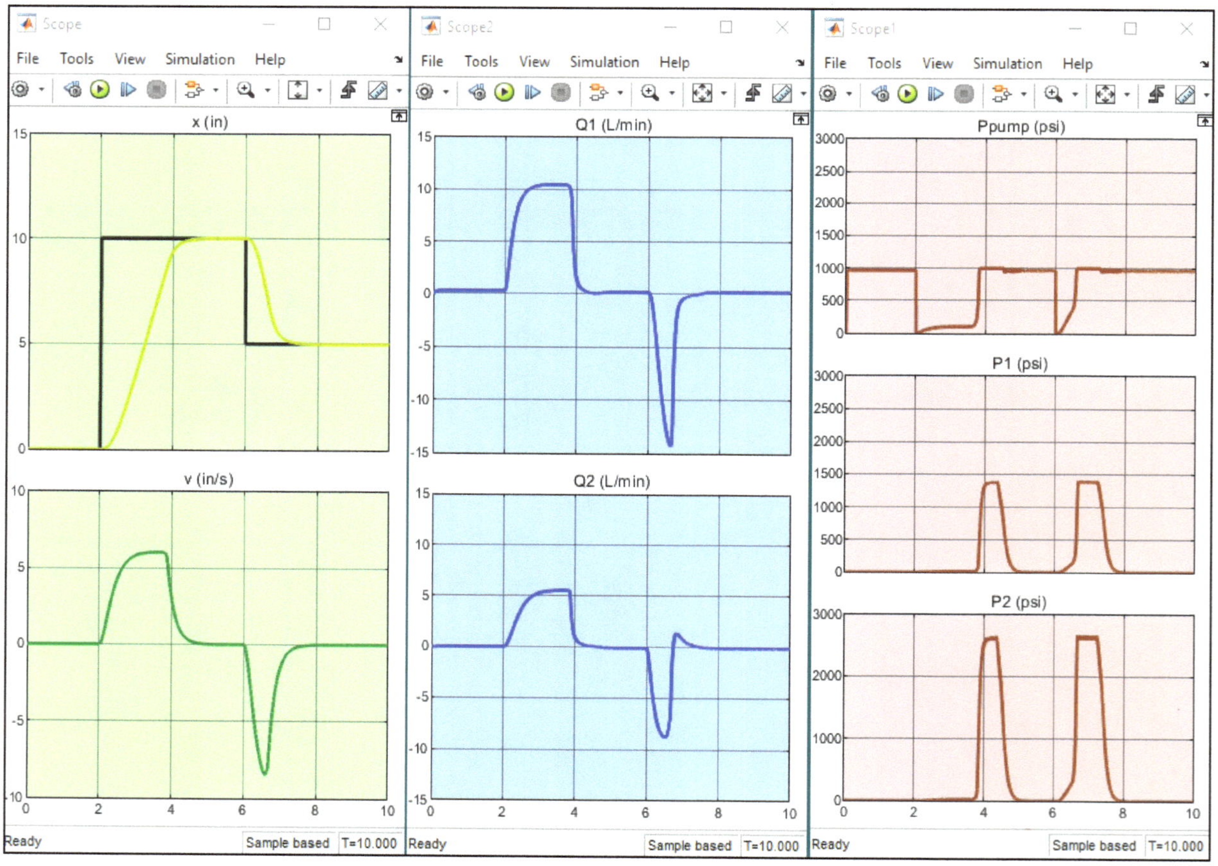

Fig. 11.4- Step Response of EH Cylinder Position Control System

11.1.2.2- Harmonic Response

The system is subjected to a sinusoidal signal and response is presented in Fig. 11.5. For an exciting signal that has a frequency of 0.5 Hz (3.14 rad/s), magnitude of 2.5 Volts, and a bias of 5 Volts, the system shows the presented phase lag and amplitude attenuation.

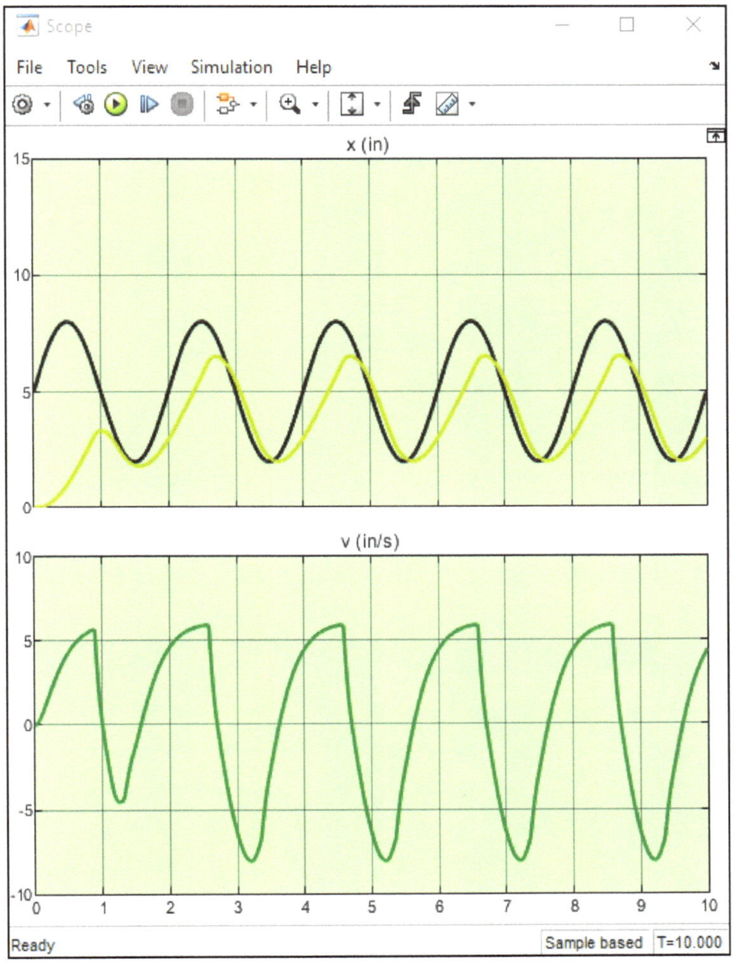

Fig. 11.5- Harmonic Response of EH Cylinder Position Control System

11.1.3- Model Validation

The system described above is built physically on the Universal Fluid Power Trainer. The system is subjected to step and harmonic input, and the actual system response is captured. Figure 11.6 and 11.7 show comparison between simulated and experimentally developed response. Agreement between results validate the model. Now the model can be used as a reference model for purposes of system design and for investigating the effect of operating conditions.

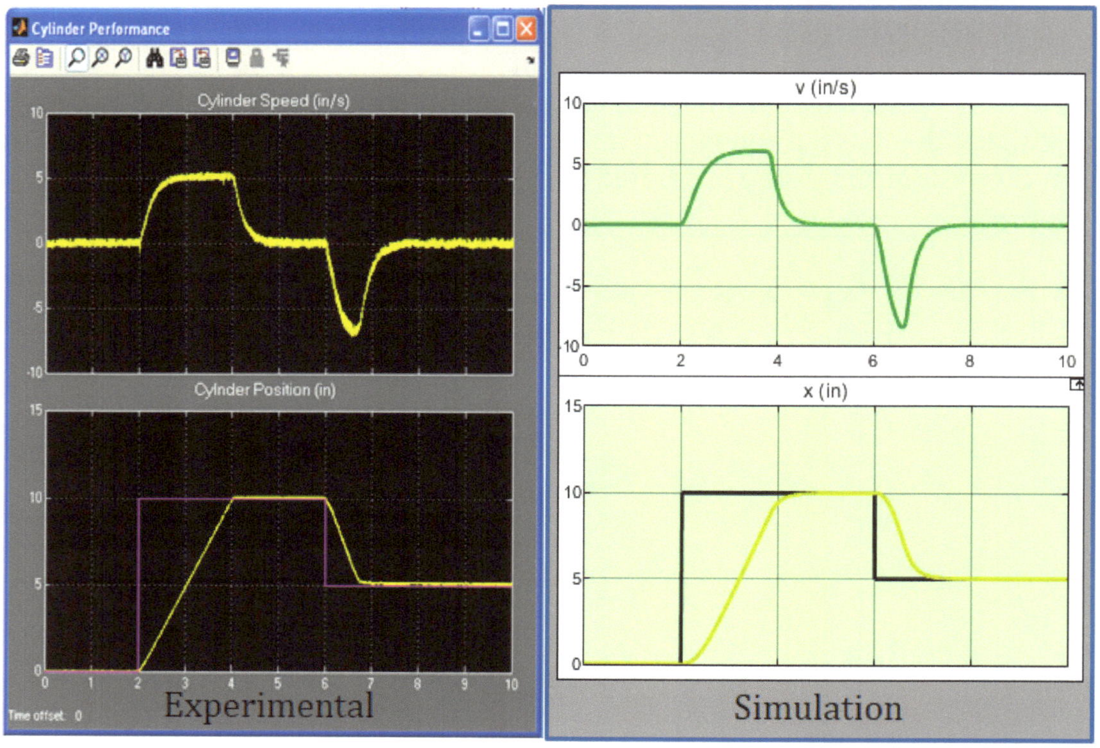

Fig. 11.6- Validation of Step Response of EH Cylinder Position Control System

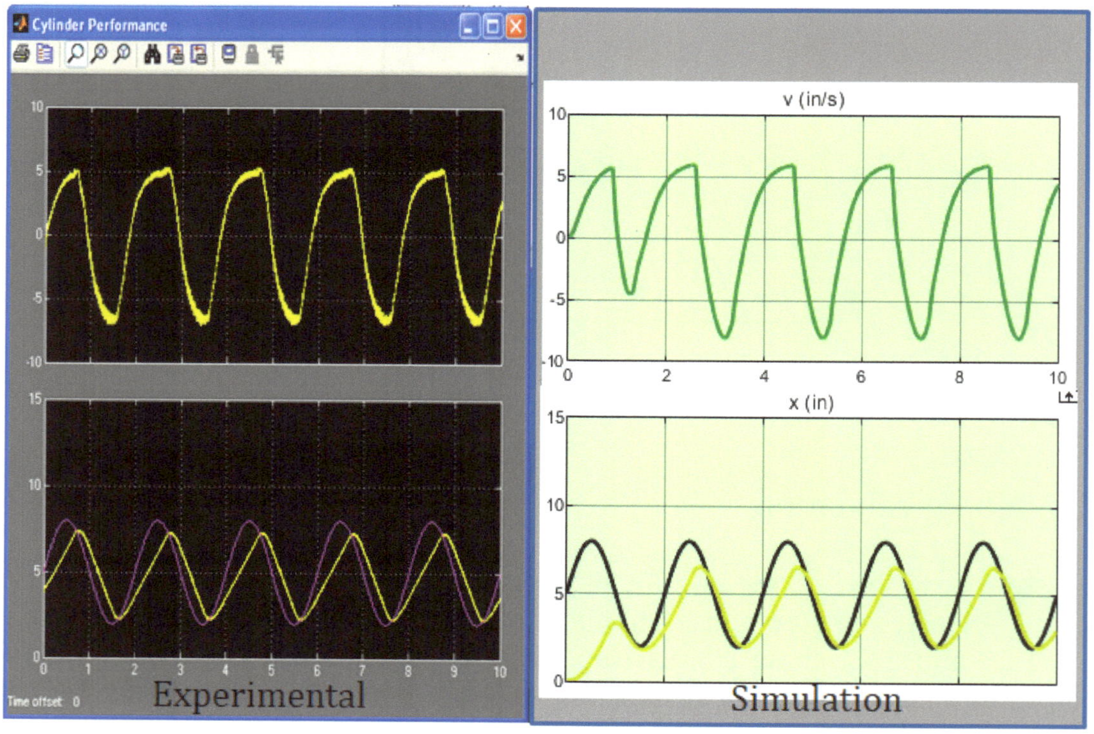

Fig. 11.7- Validation of Harmonic Response of EH Cylinder Position Control System

11.1.4- Effect of Cylinder Leakage

Figure 11.8 shows the effect of cylinder leakage on cylinder response. For a leakage factor of 2 [(lit/min)/bar] = 0.14 [(lit/min)/psi], the system response is disturbed as shown in the figure. Leakage effect will be more noticeable on frequency response.

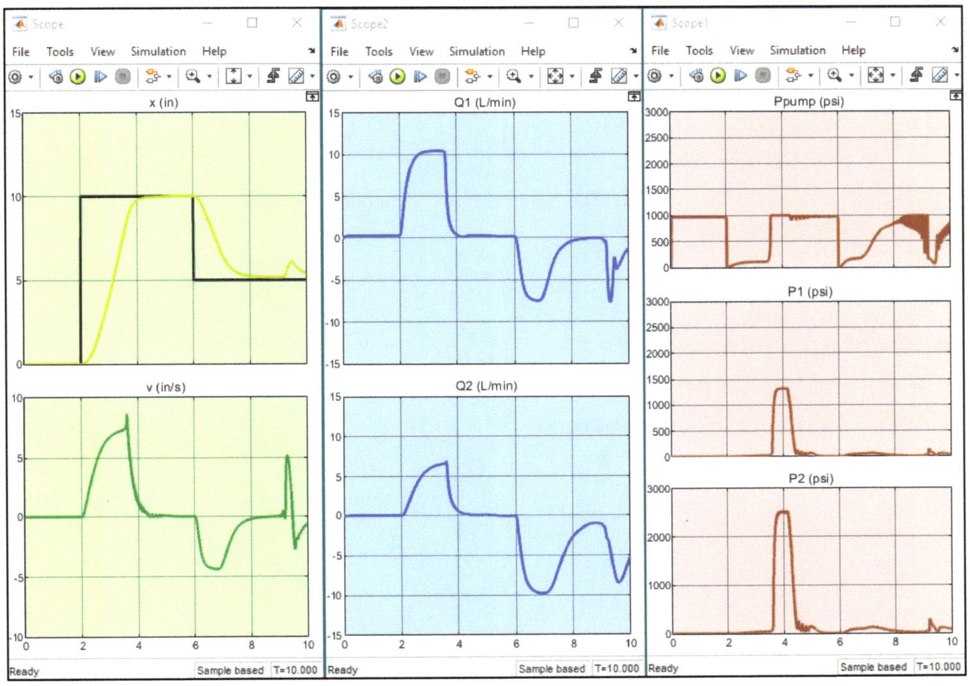

Fig. 11.8- Effect of Cylinder Leakage on Step Response of EH Cylinder Position Control System

11.1.5- Effect of Proportional Gain

Figure 11.9 shows the use of the model for proper selection of proportional gain. The figure shows that, by reducing the controller's proportional gain, the system become sluggish.

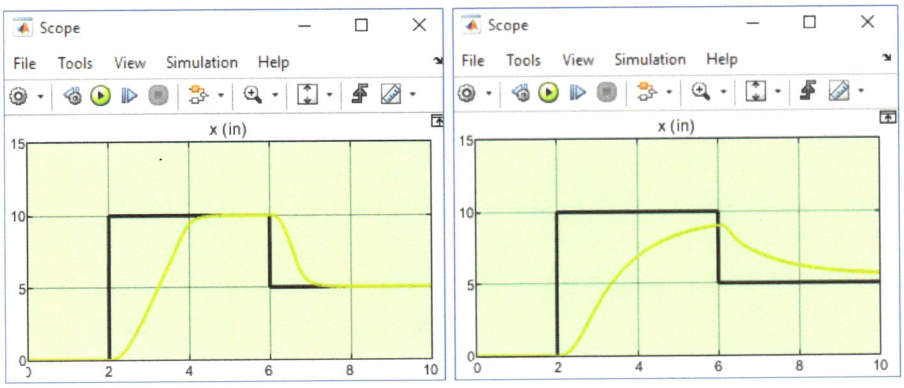

Fig. 11.9- Effect of Proportional Gain on Step Response of EH Cylinder Position Control System

11.2-Modeling of Energy Saving Loading System

In this section, a simulation model for an Energy Saving Loading System is assembled from hydraulic components lumped models that were built previously. The model is used to investigate energy saving as compared to classical system.

Figure 11.10 shows hydraulic circuit diagram for a loading system. A variable throttle valve is used to build back pressure enough to hold a load, such as a hydraulic brake system. The upper circuit uses a fixed displacement pump and a pressure relief valve. The lower circuit uses just a pressure compensated pump.

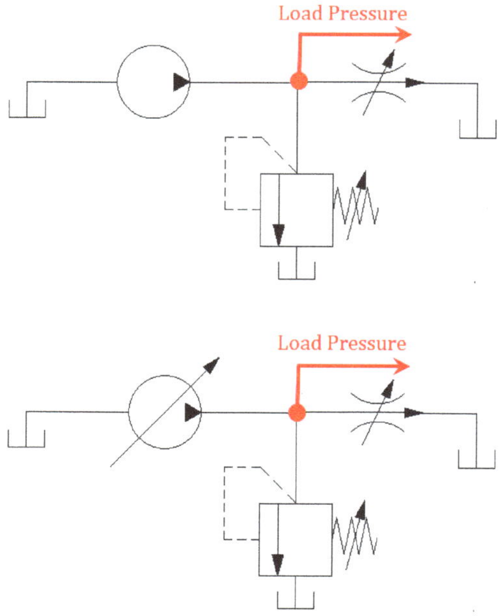

Fig. 11.10- Circuit Diagrams for Hydraulic Loading System

Figure 11.11 shows simulation model (HSV7CH11Model02R2020) for a hydraulic loading system using a fixed displacement pump (1), a PRV (2), and a variable throttle valve (3). PRV is adjusted with cracking pressure equal 100 bar and cut-off pressure equal 120 bar. System performance is simulated by ramping the % closure of the throttle valve. The figure shows that, maximum load pressure is achieved when the full pump flow is forced through the PRV. In this case, full driving power is converted into heat.

Figure 11.12 shows simulation model (HSV7CH11Model032020) for a hydraulic loading system using a pressure compensated pump instead of a fixed displacement one. A PRV is used just as a backup if the pump compensator fails. Therefore, while the cracking pressure of the pump compensator is adjusted at 100 bar, the PRV is adjusted at higher cracking pressure equal 130 bar. Simulation of this system shows that maximum load pressure is achieved with very small driving power. So, this system saves considerable amount of energy as compared to the previous one. Energy consumption in each case can be calculated by multiplying the driving power times the dwell time under maximum pressure. Energy saving is the difference between energy consumption in the two cases.

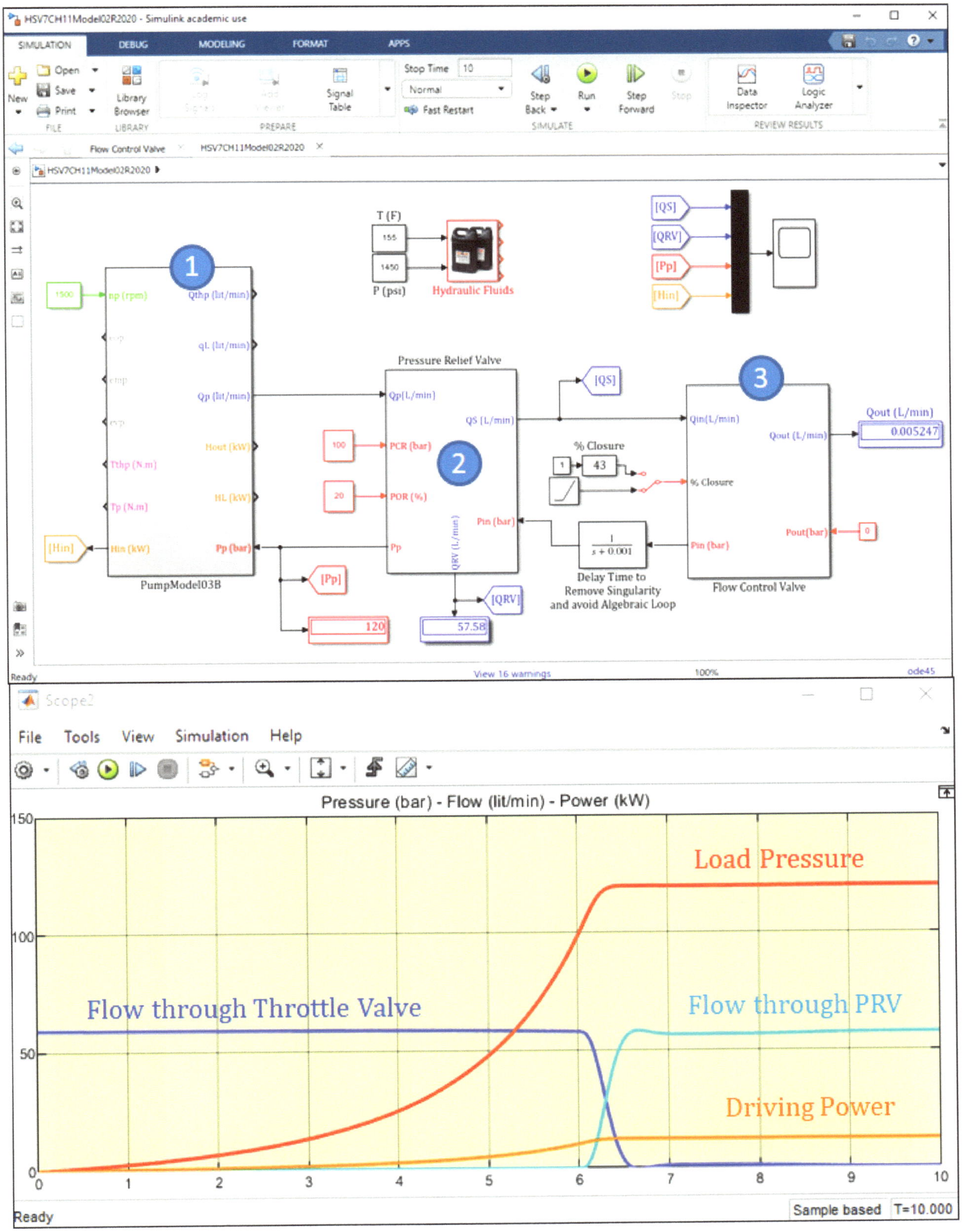

Fig. 11.11- Simulating Hydraulic Loading System using a Fixed Displacement Pump

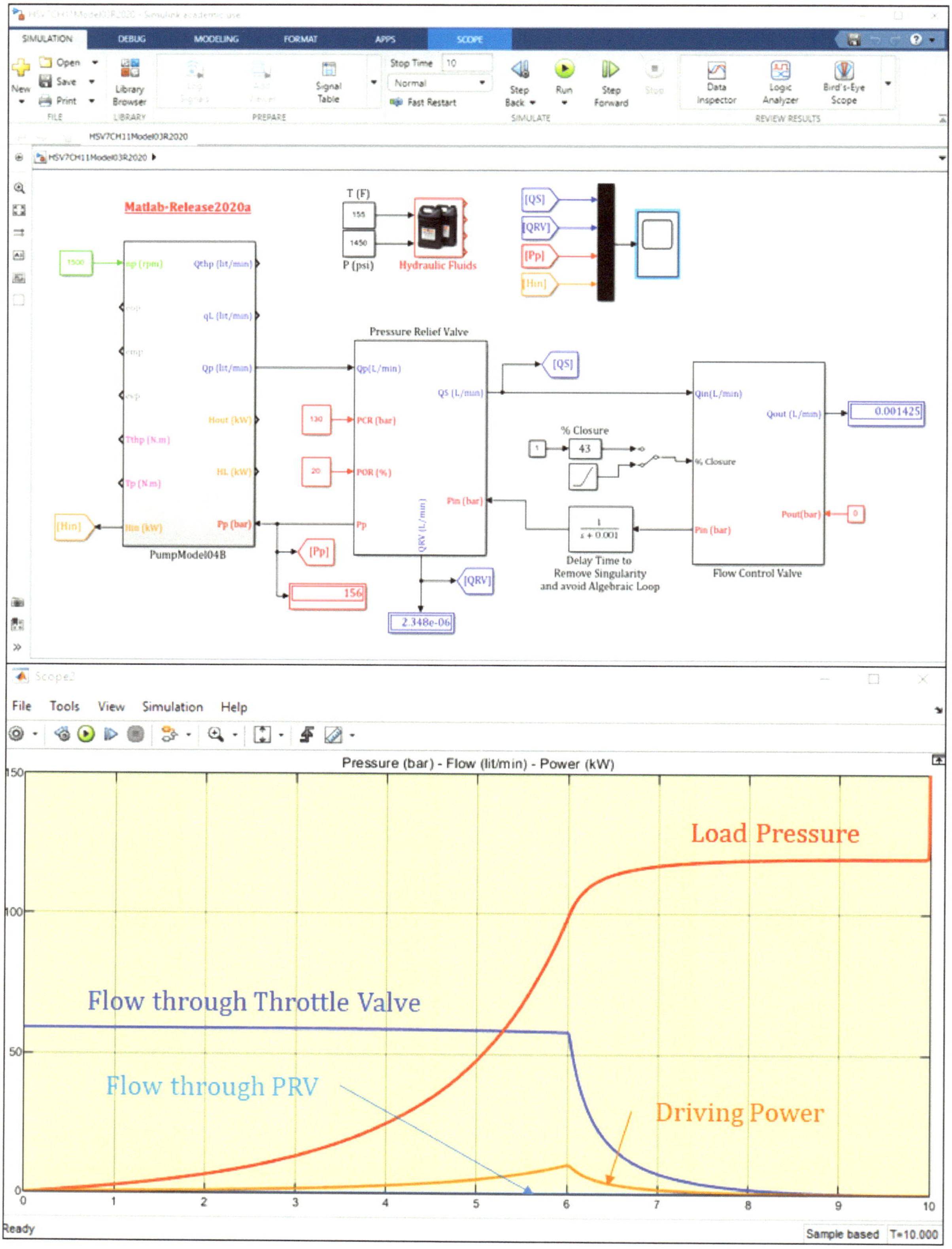

Fig. 11.12- Simulating Hydraulic Loading System using a Pressure Compensated Pump

11.3-Modeling of Electro-Hydraulic Motor Speed Control System

In this section, a simulation model for an EH *Motor Speed Control* system is assembled from hydraulic component lumped models that were built previously. System step response and harmonic response are simulated and validated based on actual system response.

11.3.1- Simulation Model

Figure 11.13 shows the hydraulic circuit diagram for an electro-hydraulic motor speed control system. Physical ports of the motor assigned as **1** on the left side, and **2** at the right side. Motor speed $\mathbf{n_m}$ is positive when the motor rotates CW, and negative when the motor rotates CCW. When the motor rotates CW, flow entering port **1** ($\mathbf{Q_1}$) and flow leaving port **2** ($\mathbf{Q_2}$) are positive. When the motor rotates CCW, flow entering port **2** ($\mathbf{Q_2}$) and flow leaving port **1** ($\mathbf{Q_1}$) are negative. Pressure readings at physical points are assigned on the figure.

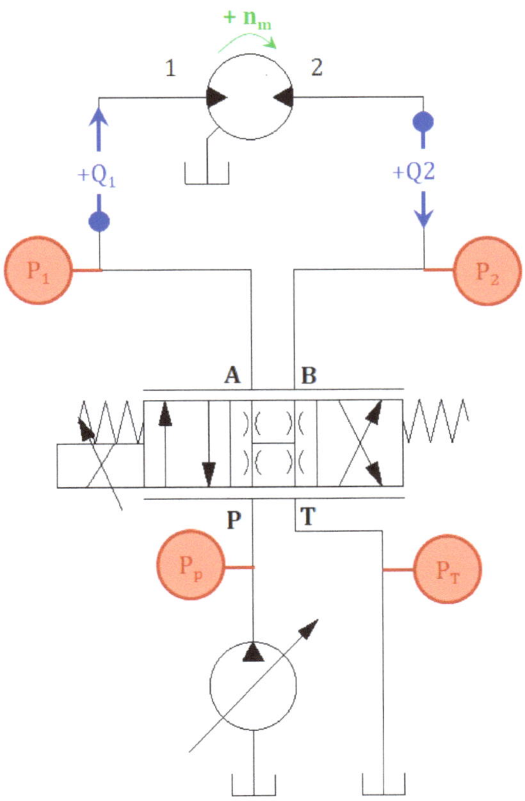

Fig. 11.13- Circuit Diagram for EH Motor Control System

Figure 11.14 shows the control diagram for the hydraulic system described above. The diagram shows a *Closed-Loop Negative-Feedback* control system. Structurally it is the same system used for cylinder position control, except that a servo valve is used with updated lookup tables. Calibration factors for the sensor and input signal are 10/765.

Figure 11.15 shows the simulation model (HSV7CH11Model04R2020) that matches the hydraulic system and the control system described above. The system contains a model for a pressure compensated pump (1), a servo valve (2), and a hydraulic motor (3).

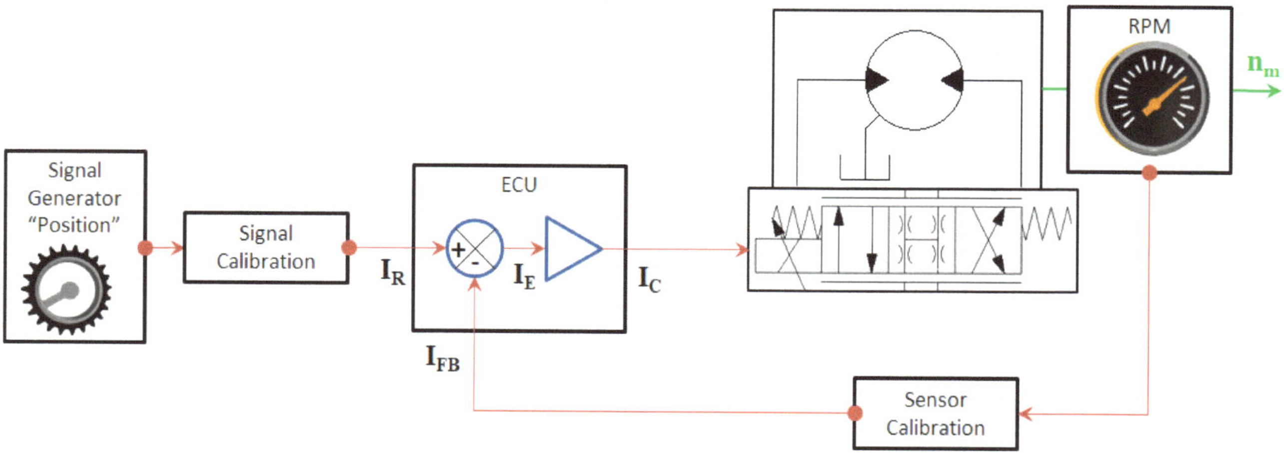

Fig. 11.14- Circuit Diagram for EH Motor Speed Control System

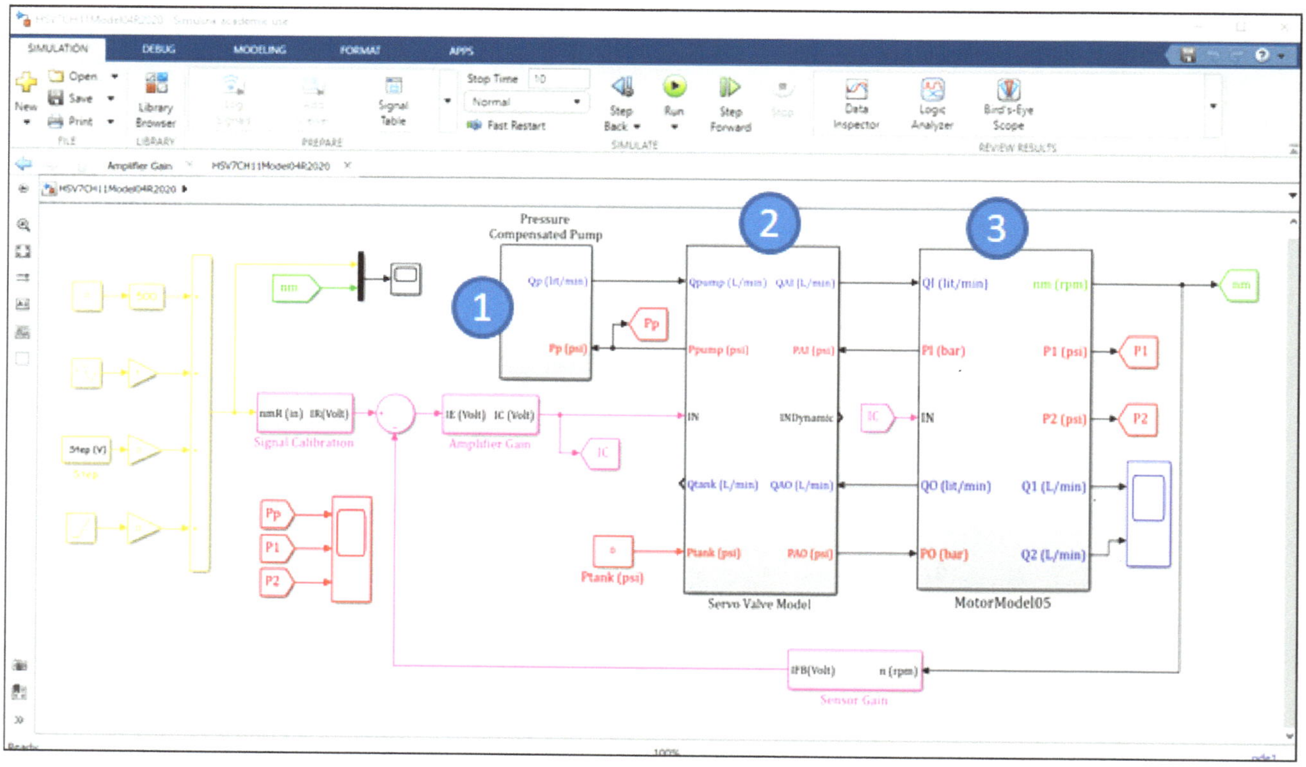

Fig. 11.15- Simulation Model for EH Motor Speed Control System

11.3.2- System Performance Simulation

11.3.2.1- Step Response

Figure 11.16 shows step response of the system. The figure shows three sections from the left, dynamics, flow, and pressure. Here is the interpretation of the system step-response.

- **Time 0-2 s.:** Control system receives 0 input signal, no flow pass through the valve, no pressure at the both motor ports, and pump is fully de-stroked at $\mathbf{P_{max}}$ = 1000 psi.
- **Time 2-6 s.:** Control system receives input signal = 500 rpm, the pump is stroked suddenly, the motor is driven and settles at the commanded rpm in approximately 1 s. Because this is a system type 0, PI controller is used. Hence, while the motor settles at the commanded rpm, control valve stays open, the pump discharges the required flow at load pressure. In this case no load condition is assumed, so $\mathbf{P_1}$ and $\mathbf{P_2}$ are due to flow paths in one leg of the valve and $\mathbf{P_p}$ is due to flow in two legs of the valve.
- **Time 6 s.:** Motor is commanded to reverse its motion at -500 rom. The valve reverses the flow, and accordingly the motor rotation.
- **Notes:** Every time the motor reverse its motion, control valve passes by its zero position, momentarily the pump is de-stroked/stroked resulting in pressure spikes.
- Flow change pattern is similar to the motor dynamic pattern, which makes sense.

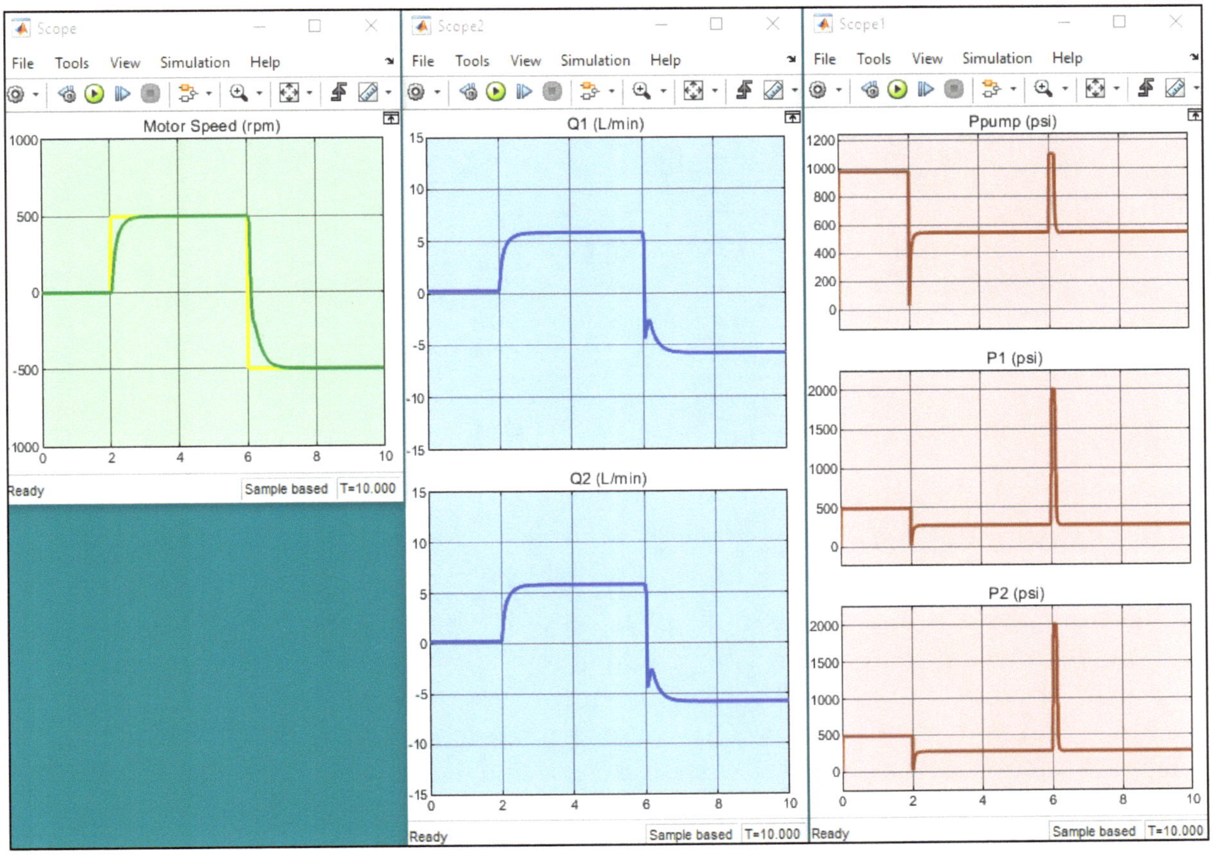

Fig. 11.16- Step Response of EH Motor Speed Control System

11.3.2.2- Harmonic Response

The system is subjected to a sinusoidal signal and response is presented in Fig. 11.17. For an exciting signal that has a frequency of 0.5 Hz (3.14 rad/s) and magnitude of 500 rom, the system shows the presented phase lag and amplitude attenuation.

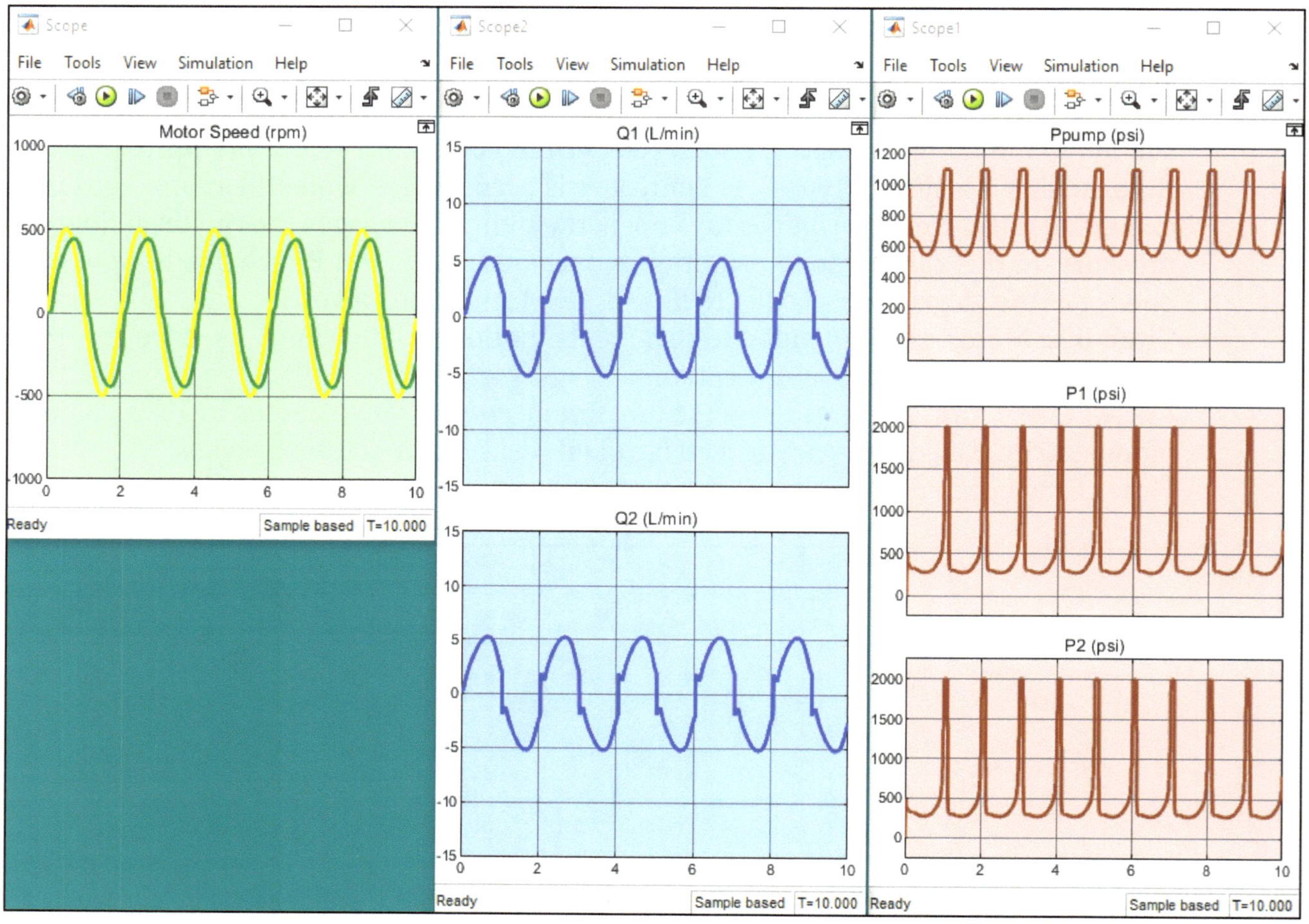

Fig. 11.17- Harmonic Response of EH Motor Speed Control System

11.3.3- Model Validation

The system described above is built physically on the Universal Fluid Power Trainer. The system is subjected to step and harmonic input, and the actual system response is captured. Figure 11.18 and 11.19 show comparison between simulated and experimentally developed response. Agreement between results validates the model. Now the model can be used as a reference model for purposes of system design and for investigating the effect of operating conditions.

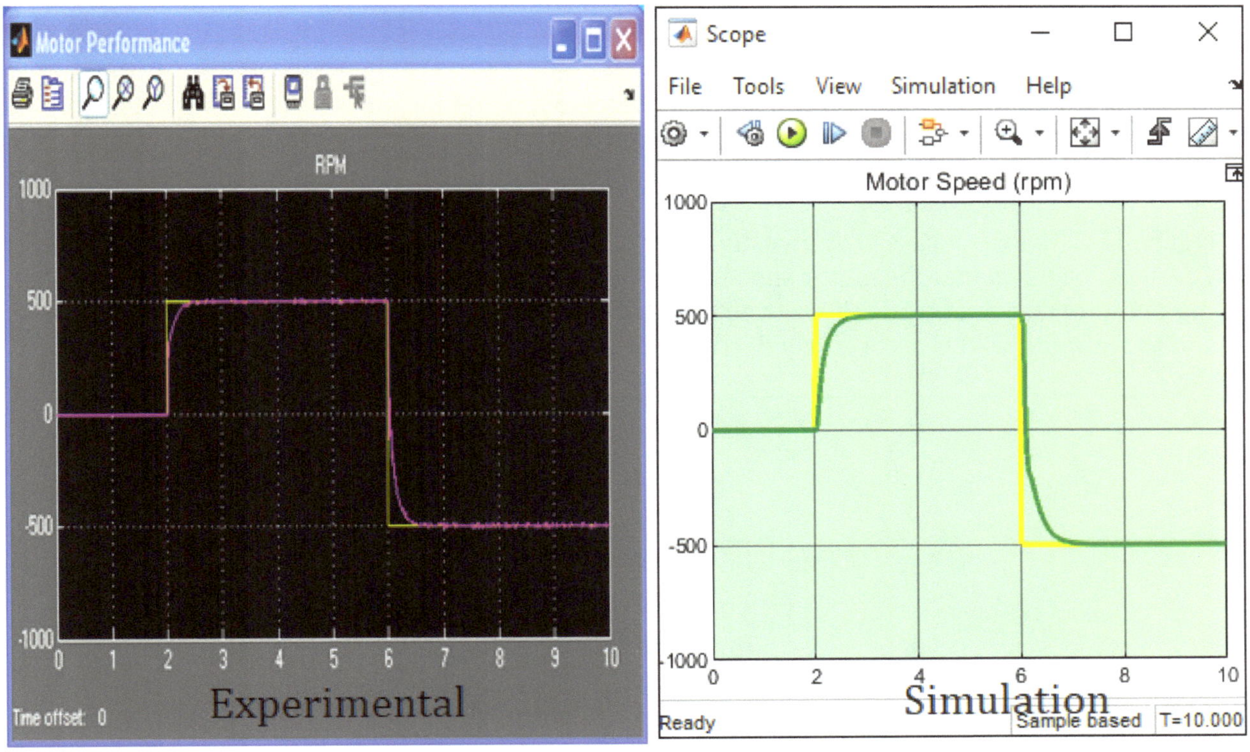

Fig. 11.18- Validation of Step Response of EH Motor Speed Control System

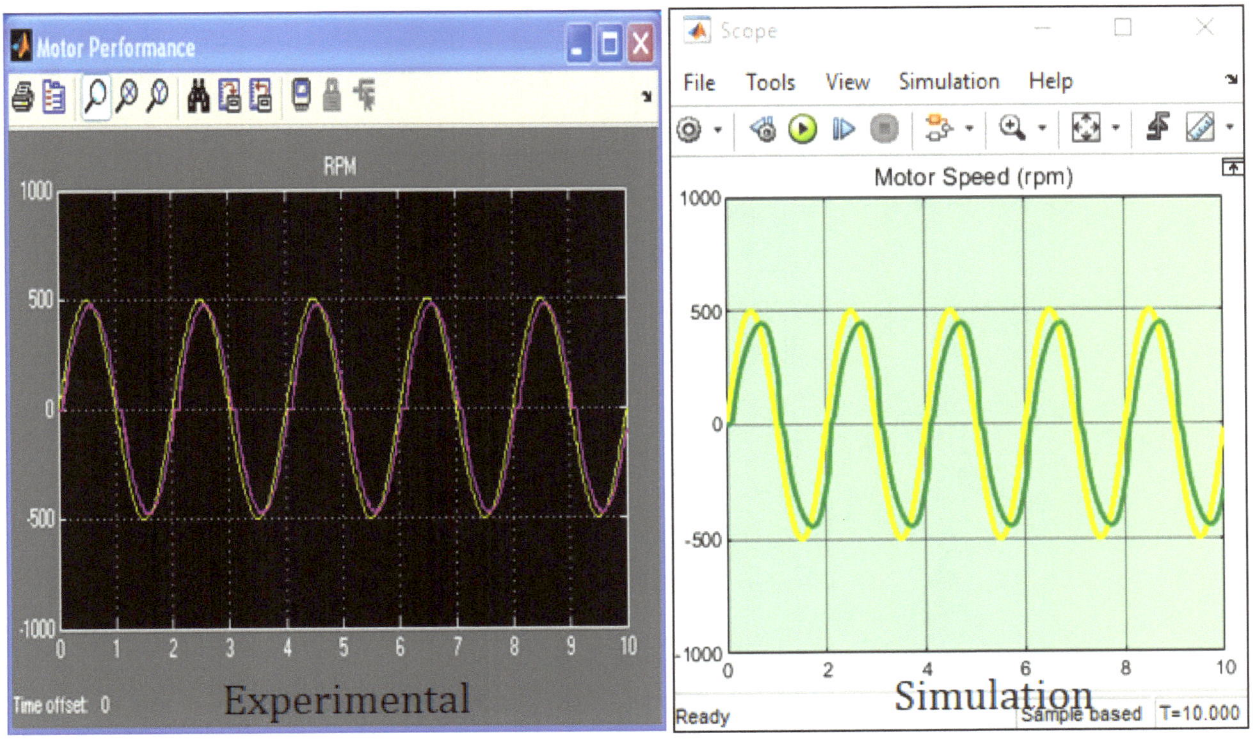

Fig. 11.19- Validation of Step Response of EH Motor Speed Control System

11.3.4- Using the Model for System Design

Figure 11.20 shows an example of using the validated model for purposes of system design. The figure shows a case when the model is used for proper controller design. The figure shows, from the left side, a higher integral gain is used, then it is lowered to improve the system performance.

Figure 11.21 shows a case that investigates system performance when the load is gradually increased. The system shows that the motor sustains the desired speed of 500 rpm for the first two seconds. When the pump pressure exceeds the setting value of the pump compensator, pump is gradually de-stroked and the motor speed is accordingly decreased.

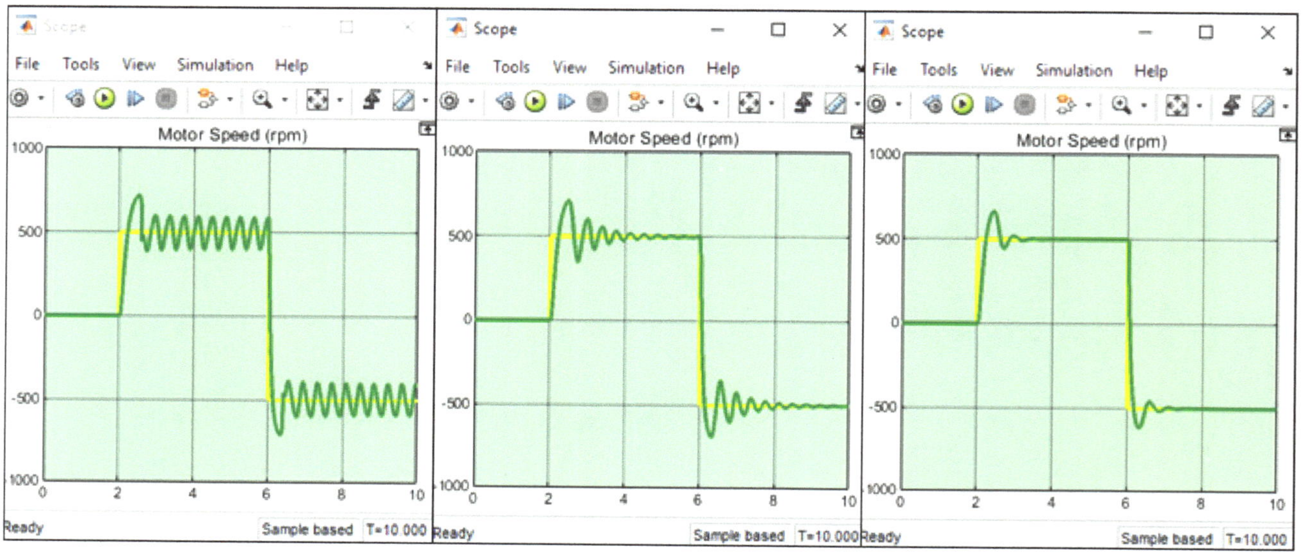

Fig. 11.20- Effect of Integral Gain

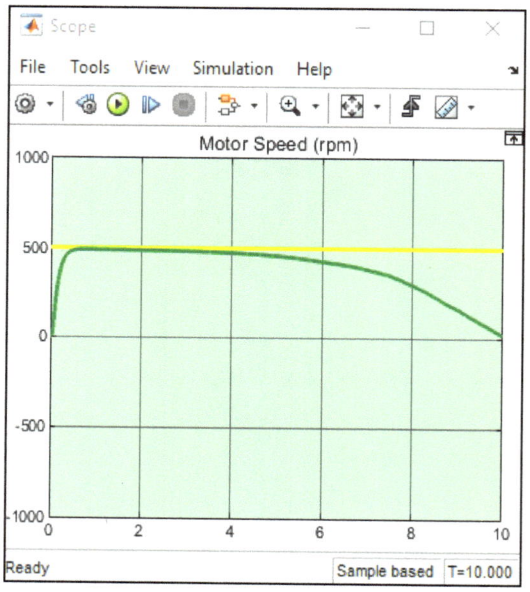

Fig. 11.21- Effect of Increasing Load

APPENDIXES

APPENDIX A: LIST OF FIGURES

Chapter 01: Introduction to Physical Systems Modeling and Simulation

Chapter 02: Modeling and Simulation of First-Order Dynamic Systems

Chapter 03: Modeling and Simulation of Second-Order Dynamic Systems

Chapter 04: Hydraulic Components Modeling Approaches

Chapter 05: Modeling of Fluid Properties

Chapter 06: Modeling of Fluid Conductors

Chapter 07: Modeling of Hydraulic Pumps

Chapter 08: Modeling of Hydraulic Motors

Chapter 09: Modeling of Hydraulic Cylinders

Chapter 10: Modeling of Hydraulic Valves

Chapter 11: Modeling of Hydraulic Control Systems

APPENDIX B: LIST OF TABLES

APPENDIX C: LIST OF MODELS AND SOFTWARE

Sign as a Member at (www.compudraulic.com) and you should be able to download sample models of the following list. Full library of the is purchased separately at (https://www.compudraulic.com/software)

These Matlab-Simulink Models are built with Release 2013b

- **HSV7CH02Model01:** First-Order System Normalized Transfer Function (NTF)
- **HSV7CH02Model02:** Simulation Model for 1st Order System Based on Differential Eq.
- **HSV7CH02Model03:** Simulation Model for 1st System Based on Transfer Function
- **HSV7CH02Model04:** Comparison of Simulation Models for 1st Order System
- **HSV7CH02Model05:** Step Response Analysis for 1st Order Systems
- **HSV7CH02Model06:** Effect of Time Constant on 1st Order System Step Response
- **HSV7CH02Model07:** Effect of Damping Coefficient on 1st Order System Step Response
- **HSV7CH02Model08:** Effect of Time Constant on 1st Order System Harmonic Response
- **HSV7CH02Model09:** Effect of Exciting Frequency on 1st Order System Harmonic System
- **HSV7CH03Model01:** Normalized Transfer Functions for 1st versus 2nd Order Systems
- **HSV7CH03Model02:** Simulation Model for 2nd Order System Based on Differential Eq.
- **HSV7CH03Model03:** Simulation Model for 2nd Order System Based on Transfer Function
- **HSV7CH03Model04:** Comparison of Simulation Models for 2nd Order System
- **HSV7CH03Model05:** Verification of a Step Response of a Second Order System
- **HSV7CH03Model06**: Effect of Design Parameters on Step Response of a 2nd Order System

- **HSV7CH03Model07:** Critically Damped 2nd Order System versus 1st Order System
- **HSV7CH03Model08:** Response of 2nd Order System Based on NTF
- **HSV7CH03Model09:** Response of 2nd Order System Based on Transfer Function
- **HSV7CH03Model10:** Response of 2nd Order System Based on Differential Equation
- **HSV7CH03Model11:** Simulated Step Response of a Hydraulic Actuator based on NTF
- **HSV7CH03Model12:** Simulated Step Response of a Hydraulic Act. based on TF
- **HSV7CH03Model13:** Simulated Step Response of a Hydraulic Actuator based on DE
- **HSV7CH05Model01:** Modeling Hydraulic Fluid Bulk Modulus
- **HSV7CH05Model02:** Modeling Hydraulic Fluid Density as Function of Temp. and Pressure
- **HSV7CH05Model03:** Modeling Specific Gravity as Function of Temp. and Pressure
- **HSV7CH05Model04:** Modeling Fluid Viscosity as Function of Temperature and Pressure
- **HSV7CH05Model05:** Lumped Model for Hydraulic Fluid Properties
- **HSV7CH06Model01:** Simulation Model for Hydraulic Transmission Line
- **HSV7CH06Model02:** Simulation Model for Hydraulic Fitting
- **HSV7CH06Model03:** Simulation Model for Hydraulic Orifice
- **HSV7CH06Model04:** Example of Modeling a Transmission Line Assembly
- **HSV7CH07Model01:** Model for Ideal Fixed Displacement Unidirectional Pump
- **HSV7CH07Model02A:** Model for a Fixed Displacement Pump Running at Constant Operating Conditions Based on Given Test Values
- **HSV7CH07Model02B:** Model for a Fixed Displacement Pump Running at Constant Operating Conditions Based on Given Efficiency Values
- **HSV7CH07Model03A:** Model for a Fixed Displacement Pump Running at Variable Operating Conditions Based on Given Test Data
- **HSV7CH07Model03B:** Model for a Fixed Displacement Pump Running at Variable Operating Conditions Based on Given Efficiency Curves
- **HSV7CH07Model04A:** Model for Pressure-Compensated Pumps Based on Given Test Data
- **HSV7CH07Model04B:** Model for Pressure-Comp. Pumps Based on Given Efficiency Curves
- **HSV7CH07Model04C:** Simplified Model for Pressure-Compensated Pumps
- **HSV7CH07Model05A:** Simplified Model for Displacement-Controlled Pumps
- **HSV7CH07Model05B**: Model for Disp.-Controlled Pumps Based on Given Efficiency Curves

- **HSV7CH07Model06A:** Simplified Model for Torque-Limited Pumps
- **HSV7CH07Model06B:** Model for Torque-Limited Pumps Based on Given Efficiency Curves
- **HSV7CH08Model01:** Model for an Ideal Fixed Displacement Motor
- **HSV7CH08Model02A:** Model for a Fixed Displacement Motor Running at Constant Operating Conditions Based on Given Test Data
- **HSV7CH08Model02B:** Model for a Fixed Displacement Motor Running at Constant Operating Conditions Based on Given Efficiency Values
- **HSV7CH08Model03A:** Model for a Fixed Displacement Motor Running at Variable Operating Conditions Based on Given Test Data
- **HSV7CH08Model03B:** Model for a Fixed Displacement Motor Running at Variable Operating Conditions Based on Given Efficiency Curves
- **HSV7CH08Model04:** Model for Variable Displacement Motors
- **HSV7CH08Model05:** Simplified Model for Valve-Controlled Fixed-Displacement Bidirectional Hydraulic Motor
- **HSV7CH09Model01:** Model for Hydraulic Cylinder
- **HSV7CH10Model01:** Model for Pressure Relief Vale Based on Linear Characteristics
- **HSV7CH10Model02:** Model for Pressure Relief Vale Based on Nonlinear Characteristics
- **HSV7CH10Model03:** Model for Flow Control Valves
- **HSV7CH10Model04:** Model for Check Valves
- **HSV7CH10Model05:** Validating the Models for Check Valve and PRV together
- **HSV7CH10Model06:** Model for Continuous Directional Control Valves

These Matlab-Simulink Models are built with Release 2020a

- **HSV7CH11Model01R2020:** Model for EH Cylinder Position Control System
- **HSV7CH11Model02R2020:** Hydraulic Loading System using a Fixed Disp. Pump
- **HSV7CH11Model03R2020:** Hyd. Loading System using a Pressure Compensated Pump
- **HSV7CH11Model04R2020:** Modeling Electro-Hydraulic Motor Speed Control System

APPENDIX D: LIST OF REFERENCES

Hydraulic Systems Volume 1- Introduction to Hydraulics for Industry Professionals
Author: Dr. Medhat Kamel Bahr Khalil, 2016.
Publisher: Compudraulic, USA.
ISBN 978-0-692-62236-0

Hydraulic Systems Volume 2- Electro-Hydraulic Components and Systems
Author: Dr. Medhat Kamel Bahr Khalil, 2016.
Publisher: Compudraulic, USA.
ISBN: 978-0-9977634-2-3

Hydraulic Systems Volume 3- Hydraulic Fluids and Contamination Control
Author: Dr. Medhat Kamel Bahr Khalil, 2016.
Publisher: Compudraulic, USA.
ISBN: 978-0-9977816-3-2

R01- Basic Electronics for Hydraulic Motion Control
Author: Jack L. Johnson, PE 1992.
Publisher: Penton Publishing Inc. 1100 Superior Avenue. Cleveland, OH 44114.
ISBN No. 0-932905-07-2.

R02- Closed Loop Electro-hydraulics Systems Manual
Author: Vickers/Eaton.
Publisher: Vickers Inc. 1992.
Training Center, 2730 Research Drive, Rochester Hills, MI 48309-3570.
ISBN 0-9634162-1-9

R03- Bosch Automation Technology
Author: Werner Gotz, Steffen Haack, Ralph Mertlick.
Publisher: Bosch.
ISBN 3-933698-05-7.

R04- Electrohydraulic Proportional and Control Systems
Publisher: Bosch Automation 1999.
ISBN 0-7680-0538-8.

R05- Proportional and Servo Valve Technology – The Hydraulic Trainer Volume 2
Author: R. Edwards, J. Hunter, D. Kretz, F. Liedhegener, W. Schenkel, A. Schmitt.
Publisher: Mannesman Rexroth AG 1988. D-8770 Lohr a. Main.
ISBN 3-8023-0266-4.

R06- Proportional Hydraulics
Author: D. Scholz.
Publisher: Festo Didactic KG, Esslingen, Germany.

R07- Electricity, Fluid Power, and Mechanical Systems for Industrial Maintenance
Author: Thomas Kissell.
Publisher: Prentice Hall, Inc. 1999, Upper Saddle River, NJ 07458.
ISBN 0-13-896473-4.

R08- Fluid Power in Plant and Field – First Edition
Author: Charles S. Hedges, R.C. Womack.
Publisher: Womack Machine Supply Co. 1968.
Womack Educational Publication, 2010 Shea Road, Dallas, TX 75235.
ISBN 68-22573 (Library of Congress Card Catalog No.).

R09- Hydraulics, Fundamentals of Service
Author: Deere and Company.
Publisher: John Deere Publishing 1999.
Almon TIAC Bldg. Suite 104, 1300-19th Street, East Moline, IL 61244.
ISBN 0-86691-265-7.

R10- Industrial Hydraulics Troubleshooting
Author: James E. Anders, Sr.
Publisher: McGraw-Hill, Inc.
ISBN 0-07-001592-9.

R11- Power Hydraulics
Author: John Ashby.
Publisher: Prentice Hall 1989. Prentice Hall International, (UK) Ltd.
66 Wood Lane End, Hemel Hempstead, Hertfordshire, HP2 4RG.
ISBN 0-13-687443-6.

R12- Fluid Power with Application
Author: Anthony Esposito.
Publisher: Prentice Hall.
ISBN 0-13-060899-8.

R13- Hydraulic Component Design and Selection
Author: E.C. Fitch.
Publisher: BarDyne Inc. 5111 North Perkins Rd. Stillwater, OK 74075.
ISBN 0-9705922-3-X.

R14- Planning and Design of Hydraulic Power Systems – The Hydraulic Trainer, Vol. 3
Author: Mannesmann Rexroth GmbH.
Publisher: Mannesman Rexroth AG 1988.
D-97813 Lhr a. Main, Jahnsrtrabe 3-5 D-97816 Lohr a. Main.
ISBN 3-8023-0266-4.

R15- Logic Element Technology: Hydraulic Trainer, Volume 4
Author: Mannesmann Rexroth GmbH.
Publisher: Mannesmann Rexroth GmbH 1989.
.Postfach 340, D 8770 Lohr am Main, Telefon (09352) 180.
ISBN 3-8023-0291-5.

R16- Hydrostatic Drives with Control of the Secondary Unit. The Hydraulic Trainer, Volume 6
Author: Dr. Alfred Feuser, Rolf Kordak, Gerold Liebler.
Publisher: Mannesmann Rexroth GmbH 1989.
Postfach 340, D 8770 Lohr am Main.

R17- Control Strategies for Dynamic Systems: Design and Implementation
Author: John H. Lumkes, Jr.
Publisher: Marcel Dekker, Inc. 2002.
Marcel Dekker, Inc. 270 Madison Avenue, New York, NY 10016.
ISBN 0-8247-0661-7.

R18- Feedback Control Of Dynamic Systems
Author: Gene F. Franklin, J. David Powell, Abbas Emami-Naeini.
Publisher: Prentice-Hall, Inc.
Upper Saddle River, New Jersey.
ISBN 0-13-032393-4.

R19- Modeling and Analysis of Dynamic Systems
Author: Charles M. Close, Dean. Frederick
Rensselaer Polytechnic Institute
Publisher: John Wiley & Sons, Inc.
ISBN 0-471-12517-2.

R20- Design of Electrohydraulic Systems For Industrial Motion Control
Author: Jack L. Johnson, PE.
Milwaukee School of Engineering.
Publisher: Parker.
Copyright © Jack L. Johnson, PE 1991.

R21- Basic Pneumatics
Author: Kjell Evensen & Jul Ruud.
Publisher: AB Mecmann Stockholm 1991.
S-125 81 Stockholm, Sweden.
ISBN 91-85800*21-X.

R22- Basic Pneumatics: The Pneumatic Trainer, Volume 1
Author: Ing. –Buro J.P. Hasebrink.
D7761 Moos.
Editor: Mannesmann Rexroth Pneumatik GmbH.
Bartweg 13, W 3000 Hannover 91.

R23- Electro-Pneumatics: The Pneumatic Trainer, Volume 2
Author: Rolf Balla.
Publisher: Mannesmann Rexroth 1990, Pneumatik GmbH.
Publication No: RE 00 262/01.92.

R24- Pneumatics Theory and Applications
Author: Bosch Automation.
Publisher: Robert Bosch GmbH 1998.
Automation Technology Division, Training (AT/VSZ)
ISBN 1-85226-135-8.

R25- Fluid Power Engineering
Author: M. Galal Rabie.
Publisher: McGraw-Hill.
ISBN 978-0-07-162246-2.

R26- Air Motors Ideas with Air
Author: GAST Mfg. Co.
Publisher: GAST Mfg. Co. 1978.
P.O. Box 97, Benton Harbor, MI 49022.
Book No: Booklet #100.

R27- Air Motor Handbook
Author: GAST Mfg. Co.
Publisher: GAST Mfg. Co. 1978.
P.O. Box 117, Benton Harbor, MI 49022.

R28- Troubleshooting Hydraulic Components: Using Leakage Path Analysis Methods
Author: Rory S. McLaren.
Publisher: Rory McLaren Fluid Power Training 1993.
562 East 7200 South, Salt Lake City, UT 84171.
ISBN No. 0-9639619-1-8.

R29- Hydraulics Theory and Application From Bosch
Author: Werner Gotz.
Publisher: Robert Bosch GmbH.
Hydraulics Division K6, Postfach 30 02 40, D-7000 Stuttgart 30.
Federal Republic of Germany, Technical Publications Department, K6/VKD2.

R30- A Complete Guide to ISO and ANSI Fluid Power Symbols
Author: Fluid Power Training Institute.
Publisher: Fluid Power Training Institute 200.
562 East Fort Union Boulevard, Midvale, Utah 84047.
R31- How to Work Safely with Hydraulics
Author: Fluid Power Training Institute.
Publisher: Fluid Power Training Institute 2004.
562 East7200 South, Midvale, Utah 84047.

R32- How to Interpret Fluid Power Symbols
Author: Rory S. McLaren.
Publisher: Fluid Power Training Institute.
Rory S. McLaren 1995.
ISBN 0-9639619-2-6.

R33- Safe Hydraulics
Editor: Gates Rubber Company.
Copyright 1995.
Denver, CO 80217.

R34- Electronically Controlled Proportional Valves. Selection and Application
Author: Michael J. Tonyan.
Publisher: Marcel Dekker, Inc. 1985.
Marcel Dekker, Inc., 270 Madison Avenue, New York, NY 10016.
ISBN 0-8247-7431-0.

R35- Introduction to Closed-Loop Oil Systems
Author: Rory S. McLaren.
Publisher: Rory McLaren Fluid Power Training Institute.
7050 Cherry Tree Lane, P.O. Box 711201, Salt Lake City, UT 84171.

R36- Industrial Hydraulic Technology, Second Edition
Author: Parker Hannifin Corporation.
Publisher: Parker Hannifin Corporation 1997.
6035 Parkland Blvd, Cleveland, OH 44124-4141.
Publication No: Bulletin 0231-B1.

R37- Basic Principle and Components of Fluid Technology – The Hydraulic Trainer, Volume 1
Author: Mannesman Rexroth.
Publisher: Mannesman Rexroth AG 1988.
D-97813 Lhr a. Main, Jahnsrtrabe 3-5 D-97816 Lohr a. Main.
ISBN 3-8023-0266-4.

R38- Safe-T-Bleed Corporation Catalog
Publisher: Safe-T-Bleed Corporation 2001.
Catalog No. STB-PC-1201-1

R39- Industrial Hydraulics Manual – EATON
Publisher: Eaton Fluid Power Training.
ISBN: 0-9788022-0-9.

R40- Vickers-Mobile Hydraulic Manual – Fourth Edition 1998
Author: Vickers.
Publisher: Vickers Inc. 1999.
Training Center, 2730 Research Drive, Rochester Hills, MI 48309-3570.
ISBN No. 0-9634162-5-1.

R41- Industrial Fluid Power Text, Volume 2
Author: Charles S. Hedges, R.C. Womack.
Publisher: Womack Machine Supply Company 1972.
Womack Educational Publications, 2010 Shea Road, Dallas, TX 75235.
ISBN 66-28254 (Library of Congress Card Catalog No.).

R42- Fluid Power Hydraulics and Pneumatics
Author: R. Daines.
Publisher: The Good-heart Willcox Company, Inc.

R43- Hydraulics in Industrial and Mobile Applications
Publisher: ASSOFLUID, Italian Association of Manufacturing and Trading Companies in Fluid Power Equipment and Components

R44- Fluid Power in Plant and Field – Second Edition
Author: Charles S. Hedges, R.C. Womack.
Publisher: Womack Machine Supply Co. 1968.
Womack Educational Publication, 2010 Shea Road, Dallas, TX 75235.
ISBN 68-22573.

R45- Mobile Hydraulics Manual
Author: Eaton.
Publisher: Eaton Corporation Training.
Eden Prairie, Minnesota.
ISBN 0-9634162-5-1.

R46- EH Control Systems
Author: F.D. Norvelle.

R47- Fluid Power Journal
Publisher: International Fluid Power Society.

R48- Fundamentals of Industrial Controls and Automation
Author: Lonnie L. Smith and Mike J. Rowlett.
Publisher: Womack Educational Publications.
Dallas, Texas.
ISBN: 0-943719-04-6.

R49- Lightning Reference Handbook
Publisher: Berendsen Fluid Power.

R50- Pneumatics Basic Level
Author: P. Croser, F. Ebel.
Publisher: Festo Didactic GmbH & Co.

R51- Electro-pneumatics Basic Level
Author: F. Ebel, G. Prede, D. Scholz.
Publisher: Festo Didactic GmbH & Co.

R52- Mechanical System Components
Author: James F. Thorpe.
Publisher: Allyn and Bacon.
Needham Heights, Massachusetts.
ISBN: 0-205-11713-9.

R53- Electrical Motor Controls for Integrated Systems, Third Edition
Author: Gary J. Rockis, Glen A. Mazur.
Publisher: American Technical Publishers, Inc.
ISBN: 0-8269-1207-9.

R54- Instrumentation, Fourth Edition
Author Franklyn W. Kirk, Thomas A. Weedon, Philip Kirk.
Publisher American Technical Publishers, Inc.
ISBN: 0-8269-3423-4.

R55- Introduction to Mechatronics and Measurement Systems, Second Edition
Author David G. Alciatore, Michael B. Histand.
Publisher McGraw-Hill, Inc.
ISBN: 0-07-240241-5.

R56- Study Guides for IFPS Certification

R57- Work Books from Coastal Training Technologies

R58- Industrial Hydraulic Manual – Fourth Edition 1999
Author: Vickers.
Publisher: Vickers Inc. 1999.
Training center, 2730 Research Drive, Rochester hills, Michigan 48309-3570.
ISBN 0-9634162-0-0.

R59- Industrial Automation and Process Control
Author: John Stenerson.
Publisher: Prentice Hall.
ISBN 0-13-033030-2.

R60- Industrial Automated Systems
Author: Terry Bartelt.
Publisher: Delmar Cengage Learning.
ISBN: 10-1-4354-888-1.

R61- Introduction to Fluid Power
Author: James L. Johnson.
Publisher: Delmar Cengage Learning.
ISBN: 10-0-7668-2365-2.

R62- Summary for Engineers
Author: Dr. Abdel Nasser Zayed.
Publisher: Dr. Abdel Nasser Zayed .
ISBN: 977-03-0647-9.

R63- Mechanics of Materials
Author: Ferdinand P.Beer, E. Russell Johnston Jr., John T DeWolf.
Publisher: McGraw Hill Publishing .
ISBN: 0-07-365935-5.

R64- Oil Hydraulic System, Principles and Maintenance
Author: S. R. Majumdar.
Publisher: McGraw Hill.
ISBN 10: -0-07-140669-7.

R65- Contamination Control in Hydraulic and Lubricating Systems
Publisher: Pall

R66- Diagnosing Hydraulic Pump Failure
Publisher: Caterpillar.

R67- Oil Service Products Catalog
Publisher: Schroder Industries.

R68- Industrial Fluid Power Volume 1
Author: Charles S. Hedges.
Publisher: Womack Educational Publication.
ISBN: 0-9605644-5-4.

R69- Industrial Fluid Power Volume 2
Author: Charles S. Hedges.
Publisher: Womack Educational Publication.
ISBN: 0-943719-01-1.

R70- Industrial Fluid Power Volume 3
Author: Charles S. Hedges.
Publisher: Womack Educational Publication.
ISBN: 0-943719-00-3.

R71- Electrical Control of Fluid Power
Author: Charles S. Hedges.
Publisher: Womack Educational Publication.
ISBN 0-9605644-9-7.

R72- Hydraulic Cartridge Valve Technology
Author: John J. Pippenger, P.E.
Publisher: Amalgam Publishing Company.
Post Office Box 617, Jenks, OK 74037 USA.
ISBN: 0-929276-01-9.

R73- Noise Control of Hydraulic Machinery
Author: Stan Skaistis.
Publisher: Marcel Dekker, 270 Madison Avenue, New York, NY 10016.
ISBN: 0-8247-7934-7.

R74-Solenoid Valves
Author: Hydraforce

R75-HF Proportional Valve Manual
Author: Hydraforce

R76-Automatic Control for Mechanical Engineers
Author: M. Galal Rabie, Professor of Mechanical Engineering
ISBN: 977-17-9869-3,2010.

R77-Fluid Power System Dynamics
Author: W. Durfee, Z. Sun

Index

A

B

C

D

E

F

G

H

I

K

L

M

N

O

P

R

S

T

U

V

Z

www.ingramcontent.com/pod-product-compliance
Lightning Source LLC
LaVergne TN
LVHW071243210926
848486LV00007B/333